GLOBAL SHIFT

FOURTH EDITION | PETER DICKEN

GLOBAL SHIFT

RESHAPING THE GLOBAL ECONOMIC MAP
IN THE 21ST CENTURY

SAGE Publications
London • Thousand Oaks • New Delhi

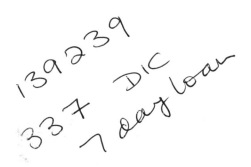

ISBN 0-7619-7149-1 (hbk)
ISBN 0-7619-7150-5 (pbk)
© Peter Dicken 2003
First published 2003
Reprinted 2004

 SAGE Publications Ltd
1 Oliver's Yard, 55 City Road
London EC1Y 1SP

SAGE Publications Inc
2455 Teller Road
Thousand Oaks, California 91320

SAGE Publications India Pvt Ltd
B–42 Panchsheel Enclave
PO Box 4109
New Delhi 110 017

British Library Cataloguing in Publication data
A catalogue record for this book is available from the British Library

Library of Congress Control Number: 2002109622

Typeset by C&M Digitals Pvt. Ltd., Chennai, India
Printed and bound in Great Britain by
Cromwell Press Limited, Trowbridge, Wiltshire

This is for Jack – the first of the new generation

SUMMARY OF CONTENTS

CONTENTS

LIST OF ABBREVIATIONS

AFTA	ASEAN Free Trade Agreement
APEC	Asia-Pacific Economic Cooperation Forum
ASEAN	Association of South East Asian Nations
CBI	Caribbean Basin Initiative
CUSFTA	Canada-United States Free Trade Agreement
ECB	European Central Bank
EMU	European Monetary Union
EU	European Union
GATS	General Agreement on Trade in Services
GATT	General Agreement on Tariffs and Trade
GDP	Gross Domestic Product
GNP	Gross National Product
HIPC	highly indebted poor country
ILO	International Labour Organization
IMF	International Monetary Fund
MAI	Multilateral Agreement on Investment
MFA	Multi-Fibre Arrangement
NAFTA	North American Free Trade Agreement
OECD	Organization for Economic Co-operation and Development
UNCTAD	United Nations Conference on Trade and Development
UNCTC	United Nations Centre on Transnational Corporations
UNDP	United Nations Development Programme
UNIDO	United Nations Industrial Development Organization
WTO	World Trade Organization

PREFACE

In the 20 years since *Global Shift* originally began as a project not only have there been immense changes in the world economy but also people's *awareness* of 'the global' has greatly intensified. From being a term that induced glazed eyes at parties, 'globalization' has become fairly common currency (though it may still induce glazed eyes). But what does it mean? What is actually going on 'out there' that matters to us in our daily lives? The problem is that, despite the proliferation of globalization literature, it is often difficult to see a clear path through the maelstrom of views and opinions. The world is a pretty confusing place. *Global Shift* aims to help to set out such a path by showing how the global economy works and what its effects are on people and communities in different parts of the world.

My basic argument is that globalization is not some inevitable kind of end-state but, rather, a complex, indeterminate set of *processes* operating very unevenly in both time and space. As a result of these processes, the nature and the degree of interconnection between different parts of the world is continuously in flux. A particular country's (or a part of a country's) position within the global system may change dramatically. Take the current (2002) case of Argentina's economic collapse. An article in the *Financial Times* carried the graphic headline: 'Crisis is sweeping Argentina back towards the world's end.'

> Argentina was perhaps the most frequently cited example of the power of globalization to bring nations together. Now Argentina is moving in the opposite direction, falling into isolation despite its best efforts. *It is a stunning example of globalization in reverse* ... The symbols of 21st century globalization are starting to look like relics, while products that disappeared a generation ago are returning to haunt the present ... International airlines have scrapped direct flights to the US and Europe or drastically reduced their frequency ... Phoning is no better: the country's telecommunications network is decaying as foreign phone operators ... halt investment. (*The Financial Times*, 22 June 2002, emphasis added)

On the one hand we need to be careful about making long-term predictions about immediate events. Similar doom-laden warnings were being made in 1997/98 when the East Asian financial crisis caught the world unawares. Yet, for the most part, that region has staged a very substantial recovery. On the other hand, some parts of the world, notably much of sub-Saharan Africa, remain very weakly integrated into the global economy.

The lesson of such cases is this: it is vital to understand the *long-term, underlying processes of global economic change* and that is the position taken in this book. In my view, these processes are worked out through the actions and interactions of two major groups of actors – transnational corporations and states – set within a volatile technological environment, of which the 'space-shrinking' technologies of transportation and, especially, communication are the most significant enabling elements. Few parts of the world are unaffected as TNCs restructure their operations, as national governments attempt to build or preserve their own economies, and as the pace of technological change quickens and its nature changes. Such changes are both geographically and sectorally uneven. It is for that reason that we will examine not only the *general* processes of change but also their *specific* expression in different places and in different kinds of activity (both manufacturing and services). But the process is not simply 'top down', from the global to the local. Conditions at a range of geographical scales, themselves play an active role in mediating and influencing the precise operation of 'global' forces. Indeed, everything that is global is also, in a very real sense, local.

The overall structure of this significantly enlarged Fourth Edition remains as before – for the simple reason that it appears to work. However, not only have all the empirical data been fully updated but also every chapter has been completely revised and extensively rewritten and a great deal of new material has been introduced. This new material includes, in particular:

- A fuller discussion of the conflicting globalization and anti-globalization debates.
- A much expanded discussion of developmental, environmental and governance issues in a substantially enlarged Part IV.
- A broadening of the sectoral cases of Part III to include the much-neglected distribution industries (notably logistics, retailing, and e-commerce).

In addition, the number of specially drawn diagrams has been increased to over 200 and all key terms are defined and cross-referenced throughout. My aim is to make the book accessible to readers without prior specialist knowledge whilst, at the same time, increasing its value to the more specialist researcher. In this respect, each chapter has detailed notes to relevant literature linked to the extremely extensive Bibliography. This is very much a multi-level, multi-disciplinary book.

I have benefited enormously from the immensely helpful comments and advice from friends and colleagues around the world – in Europe, in East Asia, in North America and in Latin America. Without such a rich network a book like this could not exist at all. As always, my accumulated intellectual (and social) debts are far too numerous to mention individually but my gratitude is nonetheless sincere. However, several groups of people merit particular mention. First, colleagues at the University of Manchester, notably Jamie Peck (now at the University of Wisconsin, Madison), Neil Coe (especially through his invaluable help with Chapter 14 and his comments on other parts of the book), Kevin Ward, Noel Castree, Jeff Henderson, Martin Hess and Jenny Johns, create a very amenable working environment, while Nick Scarle has once again worked miracles with illustrations. Second, I was fortunate to spend six months at the National University of Singapore. Not only did this facilitate my continuing collaboration with Henry Yeung but also created new ones with Philip Kelly, and Kris Olds. They provided a great intellectual atmosphere. Third, I was privileged to spend time as a Fellow at the Swedish Collegium for Advanced Study in the Social Sciences (SCASSS) in Uppsala and to strengthen long-standing links with Anders Malmberg, Sture Öberg, Ash Amin and Nigel Thrift. Fourth, my involvement with the Duxx Graduate School of Business Leadership in Monterrey, Mexico, continued to be a real source of stimulation both through Carlo Brumat and through the wonderful group of graduate students there. Fifth, a number of colleagues have continued to provide encouragment and support in a whole variety of ways, notably Roger Lee, Ray Hudson, Meric Gertler, Gary Gereffi, Amy Glasmeier, Erica Schoenberger, Adam Tickell and Yoshihiro Miyamachi (whose involvement in translating the book into Japanese is greatly appreciated). I particularly want to thank Robert Rojek at Sage Publications for being such a committed, supportive and stimulating editor, and Vanessa Harwood for doing such excellent production work. I really appreciate the atmosphere at Sage. I would also like to thank Seymour Weingarten at Guilford and the former editor there, Peter Wissoker, with whom I worked for a number of years. Of course, none of these individuals bears any responsibility for remaining weaknesses in the book. I must just try harder.

And so to the really important ones: Valerie, Christopher, Michael, Sally – and now Jack. They, as I have said before, are where it all begins and ends. The fact that Michael and Christopher still have difficulty in spelling the second word in the book's title continues to cause me distress (though they do compensate in other ways). Valerie continues to amaze me by her tolerance of my absences (in all senses) but I think she knows how much I appreciate everything she does. None of this would have been possible without her.

Peter Dicken
Manchester

ACKNOWLEDGEMENTS

We are grateful to those listed for permission to reproduce copyright material.

Figure 2.8 from *Rival States, Rival Firms: Competition for World Market Shares*, John M. Stopford and Susan Strange with John S. Henley (1991), Figure 1.6, p. 22. Reprinted with permission of Cambridge University Press.

Figure 4.12 from *Contemporary Capitalism: The Ebeddedness of Institutions*, J. Rogers Hollingsworth and Robert Boyer (1997), Figure 1.3, p. 22. Reprinted with permission of Cambridge University Press.

Figure 5.1 from *Global Politics: Globalization and the Nation-State*, Anthony C. McGrew, Paul G. Lewis et al. (eds) (1992), Figure 1.7, p. 13, Polity Press, Oxford. Reprinted with permission of Blackwell Publishers.

Table 5.1 from *The Myth of the Global Corporation*, P.N. Doremus, W.W. Keller, L.W. Pauly and S. Reich (1988), Table 2.1, p. 17. Reprinted with permission of Princeton University Press.

Figure 6.2 from *Producing Places*, R. Hudson (2001), Figure 3.3, p. 69. Reprinted with permission of Guilford Publications.

Figure 6.4 from *Between MITI and the Market: Japanese Industrial Policy for High Technology*, Daniel I. Okimoto (1989), Figure 1.5, p. 51. Reprinted with permission of the publishers, Stanford University Press. © 1989 by the Board of trustees of the Lealand Stanford Junior University.

Figure 7.2 from *The Product Life Cycle and International Trade*, Louis T. Wells, Jr (1972), p. 15. Reprinted with permission of Harvard Business School Press. © 1972 by the President and Fellows of Harvard College, all rights reserved.

Figure 8.1 from The Role of Collaborative Integration in Industrial Organization: Observations from the Canadian Aerospace Industry, Malcolm Anderson, *Economic Geography* (1995) Vol. 71, Figure 1, p. 60.

Figures 10.9, 10.10 from *A Stitch In Time: Lean Retailing and the Transformation of Manufacturing: Lessons from the Apparel and Textile Industries*, Frederick K. Abernathy (1999), Figures 3.1 and 4.1. Reprinted with permission of Oxford University Press, Inc.

Tables 12.1, 12.2, 12.3 from *Tiger Technology: The Creation of a Semiconductor Industry in East Asia*, John A. Matthews and Dong-Sung Cho (2000), Tables 3.1, 4.1 and 5.1. Reprinted with permission of Cambridge University Press.

Figure 13.4 from *Telecommunications and the Globalization of Financial Services*, B. Warf, The Professional Geographer (1989) Vol. 41, Figure 5, pp. 257–71. Reprinted with permission of Blackwell Publishers.

Figure 13.5 from *Human Resources and Corporate Strategy: Technological Change in Bank and Insurance Companies*, Olivier Bertrand and Thierry Noyelle (1988), Table 4.1. Reprinted with permission of the Organisation for Economic Cooperation and Development.

Figure 14.3 from *Managing the Global Supply Chain*, Philip B. Schary and Tage Skjott-Larsen (2001), Figure 7.4, p. 244. Reprinted with permission of Copenhagen Business School Press.

Figure 18.1 from *Globalization and Global Governance*, Vincent Cable (1999), Figure 3.1, p. 55. Reprinted with permission of Continuum.

CHAPTER 1
Introduction

The title of this book, *Global Shift: Reshaping the Global Economic Map in the 21st Century*, aims to capture the idea that there has been a fundamental redrawing of the global economic map. Its basic aim is to analyse the processes shaping and reshaping this global map and to identify their major impacts on the economic well-being of people and places occupying different positions within the global economic system. Its underlying theme is that, while there are indeed *globalizing* processes at work in *transforming* the world economy into what might reasonably be called a new geo-economy, such processes – and their outcomes – are far more diverse than we are generally led to believe. So, when we talk about globalization we must always remember that it is a set of *tendencies* and not some kind of final condition. These tendencies are both geographically and organizationally uneven. There is neither a single predetermined trajectory nor a fixed endpoint.

Equally, there is not a single transformative force at work. Although the role of the transnational corporation (TNC) is given considerable prominence in this book, I do not take the view that the TNC has rendered all other institutions – especially the state – impotent and irrelevant as economic actors or agents. On the contrary, the state continues to play a highly significant role. It seems to me to be more accurate to conceptualize the process as one of a complex interaction between TNCs and states, set within the context of a volatile technological environment. Firms and states, then, are the two major shapers of the global economy and are embedded within a triangular nexus of interactions consisting of firm–firm, state–state and firm–state relationships.

Figure 1.1 'maps' the major interconnections between the broad patterns of change within the global economy, the major processes creating such changing patterns and, most importantly, the dynamic interaction between different geographical scales – notably the global, the regional, the national and the local – as indicated by the double-headed arrows connecting the individual boxes. It will be useful to refer to Figure 1.1 from time to time throughout the book. Its central message is that the processes of globalization are not simply unidirectional, for example from the global to the local, but that all globalization processes are deeply embedded, produced and reproduced in particular contexts. Hence, the specific assemblage of characteristics of individual nations and of local communities will not only influence *how* globalizing processes are experienced but also will influence the *nature* of those processes themselves. We must never forget that all 'global' processes originate in specific places.

This conceptual framework is reflected in the way the chapters are organized. Of course, there are many alternative ways of organizing the treatment of such a broad and complex subject. The processes involved are tightly interconnected and mutually interact with one another in intricate ways. We are not dealing with a neat, linear process in which each element can be dealt with one at a time without regard for the others. Unfortunately, the constraints of language necessitate such a linear treatment for something that is not linear at all.

The book is organized into four distinct, but closely related, parts. *Part I* is concerned with *the shifting contours of the geo-economy*. It consists of two chapters. *Chapter 2* explores the conflicting currents of opinion within the globalization debate and argues that one way of understanding what is going on in the global economy is to use the concept of the network. Thinking in terms of networks forces

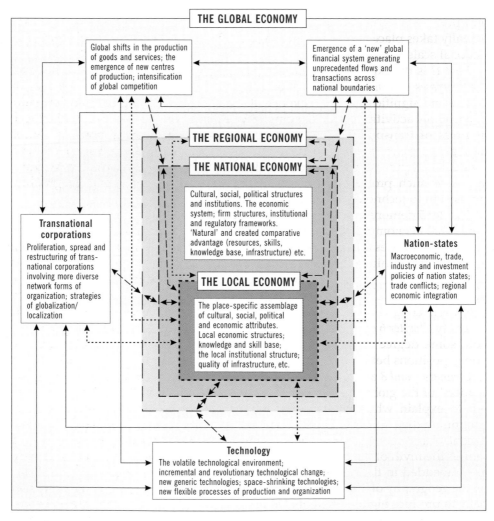

Figure 1.1   Globalizing processes as a system of interconnected elements and scales

us to emphasize the connections of economic activities and the power relationships through which such connections are established and maintained. Production networks are intrinsically spatial and territorial. They are 'grounded' in complex ways, with strong tendencies to form geographical clusters of activity. The primary purpose of production, of course, is to produce 'goods' but, in so doing, certain 'bads' are also created, most notably in terms of environmental damage.

Chapter 3 is very much an empirical chapter. Its purpose is to show how the contours of the global economic map, as measured through production, trade and foreign direct investment, have been changing very significantly in the past 50 years. In particular, it demonstrates the intensification of competition between different parts of the world, notably through the emergence of new geographical centres of economic activity. Most of the data employed in Chapter 3 are at the national scale because that is the scale at which most data are collected and published. However, we do not lose sight of the fact that economic activity actually takes place at other geographical scales, both above and below that of the national scale.

Part II is the explanatory core of the book. Its six chapters are concerned with the *processes* of *transformation* within the global economy. *Chapter 4* analyses the nature and significance of technological change in facilitating the transformation of economic activities. Particular emphasis is placed on the 'space-shrinking' technologies of transportation and communications, which make possible entirely new geographical scales and structures of organization, and on innovations in product and process technologies, which have revolutionized what is produced and how such production occurs. Again, we find a very strong geographical dimension to technological innovation.

Chapter 5 demonstrates the continuing significance of the state as a key agent in the global economy. States perform extremely important roles: as containers of distinctive cultures and practices; as regulators of trade, foreign direct investment and industry within and across their borders; and as competitors within an increasingly interconnected global economy. States are also increasingly involved in regional collaborations. But states attempt to regulate their economies in different ways, reflecting the diversity of capitalisms in today's world. To illustrate such diversity *Chapter 6* consists of a series of country case studies, which demonstrates both some degree of policy convergence but also the continuing divergence of policy positions between countries with different socio-political systems.

Chapters 7 and 8 are concerned with what are arguably the primary 'movers and shapers' of the global economy: transnational corporations (TNCs). *Chapter 7* sets out to explain why and how TNCs come into existence. It demonstrates the continuing diversity of organizational 'architectures' among TNCs as they attempt to control and coordinate their geographically dispersed operations. Chapter 7 also refutes the myth of the allegedly 'placeless' TNC, showing how closely such firms are embedded in their geographical environments. *Chapter 8* focuses specifically on the geography of TNCs' production networks, on how they are configured internally to produce highly differentiated geographies and on how they are embedded within networks of externalized relationships. The interaction between the organizational and geographical dimensions of transnational production networks creates extremely complex structures in which elements of both concentration and dispersal are apparent and which often have a strong regional expression. *Chapter 9*, the final chapter of Part II, looks at the relationships between TNCs and states as each tries to achieve its own objectives. TNCs and states are tightly intertwined in both conflictive and collaborative ways. The chapter shows how TNCs impact differentially on host and home economies along a whole range of dimensions and how the outcome depends upon the relative bargaining power of TNCs and states.

The six chapters of Part II, therefore, set out the general ways in which the major institutions and processes of change shape and reshape the global economic map. Their central theme is that the globalization of economic activity arises from the dynamic interplay between three sets of processes: the strategies of TNCs, the strategies of national governments, and the character and direction and nature of technological change. But precisely *how* these processes operate, and the *specific outcomes* produced, varies substantially between different types of economic activity. In *Part III*, therefore, we examine five sectors – three in manufacturing (textiles and garments, automobiles, semiconductors) and two in services (finance and distribution) – each of which has experienced major global shifts in recent decades. It is not suggested that these five sectors are in any way 'typical' or 'representative' of all types of economic activity. Rather, the purpose is to show how the processes of change combine to create particular organizational and geographical forms at the global scale. Each sectoral study throws a slightly different light on the processes of change.

Whereas Parts I, II and III of the book are concerned with the patterns and processes of change in the global economy, the four chapters of *Part IV* are concerned with the *impact of these processes on people and places*. Globalizing processes produce both winners and losers. *Chapter 15* provides an overview of these issues and outlines some of the major dimensions of the problem, notably those of poverty, population growth and making a living. *Chapter 16* focuses specifically on the developed countries and demonstrates that within the overall increasing affluence of the developed countries – the most obvious 'winners' in the global economy – there are significant 'losers', whether measured in terms of unemployment or of the widening income gap between different social groups. *Chapter 17* shifts the lens to the developing countries. Here we certainly find some winners – but also an awful lot of losers. There is huge heterogeneity both between and within developing countries in terms of income differentials and employment structures. But the basic problem that underlies much of the poverty is that labour force growth outstrips the creation of jobs (even 'part' jobs); one result is the increasing out-migration of labour. For the successful NIEs (newly industrialized economies), the problems are essentially those of sustaining economic growth and ensuring equity for their populations. For the poorer, less-industrialized developing countries, the problems are more fundamental: those of ensuring survival and reducing poverty. Clearly, we do not live in 'the best of all possible worlds.' How, then, might the world be made a better place in the future? The question is a moral one; the answer has to lie in political actions. In *Chapter 18*, therefore, we examine some of the key issues of global governance, particularly those relating to finance, trade, labour standards and the environment.

PART ONE

THE SHIFTING CONTOURS OF THE GEO-ECONOMY

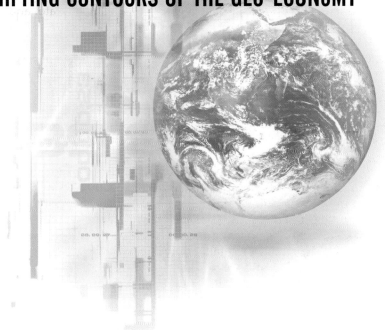

CHAPTER 2

A New Geo-Economy

Something is happening out there – but what is it?

Every age has its 'buzzword' – a word that captures the popular imagination, and becomes so widely used, that its meaning becomes confused. As the second millennium drew to a close and a new one began, the buzzword was *globalization*. Although not, by any means, a new word, its use exploded dramatically in the 1990s. A count of books and academic papers with 'global' or 'globalization' in their title[1] shows this very clearly. Between 1980 and 1984, there were only 13 such publications. In the next five-year period (1985–1989) the number had grown to 78 – a considerable increase but still a pretty modest total. Then – take-off! Between 1992 and 1996 there were almost 600 titles (in 1996 alone there were more than 200 – well over twice the number published in the whole of the 1980s).

Since then, the number of globalization titles has grown at a phenomenal rate of several hundred a year, with little sign of slackening. Populist books, with titles such as *No Logo; Global Dreams; One World Ready or Not; Turbo-Capitalism; When Corporations Ruled the World; A Future Perfect*,[2] spilled on to the bookstands and TV talk shows. Most dramatic of all was the sudden eruption of global protest movements; the explosion of street demonstrations at major international economic meetings that spanned the millennial transition. These involved a remarkable mélange of pressure groups, ranging from long-established non-governmental organizations (NGOs) to totally new groups with either very specific, or very general, foci for their protest, together with anarchist and revolutionary elements with a broad anti-capitalist agenda. First coming to real prominence at the meeting of the World Trade Organization in Seattle in November 1999, the global protest movement immediately became a permanent feature of every subsequent major meeting of international organizations. TV footage of both peaceful protest and violent confrontation was instantaneously flashed round the world. Only the cataclysmic events of 11 September 2001 in New York City produced a pause in the street demonstrations.

The explosion of interest in 'globalization' signifies a universal acceptance that something fundamental is happening in the world economy; the feeling that there are lots of 'Big Issues' that are somehow interconnected under the broad umbrella term 'globalization'. Certainly as we look around, we perceive a confusion of change, an acceleration of uncertainty; feelings intensified by our entry into the early years of a new millennium with all its symbolic promises – and threats – of epochal change. Inevitably, the commercial advertising media latch on to such uncertainties. An advertisement for Intel, the US semiconductor company in 2000, emphasized the arrival of 'the "surge economy" ... where more business is condensed into seconds than used to get done in a day, and the only constant is change'. Television news reports and specials, press headlines and the like constantly remind us of our uneasy present and our precarious future. Turbulence and change are, of course, nothing new in human affairs. But there is no doubt that today's world economy is being buffeted by extremely volatile forces. To the individual citizen the most obvious indicators of change are those that impinge most directly on her/his daily activities – making a living, acquiring the necessities of life, building a future for their children.

In the industrialized countries, there is a real fear that the dual (and connected) forces of technological change and geographical shifts in the location of economic activities are transforming employment prospects in adverse ways for many people, notably less educated and less skilled blue-collar workers, but for many others as well. Notwithstanding the remarkable economic and financial boom of the so-called 'new economy' in the United States during the 1990s – driven by the spread of information technology – the laws of economic gravity (what goes up must come down) have emphatically not been abolished. To those currently employed in a job, what matters is job security and the wages or salary received or anticipated. To those seeking a job, what matters is availability. To those self-employed in small businesses, the question is one of coping with intensifying competition. On all counts, the situation seems to have become increasingly uncertain. As consumers, the most obvious indicator of change is the vast increase in the number of products whose origins lie on the other side of the world but which are now either literally on our doorsteps (through superstores) or metaphorically in our homes (through the all-pervasive TV commercial and the more recent phenomenon of Internet shopping).

But the problems of the industrialized countries pale into insignificance compared with those of the poorest countries in what used to be called the Third World. Although there are indeed losers in the developed and affluent countries their magnitude is totally dwarfed by the poverty and deprivation of much of Africa and of many parts of South Asia and of Latin America. The development gap continues to widen: the disparity between rich and poor continues to grow. Of course, it is too simplistic to attribute these complex problems solely to 'globalization'. But there is no doubt that much of what is happening in particular places is increasingly the result of processes operating at much larger geographical scales. Not least of these are the forces of environmental change – notably the threat of global warming – that, although difficult to predict with certainty, pose enormous potential problems for societies at all levels of development.

The immediacy and longer-term impact of these major forces of change are enormously enhanced by the growing interconnections between virtually all parts of the world. Of course, internationalization is nothing new. Some products have had an international character for centuries; an obvious example being the long-established trade in spices and other exotic goods. Such internationalization was much enhanced by the spread of industrialization from the 18th century onwards. Nevertheless, until very recently, the production process itself was organized primarily within national boundaries. Today, the picture is very different: national boundaries no longer act as 'watertight' containers of the production process.

As a consequence, each of us is now more fully involved in a world economic system than were our parents and grandparents. Few, if any, economic activities now have much 'natural protection' from external competition whereas in the past, of course, geographical distance was a powerful insulator. Today, fewer and fewer activities are oriented towards local – or even national – markets. More and more have meaning only in a regional or a global context. Thus, whereas a hundred or more years ago only rare and exotic products and some basic raw materials were involved in truly international trade, today virtually everything one can think of is involved in long-distance movement. And because of the increasingly complex ways in which production is organized across national boundaries, rather than contained within them, the actual origin of individual products may be very difficult to determine.

These developments signify the emergence of a *new global division of labour*, a transformation of the old geographical pattern of specialization, in which the industrialized countries produced manufactured goods and the non-industrialized countries supplied raw materials and agricultural products to the industrialized countries and acted as a market for some manufactured goods. Such geographical specialization – structured around a *core* and a *periphery* – formed the underlying basis of much of the world's trade for many years (Figure 2.1).

Figure 2.1

The 'pre-global' international division of labour

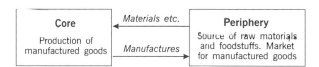

This relatively simple pattern (although it was never quite as simple as the description above suggests) no longer applies. During the past 50 years, in particular, trade flows have become far more complex. The straightforward exchange between core and peripheral areas, based upon a broad division of labour, has been transformed into a highly complex, kaleidoscopic structure involving the *fragmentation* of many production processes and their *geographical relocation* on a global scale in ways that slice through national boundaries. In addition, new centres of industrial production have emerged in the so-called *newly industrialized economies (NIEs)*. Both old and new economic activities are involved in this re-sorting of the global jigsaw puzzle in ways that also reflect the development of technologies of transportation and communications, of corporate organization and of the production process itself. Technologies of production are undergoing substantial and far-reaching change as the emphasis on large-scale, mass production assembly-line techniques has shifted towards more flexible methods. Today, the nature and the geography of the global economy are being transformed further by the shift towards an *information*, or *knowledge-driven*, economy, in which 'knowledge intervenes upon knowledge itself in order to generate higher productivity'.[3] And just as we can identify a new global division of labour in production so, too, we can identify a 'new global financial system', based on rapidly growing

24-hour transactions focused upon the world's major financial centres. But, as we saw in the financial crises of the East Asian and emerging market economies in the late 1990s, this system is highly susceptible to the dangers of contagion. No part of the world is immune.

A 'new' geo-economy?
The globalization debate

So, something is undoubtedly happening 'out there'. But precisely what that 'something' might be – and whether it really represents something new – is a subject of enormous disagreement. Three distinct, though related, questions are posed in this globalization debate, on each of which viewpoints are strongly polarized:

- What precisely is happening?
- What does it mean?
- What can or should be done about it?

Box 2.1 contrasts the views of some writers at opposite ends of the ideological spectrum.[4] The *hyperglobalists* argue that we now live in a borderless world in which the 'national' is no longer relevant. 'Globalization' is the new economic (as well as political and cultural) order. We live, it is asserted, in a world where nation-states are no longer significant actors or meaningful economic units and in which consumer tastes and cultures are homogenized and satisfied through the provision of standardized global products created by global corporations with no allegiance to place or community. Thus, the 'global' is claimed to be the *natural* order of affairs in today's technologically driven world in which time-space has been compressed, the 'end of geography' has arrived and everywhere is becoming the same. Although such a notion of a global*ized* world has become widely accepted, there are some who adopt a more *sceptical* position, arguing that the 'newness' of the current situation has been grossly exaggerated. The world economy, it is claimed, was actually more open and more integrated in the half century prior to World War I (that is, 1870–1914) than it is today.[5]

So, on the one hand, we have the view that we do, indeed, live in a new – *globalized* – world economy in which our lives are dominated by global forces. On the other hand, we have the view that not all that much has changed; that we still inhabit an *international*, rather than a globalized, world in which national forces remain highly significant. The truth, it seems to me, lies in neither of these two polarized positions.

Although, in quantitative terms, the world economy was perhaps at least as open before 1914 as it is today – in some aspects, such as labour migration, even more so – the nature of its integration was *qualitatively* very different.[6] International economic integration before 1914 – and, in fact, until only about four decades ago – was essentially *shallow integration*, manifested largely through arm's length *trade* in goods and services between independent firms and through international

Box 2.1 Contrasting positions in the globalization debate

The hyperglobalist *position*

Today's global economy is genuinely borderless. Information, capital, and innovation flow all over the world at top speed, enabled by technology and fueled by consumers' desires for access to the best and least expensive products.

Ohmae, 1995: inside front cover

What is new today is the degree and intensity with which the world is being tied together into a single globalized marketplace and village. What is also new is the sheer number of people and countries able to partake of today's globalized economy and information networks, and to be affected by them … This new era of globalization … is turbocharged.

Friedman, 1999: xvii, xviii

We are living through a transformation that will rearrange the politics and economics of the coming century. There will be no national products or technologies, no national corporations, no national industries. There will no longer be national economies, at least as we have come to understand that concept … As almost every factor of production – money, technology, factories, and equipment – moves effortlessly across borders, the very idea of an American economy is becoming meaningless, as are the notions of an American corporation, American capital, American products, and American technology. A similar transformation is affecting every other nation, some faster and more profoundly than others; witness Europe, hurtling toward economic union.

R.B. Reich, 1991: 3, 8

Globalization, as we are experiencing it, is in many respects not only new but also revolutionary … Globalization is political, technological and cultural, as well as economic.

Giddens, 1999: 10

The sceptical *position*

We do not have a fully globalized economy, we do have an international economy and national policy responses to it.

Hirst and Thompson, 1992: 394

Globalization seems to be as much an overstatement as it is an ideology and an analytical concept.

Ruigrok and van Tulder, 1995: 119

The system has … become more integrated or globalized in many respects … Nonetheless what has resulted is still very far from a globally integrated economy … In short, the world economy is considerably more globalized than 50 years ago; but much less so than is theoretically possible. In many ways it is less globalized than 100 years ago. The widespread view that the present degree of globalization is in some way new and unprecedented is, therefore, false.

Glyn and Sutcliffe, 1992: 91

movements of portfolio capital. Today, we live in a world in which *deep integration*, organized primarily within the production networks of transnational corporations (TNCs), is becoming increasingly pervasive.

Hence, although there are undoubtedly global*izing* forces at work, we do not have a fully global*ized* world economy. Globalization tendencies can occur without this resulting in an all-encompassing end-state – a global*ized* economy – in which all unevenness and difference is ironed out, market forces are rampant and uncontrollable, and the nation-state merely passive and supine.[7] The position taken in this book is that globalization is a complex of interrelated *processes*, rather than an end-state. Such tendencies are highly uneven in time and space.

In taking such a process-oriented approach to understanding the contemporary global economy it is important to distinguish between two distinct processes:

- *Internationalizing processes*. These involve the simple extension of economic activities across national boundaries. They reflect, essentially, *quantitative* changes that lead to a *more extensive* geographical pattern of economic activity.
- *Globalizing processes*. These involve not merely the geographical extension of economic activity across national boundaries but also – and more importantly – the *functional integration* of such internationally dispersed activities. They reflect, therefore, essentially *qualitative* changes in the ways economic activities are organized.

Figure 2.2 shows, in a highly stylized manner, the relationships between these two sets of transformative processes. The *geographical scope* or extent of economic activities is shown on the horizontal axis while the vertical axis displays the *intensity of functional integration*. Within this framework, *internationalizing processes* are shown as phenomena of varying spatial extent but a low level of functional integration. In contrast, *globalizing processes* are shown as embracing different degrees of both geographical and functional integration. Such processes, therefore, should be seen as *tendencies* and not as final states; they will almost certainly never produce an end-state of pure globalization.

Indeed, as Figure 2.2 indicates, a marked tendency in the contemporary world economy is for economic integration to be established at a variety of *regional* scales, above the 'national' but below the 'global'. The hypothetical 'regions' shown in Figure 2.2 reflect different degrees of functional integration and geographical extent (ranging from, for example, the highly integrated and expanding European Union to the much smaller and simpler free trade agreements found in many parts of the world). Potential connections between such geographically dispersed regions – for example through firms pursuing a multi-regional strategy – are represented by the dotted lines. Whether or not regional integration turns out to be a 'building block' towards eventual globalization or a 'stumbling block' is a matter of political-economic debate.

In sum, several sets of processes – internationalizing, regionalizing, globalizing – co-exist. In some cases, what we are seeing is no more than the continuation of a long-established international dispersion of activities. In others, however, we are undoubtedly seeing both an increasing *dispersion* and a growing *integration* of activities across national boundaries. Such developments ensure that changes originating in one part of the world are rapidly diffused to other parts. We live in a world of increasing complexity, interconnectedness and volatility; a world in which the lives and livelihoods of every one of us are bound up with processes operating at a larger geographical scale.

But although there are, indeed, powerful globalizing forces at work – they are the central focus of this book – we need to adopt a sensitive and discriminating approach to get beneath the hype and to lay bare the reality. 'Globalization is not a single, unified phenomenon, but a *syndrome* of processes and activities'.[8] Globalizing processes do not occur everywhere in the same way and at the same rate; they are intrinsically *geographically uneven*, both in their operations and in their outcomes. The particular character of individual countries and of individual localities interacts with the larger-scale processes of change to produce quite specific

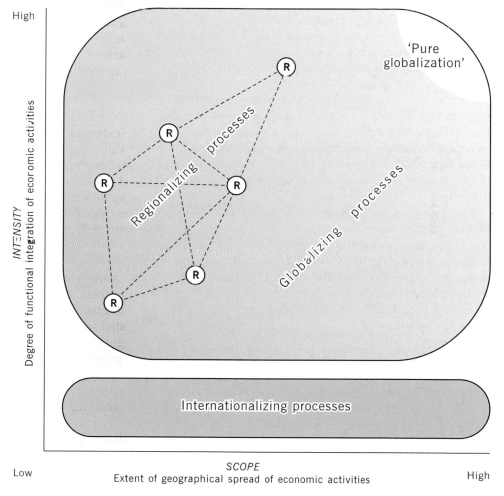

Figure 2.2   Processes of global economic transformation

outcomes. Reality is far more complex and messy than many of the grander themes and explanations would have us believe. Nevertheless, despite the need to use the term 'globalization' with care, there is no doubt that, in the popular imagination – including that of politicians – it has now taken on a life of its own. Even if its reality is far more complex than the popular view perceives, the *rhetoric* of globalization has become powerful and pervasive, particularly in terms of what the process means for society and individuals and for the ways in which national politicians employ such rhetoric to justify why they must (or cannot) adopt certain policies.

A new geo-economy: unravelling the complexity

We *are* witnessing the emergence of a new geo-economy that is *qualitatively different* from the past. The question is: how can we begin to unravel the dynamic, kaleidoscopic complexity of this geo-economy? The conventional unit of analysis is the country. Virtually all the statistical data on production, trade, investment and the like are aggregated into national 'boxes'. However, such a level of statistical aggregation is less and less useful in light of the changes occurring in the organization of economic activity. This does not mean that the national level is unimportant. On the contrary, it remains extremely significant. In any case, we have to rely heavily on national level data – 'state-istics'[9] – to explore the changing maps of production, trade and investment. But, because national boundaries no longer 'contain' production processes in the way they once did, we need to find ways of getting both below and above the national scale – to break out of the constraints of the 'national boxes' – in order to understand what is going on in the world.

One way of doing this is to use the concept of the *network*: the processes connecting 'actors' or 'agents' (firms, states, individuals, social groups, etc.) into *relational structures* at different organizational and geographical scales.[10] Adopting a network-based approach forces us to think in terms of *connections* of activities through *flows* of both material and non-material phenomena, of the different ways that networks are connected, and of the *power relationships* through which networks are controlled and coordinated.

Production chains; production networks

The production of any good or service can be conceived as a *production chain* – that is, as a *transactionally linked sequence of functions in which each stage adds value to the process of production of goods or services.*

As its name suggests, a production chain is essentially linear. It represents the *sequence* of operations required to produce and distribute a good or service (services, like any other item of consumption, have to be 'produced').

Figure 2.3*a* shows the 'stripped-down' version of a hypothetical production chain.[11] All production processes have at their core a set of four basic operations connected by a series of transactions between one element and the next: inputs are transformed into products which are distributed and consumed. But note that the processes are two-way, with:

● flows of materials, semi-finished goods and final products in one direction and
● flows of information (the demands of customers) and money (payments for goods and services) in the other direction.

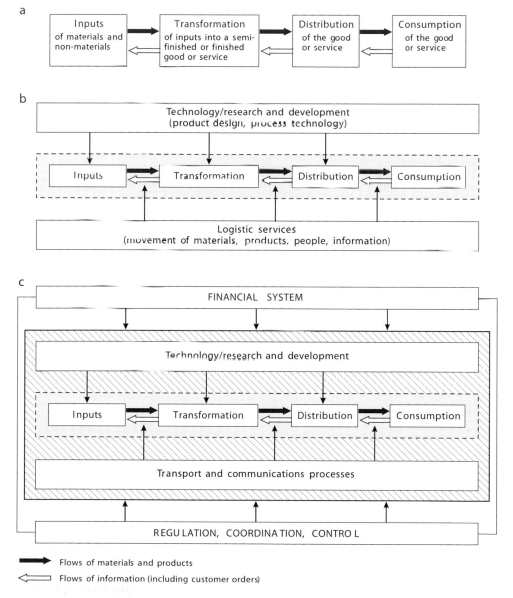

Figure 2.3 The production chain

But there is much more to it than this, as Figures 2.3(b) and 2.3(c) show.

- Each individual element in the production chain depends upon various kinds of technological inputs.
- Each production chain is embedded within a financial system that provides the necessary investment and operating capital (notably credit).
- Each has to be coordinated, regulated and controlled.

Hence, each of the individual elements in the production chain, together with their transactional links, depends upon many other kinds of inputs – particularly services – in order for the whole process to function (Figure 2.4):

> Service activities not only provide linkages between the segments of production within a [production chain] and linkages between overlapping [production chains], but they also bind together the spheres of production and circulation. Services have come to play a critical role in [production chains] because they not only provide geographical and transactional connections, but they *integrate* and *coordinate* the atomized and globalized production process.[12]

These essentially linear structures of production chains are themselves enmeshed in broader production networks of inter-firm relationships. Such networks are, in reality, extremely complex structures with intricate links – horizontal, diagonal, as well as vertical – forming multi-dimensional, multi-layered lattices of economic activity. They vary considerably between different kinds of economic activity, as we shall see in the case study chapters of Part III.

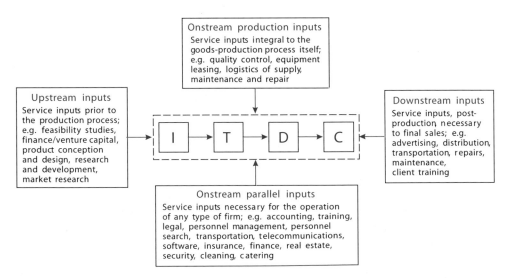

Figure 2.4   The interconnections between services and production in the production chain

Source: Based, in part, on material in UNCTC, 1988: 177

Three dimensions of production networks are especially important:

- *governance* – how they are coordinated and regulated
- *spatiality* – how they are configured geographically
- *territorial embeddedness* – the extent to which they are connected into particular bounded political, institutional and social settings.

In Figure 2.5, the two geographical dimensions are combined on the horizontal axis.

Figure 2.5

The primary dimensions of production networks

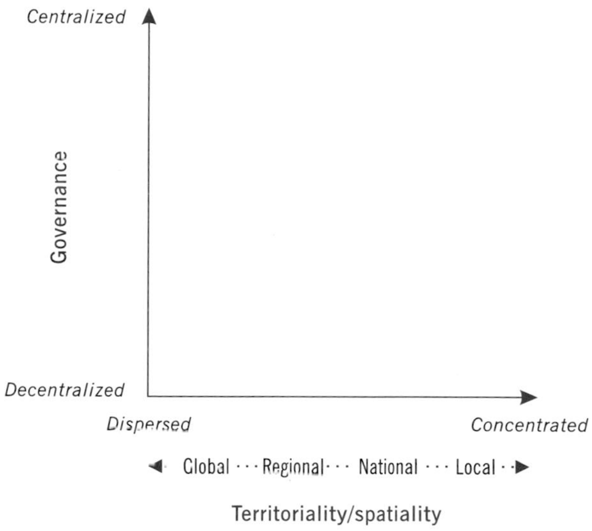

Governance of production networks[13]

In market economies (now the dominant form of economic organization in the contemporary world), production networks are coordinated and regulated primarily by *business firms*, through the multifarious forms of intra- and inter-organizational relationships that make up an economic system. As Figure 2.6 shows, economies are made up of different types of business organization – transnational and domestic, large and small, public and private – in varying combinations and interrelationships. The firms in each of the segments shown in Figure 2.6 operate over widely varying geographical ranges and perform rather different roles in the economic system.

A major theme of this book is that it is increasingly the *transnational corporation* (TNC) that plays the key role in coordinating production networks and, therefore, in shaping the new geo-economy (see Chapters 7 and 8). A broad definition of the TNC – one that goes beyond the conventional definition based upon levels of ownership of internationally based assets – is used here to capture the diversity and complexity of transnational networks. Thus, a transnational corporation is a firm that has the power to coordinate and control operations in more than one country, even if it does not own them. In fact, TNCs generally do own such assets but they are also, as we will see, typically involved in a spider's web of collaborative relationships with other legally independent firms across the globe.

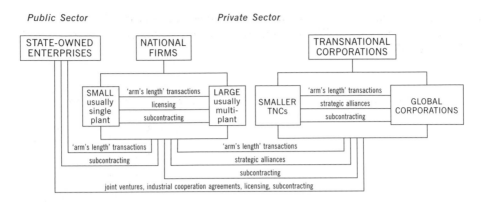

Figure 2.6 A typology of firms in an economic system

On the one hand, each function in a specific production network may be performed by separate firms. In this case, the links consist of a series of *externalized transactions* in which the transactions are organized through 'the market'. On the other hand, the whole network may be operationalized within a single firm as a *vertically integrated* system. In this case, the links consist of a series of *internalized transactions*. Here, transactions are organized 'hierarchically' through the firm's internal organizational structure.[14] In fact, this dichotomy – between externalized, market-governed transactions and internalized, hierarchically governed transactions – grossly simplifies the richness and diversity of the governance mechanisms in the contemporary economy.

The boundary between internalization and externalization is continually shifting as firms make decisions about which functions to perform 'in-house' and which to 'outsource' to other firms. What we have in reality, therefore, is a *spectrum* of different forms of coordination consisting of networks of interrelationships within and between firms structured by different degrees of power and influence. Such networks increasingly consist of a mix of intra-firm and inter-firm structures. These networks are dynamic and in a continuous state of flux. However, there will invariably be a primary coordinator driving any particular production network.

Gereffi[15] makes a useful distinction between the two types of 'driver' shown in Figure 2.7:

- *Producer-driven* production networks are characteristic of 'those industries in which transnational corporations (TNCs) or other large integrated industrial enterprises play the central role in controlling the production system (including its backward and forward linkages). This is most characteristic of capital- and technology-intensive industries like automobiles, computers, aircraft, and electrical machinery … What distinguishes 'producer-driven' production systems is the control exercised by the administrative headquarters of the TNCs'.
- *Buyer-driven* production networks tend to occur in 'those industries in which large retailers, brand-named merchandisers, and trading companies play the pivotal role in setting up decentralized production networks in a variety of exporting countries'.

In fact, as we shall see in later chapters, the governance of production networks, both between and within different economic sectors, is far more diverse than this dualistic definition implies.

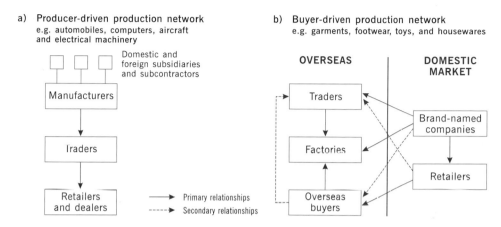

Figure 2.7 'Producer-driven' and 'buyer-driven' production networks

Source: Based on Geretti, 1994: Figure 5.1

Spatiality of production networks

Every production network has *spatiality* – the particular geographical configuration and extent of its component elements and the links between them. At the most basic level, production within a network may be organized along a spectrum from geographically concentrated to geographically dispersed (see Figure 2.5). But such terms are relative and beg the important question of geographical scale.[16] In one sense, we may simply think of geographical scale as a continuum and, therefore, of production networks as being 'more or less long and more or less connected.'[17] However, we tend to divide the continuum of scale into units that conform to some territorial division of the world, most commonly 'local', 'national', 'regional' and 'global' scales. A widely accepted view is that the geographical extensiveness of virtually all production networks has increased – in other words, we are witnessing the emergence of *global production networks* (GPNs).[18]

Thinking of the world in terms of discrete spatial scales – global, regional, national, local – is helpful in many ways. But it can too easily imply that each scale is a self-contained 'box' whereas that is not the case. For example, it has become very common to separate the global scale from the local (or the national) scale and to imply that the processes operating at these two scales are separate and distinct: in particular that the 'global' determines what happens at the local scale. In fact, the interactions are two-way: the 'local and the global intermesh, running into one another in all manner of ways'.[19] The same argument applies across all spatial scales: they are interrelated in complex ways and do not simply constitute a hierarchical structure from top (global) to bottom (local). More realistically, 'the geography of contemporary capitalism can be viewed as a polymorphic, multi-layered 'jigsaw puzzle' in which multiple forms of territorial organization … are being superimposed and intertwined even more densely.'[20]

Territorial embeddedness of production networks

'The end of geography'; 'the death of distance'.[21] These two phrases resonate, either explicitly or implicitly, throughout much of the globalization literature. Capital has become 'hyper-mobile', freed from the 'tyranny of distance' and no longer tied to 'place'. In other words, economic activity is becoming 'de-territorialized' or 'dis-embedded'. The sociologist Manuel Castells argues that the forces of globalization, especially those driven by the new information technologies, are replacing this 'space of places' with a 'space of flows'.[22] Anything can be located anywhere and, if that does not work out, can be moved somewhere else with ease.

Seductive as such ideas might seem, a moment's thought will show just how misleading they are. The world is *both* a 'space of places' *and* a 'place of flows'. Production networks don't just float freely in a spaceless/placeless world. Although transportation and communications technologies have, indeed, been revolutionized (see Chapter 4) both geographical distance and, especially, *place* remain fundamental. Every component in the production network – every firm, every economic function – is, quite literally, 'grounded' in specific locations. Such grounding is both physical, in the form of sunk costs,[23] and less tangible in the form of localized social relationships and in distinctive institutions and cultural practices. Hence, the precise nature and articulation of firm-centred production networks are deeply influenced by the concrete socio-political, institutional and cultural contexts within which they are embedded, produced and reproduced.[24]

An especially important bounded territorial form in which production networks are embedded is that of the *state*. All the elements in the production network are regulated within some kind of political structure whose basic unit is the national state but which also includes such supranational institutions as the International Monetary Fund and the World Trade Organization, regional economic groupings such as the European Union or the North American Free Trade Agreement, and 'local' states at the subnational scale. All markets are socially constructed. Even supposedly 'deregulated' markets are still subject to some kind of political regulation.

All production networks have to operate within *multi-scalar* regulatory systems. They are, therefore, subject to a multiplicity of geographically differentiated political, social and cultural influences. On the one hand, TNCs attempt to take advantage of national differences in regulatory regimes whilst, on the other hand, states attempt to minimize such 'regulatory arbitrage'. The result is a very complex situation in which firms and states are engaged in various kinds of power play: a triangular nexus of interactions comprising firm–firm, state–state and firm–state relationships (Figure 2.8).[25] In other words, the new geo-economy is essentially being structured and restructured not by the actions of either firms or states alone but by complex, dynamic interactions between the two sets of institutions.

One major advantage of adopting a network approach to understanding the global economy is that it helps us to appreciate the interconnectedness of economic activities across different geographical scales and within and across

Figure 2.8

The triangular nexus of relationships between firms and states

Source: Stopford and Strange, 1991: Figure 1.6

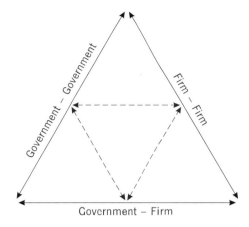

Government – Government

Firm – Firm

Government – Firm

territorially bounded spaces. The production of any commodity, whether it is a manufactured product or a service, involves an intricate articulation of individual activities and transactions across space and time. Such production networks – the nexus of interconnected functions and operations through which goods and services are produced and distributed – have become both organizationally and geographically more complex. Global and regional production networks not only integrate firms (and parts of firms) into structures that blur traditional organizational boundaries (for example, through the development of diverse forms of equity and non-equity relationships) but they also integrate national and local economies (or parts of such economies) in ways which have enormous implications for their economic development and well-being. At the same time, the specific characteristics of national and local economies influence and 'refract' the operation and form of larger-scale processes (see Figure 1.1). In that sense, 'geography matters' a lot.

The process is especially complex because, while states and local economies are essentially territorially specific, production networks themselves are not. They 'slice through' boundaries in highly differentiated ways, influenced in part by regulatory and non-regulatory barriers and local socio-cultural conditions, to create structures which are 'discontinuously territorial'. The geo-economy, therefore, can be pictured as a geographically uneven, highly complex and dynamic web of production networks, economic spaces and places connected together through threads of flows.

Figure 2.9 captures the major dimensions of these relationships. Individual production chains can be regarded as vertically organized structures configured across increasingly extensive geographical scales. Cutting across these vertical structures are the territorially defined political-economic systems which, again, are manifested at different geographical scales. It is at the points of intersection of these dimensions in 'real' geographical space where specific outcomes occur, where the problems of existing within a globalizing economy – whether as a business firm, a government, a local community or as an individual – have to be resolved.

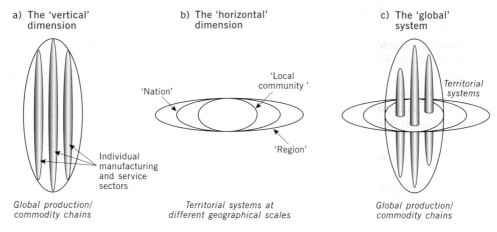

Figure 2.9 Interconnecting dimensions in a globalizing economy

Source: Based, in part, on Humbert, 1994: Figure 1

Even in a globalizing world, economic activities are geographically localized

In discussing the spatiality of production networks in the previous section we noted that economic activities may be either geographically concentrated or geographically dispersed (or anywhere in between, of course). The view of the 'hyperglobalizers' is that increasing geographical dispersal at a global scale is now the norm. If that is the case why, then, do *geographical concentrations* of economic activity not only still exist but also are the normal state of affairs? Why do 'sticky places' continue to exist in 'slippery space'?[26] When we look at the geo-economic map we find tendencies of *both* concentration *and* dispersal – but with a very strong propensity for economic activities to agglomerate into *localized geographical clusters.*[27]

Figure 2.10 identifies two types of geographical cluster: *generalized* and *specialized*. Both types are based on the notion of *externalities*, the positive 'spillovers' created when activities in a particular place are connected with one another, either directly (through specific transactions) or indirectly. Both are based on the idea that the 'whole' (the cluster) is greater than the sum of the parts because of the benefits that spatial proximity provides.

- *Generalized clusters* simply reflect the fact that human activities tend to agglomerate to form urban areas. Hence, such benefits have traditionally been labelled *urbanization economies*. General clustering of activities creates the basis for sharing the costs of a whole range of services. Larger aggregate demand in, say, a large city encourages the emergence and growth of a variety of infrastructural, economic, social and cultural facilities that cannot be provided where their customers are geographically dispersed. The larger the city, quite obviously, the greater the variety of available facilities and vice versa.

- *Specialized clusters,* on the other hand, reflect the tendency for firms in the same, or closely related, industries to locate in the same places to form what are sometimes termed 'industrial districts' or 'industrial spaces'. Such benefits have been called *localization economies.* The bases of specialized clusters arise from the geographical proximity of firms performing different – but *linked* – functions in particular production networks.

Clusters generate two types of interdependency:

- *Traded interdependencies* are direct transactions between firms in the cluster (e.g. the supply of specialized inputs of intermediate products and services). In such circumstances, spatial proximity is a means of reducing transaction costs either through minimizing transportation costs or by reducing some of the uncertainties of customer–supplier relationships.
- *Untraded interdependencies* are the less tangible benefits, ranging from such things as the development of an appropriate pool of labour, to particular kinds of institutions (such as universities, business associations, government institutions and the like) to broader socio-cultural phenomena. In particular, geographical agglomeration or clustering facilitates three important processes: face-to-face contact; social and cultural interaction; and enhancement of knowledge and innovation.[28]

But why do clusters form in the first place? Why do they arise in one place rather than another? And how do they develop over time? These are difficult questions to answer. The reasons for the origins of specific geographical clusters are highly contingent and often shrouded in the mists of history.

Figure 2.10

The bases of geographical clusters

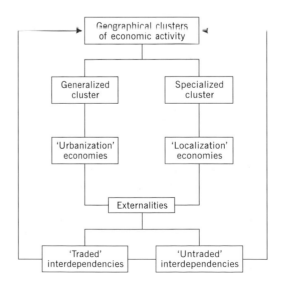

> Within broad limits the power of attraction today of a center has its origin mainly in the historical accident that something once started there, and not in a number of other places where it could equally well or better have started, and that the start met with success.[29]

Once established, a cluster tends to grow through a process of *cumulative, self-reinforcing development* involving:

- attraction of linked activities
- stimulation of entrepreneurship and innovation
- deepening and widening of the local labour market
- economic diversification
- enrichment of the 'industrial atmosphere'
- 'thickening' of local institutions
- intensification of the socio-cultural milieu
- enhanced physical infrastructures.

The cumulative nature of these processes of localized economic development suggests that the process is one of *path-dependency*. An economy becomes 'locked into' a pattern that is strongly influenced by its particular history. This may be either a source of continued strength or, if it embodies too much organizational or technological rigidity, a source of weakness. However, even for 'successful' regions, such path dependency does not imply the absolute inevitability of continued success.

> [T]he onward march of development in economically successful regions is always in practice subject to eventual cessation or reversal, not only because there *are* usually limits to the continued appropriation of external economies, but also because radical shifts in markets, technologies, skills, and so on, can undermine any given regional configuration of production. Indeed, the very existence of lock-in effects means that regions, as they develop and grow, will eventually find it difficult to adapt to certain kinds of external shocks.[30]

One way of thinking of the structure of the global economy, therefore, is to see it as the linking together of two networks: the *organizational* (in the form of production networks) and the *geographical* (in the form of localized clusters of economic activity). In Figure 2.11

> the developed areas of the world are represented as a system of polarized regional economies each consisting of a central metropolitan area and a surrounding hinterland (of indefinite extent) occupied by ancillary communities, prosperous agricultural zones, smaller tributary centers, and the like … Each metropolitan nucleus is the site of intricate networks of specialized but complementary forms of economic activity, together with large, multi-faceted local labor markets, and each is a locus of powerful agglomeration economies and increasing returns effects … These entities can be thought of as *the regional motors of the new global economy* … Equally, there are large expanses of the modern world that lie at the extensive economic margins of capitalism … Even so, underdeveloped areas are occasionally punctuated by islands of relative prosperity.[31]

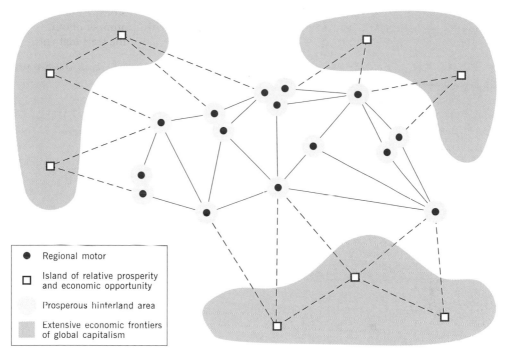

Figure 2.11 A schematic representation of the geography of the global economy

Source: Based on Scott, 1998: Figure 4.2

The geo-economy and the environment[32]

Production as a system of materials flows and balances

In our earlier description of the production process as a 'chain', in which inputs are transformed into products which are then consumed (see Figure 2.3), we ignored the fact that the inputs have to come from somewhere and that the consumption of products is not the end of the story. Ultimately, the 'somewhere' from which all inputs of materials and energy derive, and to which what is left over after production and consumption have taken place must go, is the *natural environment*. In the final analysis, '*all* production depends on and is grounded in the natural environment'.[33] Although the primary purpose of the production process shown in Figure 2.3 is the production of 'goods' for consumption (driven, in a capitalist market system, of course, by profits) the process itself – in a way unintentionally – also produces 'bads' in the form of environmental degradation. There are, in other words, unintended external effects (*negative externalities* or *spillovers*) involved in all economic activities.

Three aspects of such environmental damage are especially important:[34]

● over-use of non-renewable and renewable resources (including, for example, exploitation of fossil fuels, clearance of forests)

- overburdening of natural environmental 'sinks' (for example, the increasing concentration of greenhouse gases in the earth's atmosphere and of heavy materials in the soil)
- destruction of increasing numbers of ecosystems to create space for urban and industrial development.

The bases of such environmental damage can be understood most easily if we produce a parallel diagram to that of the production process discussed earlier. Figure 2.12 depicts the production process of Figure 2.3 in terms of *materials flows* and *materials balances*.[35] The key point of the process is that what goes in has to come out again, albeit transformed, but without being reduced. In Figure 2.12, the materials used in the production process are

> dispersed and chemically transformed. In particular, they enter in a state of low entropy (as 'useful' materials) and leave in a state of high entropy (as 'useless' materials, such as low temperature heat emissions, mixed municipal wastes, etc.) ... No material recycling processes can therefore ever be 100 per cent efficient.[36]

In effect, the economic system in general, and the production process in particular, place demands on the natural environment in two ways:

- in terms of *inputs* to the process of production which are derived from the natural environment as *resources*
- in terms of *outputs* to the natural environment in the form of pollution of various kinds.

Let's look briefly at each of these in turn.

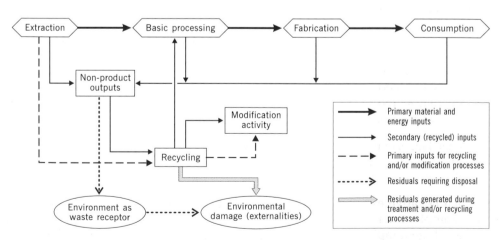

Figure 2.12   The production process and the environment

Source: Based on Turner et al., 1994: Box 1.2

The resources issue

We talk about natural resources as if they are 'naturally' resources. But that isn't strictly true. A naturally occurring element or material is a 'resource' only if it is defined as such by potential users. In other words, there must be an effective demand, there must be an appropriate technology with which to exploit it, and there must be some means of ensuring 'property rights' over its use: 'if any of these conditions ceases to hold, resources could "unbecome"'.[37] Resources, once defined as such, are of two broad kinds:

- *renewable resources* – those which, over time and with good management, can be replenished; nevertheless, most so-called renewable resources may become exhausted if they are not managed in a sustainable manner
- *non-renewable resources* – those which are fixed in overall quantity, at least under known technological conditions; the more we use today the less will be available for tomorrow.

One of the major current debates, therefore, is the extent to which non-renewable resources, in particular, are becoming increasingly scarce through excessive exploitation and, therefore, facing imminent exhaustion. Here, as in all areas of the environmental debate, views become polarized. On the one hand there is the 'Malthusian' view that resource exhaustion is inevitable; the only question is the time-scale over which such exhaustion will occur. On the other hand, there is the view that new technologies of exploration leading to discoveries of new reserves, better means of exploitation leading to more efficient use of the resource (including recycling), and the development of appropriate substitutes will put off the dreadful day. So far, at least, all of these have happened. The dire predictions of imminent resource exhaustion have not been borne out (yet). Nevertheless, there is a real danger of resource exhaustion in specific areas and of continuing localized environmental damage.

Unintended effects of production

As Figure 2.12 shows, after all efforts are made to recycle the unused energy and materials involved in production, there will still be 'things' left over in the form of residual waste and environmental damage. This is simply because the fundamental laws of thermodynamics cannot be overruled:

> the total mass of inputs to a transformation process is equal to the total mass of outputs. If inputs do not emerge as desired products, they must therefore appear as unwanted by-products or wastes.[38]

Such negative externalities are of various kinds and of varying spatial extent. For example, the negative externalities from a factory or from an airport are, at one level, geographically localized. The impact is greatest at the location of the facility itself and its immediate neighbourhood but then declines with increasing distance away from that location. On the other hand, the smoke pollution from the factory or the effect of aircraft fuel combustion may have much more extensive geographical effects, particularly on the atmosphere. The problem is that many adverse environmental effects cannot be contained within geographical boundaries.

By far the most contentious aspect of negative environmental externalities relates to potential *atmospheric damage*, that is, damage to the gaseous membrane that sustains all life on earth. The processes of material transformation involve the use of massive quantities of energy, especially of fossil fuels whose combustion products are the major source of damage to the earth's atmosphere. The problems arise because some of the key gaseous components of the earth's atmosphere – notably carbon dioxide, methane and ozone – are becoming excessively concentrated in the atmosphere. The issue is one of balance. Without these, and other, gases the earth would have a surface temperature like that of the planet Mars; that is, it would be uninhabitable. The earth's surface remains habitable precisely because of their presence in the atmosphere. In combination, they act like a 'greenhouse', preventing both excessive solar heating and excessive cooling. But it is a very delicate balance.

Most scientists believe that this balance is dangerously disturbed by human action. In the case of the main greenhouse gas, carbon dioxide, for example, there is clear evidence of a significant acceleration after around 1800:[39]

- before 1800, carbon dioxide levels in the atmosphere remained fairly stable at between 270 and 190 parts per million (ppm)
- by 1900, levels had reached 295 ppm
- by 1950, the levels had risen to between 310 and 315 ppm
- by 1995, carbon dioxide levels in the atmosphere had reached 360 ppm.

These progressive increases were closely associated with the processes of industrialization and urbanization, through the burning of fossil fuels and through deforestation. At the same time, levels of methane have increased dramatically, partly through the same fossil fuel usage and partly through agricultural practices like rice irrigation, livestock production and garbage decomposition. As a result, the atmosphere has become more efficient at trapping heat from the sun.[40]

So, the general consensus is that *global warming* is occurring, although there are some scientists who dispute its nature and rate. However, if the trend continues, it will cause potentially enormous social and economic damage in many parts of the world, notably: major dislocation of climatic zones around the world, causing flooding of many coastal and low-lying areas, altering agricultural economies; changing patterns of disease; increasing volatility of atmospheric systems (for example, an increase in severe storms). But how much, and how fast such warming will be is unclear, as is the extent to which it is primarily the result of human action rather than part of a naturally occurring climatic cycle.

The other aspect of atmospheric – or, rather, stratospheric – damage concerns the earth's *ozone layer*. Ozone is formed in the stratosphere through the chemical reaction of oxygen and sunlight. At this level, ozone is vital to the sustainability of human life on earth because it absorbs almost all the ultra-violet radiation from the sun, which would otherwise make human life impossible. Any damage to this vital protective shield poses a serious problem. Just such thinning of – or even holes in – the ozone layer (beyond natural occurrences) began to be identified in certain parts of the world in the early 1970s. One of the major effects of ozone depletion is an increase in the incidence of skin cancer. A primary cause was believed to be the chemical chlorofluorocarbon (CFC), which had become extensively used in refrigeration and aerosols. Although CFCs are now heavily restricted, the fact that the chemical is immensely stable means that the amount already in the stratosphere will still be affecting the ozone layer until about 2087.[41]

Hence, the environmental problems that are inherent in production raise serious questions about the future sustainability of economy and society as we know it. They raise big issues relating to the future of the world's economic and trading system and, indeed, to most aspects of contemporary economic life. As such, they have come to form a major element in the globalization debates and in the anti-globalization movements. This raises a major problem of global governance, which we will address in the final chapter of this book.

Conclusion

The purpose of this chapter has been to 'interrogate' the concept of globalization and to refute the popular view that it is some kind of all-embracing, inexorable, irreversible, homogenizing force. Rather, the world in which we live is constituted through, and transformed by, a complex of inter-related processes rather than by some single force called 'globalization.' The processes that are transforming the geo-economy are highly uneven in their operation and in their effects. Without doubt the world *is* a qualitatively different place from that of only sixty or seventy years ago, although it is not so much the case of being more open but of being increasingly *interconnected* in rather different ways.

One way of understanding the nature of this change is to think in terms of production chains and networks configured at a multiplicity of geographical scales, from the local through to the global. Such networks are the structures through which different parts of the world are connected together through flows of material and non-material phenomena in a system of differential power relationships. The governance of production networks has become increasingly dominated by transnational corporations, although the nature of such governance varies enormously between different kinds of economic activities. Contrary to the popular assertions of the 'end of geography', production networks are intrinsically geographical in terms of their differentiated spatial configurations and their territorial embeddedness in specific places. In particular, economic activities have a very strong propensity to cluster or agglomerate in particular kinds of location. Such clusters, once formed, have a strong tendency to develop in path-dependent ways which influence – although they do not totally determine – future geographies.

Finally, we explored briefly the connections between the production of material goods and the environmental impact of such production. Seeing the production process as a system of materials flows and balances, subject to the inexorable laws of thermodynamics, helps us to appreciate the ultimate dependence of production on the natural environment, both as a source of materials inputs in the form of renewable and non-renewable resources and as a receptor of the waste products of production. This question of the relationship between the production of 'goods' and 'bads' is at the heart of the future development of the geo-economy.

Notes

1 Based on an analysis of the *Social Sciences Citation Index*.
2 The authors are, respectively, Klein (2000), Barnet and Cavanagh (1994), Greider (1997), Korten (1995), Luttwak (1999) and Micklethwait and Wooldridge (2000).
3 Castells (1989: 10). Bell (1974) was probably the first to popularize the notion of an informational society in his influential book, *The Coming of Post-Industrial Society*.
4 Here we follow Held et al. (1999: 2–10) in distinguishing between three positions within the globalization debate: hyperglobalists, sceptics and transformationalists.
5 Kozul-Wright (1995: 139–40).
6 UNCTAD (1993: 113).
7 This argument is developed more fully in Dicken, Peck and Tickell (1997).
8 Mittelman (2000: 4).
9 Taylor (2001).
10 Castells (1996), for example, has long advocated a network approach to understanding societal change. Some of the issues involved in adopting a network approach to the global economy are discussed in Dicken, Kelly, Olds and Yeung (2001), Henderson, Dicken, Hess, Coe and Yeung (2002).
11 The 'chain' metaphor has been used in a number of different disciplinary contexts with slightly varying terminology: value chains, commodity chains, supply chains, filières, etc. Sturgeon (2001) discusses issues of definition of value chains and production networks. Raikes, Jensen and Ponte (2000) provide a comparison and critique of the global commodity chain and filière approaches. Mentzer (2001) provides a detailed treatment of supply chains in the business management and logistics context.
12 Rabach and Kim (1994: 123).
13 Humphrey and Schmitz (2001) discuss issues of governance in global value chains.
14 This 'markets and hierarchies' view of the governance of economic transactions was developed by Williamson (1975). As a concept, it derives from the work of Coase (1937), who addressed the fundamental question of why multi-function firms exist at all.
15 Gereffi (1994: 97).
16 Geographical scale is a slippery concept on which there is a large and growing literature. See, for example, Brenner (1998, 2001); Dicken, Kelly, Olds and Yeung (2001); Kelly (1999); Swyngedouw (1997, 2000).
17 Latour quoted in Thrift (1996: 5).
18 See Ernst and Kim (2001); Henderson, Dicken, Hess, Coe and Yeung (2002).
19 Thrift (1990: 181).
20 Brenner (1998: 8).
21 See, for example, O'Brien 1992; Cairncross, 1997.
22 Castells's arguments were developed initially in his book *The Informational City* (1989) and have been elaborated more recently in *The Rise of the Network Society*, Volume 1 of *The Information Age: Economy, Society and Culture* (1996).
23 Clark (1994), Clark and Wrigley (1995) and Schoenberger (1997) explore the nature and significance of sunk costs in corporate decision-making and corporate restructuring in specifically spatial contexts.
24 Granovetter (1985) is an especially influential writer in this regard. See also Dicken and Malmberg (2001); Dicken and Thrift (1992); Oinas (1997, 1999).
25 Stopford and Strange (1991).
26 Markusen (1996).
27 The pervasiveness and the significance of geographical clustering has recently been recognized by some leading economists and management theorists, notably Krugman (1995, 1998), Porter (1990, 1998, 2000) and also in national policy debates in different parts of the world. However,

economic geographers and location theorists have been pointing to the pervasiveness of this phenomenon for decades. Dicken and Lloyd (1990: chs 5 and 6) review the traditional economic-geographical and location-theoretic approaches to spatial concentration that derive from the pioneering ideas of Alfred Marshall, Alfred Weber and Edgar Hoover. The newer agglomeration literature is represented by Amin and Thrift (1992, 1994), Malmberg (1996, 1999), Markusen (1996), Scott (1995, 1998) and Storper (1997).

28 Amin and Thrift (1994: 13).

29 Myrdal (1958: 26).

30 Scott (1995: 57).

31 Scott (1998: 68, 70).

32 Very helpful discussions of the relationship between production and the environment are provided by Cairncross (1992), Hudson (2001: ch. 9), McNeill (2000) and Turner, Pearce and Bateman (1994). The Worldwatch Institute publishes an annual survey of environmental issues: see Brown (2002).

33 Hudson (2001: 300).

34 Simonis and Brühl (2002: 98).

35 Turner, Pearce and Bateman (1994: 15–23).

36 Turner, Pearce and Bateman (1994: 17).

37 Hudson (2001: 301).

38 Hudson (2001: 288).

39 McNeill (2000: 109).

40 McNeill (2000: 109).

41 McNeill (2000: 114).

CHAPTER 3

The Changing Global Economic Map

The imprint of history

For several hundred years, from the beginnings of the emergence of a capitalist market economy, the world economic map had a basically simple structure of a core and a periphery (see Figure 2.1). Of course, over time, this structure became more complex in detail and also changed in its composition. Some core economies experienced a progressive decline to semi-peripheral status during the 18th century and new economies emerged, especially in the late 19th century. In 1870, the United Kingdom was still the dominant world economic power, producing some 30 per cent of the world's total industrial output. In comparison, the newly industrializing United States produced almost 25 per cent of the total, and Germany produced 13 per cent. By 1913, the United Kingdom's share of the world total had fallen to 14 per cent while that of the United States had increased to 36 per cent.

Nevertheless, the broad contours of this long-established global economic map persisted until the outbreak of World War II in 1939. Manufacturing production remained strongly concentrated in the core: 71 per cent of world manufacturing production was concentrated in just four countries and almost 90 per cent in only 11 countries. Japan produced only 3.5 per cent of the world total. The group of core industrial countries sold some 65 per cent of its manufactured exports to the periphery and absorbed 80 per cent of the periphery's primary products.[1] A clear international division of labour persisted. International direct investment by the rapidly developing transnational corporations was also dominated by firms in these leading core nations and was most strongly concentrated in the developing countries which, on the eve of World War II, were host to 65 per cent of total foreign direct investment.

This long-established structure was shattered by World War II, which devastated the global economy. The vast majority of the world's industrial capacity (outside North America) was destroyed and had to be rebuilt. At the same time, new technologies were created and many industrial technologies were refined and improved in the process of waging war. Hence, the world economic system that emerged after 1945 was, in many ways, a new beginning. It reflected both the new political realities of the post-war period – particularly the sharp division between East and West – and also the harsh economic and social experiences of the 1930s.

The major political division of the world after 1945 was essentially that between the capitalist West (the United States and its allies) and the communist East (the Soviet Union and its allies). Outside these two major power blocs was the so-called

'Third' World, a highly heterogeneous – but generally impoverished – group of nations, many of them still at that time under colonial domination. The Third World was far from immune from the East–West confrontation. Both major powers made strenuous efforts to extend their spheres of influence with considerable implications for the subsequent pattern of global economic change. The Soviet bloc drew clear boundaries around itself and its Eastern European satellites and created its own economic system, quite separate from the capitalist market economies of the West, at least initially. In the West the kind of economic order built after 1945 reflected the economic and political domination of the United States. Alone of all the major industrial nations, the United States emerged from the war strengthened rather than weakened. It had both the economic and technological capacity and also the political power to lead the way in building a new order.

It is from this historical baseline, therefore, that recent global shifts in economic activity will be examined in this chapter. Today's world is far more complex than it was even a few decades ago. There has been a truly fundamental transformation of the world economy: a new geo-economic map has come into being, although one which, of course, bears many traces of the contours of the old. The chapter is divided into four major parts:

- First, we analyse changing aggregate trends in global economic activity, emphasizing the volatility of such trends.

- Second, we plot the changing contours of the global economic map in terms of global shifts in production and trade.

- Third, we focus on global shifts in foreign direct investment.

- Finally, we take a multi-scalar approach to demonstrate the kinds of global economic map that exist above and below the conventional scale of the nation-state.

The roller-coaster ride: aggregate trends in global economic activity

> The world economy performed better in the last half century than at any time in the past. World GDP increased six-fold from 1950 to 1998 with an average growth of 3.9 per cent a year compared with 1.6 from 1820 to 1950, and 0.3 per cent from 1500 to 1820.[2]

However, even within this broad surge of economic growth there were some pretty large interruptions to the upward curve. In fact, the path of economic change is best seen as being like a roller-coaster. Sometimes the ride is relatively gentle with just minor ups and downs; at other times the ride is truly stomach-wrenching, with steep upward surges separated by vertiginous descents to what seem like bottomless depths (Figure 3.1).

The years immediately following World War II were ones of basic reconstruction of war-damaged economies throughout the world. At the time it was felt that growth rates would then slacken in the 1960s. This did not happen. Instead, rates of economic growth reached unprecedented levels:

- between 1948 and 1953 world trade increased at an average annual rate of 6.7 per cent
- between 1958 and 1963 the rate rose to 7.4 per cent
- between 1963 and 1968 it accelerated further to 8.6 per cent.

The period between the early 1950s and the early 1970s became known as the 'golden age'. People began to believe that the roller-coaster days were over. Growth, in the Western industrialized economies at least, came to be an expectation rather than a hope, even though such growth was extremely variable from one country to another and between different parts of the same country. In fact, 'the golden age was only partly golden: it was more golden in some places than others, for some people than others'.[3] But then, in the early 1970s, the sky fell in. The long boom suddenly went 'bust', the 'golden age' of growth became tarnished. As Figure 3.1 shows, growth rates declined with each successive decade. Throughout the 1980s and 1990s annual rates of growth became very variable indeed. The roller-coaster had come back with a vengeance.

The most obvious explanation for the sudden end of almost continuous growth from 1950 to 1973, at the time, seemed to be the decision by the Organization of Petroleum Exporting Countries (OPEC) to quadruple oil prices. Certainly this was a massive shock to the system but, with hindsight, it became clear that this was merely the final piece in a jigsaw puzzle which had been taking shape since the late 1960s. Commodity prices, other than oil, had been rising steeply. Labour costs in all the industrialized nations had begun to accelerate as the level of wage

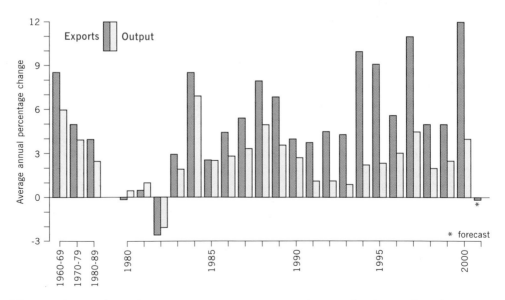

Figure 3.1   The roller-coaster of world merchandise production and trade

Source: Based on GATT (1990) *International Trade, 1989–1990*, vol. II: Chart I.1; WTO (2001) *International Trade Statistics, 2001*: Table I.1

settlements rose. The international monetary system created at Bretton Woods in 1944 became increasingly unstable as national currencies became more and more out of line with the fixed exchange rates. In 1971 the United States moved to a floating exchange rate, other countries followed, and the Bretton Woods system was no more. As the recession deepened during the second half of the 1970s and into the early 1980s emphasis shifted towards the view that the changes were the result of more deep-seated and fundamental processes.

Rates of growth during the 1980s were extremely variable, ranging from the negative growth rates of 1982 through to two years (1984 and 1988) when growth of world merchandise trade reached the levels of the 1960s once again. Overall, growth – albeit uneven growth – reappeared. But then, in the early 1990s, recession occurred again. In 1994 and 1995, strong growth reappeared, especially in exports. A similarly volatile pattern characterized the last years of the century. There was spectacular growth in world trade in 1997, followed by far slower growth in 1998 and 1999 (partly related to the Asian financial crisis and to its contagious effects on other parts of the world). Then, once again, there was spectacular growth in world trade in 2000 (12 per cent). At the time of writing, however, it is clear that the growth bubble of 2000 has burst again, a problem certainly exacerbated (though not caused) by the events of 11 September 2001.

Thus, one major characteristic of world economic growth is its inherent volatility: periods of very rapid growth being interspersed with periods of very slow – or even negative – growth. A second is the faster growth of trade than of production, a clear indication of the greater *interconnectedness* of the global economy. This second characteristic is shown very clearly by the widening gap between the two curves of trade and production in Figure 3.2. Between 1950 and the end of the 20th century, world merchandise trade increased almost twenty-fold while world merchandise production increased just over six-fold.

Figure 3.2

The growing interconnectedness of the world economy: the widening gap between trade and production

Source: Based on WTO (1995)
International Trade, 1995: Chart II.1

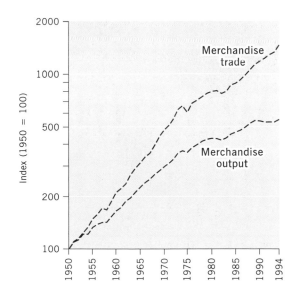

Changing contours of the global economic map: global shifts in production and trade

Aggregate figures at the global scale mask very significant geographical variations. However, mapping the changing contours of the global economic map is a difficult and challenging task, partly because of the problems of acquiring appropriate and up-to-date statistical data and partly because of the intrinsic complexity of the processes involved. We will begin by providing a snapshot of the current global maps of production and trade, and of recent changes in these maps, before looking at the differentials within and between the major regions of the world.

Manufacturing production

Despite the vastly increased importance of services in virtually all economies in recent decades, for most of the past 50 years it has been manufacturing production – and especially manufacturing trade – that has been the primary driver of the global economy. This has been especially true for those developing countries that have become increasingly integrated into the global economy. But although there are virtually no countries in the world without at least some manufacturing activity, the map of world manufacturing production (Figure 3.3) is incredibly uneven. The overwhelming majority of manufacturing production is concentrated in a relatively small number of countries.

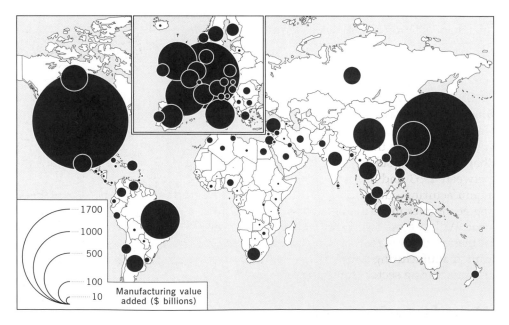

Figure 3.3   The map of world manufacturing production

Source: Based on material in UNIDO (1997) *Industrial Development: Global Report, 1997*

These features can be seen more clearly in Table 3.1. The 15 countries listed produce 84 per cent of total world manufactured output. However, a mere three countries – the United States, Japan and Germany – account for almost 60 per cent of the total. Only four of the countries listed in Table 3.1 can be regarded as developing countries. The vast majority of developing countries have only a very small manufacturing base; the 'manufacturing tail' of the world economy is very long indeed.

Table 3.1 The world 'league table' of manufacturing production

Rank 1996	Country	Manufacturing value added (US$m)	Percentage of world total
1	United States	1,696,955	24.8
2	Japan	1,365,523	20.0
3	Germany	808,320	11.8
4	France	297,536	4.3
5	Brazil	277,242	4.1
6	United Kingdom	266,606	3.9
7	South Korea	196,400	2.9
8	China	186,952	2.7
9	Italy	156,300	2.3
10	Canada	119,348	1.7
11	Spain	95,386	1.4
12	Taiwan	77,097	1.1
13	Switzerland	69,457	1.0
14	Australia	65,859	1.0
15	Netherlands	56,417	0.8
	Total		83.8

Source. Based on data in UNIDO (1997) *Industrial Development: Global Report, 1997*

If we compare the current situation with that of only 50 years ago we can see that some very substantial global shifts have occurred. Figure 3.4 shows that between 1953 and the late 1990s the industrialized economies' share of world manufacturing output declined from 95 to 77 per cent while that of the developing economies more than quadrupled, albeit from a very low base level, to 23 per cent. Within these two broad categories, however, there has been wide variation in manufacturing growth. Only a few developing countries have become significant manufacturing producers. For the vast majority, as Figure 3.3 shows, the manufacturing sector is minuscule.

Manufacturing trade

Very rapid growth of manufacturing trade was a distinctive feature of the postwar period (see Figures 3.1 and 3.2). Manufacturing has come to account for an

Figure 3.4

The changing distribution
of world manufacturing
production between
industrialized and developing
countries

Source: Based on UNIDO (1986) *World
industry: a statistical review, 1985,
Industry and Development*, 18:
Figure 1; UNIDO data base

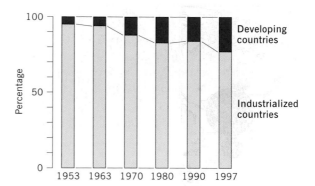

increasingly larger proportion of total exports in both developed and developing
market economies. However, the shift was particularly strong in the latter, where
manufacturing accounted for only 20 per cent of total exports in 1960 but accounts
for more than 50 per cent today. More significantly, by the end of the 1970s – for
the very first time – the value of manufactured products exported from the developing economies exceeded that of their food and raw materials exports. From the
early 1970s, exports of manufactured goods from developing countries as a whole
grew at twice the rate of exports of raw materials. Without doubt the old international division of labour has been displaced.

Viewed in isolation international trade figures can be misleading. For example,
other things being equal, international trade will be far more important relatively
speaking for a small economy than for a larger one (an obvious example would be
to contrast the United States with Singapore or Hong Kong). Similarly, although in
general there is a relationship between manufacturing production and trade, the
relationship is not exact. Some countries which are very significant as producers
are less significant as exporters (again, the United States is an example, as is Brazil).

Figure 3.5 shows the global map of merchandise exports. Although the trade
map is broadly similar to that of production (see Figure 3.3), the major difference
is that the East Asian economies are more important as exporters of manufactured
goods than they are as producers, while the converse is true of the United States
and Latin America.

Table 3.2 (p. 42) is the 'league table' of trade by country and should be compared
with Table 3.1. Such a comparison reveals some interesting differences:

- The origins of manufactured exports are less concentrated than those of
  production (69 per cent compared with 84 per cent).

- Whereas the top three producers (the United States, Japan and Germany)
  account for 57 per cent of world manufacturing production the same three
  countries account for only 29 per cent of world manufacturing exports.

- The share of world manufactured imports accounted for by the top 15
  importers increased dramatically between 1963 and 2000, from 46 per cent
  to 71 per cent.

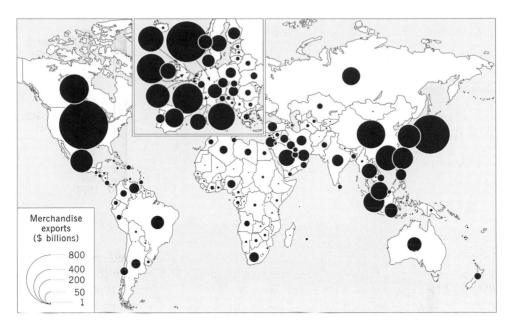

Figure 3.5 The map of world merchandise exports

Source: Based on data in WTO (2001) *International Trade Statistics, 2001*: Table A5

- Whereas only four developing countries are in the top 15 producers of manufactures, there are six in the top 15 exporters. Only one of the six (Hong Kong) was in the top 15 exporters in 1963.
- Although the United States and Germany were the top two exporters of manufactured products in both 1963 and 2000, their combined share has fallen from 33 per cent to 21 per cent.
- Conversely, their share of manufactured imports increased – from around 15 per cent in 1963 to almost 27 per cent in 2000. Virtually all of that increase was accounted for by the United States, whose share of world manufactured imports increased from less than 9 per cent of the world total in 1963 to 19 per cent in 2000. Clearly, the United States had become the 'importer of last resort' for the world economy.

An important aspect of a country's trade is the net balance between exports and imports. Figure 3.6 shows these net differences and reveals the marked contrast between the large manufacturing trade surplus of Japan and Germany, on the one hand, and the massive trade deficit of the United States on the other. Most Latin American countries (apart from Argentina, Colombia, Ecuador and Venezuela) have manufacturing trade deficits. Conversely, all the East Asian economies (apart from Hong Kong) have large surpluses. South Asia, on the other hand, is uniformly a deficit region. The situation in Europe is mixed. While Germany, the Scandinavian countries, Ireland, Italy and the Netherlands have surpluses, the United Kingdom (in particular), France, Portugal, Spain and Switzerland have deficits.

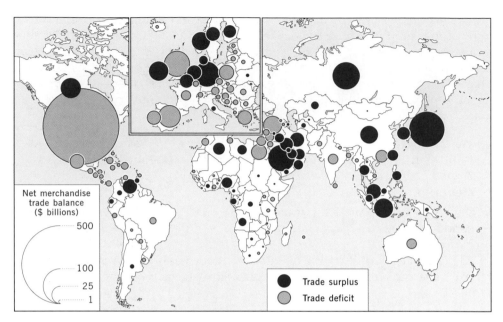

Figure 3.6 The pattern of merchandise trade surpluses and deficits

Source: Based on data in WTO (2001) *International Trade Statistics, 2001*: Table A6

Table 3.2 The world 'league table' of manufacturing trade

	\multicolumn{4}{c}{Exports}		\multicolumn{4}{c}{Imports}
	\multicolumn{2}{c}{2000}	\multicolumn{2}{c}{1963}		\multicolumn{2}{c}{2000}	\multicolumn{2}{c}{1963}
	%	Rank	%	Rank		%	Rank	%	Rank
USA	12.3	1	17.4	1	USA	18.9	1	8.6	1
Germany	8.7	2	15.6	2	Germany	7.5	2	6.2	2
Japan	7.5	3	6.1	5	Japan	5.7	3	1.9	13
France	4.7	4	7.0	4	UK	5.1	4	4.6	6
UK	4.5	5	11.4	3	France	4.6	5	4.7	5
Canada	4.3	6	2.6	12	Canada	3.7	6	5.0	4
China	3.9	7	NA	NA	Italy	3.5	7	4.1	8
Italy	3.7	8	4.7	6	China	3.4	8	NA	NA
Netherlands	3.3	9	3.3	9	Hong Kong	3.4	9	NA	NA
Hong Kong	3.2	10	0.9	15	Netherlands	3.0	10	4.3	7
Belgium	2.9	11	4.3	7	Mexico	2.7	11	NA	NA
South Korea	2.7	12	0.0	NA	Belgium	2.7	12	NA	NA
Mexico	2.6	13	NA	NA	South Korea	2.4	13	0.3	NA
Taiwan	2.3	14	0.2	NA	Spain	2.3	14	1.2	NA
Singapore	2.2	15	0.4	NA	Taiwan	2.1	15	0.2	NA
Total	68.8		76.7		Total	71.0		46.1	

NA, data not available.

Source: Based on data in WTO (1996) *Annual Report, 1996: Volume II*: Tables II.11, II.12; WTO (2001) *International Trade Statistics*: Table I.5

The geography of trade in manufactures is far more complicated than that of production, simply because trade consists of *flows* between areas. However, there is a strong tendency for countries to trade most with their neighbours: a considerable proportion of world trade is intra-regional, although this does not imply that such regionalizing tendencies dominate in all cases.[4] Figure 3.7 provides a broad-brush picture of this global network of merchandise trade. Several features merit attention:

- Western Europe is the world's major trading region. Two-thirds of that trade is intra-regional, that is, between Western European countries themselves. Around 11 per cent of Western Europe exports go to North America and about 8 per cent to Asia.

- Asia is the second most significant trade region, with half of its trade conducted internally. Just over one-quarter of Asia's external trade goes to North America and a further 17 per cent to Western Europe. Within Asia, 43 per cent of Japan's trade is within the region and no less than 65 per cent of Australia and New Zealand's trade is within Asia (a very clear indication of

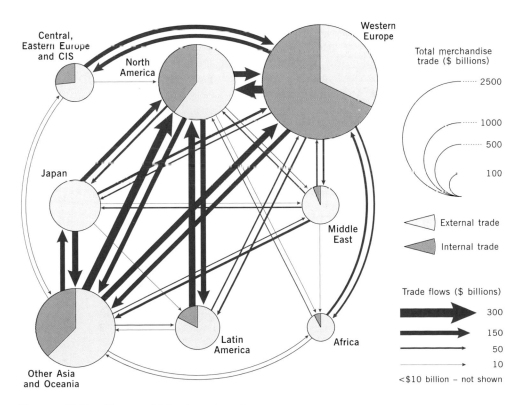

Figure 3.7   The world trade network

Source: Based on data in WTO (2001) *International Trade Statistics, 2001*: Table A3

their geopolitical re-orientation away from their traditional markets in Europe).

● North America conducts around 40 per cent of its trade internally. Its external trade is distributed fairly evenly between Asia (22 per cent), Western Europe (18 per cent) and Latin America (16 per cent).

Production of services

To focus only on manufacturing production is to tell only a part of the story. The growth of international manufacturing activity and of international trade in manufactured goods can occur only with the parallel development of circulation activities within the production chain (see Figures 2.3 and 2.4), notably the whole variety of commercial, financial and business services. These are, in themselves, becoming increasingly internationalized. In effect, the production – and trade – of goods and services are becoming increasingly interlinked.

In all the industrialized economies, the services sector accounts for a larger share of Gross Domestic Product (GDP) than manufacturing (Table 3.3). However, a large service sector is not confined to the developed economies, although it is certainly true that the relative importance of services does vary more or less directly with a country's income level. What does differ is the kind of service activity involved. Much of the service sector in both developing and industrialized countries is either low-skill, low-technology, private service activity (including wholesale and retail activity) or activities within the public sector (for example, health, education, welfare services, etc.). The service sector is exceptionally diverse. It is the business services which tend to be much more strongly developed in the industrialized countries and far less developed in countries lower down the development scale.

Table 3.3 The contribution of the service sector to Gross Domestic Product

Country group	Percentage of GDP		
	1965	*1987*	*2000*
Low income	30	32	43
Lower-middle income	50	46	46
Upper-middle income	42	50	61
High income	54	61	64

Source: Based on data in World Bank, *World Development Report*, various issues

Trade in services

Although many service activities are not 'tradable' – that is, they have to be provided to customers on a face-to-face basis – the growth of international trade in commercial services has accelerated. This is especially true of telecommunications, financial services, management, advertising, professional and technical services. Overall, between 1989 and 2000, services exports grew at 10.7 per cent per year while manufacturing exports grew at an annual average rate of 9.6 per cent.

Figure 3.8 is the global map of commercial services trade. Like the map of manufacturing exports, it shows great geographical unevenness, although services trade is less concentrated than manufacturing. Around 63 per cent of services exports and 67 per cent of imports are generated by the top 15 countries. As in manufacturing, the United States is the dominant player, with 12 per cent of services exports and 14 per cent of imports. The United Kingdom is also far more important as a generator of services exports than it is of manufactures. Conversely, Germany's position is substantially weaker in services, as is Japan's. Regionally, Western Europe dominates, with 45 per cent of total services exports, followed by the United States with 22 per cent and Asia with 21 per cent. Together, the three regions account for 88 per cent of services exports compared with 84 per cent of manufacturing exports.

The pattern of trade surpluses and deficits in services is rather different from that in manufacturing (compare Figure 3.9 with Figure 3.6). At the regional scale, North America and Western Europe have a substantial surplus in services trade (compared with their deficits in manufacturing). Conversely, Asia has a large deficit in its services trade in contrast with its large manufacturing surplus. These

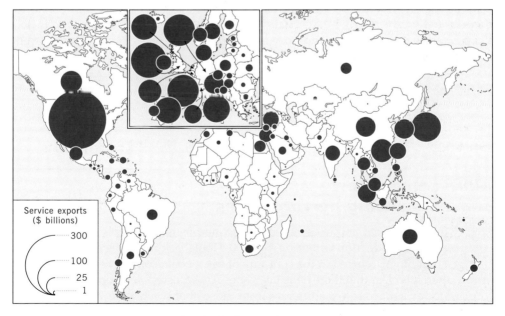

Figure 3.8 The map of world services exports

Source: Based on data in WTO (2001) *International Trade Statistics, 2001*: Table A7

contrasts are reflected at the level of individual countries. Among the industrialized countries, the United States, the United Kingdom and France have trade surpluses in services and deficits in manufacturing while Germany and Japan have services deficits but manufacturing surpluses. Of the Asian developing countries, Hong Kong and Singapore have services surpluses but most of the others have deficits. Nevertheless, the East Asian countries, in particular, have become increasingly important as service exporters. As Table 3.4 shows, Hong Kong, China, South Korea and Singapore all appear in the list of the world's top 15 services exporters.

Table 3.4   The world 'league table' of commercial services trade, 2000

	Exports			Imports	
	%	Rank		%	Rank
USA	12.3	1	USA	13.8	1
UK	7.0	2	Germany	9.2	2
France	5.7	3	Japan	8.1	3
Germany	5.6	4	UK	5.7	4
Japan	4.8	5	France	4.3	5
Italy	4.0	6	Italy	3.9	6
Spain	3.7	7	Netherlands	3.6	7
Netherlands	3.6	8	Canada	2.9	8
Hong Kong	2.9	9	Belg.–Lux.	2.7	9
Belg.–Lux.	2.9	10	China	2.5	10
Canada	2.6	11	South Korea	2.3	11
China	2.1	12	Spain	2.1	12
Austria	2.1	13	Austria	2.0	13
South Korea	2.0	14	Ireland	2.0	14
Singapore	1.9	15	Hong Kong	1.8	15
Total	63.2		Total	66.9	

Source: Based on data in WTO (2001) *International Trade Statistics, 2001*: Table I.7

More competitors – and some surprising performances

So far we have looked at the current global economic map in terms of its individual components of production and trade. But if we look back more systematically at the changes that have occurred in the contours of the global economic map since World War II, then it is clear that there have been some unexpected developments. The contours of today's map are very different from those of only a few decades ago. In particular, there is a much wider spread of competition in the global economy than in the past when a few older industrialized countries dominated. Seven broad changes stand out above others:

- The rise of Japan to become the second biggest economy in the world in terms of production and the third biggest in terms of exports.
- The continued dominance of the United States as the world's largest economy despite experiencing fierce competition from Japan and from the newer East Asian economies.
- The uneven economic performance of the West European economies.
- The emergence of a number of newly industrialized economies in East Asia to become major global players, especially as exporters of manufactures.
- The very recent, and especially rapid, emergence of China as a major force in the world economy.
- The weak economic performance of most Latin American economies – indeed, of most developing countries outside East Asia. Weakest of all, of course, is sub-Saharan Africa.
- The sudden appearance of a new group of 'transitional economies' in Central and Eastern Europe and the former Soviet Union with the collapse of the Soviet-led system in 1989.

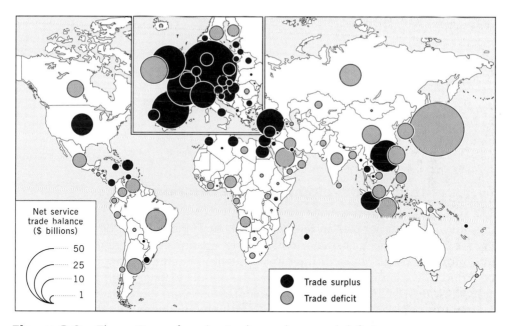

Figure 3.9   The pattern of service trade surpluses and deficits

Source: Based on data in WTO (2001) *International Trade Statistics, 2001*: Table A8

The changing positions of leading industrialized countries

Table 3.5 shows growth rates for the United States, Japan and the four leading West European economies. The changing relative performance of the United States and Japan is especially interesting. In the early 1960s, Japan was the fifth largest economy and the United States the first largest. By the late 1980s, Japan had reached second position, which it has retained since then. However, after the remarkably high growth rates during the 1960s, 1970s and 1980s, Japan's growth rate collapsed during the 1990s. Conversely, the United States' growth rate was around one-third of that of Japan to the end of the 1980s but then surged to a level almost five times greater during the long US boom of the 1990s. In contrast, both the United Kingdom and Italy declined in relative standing while Germany and France sustained their positions.

Table 3.5   Changing manufacturing performance of leading industrialized countries

	1963		1987		1999		Average annual % change			
		% share		% share		% share	1960/ 70	1970/ 81	1980/ 87	1990/ 99
United States	1	40	1	24	1	25	5.3	2.9	3.9	4.9
Japan	5	6	2	19	2	20	13.6	6.5	6.7	1.1
Germany	2	10	3	10	3	12	5.4	2.1	1.0	NA
United Kingdom	3	7	5	3	6	4	3.3	−0.5	1.3	NA
France	4	6	4	5	4	4	7.8	3.2	−0.5	0.6
Italy	6	3	6	2	9	2	8.0	3.7	0.9	0.9

NA, data not available.

Source: Based on data in World Bank, *World Development Report*, various issues; UNIDO, *Industrial Development: Global Report*, various issues

Shifts within the developing market economies

The developing countries as a whole increased their share of world manufacturing output from 5 per cent in the early 1950s to 23 per cent by the late 1990s (see Figure 3.4). For the vast majority, however, manufacturing remains relatively unimportant. In fact, most of the rapid growth in manufacturing production has occurred in what the World Bank terms 'the middle income group' of developing countries. But there are pronounced geographical variations in this process, as Figure 3.10 very clearly shows. Although growth rates have fluctuated (the roller-coaster effect at a regional scale), the rates of growth were highest of all in East Asia and lowest in sub-Saharan Africa. Growth rates in Latin America and in South Asia (both important manufacturing locations in absolute terms) were between these two extremes but with variability in manufacturing growth being much greater in Latin America.

The increasingly dominant position of the East Asian developing countries compared with those in other parts of the world is shown in more detail in Table 3.6. The gap between the developing economies of East Asia, on the one hand, and those of Latin America and South Asia on the other has widened enormously. Throughout the 1960s, 1970s and 1980s, the four 'tiger' economies of South Korea,

Taiwan, Singapore and Hong Kong dominated the picture. More recently, as Figure 3.10 and Table 3.6 show, other East Asian countries – notably Malaysia, Thailand, Indonesia and China – have grown extremely rapidly. These East Asian manufacturing growth rates should be contrasted with those of the leading developed market economies discussed earlier (see Table 3.5).

The relative importance of the East Asian economies is especially marked in the sphere of exports, as the bottom part of Table 3.6 shows. Indeed, in the global

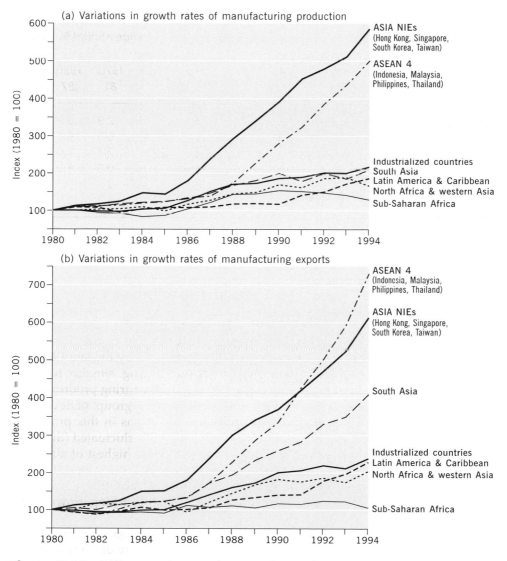

Figure 3.10 Differences in growth rates of manufacturing production and exports between country groups

Source: Based on data in UNIDO (1996) *Industrial Development: Global Report, 1996*

Table 3.6   Growth of manufacturing production and exports in newly industrializing economies

	Share of world output (%)			Average annual % change			
	1963	*1980*	*1999*	*1960/70*	*1970/81*	*1980/87*	*1990/99*
South Korea	0.1	0.6	2.9	17.6	15.6	10.6	6.2
Taiwan	0.1	0.5	1.1	16.3	13.5	7.5	NA
Hong Kong	0.1	0.2	0.2	NA	10.1	NA	NA
Singapore	0.1	0.1	0.4	13.0	9.7	3.3	7.9
Malaysia	NA	0.1	0.4	NA	11.1	6.3	9.4
Thailand	NA	0.3	0.8	NA	NA	NA	6.7
Indonesia	NA	0.1	0.5	NA	NA	NA	7.8
Philippines	NA	0.2	0.2	NA	NA	NA	3.4
China	NA	2.9	2.7	NA	NA	NA	14.4
India	NA	0.4	0.5	4.7	5.0	8.3	6.7
Brazil	1.6	2.4	4.1	NA	8.7	1.2	3.2
Argentina	NA	0.8	0.8	5.6	0.7	0.0	4.8
Mexico	1.0	1.4	0.8	10.1	7.1	0.0	3.6

	Share of world exports (%)		Average annual % change			Manufactures as % of total exports	
	1963	*2000*	*1970/80*	*1980/90*	*1990/99*	*1980*	*1998*
South Korea	0.01	2.7	22.7	12.0	15.6	89	91
Taiwan	0.2	2.3	16.5	NA	NA	88	94
Hong Kong	0.8	3.2	9.9	14.4	8.4	88	95
Singapore	0.4	2.2	NA	NA	NA	43	86
Malaysia	0.1	1.5	3.3	10.9	11.0	19	79
Thailand	NA	1.1	8.9	14.1	9.4	25	74
Indonesia	NA	1.0	6.5	2.9	9.2	2	45
Philippines	NA	0.6	NA	3.5	9.6	NA	90
China	NA	3.9	8.7	19.3	13.0	48	87
India	0.8	0.7	5.9	5.9	11.3	51	74
Brazil	0.1	0.9	8.6	7.5	4.9	37	55
Argentina	NA	0.4	8.9	3.9	8.7	23	35
Mexico	0.2	2.6	5.5	7.0	14.3	10	85

NA, data not available.

Source: Based on data in OECD, 1979: Table 1; *World Bank, World Development Report*, various issues; UNIDO, *Industrial Development: Global Report*, various issues; WTO *Annual Report*, various issues

reorganization of manufacturing trade the increased importance of East Asia as an exporter of manufactures is unique in its magnitude. The nine countries listed in the top half of Table 3.6 increased their collective share of total world manufactured exports from a mere 1.5 per cent in 1963 to almost 20 per cent in 1999 (and remember that this period includes the East Asian financial crisis of 1997/98, which had a devastating effect on most of the East Asian economies). China's growth has been especially marked. During the 1980s, its exports grew at more than 19 per cent per year; during the 1990s, they grew at 13 per cent per year.

So, it is especially in their role as exporters that the East Asian economies are particularly significant. In some cases the transformation has been nothing short of spectacular, as the figures for the share of manufactures of total exports shows (Table 3.6). For example, in 1980, less than 20 per cent of Malaysia's exports were of manufactures; by 1998 the figure was 79 per cent. Indonesia provides an even more striking experience. In 1980 a mere 2 per cent of the country's exports were of manufactures; in 1998 almost half was in that category. China, Singapore and Thailand also show a similar transformation. In comparison, the structural change in Argentina and Brazil was far smaller, although Mexico's transformation mirrors that of the East Asian economies in this respect (Mexico's export position has been transformed by its increasing integration with the United States through the North American Free Trade Agreement – see Chapters 5 and 6).

The basis of these transformations of the East Asian economies was the incredibly high annual growth rates that were sustained over a very long period of time, as Table 3.6 shows. In comparison, the manufacturing export performance of most other developing countries was relatively modest. As a result, the gap between the East Asian economies and other developing countries widened substantially. By any criterion, the first generation of East Asian newly industrializing economies are now, in effect, industrialized – rather than industrializing – countries. Both Singapore and Hong Kong now have per capita GNP higher than a number of industrialized countries, including the United Kingdom, Italy and Canada. Apart from South Korea, they are relatively small countries in terms of population. However, the newly emerging group of East Asian NIEs are very much larger. China, of course, is in a size league of its own, with its population of 1.3 billions. But Indonesia, too, has a population larger than any industrialized country other than the United States.

Import penetration of developed country markets

One of the most controversial aspects of the spectacular export growth of NIEs is their growing penetration of the domestic markets of developed economies as well as their success in competing with industrialized nations elsewhere. Figure 3.11 reveals the increasing difference in penetrating developed country markets between the East Asian NIEs and those from other parts of the world, especially Latin America. Not only has the overall penetration of developed country markets by East Asian NIEs increased far more than that of other NIEs but also they are far more concentrated in those products for which demand is growing most rapidly.

> The East Asian economies have been singularly successful in increasing their exports of products that are growing in importance in international trade. In 1990 about three-quarters of their exports were in goods for which the share in total OECD

imports had been expanding over the previous three decades. In contrast, the proportion for Latin America was only 38 per cent, and only 24 per cent if Mexico is excluded. There was, however, some difference between the first-tier and second-tier NIEs … the second-tier NIEs, in this respect, resemble more closely non-Asian NIEs, such as Mexico, Tunisia, Morocco and Turkey … [T]he East Asian economies have been far more successful than other developing economies in entering markets for products with high income elasticities of demand. While 9 of the 10 leading exports from the first-tier NIEs to OECD in 1993 were income-elastic, there were only 2 in Latin America.[5]

Figure 3.11

Shares of NIEs in manufactured imports to developed countries

Source: Based on data in UNCTAD, 1996a: Chart 5

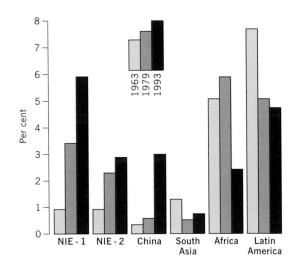

The 'transitional economies' of Eastern Europe and the former Soviet Union

Since 1989, there has been a further significant development in the changing geography of the global economy. The political collapse of the Soviet-led group of countries, and, indeed, of the Soviet Union itself, produced a group of so-called 'transitional economies': former centrally planned economies now in various stages of transition to a capitalist market economy. The process of transition, from a centrally planned economic system, with a heavy emphasis on basic manufacturing industries, to a capitalist market system, has been painful in many cases. The kinds of industries favoured in the centrally planned system are less viable in the context of today's highly competitive global economy, as are the kinds of industrial organization themselves. In 1985, for example, the USSR accounted for almost 10 per cent of world manufacturing output; by the mid-1990s, the share of the Russian Federation was around 1 per cent.

There are big differences within the group of transitional economies both in terms of their scale and in terms of their potential for growth within a market, as opposed to a centrally planned system. Table 3.7 provides some idea of this variation. The four most significant transitional economies are the Russian Federation,

Poland, the Czech Republic and Hungary. Together, these four account for most of the manufacturing production and exports. Clearly, their current importance remains quantitatively small although their location adjacent to the European Union (to which several of these countries have applied for membership) gives them an important strategic location for firms seeking lower-cost production sites from which to serve the EU market. But, as Table 3.7 shows, these economies remain very volatile in growth terms.

Table 3.7   The transitional economies

	Share of world output %	Share of world exports %	Annual % change in exports				Manufactures as % of total exports	
			1997	1998	1999	2000	1990	1999
Russian Federation	0.8	1.7	0	−15	1	39	NA	28
Poland	0.4	0.5	5	3	−3	15	59	77
Czech Republic	0.1	0.5	4	16	2	11	NA	88
Hungary	0.1	0.4	22	20	9	12	63	82
Ukraine	NA	0.2	−1	−11	−8	26	NA	NA
Slovak Republic	0.03	0.2	−7	11	−4	16	NA	84
Romania	0.1	0.2	4	2	2	22	73	81
Kazakhstan	NA	0.1	10	−16	3	64	NA	23
Belarus	NA	0.1	29	−3	−16	25	NA	76
Bulgaria	0.1	0.1	1	−13	−6	20	NA	61

Source: Based on data in World Bank (2001) *World Development Report, 2001;* UNIDO (1997) *Industrial Development: Global Report;* WTO (2001) *International Trade Statistics, 2001*

Changing contours of the global economic map: global shifts in foreign direct investment

The changing global maps of production and of trade are important indicators of the increasing globalization of economic activity. A third, intimately related, indicator is the growth in the scale and complexity of foreign investment. Here, it is *foreign direct investment* (FDI), rather than portfolio investment, which is especially important. 'Direct' investment is an investment by one firm in another with the intention of gaining a degree of control over that firm's operations. 'Foreign' direct investment is simply direct investment which occurs across national boundaries, that is, when a firm from one country buys a controlling investment in a firm in another country or where a firm sets up a branch or subsidiary in another country.

It differs from 'portfolio' investment which refers to the situation in which firms purchase stocks/shares in other companies purely for financial reasons. But, unlike direct investment, portfolio investments are not made to gain control.

In this section we examine global trends in foreign direct investment over the past few decades. FDI statistics constitute the most comprehensive single indicator of the activities of TNCs and of the growth of international production. However, FDI is only one measure – albeit a very important one – of TNC activity. It does not capture the increasingly diverse ways in which firms engage in international operations, for example, through various kinds of collaborative ventures and alliances, or through their coordination and control of production chain transactions. We will look at these issues in Chapter 8.

Aggregate trends in FDI[6]

Although there was very considerable growth and spread of foreign direct investment during the first half of the 20th century, that was as nothing compared with its spectacular acceleration and spread after the end of World War II. The post-war surge of FDI was an integral part of the 'golden age' of economic growth of the 1960s discussed earlier. In fact, during the 1960s, FDI grew at twice the rate of global gross national product and 40 per cent faster than world exports.

Figure 3.12 shows that during the 1970s and into the first half of the 1980s the trend lines of both FDI and exports ran more or less in parallel. Then, from 1985 to 1990 the rate of growth of FDI and of exports and GDP diverged rapidly. Between 1986 and 1990 FDI outflows grew at an average annual rate of 26 per cent and cumulative FDI stocks at a rate of 20 per cent a year compared with a growth rate of world exports of 15 per cent. FDI during the 1980s grew more than four times faster than world GNP. The recession of the early 1990s reduced the FDI growth rates significantly but by the mid-1990s the upward trend had resumed. Indeed, as the figures in Figure 3.12 show, the second half of the 1990s saw unprecedented rates of growth of FDI.

This divergence in growth trends between FDI and trade is extremely significant. In discussing Figure 3.2 earlier, we observed that the divergence between the growth of trade and of production was an indicator of the increasing interconnectedness within the global economy. The fact that, after the mid-1980s, FDI grew much faster than trade suggests that the primary mechanism of integration has shifted from trade to FDI. Of course, these trends in the growth of FDI, trade and production are not independent of one another. The common element is the transnational corporation (TNC).

TNCs account for around two-thirds of world exports of goods and services of which a significant share is *intra-firm trade*. In other words, it is trade within the boundaries of the firm – although across national boundaries – as transactions between *different parts of the same firm*. Unlike the kind of trade assumed in international trade theory – and in the trade statistics collected by national and international agencies – intra-firm trade does not take place on an 'arm's-length' basis. It is, therefore, not subject to external market prices but to the internal decisions of TNCs. Unfortunately, there are no comprehensive and reliable statistics on intra-firm trade. The 'ball park' figure is that approximately one-third of total world trade is intra-firm although, again, that could well be a substantial underestimate.

According to UNCTAD, the number of TNCs has grown exponentially over the past three decades. Today there are around 60,000 parent company TNCs controlling around 700,000 foreign affiliates. Of course, this is only a very small proportion of the total number of business firms in the world, the vast majority of which are small and entirely domestically oriented. There are two ways in which a firm may become a TNC in the first place or subsequently to extend its existing transnational operations:

- by building entirely new facilities ('greenfield' investment)
- by merging with, or acquiring, a firm or firms in other countries (M&A).

It is impossible to measure accurately the contribution of these two mechanisms to the overall growth of FDI. However, in recent years, most of the growth in transnational production has been through merger and acquisition, rather than through greenfield investment.[7] Figure 3.13 shows the rapid increase in cross-border M&A activity during the second half of the 1990s, although there was a decline in the intensity of such activity in 2001.

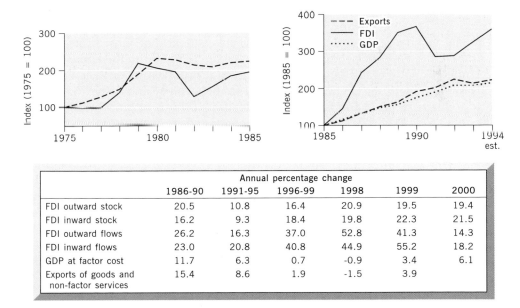

	Annual percentage change					
	1986-90	1991-95	1996-99	1998	1999	2000
FDI outward stock	20.5	10.8	16.4	20.9	19.5	19.4
FDI inward stock	16.2	9.3	18.4	19.8	22.3	21.5
FDI outward flows	26.2	16.3	37.0	52.8	41.3	14.3
FDI inward flows	23.0	20.8	40.8	44.9	55.2	18.2
GDP at factor cost	11.7	6.3	0.7	-0.9	3.4	6.1
Exports of goods and non-factor services	15.4	8.6	1.9	-1.5	3.9	

Figure 3.12 Growth of foreign direct investment compared with trade and production

Source: Based on material supplied by UNCTAD

Figure 3.13

Growth in cross-border merger and acquisition activity

Source: Based on UNCTAD, 2000b: Figure I.4

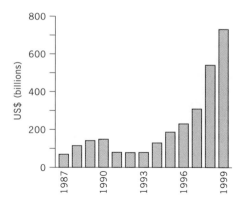

Where does FDI come from?

Although the aggregate growth of FDI has been extraordinarily rapid and sustained, especially since the mid-1980s, its *geography* is extremely uneven. This is true of both its origins and its destinations. Figure 3.14 maps the distribution of FDI across the world. It shows both the level of outward investment from each country (the left-hand segment of each circle) and the inward stock (the right-hand segment). Although the number of source countries has grown in the past three decades, FDI originates overwhelmingly from the developed market economies.

Figure 3.15 and Table 3.8 illustrate FDI trends among the leading developed country sources. The most significant feature of Figure 3.15 is the difference in relative growth rates for individual source nations and the consequent shifts in their relative shares of world foreign direct investment. For most of its history, world FDI has been overwhelmingly dominated by TNCs from the United States, the United Kingdom and one or two continental European countries. From the 1950s to the mid-1970s, US firms accounted for between 40 and 50 per cent of the world total. In 1960, US and UK TNCs made up two-thirds of the world total. But although TNCs from both countries have continued to invest heavily overseas, other countries' outward investment has increased more rapidly, as Figure 3.15 shows. By 1985 the combined US–UK share of the world total had fallen to around half. Conversely, the German share of the total increased from 1.2 to 8.5 per cent while Japan's share had grown even more sharply, from 0.7 to 6.2 per cent. From being a very minor player in terms of foreign direct investment in 1960, and not especially important in 1975, by 1985 Japan had surged up the league table to fifth place.

The year 1985 marked a major acceleration in the growth of world FDI to unprecedented levels. In that acceleration, Japan was undoubtedly a leading player, at least initially. Japanese outward direct investment had grown from 6 per cent of the world total in 1975 to 12 per cent by 1990. However, there was a pronounced slackening in the rate of Japanese FDI growth in the 1990s so that, by 2000, its share of the world total had fallen to less than 5 per cent. Conversely, the UK's position as the second most important source of global FDI remained unchallenged throughout the 40-year period. In 2000, 15 per cent of world FDI

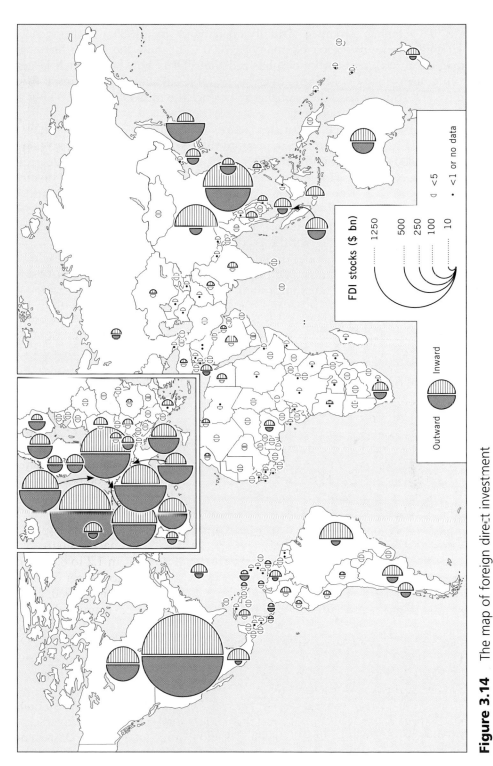

Figure 3.14 The map of foreign direct investment

Source: Based on data in UNCTAD, 2001: Annex Tables B3, B4

originated from the UK compared with 18 per cent in 1960. Germany's share of the world FDI total has remained virtually unchanged since 1985 at just under 9 per cent, while France overtook Germany in 2000 to become the fourth largest FDI source. Of course, it must be emphasized that FDI is now very much larger in absolute terms than ever before. Even so, as Table 3.8 shows, the share of the ten leading FDI sources has fallen substantially since the mid-1970s: from 93 per cent of the world total to 74 per cent.

Clearly, the number of countries acting as significant sources of FDI has increased. Thirty-three countries had more than $10 billion of direct investment abroad in 2000 compared with only 10 in 1985. Many of these are, of course, other developed economies. However, a most significant development has been the emergence of TNCs from developing countries. In 1960, 99 per cent of world FDI came from the developed economies. Fifteen years later, in 1985, the developing countries' share was only around 3 per cent. By 2000, however, this had quadrupled to 12 per cent of the world total (Figure 3.16).

As yet, only a small number of developing countries are involved. Eighty per cent of all developing country FDI originates from just seven countries, of which Hong Kong is by far the most important. Indeed, Hong Kong alone accounts for

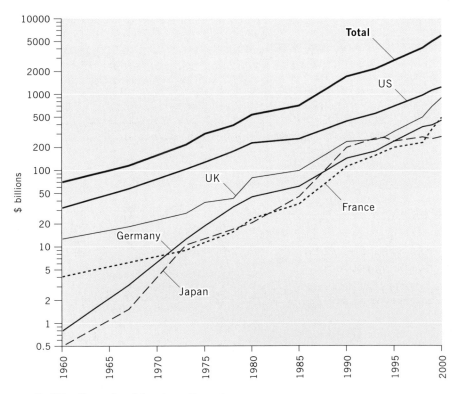

Figure 3.15 Growth of foreign direct investment

Source: Based on data in UNCTAD, *World Investment Report*, various issues

two-thirds of the total of the seven countries shown in Figure 3.17. Six of the seven are East Asian NIEs. However, although modest in scale, this is undoubtedly the harbinger of an important new development. Again, therefore, we see clear signs that the relatively simplistic division of the global economy has disappeared. The world's population of TNCs is not only growing very rapidly but also there has been a marked increase in the geographical diversity of its origins in ways that cut across the old international division of labour.

Table 3.8 World foreign direct investment: changing relative importance of leading source countries

Country of origin	Percentage of world total of outward direct investment stock					
	1960	1975	1985	1990	1995	2000
United States	47.1	44.0	35.5	25.1	24.3	20.8
United Kingdom	18.3	13.1	14.2	13.4	10.6	15.1
Japan	0.7	5.7	6.2	11.7	8.3	4.7
Germany	1.2	6.5	8.5	8.6	9.0	7.4
France	6.1	3.8	5.2	7.0	7.2	8.3
Netherlands	10.3	7.1	6.8	6.0	5.8	5.5
Canada	3.7	3.7	6.1	4.9	4.1	3.4
Switzerland	3.4	8.0	3.5	3.9	4.9	3.9
Italy	1.6	1.2	2.4	3.3	3.8	3.0
Sweden	0.6	1.7	1.5	2.9	2.5	1.9
Total	93.0	94.8	89.9	86.8	80.5	74.0

Source: Based on data in UNCTAD, *World Investment Report*, various issues

Figure 3.16

The increasing share of foreign direct investment originating from developing countries

Source: Based on data in UNCTAD, *World Investment Report*, various issues

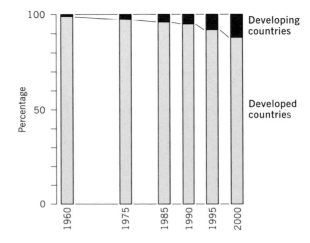

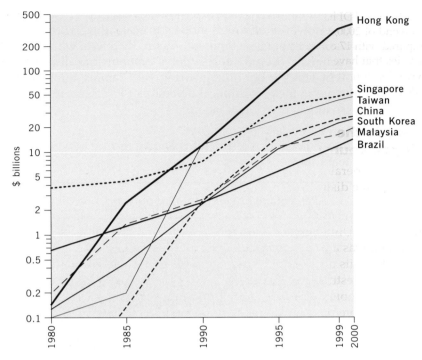

Figure 3.17 Foreign direct investment from leading developing countries

Source: Based on data in UNCTAD, *World Investment Report*, various issues

Where does FDI go to?

It is perhaps not very surprising that the majority of the world's FDI originates from developed economies. What is more surprising is that most of the world's FDI also *goes to* developed countries. In fact, 67 per cent of world FDI stock in year 2000 went to developed economies and rather less than one-third went to developing countries. This is an almost exact reversal of the position just prior to World War II: in 1938, some 65 per cent of the world's FDI was located in developing countries.

At one time, it was reasonable to distinguish between those countries that were primarily sources of FDI and those that were primarily destinations for FDI. But, as Figure 3.14 shows, this is far less possible today, particularly for the developed economies. In most cases the sizes of the two segments in Figure 3.14 are very similar. If we measure this as a simple FDI ratio (outward FDI/inward FDI) then a value of 1.0 would signify an exact balance between outward and inward investment. With just four exceptions, all the major countries of Western Europe and North America had FDI ratios within the range 0.9 and 1.5 in 2000. The exceptions were France (1.9), the United Kingdom (1.9) and Switzerland (2.7). What these patterns imply, in fact, is a high degree of *cross-investment* between the major developed market economies: each is investing in each other's home territory.

Nevertheless, FDI has become more geographically dispersed since the mid-1980s: by the end of 2000, 51 countries reported inward FDI stocks of more than $10 billion, compared with 17 countries in 1985 … [there has been] … growth in the number of countries that have become major recipients … of FDI. Among them, the number of developing countries had risen from 7 in 1985 to 24 in 2000 in the case of inward FDI stocks.[8]

Changes in the relative position of developed countries as FDI destinations

Within the general pattern of interpenetration of FDI between developed economies, three distinctive features stand out:

- *The United States has strengthened its position as a favoured destination for FDI.* Although the United States has attracted FDI for many decades, such inward investment was a tiny fraction of the country's outward direct investment. Even in 1975, its outward investment was four-and-a-half times greater than its inward investment. Since then, however, the United States has become significantly more important as an FDI destination as firms from Europe, Japan and, more recently, some East Asian NIEs reoriented the geographical focus of their overseas direct investments. By 2000, the US FDI ratio was almost exactly 1.0: a perfect balance in terms of outward and inward stocks.

- *Europe remains a major magnet for inward investment.* Western Europe's share of total world inward investment fell quite substantially between 1975 and 1985 (from 41 per cent of the world total to 33 per cent). But by 2000 its share had risen again to 40 per cent. Much of this resurgence is explained by the programme to complete the Single European Market in 1992 and by further economic integration through the 1990s, culminating in the creation of the eurozone (see Chapter 5). Non-European companies made major efforts to gain a direct presence in the single market spurred on by the fear of a potential 'fortress Europe'. Both statistical and anecdotal evidence suggest a major upsurge in direct investment in Europe by both US and, especially, Japanese and other Asian companies. At the same time, there continues to be a very high level of intra-European FDI. For all the major European countries (excluding the United Kingdom), more than half of their FDI outflows are to other European countries. In most cases, this regional orientation has actually increased. For example, in 1985 49 per cent of Italy's outward FDI was in Europe. By the end of the 1990s, this share had grown to 69 per cent. The comparable figures for Germany were 44 per cent and 60 per cent; for the Netherlands 40 per cent and 55 per cent, for France 58 per cent and 60 per cent. For the United Kingdom, in contrast, the figures were only 28 per cent and 38 per cent. The transformation of the political situation in Eastern Europe is also leading to the growth of FDI in these economies although, as yet, the volumes are relatively small. In 2000, only 2 per cent of world inward FDI was located in Eastern Europe. However, this represented a ten-fold increase over the position in 1990.

● *Japan's anomalous FDI position continues – though rather less so.* While the US direct investment position has been transformed from one of being overwhelmingly a home country for FDI to one in which the ratio of outward to inward investment is in balance, the same cannot be said of Japan. While Japanese outward investment has grown spectacularly, there has been only very limited growth of inward investment. Indeed, the outward/inward ratio actually increased – from an already exceptionally high level of 10.7 in 1975 to 20.5 in 1990. Whereas in 1990 Japan accounted for 11.7 per cent of total world outward direct investment (see Table 3.8), it was the host to only 0.5 per cent of world inward investment. Along with trade frictions, this huge imbalance in the Japanese direct investment account caused major concern among businesses and policy-makers in the West, especially in the United States. However, in the light of Japan's economic crisis throughout the 1990s, the FDI imbalance has changed considerably, although Japan still has the most unbalanced FDI position of all the major developed economies (an FDI ratio of 5.2 in 2000).

The highly uneven distribution of FDI among developing countries

As we have seen, the developing countries as a whole are host to less than one-third of total world FDI. Within that relatively small share there is a very high level of concentration in a small number of countries. Figure 3.14 shows that inward FDI is minuscule in the majority of developing countries. There is a clear regional dimension to this. Africa's share of the developing countries' total had declined to a mere 5 per cent by 2000. Latin America's share had fallen to 31 per cent. In contrast, the Asian share had increased more than two-and-a-half times, from 21 per cent in 1975 to 64 per cent in 2000. Table 3.9 shows that ten countries account for 70 per cent of all FDI in developing countries. Six of these are in East Asia, including by far the largest developing country recipient of inward FDI – China (with Hong Kong). Indeed, the extent to which FDI in China has grown since the early 1980s is nothing short of spectacular.

How important is inward FDI to host countries?

Statistics on the scale of FDI are important in showing us the marked geographical variations in destinations. But they tell us nothing about how important such investment is to an individual host economy. In fact, such importance varies enormously from one country to another. One measure of relative domestic importance of inward FDI is to compare it to a country's gross domestic product (GDP). Table 3.10 shows that, in every case, the relative importance of FDI increased between 1990 and 1999. In some cases, the increase was quite spectacular. For developed countries as a whole, inward FDI constituted around 15 per cent of GDP in 1999 compared with 8 per cent in 1990. Among developed economies, the relative importance of FDI to domestic economies is high throughout Western Europe. It is particularly high in Ireland and the Netherlands. In North America, there is a pronounced contrast between Canada (where FDI contributes

Table 3.9 The concentration of foreign direct investment in developing countries, 2000

Country	FDI stock ($bn)	% of developing country total
Hong Kong	469,776	23.7
China	346,694	17.5
Brazil	197,652	10.0
Mexico	91,222	4.6
Singapore	89,250	4.5
Argentina	73,441	3.7
Indonesia	60,638	2.9
Malaysia	54,315	2.7
Chile	42,933	2.2
South Korea	42,329	2.1
Total, 10 developing countries		70.4

Source: Based on data in UNCTAD, 2001: Annex Table B3

Table 3.10 Inward foreign direct investment as a share of Gross Domestic Product

	Percentage share of GDP				
	1990	*1999*		*1990*	*1999*
Developed countries	**8.4**	**14.5**	**Developing countries**	**13.4**	**28.0**
Ireland	12.2	50.7	Singapore	76.3	97.5
Netherlands	23.6	50.1	Malaysia	24.1	65.3
Sweden	5.4	32.7	Chile	33.2	55.2
Canada	19.7	27.9	Indonesia	34.0	46.2
United Kingdom	20.8	26.8	China	7.0	30.9
Portugal	15.3	21.2	Argentina	6.4	22.2
Denmark	6.9	20.9	Brazil	8.0	21.6
Spain	13.4	20.5	Thailand	9.6	17.5
Greece	16.9	17.7	Pakistan	4.8	17.2
France	8.4	17.1	Mexico	8.5	16.4
Finland	3.8	14.5	Philippines	7.4	14.9
Germany	7.3	13.7	Taiwan	6.1	8.0
United States	7.1	11.1	South Korea	2.0	7.9
Italy	5.3	9.4	India	0.6	3.6
Japan	0.3	1.0			

Source: Based on data in UNCTAD, 2001: Annex Table B6

28 per cent of GDP) and the United States (11 per cent). Not surprisingly we find the lowest contribution of all in the case of Japan (1 per cent).

In the case of developing countries as a whole, inward FDI is twice as significant to the domestic economy as in developed countries (28 per cent in 1999). But, again, there are marked differences between individual countries. Within Asia, FDI is overwhelmingly important to Singapore's economy, in particular, but also to Malaysia's, Indonesia's and China's. In Latin America foreign firms are especially important to Chile (55 per cent of GDP). In contrast, FDI is far less important relatively to Taiwan (8 per cent), South Korea (8 per cent) and India (4 per cent).

Which sectors attract FDI?

Just as FDI as a whole is unevenly distributed geographically so, too, it tends to be concentrated rather more in some types of economic activity than in others. At the broad level, two major trends have been, first, the relative decline in the proportion of outward FDI in the primary sector and, second, the corresponding increase in importance of the services sector. Within each of these broad categories, however, important transformations have also been taking place. Unfortunately, there are no comprehensive data on the sectoral composition of world FDI so it is difficult to make accurate general statements. Table 3.11 helps a little: it shows the sectoral focus of the world's largest 100 TNCs during the 1990s, together with the extent to which each sector is 'transnational'.

Historically, the major proportion of FDI was concentrated in the *natural resource-based sectors*: the mining of geographically localized minerals and ores, the operation of large-scale plantations for the production of commercial foodstuffs for export, and the like. The major reasons for such a concentration have been the lack of domestic financial and technical resources in firms in developing countries. TNCs are still heavily involved in such commodity sectors, the most obvious example being the giant petroleum companies. In some extractive activities, national governments have acquired control of their natural resource operations but, even so, TNCs often remain in control of the commodities' marketing channels.

Although foreign direct investment in the extractive industries remains extremely important, the emphasis has certainly shifted. Until the early 1980s, the bulk of FDI was in *manufacturing industry* and was an integral part of the rapid growth and internationalization of economic activity which, as we have seen, helped to transform the world economy. Three broad types of manufacturing industry have an especially large TNC involvement:

- *technologically more advanced sectors* – for example, pharmaceuticals, computers, scientific instruments, electronics, synthetic fibres
- *large-volume, medium-technology consumer goods industries* – for example, motor vehicles, tyres, televisions, refrigerators
- *mass-production consumer goods industries supplying branded products* – for example cigarettes, soft drinks, toilet preparations, breakfast cereals.

These are the industries in which a high level of technological expertise and resource (both human and financial) are required, in which demand is strongly income-elastic and for which the extensive operations of TNCs are especially suitable. In fact, there has been a gradual move of FDI away from labour-intensive, low-cost, low-skill manufacturing and towards more capital-, knowledge- and skill-intensive industries.

However, it is in the *service industries* that the most significant relative change has occurred. FDI in services tends to be concentrated in certain key sectors, notably:

- *financial services* (banking, insurance, accounting)
- *trade-related services* (wholesaling, marketing, distribution)
- *telecommunication services*

Table 3.11 Sectoral focus of the world's 100 largest TNCs

Sector	Number of entries		Average TNI[a] per sector	
	1990	*1999*	*1990*	*1999*
Media	2	2	82.6	86.9
Food / beverages / tobacco	9	10	59.0	78.9
Construction	4	2	58.8	73.2
Pharmaceuticals	6	7	66.1	62.4
Chemicals	12	7	60.1	58.4
Petroleum	13	13	47.3	53.3
Electronics / electrical equipment / computers	14	18	47.4	50.7
Motor vehicles and parts	13	14	35.8	48.4
Metals	6	1	55.1	43.5
Diversified	2	6	29.7	38.7
Retailing	—	4	—	37.4
Utilities	—	5	—	32.5
Telecommunications	2	3	46.2	33.3
Trading	7	4	32.4	17.9
Machinery / engineering	3	—	54.5	—
Other	7	4	57.6	65.7
Total / average	100	100	51.1	52.6

[a]TNI is the 'transnationality index'. Calculated as the average of three ratios: foreign assets to total assets; foreign sales to total sales; foreign employment to total employment.

Source: Based on UNCTAD, 2001: Table III.6

- *business services* (consulting, advertising, hotels, transportation, construction)
- *some consumer services* (retailing, fast-food).

A closer look at the geography of FDI: some case examples

TNC activity as a whole is very unevenly distributed, both geographically and sectorally. However, there are significant differences between individual source countries. In this section we look, very briefly, at the broad trends over the past decade in FDI from four leading developed countries (the United States, the United Kingdom, Germany, Japan) and from two East Asian NIEs (South Korea, Singapore). The aim is to provide no more than a brief 'thumbnail' sketch of each case. The reasons for doing this are two-fold. First, it helps to add some individual detail to the broad picture painted in this chapter so far. Second, one of the themes of this book is that TNCs are not all the same. Although, as profit-seeking enterprises operating within a capitalist market system they do, indeed, share some common characteristics, they are far from being homogeneous. There are many reasons for such differences, as we shall see in later chapters. One very important reason is, without doubt, the influence of the TNC's *home country* environment, notably its political, social, cultural and economic characteristics (this is an issue we will discuss in Chapter 7).

US foreign direct investment

In 1986, *The Economist* published an article headed 'American multinationals: the urge to go home' and asserted that American firms were 'turning inward'. This turned out to be a rash prediction. In six of the ten years shown in Figure 3.18, the growth rate was in double figures. Although the United States' share of the world FDI total has fallen, it should be remembered that the world total itself has expanded dramatically. In absolute, if not in relative terms, US TNCs are more significant today than they were in the past. They constitute the largest and most extensive network of international production facilities in the world.

The *sectoral* distribution of US FDI has changed substantially in recent years. In 1985, 27 per cent was in the primary sector (predominantly in oil), 41 per cent in manufacturing and 31 per cent in services. In manufacturing, five industries (food, beverages, tobacco; chemicals; mechanical equipment; electric and electronic equipment; motor vehicles) accounted for three-quarters of the total. In services, nine-tenths was in finance, insurance and business services and in wholesale and retail trade. By the late 1990s, the picture had changed in some important respects. The share of the primary sector had fallen to 6 per cent, that of manufacturing to 30 per cent, while the service sector had grown to 63 per cent. Within manufacturing, the major increases were in food, drink and tobacco and in chemicals set against a major decline in the share of mechanical equipment. The shares of FDI in motor vehicles and in electric and electronic equipment increased only slightly. The biggest relative shift was in the services sector where FDI in finance, insurance and business services had grown from 54 to 69 per cent of the sector total.

Geographically, 51 per cent of US FDI was located in Europe in the late 1990s. The United Kingdom alone was host to 37 per cent of this investment. The Netherlands, the second most important European host country for US FDI, had 18 per cent, and Germany 9 per cent. Even so, three countries host more than half of all US FDI in Europe. In the Americas, there was a marked contrast between Canada (whose share of US FDI declined from 20 per cent in 1985 to 10 per cent) and Latin America, whose share grew from 12 per cent to 17 per cent.

Asia contained only around 13 per cent of US FDI in the late 1990s (up from 11 per cent in 1985). Within Asia, the major foci were Japan (33 per cent of the regional total compared with 38 per cent in 1985), Hong Kong (14 per cent compared with 13 per cent) and Singapore (17 per cent compared with 8 per cent). China, South Korea and Taiwan together contained 16 per cent of US FDI in Asia in the late 1990s, up from 7 per cent in 1985. The spread of US FDI outside Europe,

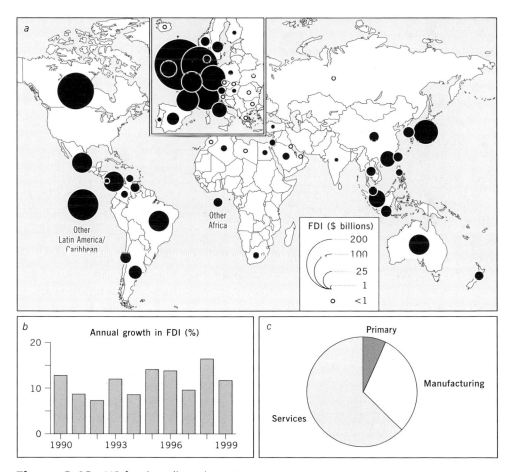

Figure 3.18 US foreign direct investment

Source: Based on data in OECD (2000) *International Direct Investment Statistics Yearbook.* Paris: OECD

Latin America and Asia was very thin indeed. The whole of Africa, for example, contained only a little over 1 per cent of the total.

UK foreign direct investment

UK FDI shares at least one common feature with that of the United States: both have experienced a relative decline in their share of world direct investment in the past two or three decades. However, the year-by-year trend has been rather different, as Figure 3.19 shows. The United Kingdom's growth pattern has been far more volatile than that of the United States, with low growth in 1990 and 1991, double-digit growth in 1992, 1993, 1995 and 1997 interspersed with an actual decline in 1996. But then UK FDI surged by more than 30 per cent a year in 1998 and 1999.

At the broad level of aggregation, the *sectoral* structure of UK FDI changed only slightly between 1987 and the late 1990s (Figure 3.19). The share of the primary

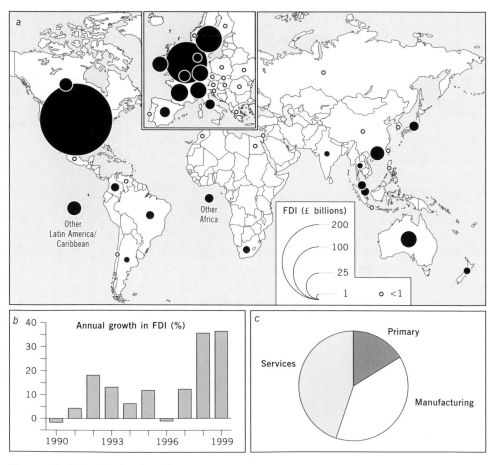

Figure 3.19 UK foreign direct investment

Source: Based on data in OECD (2000) *International Direct Investment Statistics Yearbook*. Paris: OECD

sector fell substantially, that of the manufacturing sector increased slightly, to just under 40 per cent of the total. The share of services increased to 45 per cent. Significant changes occurred within both the manufacturing and services categories. Within manufacturing, two industries – food, drink, tobacco and petroleum and chemical products – account for no less than 60 per cent of the UK FDI total. The biggest change was the dramatic fall in the share of electrical and electronic equipment: from 11 per cent to 0.1 per cent. Within services, predictably, it was finance, insurance and business services which grew fastest to account for 61 per cent of the sector total in the late 1990s (compared with 35 per cent in 1987).

UK FDI has always displayed a particularly extensive *geographical* spread. Historically, that was clearly related to the extensiveness of its colonial possessions. But the geographical emphasis has certainly changed, with a much greater emphasis on Europe. However, it is noticeable that the UK's FDI in Africa is considerably greater than that of the United States. There is also still important UK investment in Australia, although it is relatively less important than it used to be. The major shift has been towards Europe, where almost 40 per cent of UK FDI is now located (compared with 28 per cent in 1987). Within Europe, 40 per cent is in the Netherlands, 17 per cent in Sweden, 7 per cent in France and 6 per cent in Germany.

Outside Europe, UK FDI is heavily oriented towards the United States (47 per cent of the world total) but Canada's significance as a host country has declined. Asia is far less important to the United Kingdom than to the United States. Only 4 per cent of UK FDI is in what has been the most dynamic of all world regions. Almost 20 per cent of the regional total is in Japan (little changed from 1987), but the major focus otherwise is on the former colonies of Hong Kong, Singapore and Malaysia. Together, these account for 70 per cent of all UK FDI in Asia.

German foreign direct investment

German FDI grew especially rapidly during the 1960s and 1970s. More recently, however, its relative growth has been less pronounced, although in the second half of the 1990s, growth rates were consistently in double figures but with considerable fluctuation (Figure 3.20). In *sectoral* terms, German FDI shows a number of distinctive features. FDI in the primary sector is virtually non-existent whereas the oil industry's significance in the United States and the United Kingdom is very marked. The significance of manufacturing halved between 1985 and 1994, from 60 per cent to 33 per cent. Within manufacturing, again, the German profile is distinctive. The chemicals industry dominates the sector (35 per cent) but motor vehicles (17 per cent) and metal and mechanical products (14 per cent) are also very significant. There is far more of a 'high-tech' flavour about German FDI in manufacturing. Finance, insurance and business services are highly significant, constituting 77 per cent of the sector total.

The *geographical* distribution of German FDI has become overwhelmingly oriented towards Europe. In the late 1990s, 60 per cent of the world total was located in Europe (up from 44 per cent in 1985). Although the largest concentration is in Belgium–Luxembourg and the Netherlands, German FDI in Europe is rather more evenly spread than that of the United Kingdom. In addition, however, German firms have a far stronger orientation towards Eastern Europe, particularly Hungary, the Czech Republic and Poland. As the emphasis on Europe has increased the relative importance of North America has declined – from 33 per cent in 1985 to 27 per cent. The Latin American focus is also rather less significant than it was (down from 9 per cent to 5 per cent) but almost 50 per cent of the Latin American FDI is in Brazil. Asia hosts only 5 per cent of German FDI, a remarkably low figure. Of this, almost 30 per cent is located in Japan with a further 12 per cent in Singapore. But the biggest change involves China. By the late 1990s, almost one-fifth of German FDI in Asia was located there, a measure of German firms' determined drive to enter the Chinese market.

Japanese foreign direct investment

Until relatively recently, FDI from Japan was extremely small. However, it grew with great speed from the 1980s, only recently slowing down significantly.

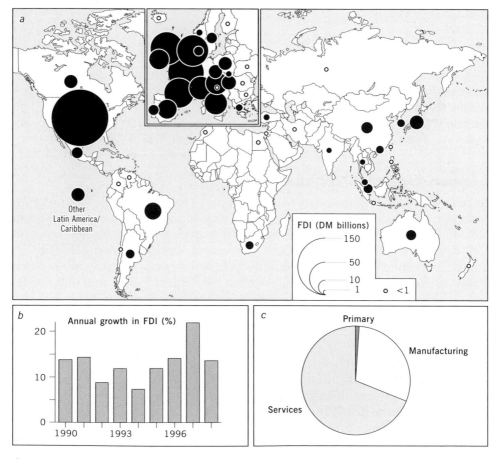

Figure 3.20   German foreign direct investment

Source: Based on data in OECD (2000) *International Direct Investment Statistics Yearbook.* Paris: OECD

From 1985 to 1990, annual rates of Japanese FDI growth were between 22 and 36 per cent, by far the highest of any country. Only after 1991 was there a marked slowdown but, even so, Japanese FDI stock still grew at almost 10 per cent a year up to 1994 (Figure 3.21). Since then, growth rates have been far more volatile; an increase of 18 per cent in 1997 being followed by a steep decline of 12 per cent in 1998.

The *sectoral* trajectory of Japanese FDI has been quite different from that of the other industrialized countries. In the early stages (the 1960s) the emphasis was overwhelmingly on FDI in the natural resources sector – a reflection of Japan's almost total lack of indigenous industrial raw materials. By the late 1990s, however, the dominant sector was services (66 per cent) – far higher than that of the United States, the United Kingdom and Germany. In fact, contrary to much popular opinion, manufacturing has never been the dominant component of Japanese FDI. Within manufacturing, there is a strong focus upon electric and

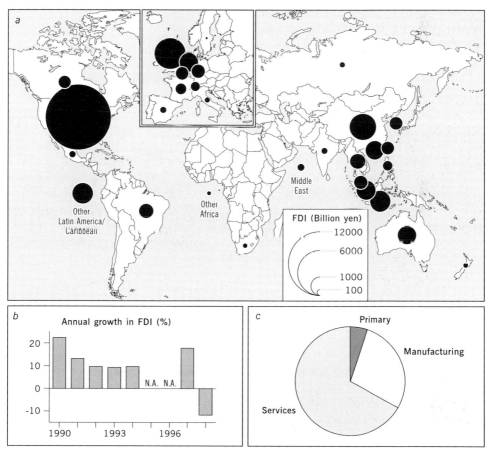

Figure 3.21 Japanese foreign direct investment

Source: Based on data in OECD (2000) *International Direct Investment Statistics Yearbook*. Paris: OECD

electronic equipment, motor vehicles and chemicals. The services sector is also distinctive. FDI in distribution services has been especially significant.

Between 1985 and the late 1990s, several major changes occurred in the *geography* of Japanese FDI. The relative importance of Asia increased (from 23 per cent to 26 per cent of the total), as did Latin America (from 19 per cent to 6 per cent). Conversely, both North America and Europe increased very substantially in importance. The North American share increased from 32 per cent to 42 per cent; the European share from 13 to 19 per cent. Japanese FDI in Europe is especially heavily biased geographically towards the United Kingdom, which contains almost 50 per cent of all Japanese FDI in Europe, followed – a long way behind – by the Netherlands (18 per cent) and Germany (9 per cent). In Asia, the major change has been a huge increase in the relative importance of China as a host country (up from 2 per cent in 1985 to 25 per cent in the late 1990s) as well as – more modestly – Thailand (4 per cent to 8 per cent). On the other hand, Indonesia's relative importance has waned (from 43 per cent in 1985 to 16 per cent).

A new wave? The emergence of FDI from newly industrializing economies

As shown earlier, by 2000 around 12 per cent of world FDI originated from developing countries, primarily from NIEs in East Asia. Contrary to much of the conventional literature on the so-called 'Third World multinationals', the sectoral and geographical distribution of FDI from Asian NIEs is very diverse. In particular, the idea that such FDI is predominantly located in developing countries is far from being universally true.[9] The following brief cases provide at least a hint of such diversity.

- *South Korea*. Korean outward FDI began in the late 1960s with investments in Indonesia to procure timber for the Korean plywood industry. The first overseas manufacturing plant, to produce food seasonings, was also located in Indonesia in the early 1970s. By the late 1970s, Korean outward investment was picking up speed and, like Taiwanese FDI, accelerated during the 1980s. The two share some similar features in the geography of their investment. Like Taiwan, South Korean FDI, from at least the late 1980s, has had a strong orientation towards North America. In 1985, 32 per cent of the total was located there compared with 22 per cent in Asia and 11 per cent in Europe. However, by the late 1990s, the pattern had changed somewhat, as Figure 3.22 shows. The Asian share of the total had increased dramatically to 45 per cent. Within Asia, China is the primary destination by a very wide margin. Over 40 per cent of Korean FDI in Asia is located in China with a further 14 per cent in Indonesia. The European share of the world total had grown to 15 per cent. Roughly 40 per cent of Korean FDI in Europe is located in the United Kingdom (25 per cent) and Germany (14 per cent). Korean firms also have some significant investments in Eastern Europe, notably in Poland. However, the restructuring problems facing the huge Korean conglomerate firms (the *chaebol*) in the aftermath of the 1997/98 crisis have led to the abandonment, or at least postponement, of some very large investments in the United Kingdom. Korean FDI in North America has remained stable in relative terms.

● *Singapore.* In common with the other Asian NIEs, FDI from Singapore grew quite rapidly during the early 1980s. The 1985 recession slowed down the rate of growth but from the early 1990s growth has been extremely rapid, as Figure 3.23 shows. One-third of Singapore's outward FDI is in the manufacturing sector and 42 per cent in the financial sector. Geographically, Singaporean FDI is very strongly oriented towards other East Asian countries. More than half of Singapore's total FDI goes to Asia. There have been some substantial changes in the geographical pattern within Asia during the 1990s, although around half the total still goes to the ASEAN countries. Within ASEAN, Malaysia's dominance has been reduced (from 40 per cent of the Asian total in 1990 to 24 per cent in the late 1990s). Conversely, Indonesia became more important as a destination for Singaporean FDI (up from 3 per cent to 15 per cent). But the biggest relative shift involved China. In 1990, only 3 per cent of Singapore's FDI in Asia went to China. By the late 1990s, China's share had risen to 24 per cent. Outside Asia, Europe is a far more significant FDI destination than the United States. Whilst only 4 per cent

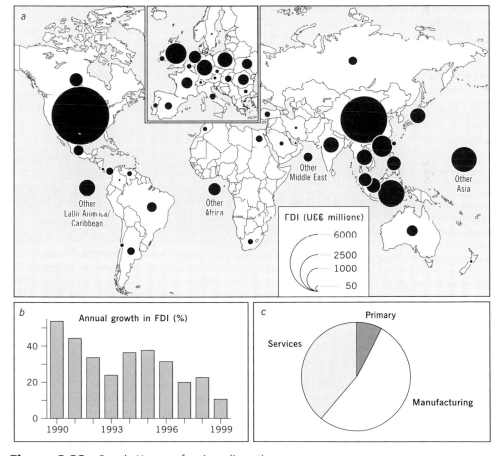

Figure 3.22   South Korean foreign direct investment

Source: Based on data in OECD (2000) *International Direct Investment Statistics Yearbook.* Paris: OECD

of Singapore's FDI goes to the United States, Singapore's FDI in Europe increased from 7 per cent to 16 per cent of its world total. Two-thirds of the investment in Europe was located in the United Kingdom.

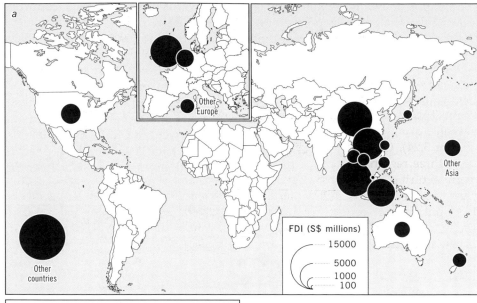

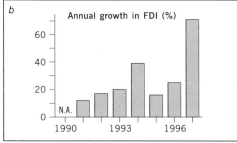

Figure 3.23   Singaporean foreign direct investment

Source: Based on data provided by Singapore Department of Statistics

Changing the lens: the macro-, micro- and meso-scale geography of the global economy

Our analytical lens throughout this chapter has been that of the *national* unit. This is primarily because the data we need to explore economic change at the global scale are only available at that level of spatial aggregation. But although it may be difficult to obtain comprehensive data at other spatial scales we can at least conceive of other highly relevant scales at which economic activities operate. Here, we look at three scales: the macro-, the micro- and the meso-scales.

The macro-scale: a global triad?

Almost 20 years ago, the Japanese management writer, Kenichi Ohmae, coined the term *global triad* to capture the macro-scale trends in the world economy.[10] According to this view, the world economy is now essentially organized around a tripolar, macro-regional structure whose three pillars are North America, Europe and East Asia. It is a view that has received very wide acceptance, especially within business circles. Certainly, if we look at the statistical data on production, trade and foreign direct investment the case looks pretty convincing, as Figures 3.24 and 3.25 show. Together, these three macro-regions contain 85 per cent of total world manufacturing output and generate 81 per cent of world merchandise exports (Figure 3.24). Over the past 20 years, the degree of economic concentration in these three regions increased. In 1980, the triad contained 76 per cent of world manufacturing output and 71 per cent of merchandise trade. Not surprisingly, in view of our earlier discussion, East Asia experienced the most marked increase. In 1980, only 16 per cent of total world manufacturing output was located there. Today, the region's share stands at 29 per cent whilst its share of total world merchandise trade has increased from 17 per cent to 26 per cent. The triad appears

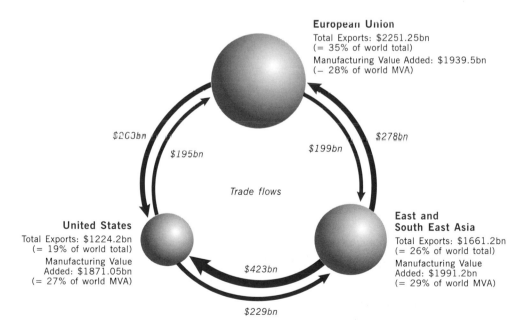

Figure 3.24   The global triad – concentration of manufacturing production and merchandise trade

Source: Based on material in UNIDO (1997) *Industrial Development: Global Report, 1997*; WTO (2001) *International Trade Statistics, 2001*

to be equally dominant in terms of foreign direct investment. Figure 3.25 shows that around 80 per cent of the world stock of FDI originates in the three regions. It also reveals very clearly the asymmetrical nature of FDI flows between the three macro-regions. The biggest flows by far involve the United States.

These trends seem to imply that the global triad is, in effect, 'sucking in' more and more of the world's production, trade and direct investment. The triad appears to sit astride the global economy like a modern three-legged Colossus, constituting the world's 'mega-markets'. However, whether it is 'real' or more of a statistical artefact is subject to debate. Two of the three pillars – Europe and North America – are both more formally organized politically into regional trading blocs (the European Union and the North American Free Trade Agreement. respectively).

In fact, if patterns of trade and FDI *flows* are examined more carefully and analysed statistically the evidence of the world crystallizing into just three mega-blocs is less clear:

> In analysing trade and FDI intensities over the 1985–1995 period, evidence of the triad appears to be weak … On one hand, trade interactions between countries are more regionalized and organized along continental lines than FDI … On the other, while sources of FDI are less dispersed than those of trade, and also more spatially concentrated in developed countries, FDI interactions between core and their regional members are far more internationalised and spatially dispersed.[11]

However, if the triad does represent a functional reality (actual or potential), with internally oriented production and trade systems, then it poses major problems for those parts of the world – notably the least developed countries – which are not integrated into the system.

Figure 3.25

The global triad – concentration of foreign direct investment

Source: Based on data in UNCTAD, 2001

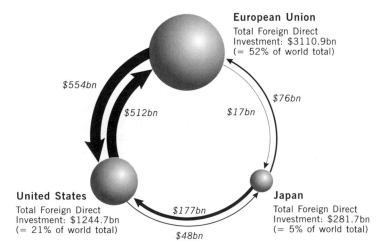

European Union
Total Foreign Direct Investment: $3110.9bn
(= 52% of world total)

$554bn

$512bn

$76bn

$17bn

United States
Total Foreign Direct Investment: $1244.7bn
(= 21% of world total)

$177bn

$48bn

Japan
Total Foreign Direct Investment: $281.7bn
(= 5% of world total)

The micro-scale: a 'fine-grained mosaic'

The global triad is one view of the world. Insofar as both business decision-makers and politicians believe in its existence it will continue to have a major influence on decisions that affect all our lives and livelihoods, regardless of where we actually live. But, of course, as individuals we do not think of ourselves, first and foremost, as living in one or other part of the global triad. We live in specific localized communities – cities, towns, or villages. If you could observe the earth from a very high altitude and look at its 'economic surface' you certainly would not see the kinds of national economic boxes we have had to use as the basis of our discussion in this chapter. Particularly if you were making the observation at night what you would see are distinctive *clusters*, picked out by the lights of localized agglomerations of activity.

Unfortunately, data disaggregated in this way, showing details of production and trade, are simply not available. We have to resort to surrogate measures or individual case studies. But it is vital to stress this most fundamental fact of economic life: the place-specific and clustered nature of most economic activity. The most widely available micro-scale indicator of the localized clustering of economic activity is the map of the world's cities (Figure 3.26). Virtually all manufacturing and business service activity is located in urban places. In some countries, just one, or perhaps two, major cities dominate; in other countries there is a 'flatter' urban hierarchy and a wider spread of activity among more evenly sized cities.

It is these cities, and their associated local regions, which contain a nation's economic activity, not some statistical box. Within any individual country, there will almost certainly be considerable diversity between cities / local regions not only in terms of their particular economic specializations but also in terms of their growth rates. In most cases, this reflects their specific historical trajectory – the 'path-dependency' idea introduced in Chapter 2. In others, however, such differentials may be the outcome of very specific political decisions to develop one particular part of a country rather than another. For example, China's recent spectacular economic growth has been articulated, quite deliberately, by the Chinese government around a limited number of Special Economic Zones and Coastal Cities (see Chapter 6).

The meso-scale: transborder clusters and corridors

Between the macro-scale of the global triad and the highly localized agglomerations of economic activity lies a meso-scale of economic-geographic organization which crosses, or sometimes aligns with, national boundaries. In some cases, this scale of organization is actually defined and created by the existence of the political boundary itself. In others it develops in spite of such boundaries and simply extends across them in a functionally organized manner. Three examples, one drawn from each of the three global triad regions, illustrate this meso-scale phenomenon.

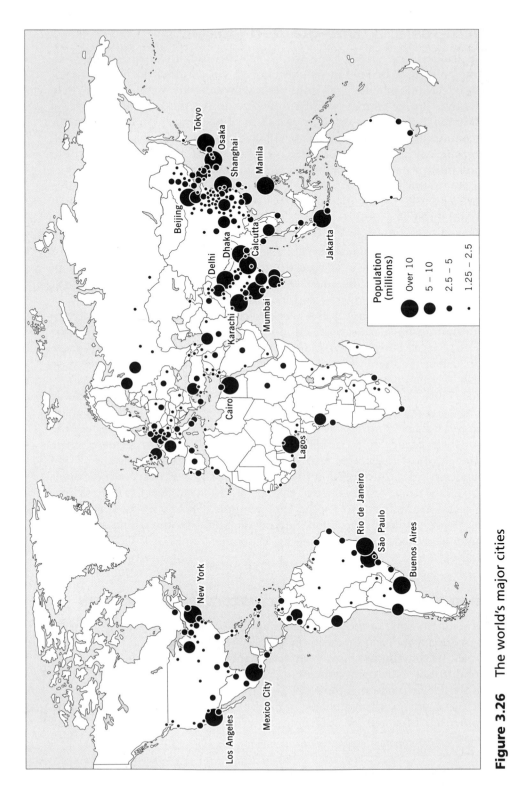

Figure 3.26 The world's major cities

Source: Based on United Nations Centre for Human Settlements, 2001: Table A1

● *Europe's major economic growth axis.* Within Europe, the pattern of economic activity is extremely uneven, both within and between individual countries. But as Figure 3.27 shows, we can also identify a distinctive 'growth axis' which runs north-west to south-east across the core area, cutting across national boundaries.

> The most advanced areas of Europe and most of Europe's major international cities lie on or near an axis extending from the north-west of London through Germany to Northern Italy … Along this axis lie two major foci: in the north-west are found the historic capitals of Europe's major colonial powers (Paris, London and Randstad-Holland) … in the south-east are cities and regions whose faster recent economic growth has pulled the axis' centre of gravity to the south-east. A parallel axis extends from Paris to the Mediterranean – and a south-western extension stretches down to the major cities in Iberia … another parallel axis may emerge in the east extending from Hamburg to Berlin, Leipzig, Prague and Vienna.[12]

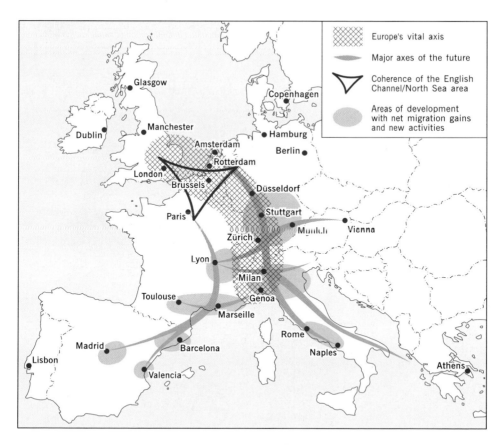

Figure 3.27   Europe's major growth axis

Source: Based on Dunford and Kafkalas, 1992: Figure 1.4

• *Emerging urban corridors in Pacific Asia.* The European 'growth axis' has evolved in a part of the world which has been strongly industrialized for a very long time and also in a region which has had a strong political manifestation through the European Community/European Union. However, we can see a similar phenomenon, at least in embryonic form, developing in Pacific Asia. Such 'growth triangles' include the Singapore–Batam–Johor triangle and the Southern China–Hong Kong–Taiwan triangle, both of which are focused upon a distinctive major city. But some Asian urban scholars argue that much larger urban corridors are becoming evident, as Figure 3.28 suggests.

Figure 3.28

Emerging urban corridors in Pacific Asia

Source: Based on Yeung and Lo, 1996: Figure 2.8

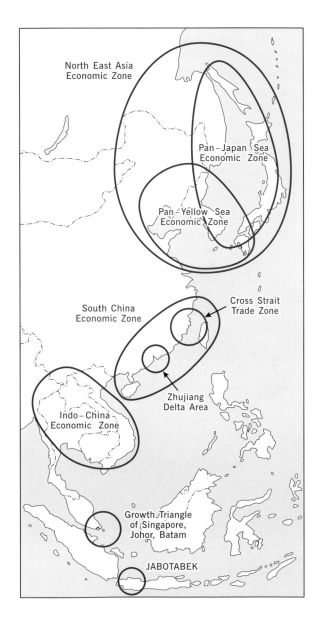

These urban corridors are at varying stages of formation, some exhibiting but incipient development whereas others are quite advanced in form and connectivity among the mega-cities. The best illustration of a mature urban corridor is ... an inverted S-shaped 1,500 km urban belt from Beijing to Tokyo via Pyongyang and Seoul ... [which] ... connects 77 cities of over 200,000 inhabitants each. More than 97 million urban dwellers live in this urban corridor, which, in fact, links four separate megalopolises in four countries in one.[13]

● *The United States–Mexico border zone.* The two previous examples illustrate the development of meso-scale regions which cut across national boundaries and create 'trans-national regions'. But there are other cases where the form of economic and urban development is actually defined and created by the existence of a border between countries. Where there is a very marked differential between two adjacent countries – for example, in taxation rates or production costs – there is often a strong incentive for development to occur on one side of the border to take advantage of benefits on the other side. One of the best examples of this is the United States–Mexico border, the sharpest geographical interface between an extremely wealthy economy and a much poorer developing economy. Far from being just a line on the map, the US–Mexico border is defined in the starkest of physical terms by a whole string of towns and concentrations of manufacturing activity along its entire length. These are not simply spontaneous clusters of activity, however. They owe their existence to political decisions taken initially in 1965 when the Mexican government instituted its Border Industrialization Program to help alleviate the severe economic and social problems of the northern border towns. Its focus was the 'in-bond' assembly (*maquiladora*) plant, which was

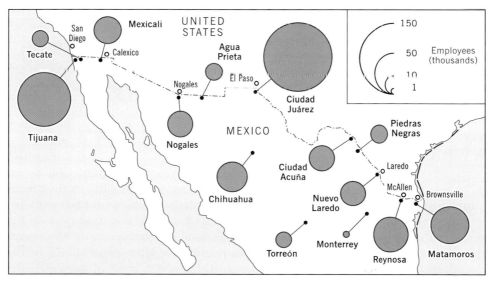

Figure 3.29   Employment in the major *maquiladora* centres

Source: Based on data in Instituto Nacional de Estadistica, Geografia e Informatica (2000) *Industria Maquiladora de Exportación.* Mexico: INEGI

allowed to import materials and components from the United States free of duty provided the end-products were then exported.[14] For United States manufacturers, the attraction of using the *maquiladoras* lay in their very cheap labour and the less stringent environmental controls south of the border. The result was spectacular growth in both employment and population in towns on both sides of the border, but especially on the Mexican side. Figure 3.29 shows the current scale of employment in the major *maquiladora* centres. Almost 600,000 workers are employed in these plants compared with 310,000 in 1988. The associated growth of the main towns has been very substantial. In 1950, around 390,000 people lived in the six major Mexican cities along the border. Today there are around 3 million.

Conclusion: a multi-polar, kaleidoscopic global economy

In these first years of the new millennium, the global economic map is vastly more complicated than that of only a few decades ago. Although there are clear elements of continuity, dramatic changes have occurred. The overall trajectory of world economic growth has become increasingly volatile; the pattern is one of short-lived surges in economic growth punctuated by periods of downturn or even recession.

Within this uneven trajectory, however, there has been a substantial reconfiguration of the global economic map. Although the world economy continues to be dominated by a small group of core economies, manufacturing production – and particularly trade – is no longer almost exclusively a core–region activity as it had been for the previous 200 years. Although a handful of core economies still dominates international trade flows, the most spectacular growth rates – apart from that of Japan – have been achieved by the East Asian NIEs. Although foreign direct investment is still dominated by the core economies, we are now seeing very substantial growth in such investment from some of the NIEs. Although most transnational investment originates from, and flows to, the core economies, the pattern of the investment flows has become increasingly complex. Today, there is a great deal of *interpenetration* of investment between national economies.

Without doubt, then, the most important single global shift of recent times has been the emergence of East Asia – including the truly potential giant, China – as a dynamic growth region. During the last three decades of the 20th century, it certainly appeared that the geographical centre of gravity of the world economy had begun to shift towards the Asia-Pacific region. Of course, the sudden crisis that broke in the region in 1997 raised question marks about that scenario. At the same time, the resurgence of the United States during the later 1990s raised further questions about the durability of such a shift. However, most of the East Asian economies have staged a considerable economic recovery while the United States' remarkable economic boom faltered in 2000–2001. Of course, in an interconnected global economy what happens in one part of the world has repercussions in other parts of the world. So, for example, the continued recovery of the East Asian economies depends to a large extent on continued growth of its export markets – and the most important of these is the United States!

So, there have been big changes in the contours of the global economic map. But the fact remains that the actual extent of global shifts in economic activity is extremely uneven. Only a small number of developing countries have experienced substantial economic growth; a good many are in deep financial difficulty whilst others are at, or even beyond, the margins of survival. Thus, although we can indeed think in terms of a new international division of labour, its extent is far more limited than is sometimes claimed. What is clear is that a relatively simple international division of labour no longer exists. It has been replaced by a far more complex, multi-scalar, structure. The global economy can perhaps best be described as 'a mosaic of unevenness in a continuous state of flux'.[15] That mosaic is, however, made up of processes which operate – and are manifested – at different, but interrelated spatial scales.

Notes

1   League of Nations (1945).
2   Maddison (2001: 125).
3   Webber and Rigby (1996: 6).
4   See Poon (1997).
5   UNCTAD (1996a: 124, 126).
6   The *World Investment Report*, compiled by UNCTAD on an annual basis, is the most comprehensive source of data on foreign direct investment. Historical trends are discussed by Dunning (1993) and Kozul-Wright (1995).
7   UNCTAD (2000b: 10). See also Kang and Johansson (2000).
8   UNCTAD (2001: 47).
9   For a discussion of these issues, see Yeung (1999a: xiii–xlvi).
10   Ohmae (1985).
11   Poon, Thompson and Kelly (2000: 440).
12   Dunford and Kafkalas (1992: 25, 27).
13   Young and Lo (1996; 39, 41).
14   Sklair (1989) provides a very thorough discussion of the *maquiladora* programme.
15   Storper and Walker (1984: 37).

PART TWO
PROCESSES OF GLOBAL SHIFT

CHAPTER 4

Technology: The 'Great Growling Engine of Change'

Technology and economic transformation

Technology is one of the most important processes underlying the globalization of economic activity. But, in saying that, we must avoid adopting a technologically deterministic position. It is all too easy to be seduced by the notion that technology 'causes' a specific set of changes, makes particular structures and arrangements 'inevitable', or that the path of technological change is linear. In fact, technology in, and of, itself does not cause particular kinds of change. It is, essentially, an *enabling* or *facilitating* agent. It makes possible new structures, new organizational and geographical arrangements of economic activities, new products and new processes, while not making particular outcomes inevitable. On the other hand, in a highly competitive environment, once a particular technology is in use by one firm, then its adoption by others may become virtually essential to ensure competitive survival.

Technological change, then, lies at the very heart of the processes of economic growth and development. As Schumpeter pointed out,

> the fundamental impulse that sets and keeps the capitalist engine in motion comes from the new consumers' goods, the new methods of production or transportation, the new markets, the new forces of industrial organization that capitalist enterprise creates.[1]

Technological change has been variously described as the 'prime motor of capitalism'; the 'great growling engine of change';[2] the 'fundamental force in shaping the patterns of transformation of the economy';[3] the 'chronic disturber of comparative advantage'.[4] Certainly, the so-called 'new economy' that was alleged to be emerging, especially in the United States in the 1990s, was driven by technological developments in information and communications technologies (ICT). More broadly, 'knowledge' has become the key source of wealth creation in contemporary societies. One estimate is that, by the mid-1990s, 70 per cent of value added in US manufacturing was knowledge-related compared with only 20 per cent in the 1950s. Conversely the share contributed by primary-processed materials and products fell from 80 per cent in the mid-1950s to 30 per cent.[5] Technological change and knowledge creation also have a distinctive geography, one characterized by both concentration and dispersal. Despite the transformative influences of those changes in communications technologies that facilitate the rapid and geographically extensive diffusion of innovations and knowledge, there continues to be a strong geographical localization of innovative activity.

In this chapter we focus only on those aspects of technological change that specifically influence the globalization of economic activity. The chapter is divided into four major parts:

- First, we discuss some of the broad characteristics of technological change in order to identify the key technologies and their evolution over time.
- Second, we focus on the 'space-shrinking' technologies of transportation and communication which are obviously central to the processes of internationalization and globalization.
- Third, we look at technological changes in both products and processes and explore the extent to which totally new forms of production technology and organization are occurring.
- Fourth, we focus explicitly on the geography of innovation, on the different scales – national and local – at which innovation processes operate.

Processes of technological change: an evolutionary perspective[6]

Technological change is a form of *learning* – of how to solve specific problems in a highly differentiated, and volatile, environment. However, it is more than a narrowly 'technical' process. Technology is not independent or autonomous; it does not have a life of its own.

> Specific choices within the frontier of technological possibilities are not the product of technological change; they are, rather, the product of those who make the choices within the frontier of possibilities. *Technology does not drive choice; choice drives technology.*[7]

Technology is a socially and institutionally embedded process. The ways in which technologies are used – even their very creation – are conditioned by their social and their economic context. In the contemporary world this means primarily the values and motivations of capitalist business enterprises, operating within an intensely competitive system. Choices and uses of technologies are influenced by the drive for profit, capital accumulation and investment, increased market share, and so on.

Types of technological change

Four broad types of technological change can be identified,[8] each of which is progressively more significant and far-reaching in its impact:

- *Incremental innovations*: the small-scale, progressive modifications of existing products and processes, created through 'learning by doing' and 'learning by using'. Although individually small – and, therefore, often unnoticed – they accumulate over a period of time to create very significant changes.
- *Radical innovations*: discontinuous events that drastically change existing products or processes. A single radical innovation will not, however, have a

widespread effect on the economic system; what is needed is a 'cluster' of such innovations.

- *Changes of technology system*: extensive changes in technology that impact upon several existing parts of the economy, as well as creating entirely new sectors, are based on a combination of radical and incremental technological innovations, along with appropriate organizational innovations. Changes of technology system tend to be associated with the emergence of key generic technologies (for example, information technology, biotechnology, materials technology, energy technology, space technology).

- *Changes in the techno-economic paradigm*: the truly large-scale revolutionary changes, embodied in new technology systems. They

    have such pervasive effects on the economy as a whole that they change the 'style' of production and management throughout the system. The introduction of electric power or steam power or the electronic computer are examples of such deep-going transformations. A change of this kind carries with it many clusters of radical and incremental innovations, and may eventually embody several new technology systems. Not only does this fourth type of technological change lead to the emergence of a new range of products, services, systems and industries in its own right – it also affects directly or indirectly almost every other branch of the economy … the changes involved go beyond specific product or process technologies and affect the input cost structure and conditions of production and distribution throughout the system.[9]

Long-waves[10]

The notion that economic growth occurs in a series of long-waves – 'Kondratiev' waves – of more or less 50 years' duration goes back to the 1920s. Figure 4.1 outlines the kind of long-wave sequence commonly envisaged. Four complete K-waves are identified; we are now in the midst of a fifth. Each wave may be divided into four phases: prosperity, recession, depression and recovery. Each wave tends to be associated with particularly significant technological changes around which other innovations – in production, distribution and organization – cluster and ultimately spread through the economy. Such diffusion of technology stimulates economic growth and employment, although technology alone is not a sufficient cause of economic growth. Demographic, social, industrial, financial and demand conditions also have to be appropriate. At some point, however, growth slackens: demand may become saturated or firms' profits become squeezed through intensified competition. As a result, the level of new investment falls, firms strive to rationalize and restructure their operations and unemployment rises. Eventually, the trough of the wave will be reached and economic activity will turn up again. A new sequence will be initiated on the basis of key technologies – some of which may be based on innovations that emerged during recession itself – and of new investment opportunities. Although there is disagreement over the precise mechanisms and timing involved, each of the waves is generally associated with changes in the techno-economic paradigm, as one set of techno-economic practices is displaced by a new set. This is not a sudden process but one that occurs gradually and involves the ultimate 'crystallization' of a new paradigm.

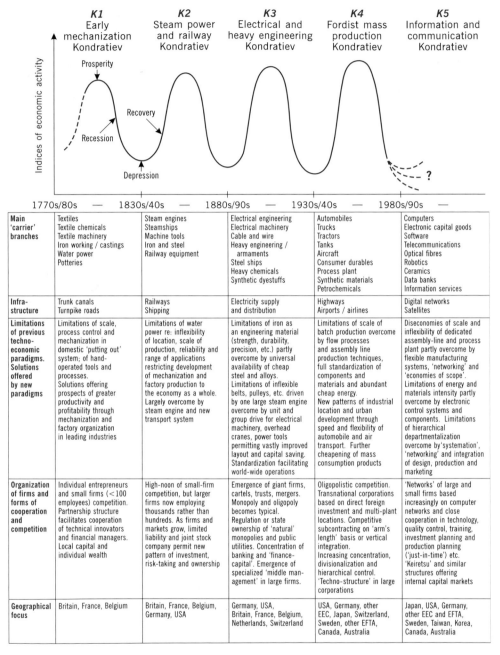

	K1 Early mechanization Kondratiev 1770s/80s	K2 Steam power and railway Kondratiev 1830s/40s	K3 Electrical and heavy engineering Kondratiev 1880s/90s	K4 Fordist mass production Kondratiev 1930s/40s	K5 Information and communication Kondratiev 1980s/90s
Main 'carrier' branches	Textiles Textile chemicals Textile machinery Iron working / castings Water power Potteries	Steam engines Steamships Machine tools Iron and steel Railway equipment	Electrical engineering Electrical machinery Cable and wire Heavy engineering / armaments Steel ships Heavy chemicals Synthetic dyestuffs	Automobiles Trucks Tractors Tanks Aircraft Consumer durables Process plant Synthetic materials Petrochemicals	Computers Electronic capital goods Software Telecommunications Optical fibres Robotics Ceramics Data banks Information services
Infra-structure	Trunk canals Turnpike roads	Railways Shipping	Electricity supply and distribution	Highways Airports / airlines	Digital networks Satellites
Limitations of previous techno-economic paradigms. Solutions offered by new paradigms	Limitations of scale, process control and mechanization in domestic 'putting out' system; of hand-operated tools and processes. Solutions offering prospects of greater productivity and profitability through mechanization and factory organization in leading industries	Limitations of water power re: inflexibility of location, scale of production, reliability and range of applications restricting development of mechanization and factory production to the economy as a whole. Largely overcome by steam engine and new transport system	Limitations of iron as an engineering material (strength, durability, precision, etc.) partly overcome by universal availability of cheap steel and alloys. Limitations of inflexible belts, pulleys, etc. driven by one large steam engine overcome by unit and group drive for electrical machinery, overhead cranes, power tools permitting vastly improved layout and capital saving. Standardization facilitating world-wide operations	Limitations of scale of batch production overcome by flow processes and assembly line production techniques, full standardization of components and materials and abundant cheap energy. New patterns of industrial location and urban development through speed and flexibility of automobile and air transport. Further cheapening of mass consumption products	Diseconomies of scale and inflexibility of dedicated assembly-line and process plant partly overcome by flexible manufacturing systems, 'networking' and 'economies of scope'. Limitations of energy and materials intensity partly overcome by electronic control systems and components. Limitations of hierarchical departmentalization overcome by 'systemation', 'networking' and integration of design, production and marketing
Organization of firms and forms of cooperation and competition	Individual entrepreneurs and small firms (<100 employees) competition. Partnership structure facilitates cooperation of technical innovators and financial managers. Local capital and individual wealth	High-noon of small-firm competition, but larger firms now employing thousands rather than hundreds. As firms and markets grow, limited liability and joint stock company permit new pattern of investment, risk-taking and ownership	Emergence of giant firms, cartels, trusts, mergers. Monopoly and oligopoly becomes typical. Regulation or state ownership of 'natural' monopolies and public utilities. Concentration of banking and 'finance-capital'. Emergence of specialized 'middle man-agement' in large firms.	Oligopolistic competition. Transnational corporations based on direct foreign investment and multi-plant locations. Competitive subcontracting on 'arm's length' basis or vertical integration. Increasing concentration, divisionalization and hierarchical control. 'Techno-structure' in large corporations	'Networks' of large and small firms based increasingly on computer networks and close cooperation in technology, quality control, training, investment planning and production planning ('just-in-time') etc. 'Keiretsu' and similar structures offering internal capital markets
Geographical focus	Britain, France, Belgium	Britain, France, Belgium, Germany, USA	Germany, USA, Britain, France, Belgium, Netherlands, Switzerland	USA, Germany, other EEC, Japan, Switzerland, Sweden, other EFTA, Canada, Australia	Japan, USA, Germany, other EEC and EFTA, Sweden, Taiwan, Korea, Canada, Australia

Figure 4.1   Kondratiev long-waves and their basic characteristics

Source: Based, in part, on material in Freeman and Perez, 1998: Table 3.1

As Figure 4.1 shows, however, the process of change involves more than just technical change. Each phase is also associated with characteristic forms of economic organization, cooperation and competition. Organizational change has followed a path from an early focus on individual entrepreneurs in K1, through small firms, but of larger average size, in K2, to the monopolistic, oligopolistic and cartel structures of K3, the centralized, hierarchical TNCs of K4 and, it is argued, the 'network' and alliance organizational forms of K5. (These are issues we will explore in Chapter 8.)

Each successive K-wave also has a *specific geography* as technological leadership shifts over time. The technological leaders of K1 were Britain, France and Belgium. In K2 these were joined by Germany and the United States. K3 saw leadership firmly established in Germany and the United States, although the other earlier leaders were still prominent and had been joined by Switzerland and the Netherlands. By K4 Japan, Sweden and the other industrialized countries were in the leadership group. K5 has seen a more prominent role in technological leadership by Japan and, more unexpectedly, by the emergence of two of the East Asian NIEs – Taiwan and South Korea – to prominent technological positions in specific areas. There is also a clear micro-geography characteristic of each stage. In effect, 'the locus of the leading-edge innovative industries has switched from region to region, from city to city'.[11]

Information technology: a key generic technology[12]

The fifth Kondratiev cycle is associated primarily with *information technology* (IT).

> For the first time in history, information generation, processing and transmission have become the main commodities and sources of productivity and power and not only a means of achieving better ways of doing things in the production process. New information technologies are not simply tools to be applied but processes to be developed.[13]

Information technology, in itself, is nothing new. But the current generation of information technologies has one very special characteristic, as Figure 4.2 shows. It is based upon the *convergence* of two initially distinct technologies: *communications technologies* (concerned with the transmission of information) and *computer technologies* (concerned with the processing of information). It is this convergence, especially through the transition from analogue to digital systems, that is so important for developments in the global economy.

The 'space-shrinking' technologies[14]

The processes of economic globalization have one fundamental requirement: the development of transportation and communications technologies that overcome the frictions of space and time. Neither of these technologies can be regarded as the cause of globalization. However, without them, today's complex global economic system simply could not exist. Transportation and communication technologies perform two distinct, though closely related and complementary, roles.

● *Transportation systems* are the means by which materials, products and other tangible entities (including people) are transferred from place to place.

● *Communications systems* are the means by which information is transmitted from place to place in the form of ideas, instructions, images, and so on.

For most of human history, transportation and communications were effectively one and the same. Prior to the invention of electric technology in the nineteenth century, information could move only at the same speed, and over the same distance, as the prevailing transportation system allowed. Electric technology broke that link, making it increasingly necessary to treat transportation and communication as separate, though intimately related, technologies. Developments in both have transformed our world, permitting unprecedented mobility of materials and products and a massive geographical expansion of markets.

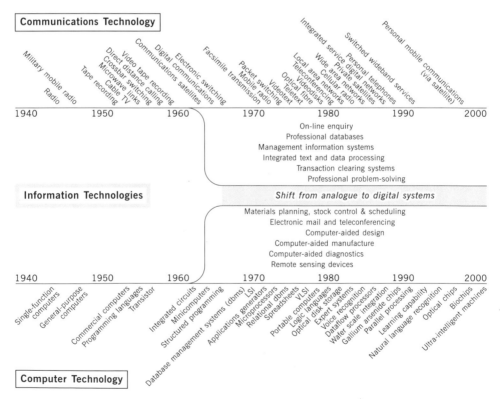

Figure 4.2 Information technology: the convergence of communications technologies and computer technologies

Source: Based on Freeman, 1987: Figure 2

Major developments in transportation technologies

In terms of the time it takes to get from one part of the world to another, there is no doubt that the world has 'shrunk' dramatically (Figure 4.3). For most of human history, the speed and efficiency of transportation were staggeringly low and the costs of overcoming the friction of distance prohibitively high. Movement over land was especially slow and difficult before the development of the railways. Indeed, even as late as the early 19th century, the means of transportation were not really very different from those prevailing in biblical times. The major break-through came with two closely associated innovations: the application of steam power as a means of propulsion and the use of iron and steel for trains, railway tracks and ocean-going vessels. These, coupled with the linking together of over-land and oceanic transportation (for example, with the cutting of the canals at Suez and Panama), greatly telescoped geographical distance on a global scale. The railway and the steamship introduced a new, and much enlarged, scale of human activity. Flows of materials and products were enormously enhanced and the possibilities for geographical specialization greatly stimulated. Such innovations were a major factor in the massive expansion in the global economic system during the nineteenth century.

The past few decades have seen an acceleration of this process of global shrink-age. In economic terms, there have been two particularly important develop-ments, both of them appearing during the 1950s. One development was the introduction of *commercial jet aircraft*, whose 'take-off' in the late 1950s was extra-ordinarily rapid, as Figure 4.4 shows. The other development was the introduc-tion of *containerization* for the movement of ocean and land freight, an innovation that vastly simplified transhipment of freight from one mode of transportation to another, increased the security of shipments, and greatly reduced the cost and time involved in moving freight over long distances. The first container ship, launched in 1956 to move goods from Newark, New Jersey to Houston, Texas, through the Gulf of Mexico, was merely a conventional oil tanker strengthened to take 58 9-metre boxes. By the late 1990s, roughly 90 per cent of total world trade was moved in containers on purpose-built ships.[15] According to one observer,

> Container shipping certainly is the great hidden wonder of the world, a vastly underrated business … It has shrunk the planet and brought about a revolution because the cost of shipping boxes is so cheap. People talk about the contribution made by the likes of Microsoft. But container shipping has got to be among the 10 most influential industries over the past 30 years… Before container shipping, seaborne trade was slow and unreliable. In the early Sixties, unloading a ship at, say, Liverpool docks could take weeks, even months. And during that time a substantial proportion of the goods could fall prey to thieves, or the weather. Today, the goods are protected in a container, during passage and in port. With cranes specially built to lift the containers, a ship can be in and out of a port in 10 hours, saving thousands in port charges and speeding trade.[16]

In the process, of course, hundreds of thousands of jobs were lost in port indus-tries throughout the world, a reminder that technological change produces losers as well as winners.

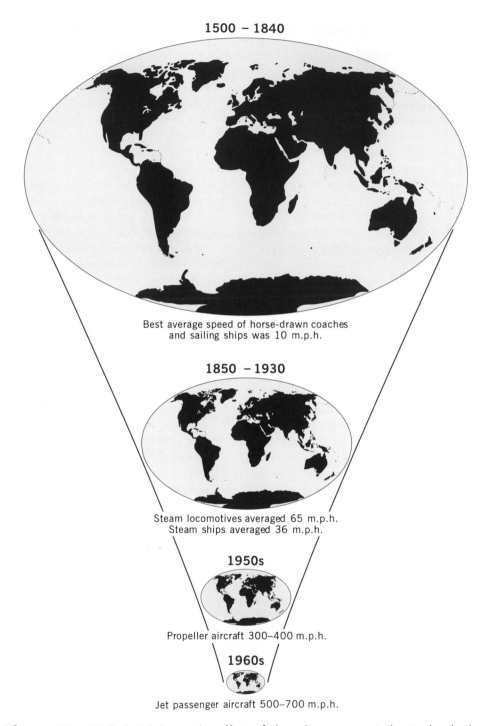

Figure 4.3 Global shrinkage: the effect of changing transportation technologies on 'real' distance

Source: Based on McHale, 1969: Figure 1

Figure 4.4

The 'take-off' of air traffic after the introduction of the jet aircraft

Source: Based on C. Cherry (1978) *World Communications: Threat or Promise?* Chichester: Wiley: Figure 3.16

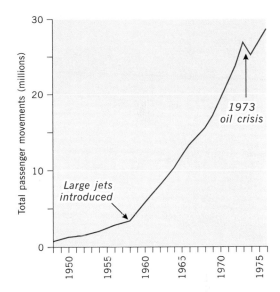

Although the world has indeed shrunk in relative terms, however, we need to be aware that, contrary to the impression given by Figure 4.3, such shrinkage has been, and continues to be, highly uneven. Technological developments in transportation have a very strong tendency to be geographically concentrated. *Time–space convergence* affects some places more than others. While the world's leading national economies and the world's major cities are being pulled closer together, others – less industrialized countries or smaller towns and rural areas – are, in effect, being left behind. Figure 4.5 shows this process of differential geographical shrinkage for just one country, Britain, in the 1960s. It shows the fastest travel times from London to other parts of the country. The effects of the major investments in transportation technologies of that time were to pull the bigger cities, like Manchester, Birmingham, Glasgow and Edinburgh, closer to London But the effect of differential investment in transportation technologies was to make other places less accessible. For example, the former cotton textiles town of Burnley, though only 25 miles north of Manchester, became twice as far away from London as Manchester. The small Welsh town of Pwllheli was almost as far away from London as were New York or Montreal.

The same process of geographically uneven shrinkage applies at the global scale. The time–space surface is highly plastic; some parts shrink whilst other parts become, in effect, extended. By no means everywhere benefits from technological innovations in transportation.

Major developments in communications technologies

Both the time and the relative cost of transporting materials, products and people have fallen dramatically as the result of technological innovations in the

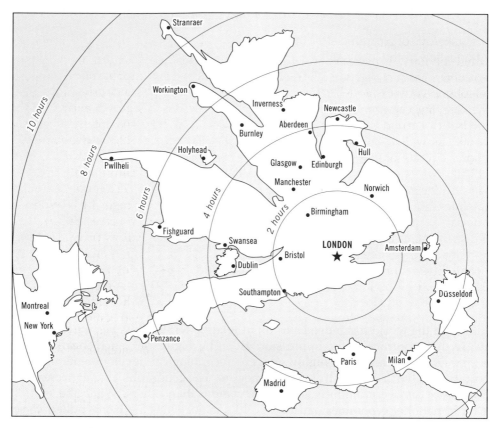

Figure 4.5 The unevenness of time–space convergence

Source: Based on P. Dicken and P.E. Lloyd (1981) *Modern Western Society*. London: Harper & Row: Figure 2.7

transportation media. Such developments have depended on parallel developments in communications technology. However, the communications media are fundamentally significant in their own right. Indeed, communications technologies must now be regarded as *the* key technologies transforming relationships at the global scale.

> The new telecommunications technologies are the electronic highways of the informational age, equivalent to the role played by railway systems in the process of industrialization.[17]

Transmission channels: satellites and optical fibres

Global communications systems have been transformed radically during the past 20 or 30 years through a whole cluster of significant innovations in information technologies. Two innovations have been especially significant in the development of the global economy: satellite communications and optical fibre technologies.

Satellite technology revolutionized global communications from the mid-1960s when the Early Bird or Intelsat I satellite was launched. This was the first geo-stationary satellite,[18] located above the Atlantic Ocean and capable of carrying 240 telephone conversations or two television channels simultaneously. Since then, the carrying capacity of the communications satellites has grown exponentially. Intelsat IV carried 6,000 simultaneous telephone conversations; Intelsat V carried 12,000, as well as television channels; Intelsat VIII carries 22,500 two-way telephone circuits and three television channels. Its capacity can be increased to 112,500 two-way telephone circuits with the use of digital multiplication equipment.[19] The Intelsat system is a multi-nation consortium of 122 countries whose satellites are positioned to provide complete global coverage. In addition to Intelsat, there are regional systems (such as Eutelsat, which serves Europe, and others in Asia, Latin America and the Middle East) as well as private satellite systems. Today, there are around 100 geo-stationary satellites in orbit.

Satellite technology made possible remarkable levels of global communication of both conventional messages and the transmission of data. A message could be transmitted in one location and received in another on the other side of the world virtually simultaneously. But not quite simultaneously. A geo-stationary satellite system involves a short delay in the transmission and receipt of signals. One way of getting round these problems is to use a whole series of satellites that operate together in low earth orbit (LEO systems).

In the past two or three decades, satellite communications have been increasingly challenged by *optical fibre* technology. The first commercially viable optical fibre system was developed (by Corning) in the United States in the early 1970s. Since then, the speed, carrying capacity and cost of optical fibre transmission cables have changed dramatically. Optical fibre systems have a huge carrying capacity, and transmit information at very high speed and, most importantly, with a high signal strength. By the end of the 1990s, for example,

> a single pair of optical fibres, each the thickness of a human hair ... [could] ...carry North America's entire long-distance communications traffic. Gemini, a transatlantic undersea cable... completed... [in 1998]... ha[d] more capacity than all existing transatlantic cables combined.[20]

Since then, technological developments in optical fibres have continued to accelerate, vastly increasing the speed and capacity of communications networks. At the same time, the geographical spread of optical fibre systems has increased. For example, the privately financed FLAG (fibre link around the globe) system has three transoceanic systems in operation that link Europe, Asia and North America. FLAG Europe–Asia, for example, is a 27,000 km system that connects north-west Europe and Japan via 11 'landing points' in southern Europe, the Middle East, India and South East Asia. The FLAG system is a 'global carriers' carrier', providing transmission services for large numbers of international carriers.

Figure 4.6 shows the enormous increase in the carrying capacity of submarine oceanic cable systems created by developments in optical fibre technologies. Figure 4.7 shows the extent to which cable systems have overtaken satellite systems in the transmission of voice traffic across the Atlantic. However, communications systems consist of more than just pushing more and more signals along optical fibre cables. At present, a complete 'all-optical, end-to-end' network is not possible. In particular, getting the signals direct to individual consumers remains difficult because control of the so-called 'last mile' of communications systems remains in the hands of local, often formerly state-owned, telecommunications companies.

Nevertheless, technological developments in satellite and cable technologies have transformed the relationship between geographical distance and the cost of transmitting and receiving information. Two examples demonstrate this:

- In the 1960s the annual cost of an Intelsat telephone circuit, giving a connection from one point on the earth's surface to any other, was more than $60,000. In the late 1980s the same facility cost only $9,000. Between 1973 and 1993, the price of a 3-minute international telephone call from London to New York fell by almost 90 per cent in real terms.[21]

- In 1970 the cost of transmitting the *Encyclopaedia Britannica* electronically from the east to west coasts of the United States would have been $187. In 2000, 'the entire content of the Library of Congress could be sent across America for just $40. As bandwidth expands, costs will fall further.'[22]

Even so, the cost of international telecommunications remains far higher than should be the case in purely technological terms. This is because the

Figure 4.6

The growth in the information carrying capacity of submarine cable systems

Source: Based on material in *The Financial Times*, 15 November 2000

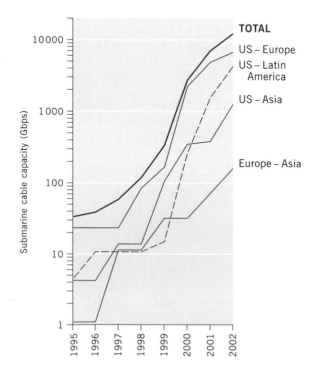

telecommunications industries have been very highly regulated at the national scale, despite the moves towards deregulation in many parts of the world.

Figure 4.7

Cable overtakes satellite across the Atlantic

Source: Based on Graham and Marvin, 1996: Figure 1.6

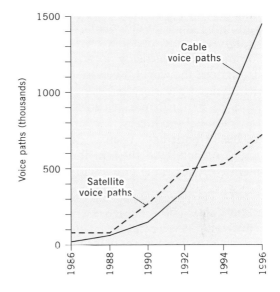

The electronic mass media

Developments in the communications media have revolutionized the potential for large organizations to operate over vast geographical distances. But there is another sphere – that of the *mass media* – in which innovations have transformed the global economy and are facilitating the globalization of markets. Large business firms require large markets to sustain them; global firms aspire to global markets. The existence of such markets obviously depends on income levels, but it depends, too, on potential customers becoming aware of a firm's offerings and being persuaded to purchase them. Even where consumer incomes are low, the ground may be prepared for possible future ability to purchase by creating an aspirational image. The mass media are particularly powerful means both of spreading information and of persuasion, hence their vital importance to the advertising industry and, in particular, to branded products.

The electronic media – particularly radio and television – are the most significant global media. In part, this is because of their vividness and sense of immediacy and involvement but it is also because they make no demands on literacy, something that even the most primitive news-sheet makes. Perhaps more than any other innovation in the mass media, it was the development of the transistor radio receiver that initially had the most revolutionary effects, especially in developing countries. Not only is it portable but also it is relatively cheap. Largely because of this, the sequence of development of the mass media has been rather different in the developing countries.

In the Western world, the pattern has been newspapers, radio, television. In Africa and much of Asia, the first contact the ordinary man [*sic*] has with any means of mass communication is the radio. It is the transistor which is bringing the people of remote villages and lonely settlements into contact with the flow of modern life.[23]

Today, television is the medium that has the most dramatic impact on people's awareness and perception of worlds beyond their own direct experience. Although the electronic media transmit messages of all kinds a very large proportion of these messages are *commercial* messages aimed at the consumer. Commercial advertising is a feature of most radio and television networks throughout the world. Even in state-controlled systems some advertising is often included. In a variety of ways, therefore, the communications media open the doors of national markets to the heavily advertised, branded products of the transnational producers.

These trends have been under way for several decades. The 1980s, however, saw a major 'phase shift' in the mass media with the appearance of cable and satellite broadcasting and a widespread deregulation of the media. As a result,

> In the US the number of independent TV stations grew during the 1980s from 62 to 330. Cable systems in major metropolitan areas feature up to 60 channels, mixing network TV, independent stations, cable networks, most of them specialized, and pay TV. In the countries of the European Union, the number of TV networks increased from 40 in 1980 to 150 by the mid-1990s, one-third of them being satellite broadcasted.[24]

Prior to the diversification wave of the 1980s, there was a high level of standardization in the kinds of TV programme available. It was this kind of 'mutual experience' that led Marshall McLuhan to coin the metaphor of the *global village* in which certain images are shared and in which events take on the immediacy of participation. Although in one sense the world may not have shrunk for the rural peasant or the urban slum dweller with no adequate means of personal transportation, it had undoubtedly shrunk in an indirect sense. It was now possible to be aware of distant places, of lifestyles, of consumer goods through the vicarious experience of the electronic media. But the increasing segmentation of TV messages made possible by the communications revolution of the 1980s means that the global village idea may no longer be an accurate picture of reality:

> the fact that not everybody watches the same thing at the same time, and that each culture and social group has a specific relationship to the media system, does make a fundamental difference vis-à-vis the old system of standardized mass media ... While the media have become indeed globally interconnected, and programs and messages circulate in the global network, *we are not living in a global village, but in customized cottages globally produced and locally distributed.*[25]

The Internet: the 'skeleton of cyberspace'[26]

Within the space of just a few years, these developments in the communications media have been joined by an even more revolutionary technology that provides the potential for *interactive* communications on a global basis. The 'skeleton' of this emerging cyberspace is the Internet, whose origins go back to the early 1970s and are to be found within the US Department of Defense. It spread initially through the linking of more specialized computer networks and now consists of a complex multi-level geographical structure – a series of 'backbone' networks into which regional and local networks are linked. As Figure 4.8 shows, growth of the Internet – measured in terms of the number of registered computers ('hosts') – has been exponential.

Although much of the discussion of the Internet is very heavy on hype, there is no doubt that the incredibly rapid development of a mass user, computer-based communications system is indeed revolutionary. Almost overnight, it seems, anybody who has a PC and telephone line can link into a global communications network that provides both inter-personal and inter-organizational communication through e-mail and also access to a cornucopia of information through the World Wide Web. The popular view is that this heralds the true arrival of a 'placeless' world in which anybody can be anywhere in 'communication space'. And yet when we look at the map of the Internet (even allowing for problems of measurement) we find a very distinctive – and very uneven – geography.[27]

Figure 4.8

Exponential growth of the Internet

Source: Based on data compiled by Internet Software Consortium [http://www.isc.org]

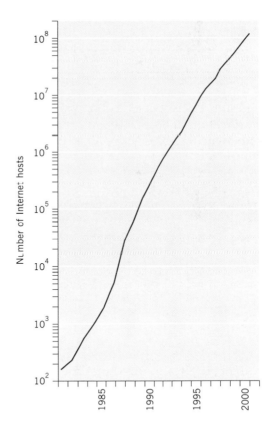

Figures 4.9 and 4.10 show some of the key features of Internet geography. More than half of all the registered domain names are in the United States alone. Germany and the United Kingdom account for a little less than 7 per cent each while the next more significant European country on this measure, France, has

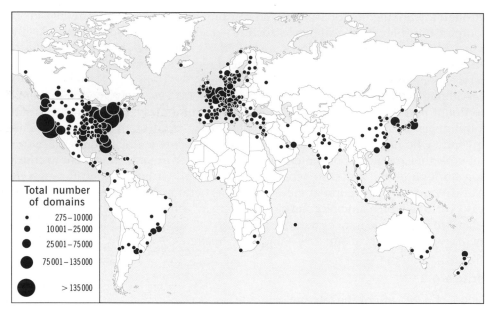

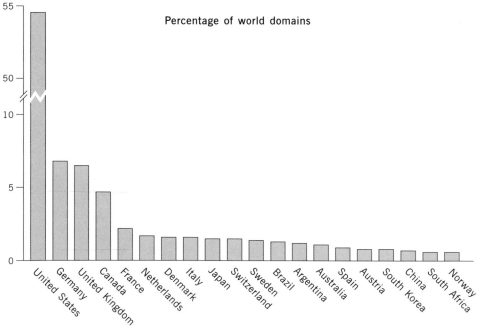

Figure 4.9 The geography of the Internet

Source: Based on Zook, 2001: Table 1, Figure 1

only 2.2 per cent of world domain names. Even more strikingly, Japan has only 1.5 per cent of the world total. These very strong geographical concentrations are further reflected in the pattern of Internet capacity between world regions (Figure 4.10).

In other words, the Internet is far from being the placeless/spaceless phenomenon so often envisaged. In particular, as far as provision of its basic infrastructure is concerned it is overwhelmingly an *urban* phenomenon:

> This is partly an historical accident … the Internet's fibre-optic cables often piggyback on old infrastructure where a right-of-way has already been established: they are laid alongside railways and roads or inside sewers … Building the Internet on top of existing infrastructure in this way merely reinforces real-world geography. Just as cities are often railway and shipping hubs, they are also the logical places to put network hubs and servers, the powerful computers that store and distribute data. This has led to the rise of 'server farms', also known as data centres or web hotels – vast warehouses that provide floorspace, power and network connectivity for large numbers of computers, and which are located predominantly in urban areas.[28]

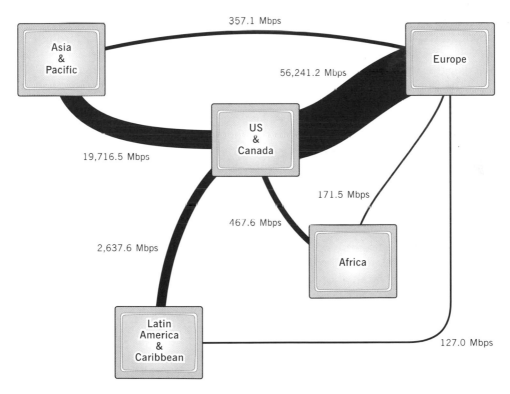

Figure 4.10 Inter-regional Internet bandwidth

Source: Based on material in *Telegeography* (2002)

It is too early to assess the eventual impact of the Internet and it is already clear that the system has grown so fast that there are major problems of effective usage. The instantaneous character of the Internet is often contradicted by long connection delays because of the sheer number of people trying to use it.

Uneven access to communications technologies

Technological developments in communications media have transformed space–time relationships between all parts of the world. In general terms it is clear that we now live in a *digital age*, characterized by certain key trends:[29]

- convergence between computer technologies and communications technologies
- continuing increase in the speed and capacity of these technologies and a continuing decline in their cost
- increasing ease of interfacing between different parts of the Internet
- growth of broadband communications technologies
- growth of mobile telecommunications.

Although we need to beware of the hype that surrounds the 'information revolution', there is no doubt that epochal changes are occurring through the development of digital technologies. However, not all places are equally connected; the 'time–space convergence' process is geographically uneven. In general, the places that benefit most from innovations in the communications media are the already 'important' places. New investments in communications technology are market-related; they go to where the returns are likely to be high. The cumulative effect is to reinforce both certain communications routes at the global scale and to enhance the significance of the nodes (cities/countries) on those routes. For example, although developing countries contain around 75 per cent of the world's population they have only around 12 per cent of the world's telephone lines.

Table 4.1 summarizes some of these global inequalities in access to the communications media. The range is enormous. In the case of telephones, for example, there are around 552 lines per 1000 population in the high-income group of countries (644 in the United States). In the low-income group there are only 32 lines per 1000 population on average, but in many African and some Asian countries the levels are in low single figures. The differentials are even greater in the case of mobile telephones, personal computers and the Internet. In the latter case, the high-income country group has 470 Internet hosts per 10,000 population (the United States has 1,131) while the low-income group has only 0.17 per 10,000. Such figures do, indeed, reflect the extreme inequalities that exist in access to communications technologies. But the figures do need to be treated with a little caution because, in most developing countries, there is a large amount of 'shared' viewing and listening in communal places. Also, the development of mobile telephonic systems helps to avoid the need to provide very expensive terrestrial-based systems with high fixed costs.

There is also an additional factor limiting the universal spread of new communications technologies. In virtually all countries of the world, governments regulate the communications industries within their borders. Today, however, there is a strong trend towards the deregulation of telecommunications in many countries. Within this geographically uneven communications surface there is also a social dimension. Not everybody – whether they are business firms or private individuals – have equal access. Despite the general decline in communications costs driven by technological change, the costs of usage are far from trivial. The major beneficiaries are the big corporations.

Table 4.1 The uneven access to communications media

Region	Radios/ 1000 pop'n	Television sets/1000 pop'n	Telephone lines/1000 pop'n	Mobile phones/ 1000 pop'n	Personal computers/ 1000 pop'n	Internet hosts/ 1000 pop'n
High income	**1300**	**664**	**552**	**188**	**269.4**	**470.12**
United States	2115	847	644	206	406.7	1131.52
United Kingdom	1445	641	540	151	242.4	240.99
Germany	946	570	550	99	255.5	160.23
Japan	957	708	479	304	202.4	133.53
Sweden	907	531	679	358	350.3	487.13
Middle income	**383**	**272**	**136**	**24**	**32.4**	**10.15**
Lower middle	327	247	108	11	12.2	4.91
Upper middle	469	302	179	43	45.5	19.01
Low income	**147**	**162**	**32**	**5**	**4.4**	**0.17**
Low and middle income	**218**	**194**	**65**	**11**	**12.3**	**3.08**
East Asia and Pacific	206	237	60	15	11.3	1.66
Europe and Central Asia	412	380	189	13	17.7	13.00
Latin America and Carib.	414	263	110	26	31.6	9.64
Middle East and N. Africa	265	140	71	6	9.8	0.25
South Asia	99	69	18	1	2.1	0.14
Sub-Saharan Africa	172	44	16	4	7.2	2.39
World	**380**	**280**	**144**	**40**	**58.4**	**75.22**

Source: Based on World Bank, 2000: Table 19

Technological changes in products and processes

Product innovation and the product life cycle[30]

In an intensely competitive environment, the introduction of a continuous stream of new products is essential to a firm's profitability and, indeed, survival. 'Long-run growth requires either a steady geographical expansion of the market area or the continuous innovation of new products. In the long run only product innovation can avoid the constraint imposed by the size of the world market for a given product.'[31] The idea that the demand for a product will decline over time is captured in the concept of the *product life cycle* (PLC).

Its essence is that the growth of sales of a product follows a systematic path from initial innovation through a series of stages: early development, growth, maturity and obsolescence (Figure 4.11). When a new product is first introduced on the market the total volume of sales tends to be low because customers' knowledge is

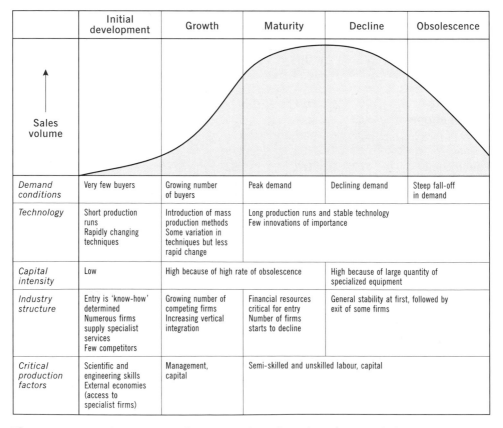

Figure 4.11 The product life cycle and its changing characteristics

Source: Based, in part, on Hirsch, 1967: Table II (1)

limited and also they tend to be uncertain about the product's quality and reliability. Assuming that the new product gains a foothold in the market (and very many do not get beyond this initial stage), it then enters a phase of rapid growth as overall demand increases. Such growth is likely to have a ceiling, however; the product attains maturity in which demand levels out. Eventually, demand for the product will slacken as the product becomes obsolescent.

The kind of development path suggested by the product life cycle concept has very important implications for the growth of firms and for their profit levels. The product life cycle implies that all products have a limited life; that obsolescence is inevitable. Of course, the rate at which the cycle proceeds will vary from one product to another. In some highly ephemeral products the cycle may run its course within a single year or even less. In others the cycle may be very long. However, product cycles are becoming shorter. In order to continue to grow, and to make profits, firms need to innovate on a regular basis (or to acquire innovations from other firms). There are three major ways in which a product's sales may be maintained or increased:

- to introduce a new product as the existing one becomes obsolete so that 'overlapping' cycles occur
- to extend the cycle for the existing product, either by making minor modifications in the product itself to 'update' it or by finding new uses for it
- to make changes to the production technology itself to make the product more competitive.

Whichever strategy is pursued, innovation and technological change are fundamental. In so far as product cycles are shortening in many sectors, this implies increasing pressure on firms to develop new products.

The production process and technology

In today's intensely competitive global environment product innovation alone is inadequate as a basis for a firm's survival and profitability. Firms must endeavour to operate the *production process* as efficiently as possible. Recent developments in technology – and, especially, in information technologies – are having profound effects upon production processes in all economic sectors. Three major, and closely interrelated, decisions are involved in the production process:[32]

- *Technique of production.* This decision concerns both the particular technology used and also the way in which the various inputs or factors of production are combined. It is almost always possible to vary the precise combination of, say, labour and capital according to their relative availability and cost. However, there are limits to such substitution of factors. Some production processes are intrinsically more capital intensive than others and vice versa. Closely related to the question of technique is that of –
- *Scale of production.* In general, the average cost of production tends to decline as the volume of production increases. The extent of such economies of scale

varies considerably from one industry to another. They are much greater, for example, in automobile production than in the manufacture of fashion garments. Technique and scale are, themselves, closely related.

- *Location of production.* The geographical location of production is intimately related to questions of technique and scale. Large-scale operations require access to large markets; highly labour-intensive production processes need access to appropriate pools of labour. Clearly, therefore, a firm seeking to reduce its production costs, or to increase its efficiency and productivity, can seek such economies at different points in the production process. It can attempt to purchase lower-cost inputs. In the case of material inputs this has increasingly involved a shift to supplies in developing countries. In the case of labour, a relatively immobile factor of production, the search for lower costs may involve the physical relocation of production to a cheaper labour location.

Production processes and the product life cycle

The nature of the production process itself also tends to vary systematically according to stage in the product life cycle (Figure 4.11). Each stage will tend to vary in terms of technology, capital intensity, labour force characteristics and industry structure.

- During the *early stage* of the cycle, production technology tends to be volatile with frequent changes in product specification. Production tends to be in short runs or batches. There is a tendency to rely on specialist suppliers and subcontractors. Capital intensity is relatively low. The most important type of labour is scientific and engineering. Entry into the new industry is determined largely by 'know-how' rather than by financial resources.

- By the time the product has progressed to the *growth stage* (assuming that it does) some important changes have occurred in the production process. The rapidly growing demand for the product (see Figure 4.11) permits the introduction of mass production and assembly line techniques. Even though the technology may be evolving, it is less volatile than in the earlier stage of the cycle. Capital intensity at the growth stage is considerably higher and the key type of labour is now managerial. The need is for administrative and marketing, rather than scientific, skills. In the growth stage also the number of firms engaged in the industry tends to be increasing but with a high casualty rate. There is a tendency towards increased vertical integration as firms seek to ensure the stability of component supplies and to exert greater influence over distribution of their product, often through acquisition and merger.

- In the *mature stage* of the cycle, demand has reached its peak (see Figure 4.11) and the market is becoming saturated. The major emphasis is on reducing the costs of production through long production runs. By this stage the technology is stable with few important changes. Long runs require the installation of high-volume specialist equipment that, in turn, increases the capital intensity of the industry. Stable technology and the application of a finer division of labour alter the labour force focus. In the mature stage the emphasis is on semi-skilled and unskilled labour performing routine, repetitive tasks in a mechanized manner. Labour costs become an increasingly significant element in production costs. The major barrier to entry into the industry is finance. Acquisition and merger are important mechanisms of entry and exit.

Hence, each stage in the product life cycle has a significant influence on the nature of the production process. Each stage has particular production characteristics. One of the most important of these is the way in which the relative importance of the major production factors changes. In general, as the cycle proceeds, the emphasis shifts from product-related technologies to process technologies and, in particular, to ways of minimizing production costs. In this respect, the relative importance of labour costs – especially of semi-skilled and unskilled labour – increases. More generally, different types of geographical location are relevant to different stages of the product cycle.

This view of systematic changes in the production process as a product matures is appealing and has some validity. There undoubtedly are important differences in the nature of the production process between a product in its very early stages of development and the same product in its maturity. But this linear, sequential notion of change in the production process is overly simplistic and deterministic. At any stage, the production process may be 'rejuvenated' by technological innovation. There may not necessarily be a simple sequence leading from small-scale production to standardized mass production.

Flexibility: the world 'after Fordism'

This leads us to consider the major recent developments that have been occurring in the technology of production processes and, particularly, those associated with the techno-economic paradigm of information technology. Most technological developments in production processes are, as we observed earlier, gradual and incremental, the result of 'learning by doing' and 'learning by using'. But periods of radical transformation of the production process have occurred throughout history. We are now in the midst of such a radical transformation, although precisely what is involved is the subject of much argument.

Over the long time-scale of the development of industrialization, the production process has developed through a series of stages each of which represents increasing efforts to mechanize and to control more closely the nature and speed of work. Five stages are generally identified:

- *Manufacture*: the collecting together of labour into workshops and the division of the labour process into specific tasks.
- *Machinofacture*: the application of mechanical processes and power through machinery in factories. Further division of labour.
- Scientific management ('Taylorism'): the subjection of the work process to scientific study in the late 19th century. This enhanced the fineness of the division of labour into specific tasks together with increased control and supervision.

- *'Fordism'*: the development of assembly line processes that controlled the pace of production and permitted the production of large volumes of standardized products.
- *'After-Fordism'*: the development of new flexible production systems based upon the deep application of information technologies.

These stages in the production process map fairly closely on to the long-wave sequence shown earlier in Figure 4.1. The first Kondratiev wave was associated with the transition from manufacture to machinofacture. The application of scientific management principles to the production process emerged in the late phase of K2 and developed more fully in K3. The bases of Fordist production were established during K3 but reached their fullest development during K4. The fifth Kondratiev is seen as marking the transition from Fordism to a new regime, the crossing of what Piore and Sabel termed the 'Second Industrial Divide'.[33] However, there is considerable disagreement over the precise form of this new regime, hence the use here of the neutral term 'after-Fordism'.

The Fordist system was epitomized by very large-scale production units, using assembly line manufacturing techniques and producing large volumes of standardized products for mass market consumption. It was a type of production especially characteristic of particular industrial sectors, notably automobiles (see Chapter 11). Not all sectors, nor all production processes, lent themselves to such a system of mass production but it was seen to be the main characteristic of the K4 phase. Many now argue that this Fordist system of production (and its associated organizational structures) entered a period of 'crisis' from about the mid-1970s and that it has been replaced by new modes of production. The most important characteristic of this new system is claimed to be *flexibility*: of the production process itself, of its organization within the factory, and of the organization of relationships between customer and supplier firms (see Chapter 8).

The key to production flexibility lies in the use of *information technologies* in machines and operations. These permit more sophisticated control over the production process. With the increasing sophistication of automated processes and, especially, the new flexibility of electronically controlled technology, far-reaching changes in the process of production need not necessarily be associated with increased scale of production. Indeed, one of the major results of the new electronic and computer-aided production technology is that it permits rapid switching from one part of a process to another and allows – at least potentially – the tailoring of production to the requirements of individual customers. 'Traditional' automation is geared to high-volume standardized production; the newer 'flexible manufacturing systems' are quite different, allowing the production of small volumes without a cost penalty.

The buzzwords now are *mass customization* – the ability of producers to make products to suit individual customers' needs, without sacrificing the lower costs conventionally associated with mass production. One of the best examples of the success of mass customization is in the personal computer industry where Dell has built its entire growth on this principle. A major benefit is that it can save huge amounts of capital that would otherwise be tied up in warehouse inventories of materials, components and finished products waiting for consumers to buy them (see Chapter 14). However, just how far mass customization will spread through the entire economy is yet to be seen.

Nevertheless, the potential of such flexible technologies is immense, and their implications enormous, for the nature and organization of economic activity at all geographical scales, from the local to the global. They involve three major tendencies:[34]

- a trend towards information intensity rather than energy or materials intensity in production
- a much enhanced flexibility of production that challenges the old best-practice concept of mass production in three central respects:
  - o a high volume of output is no longer necessary for high productivity; this can be achieved through a diversified set of low-volume products
  - o because rapid technological change becomes less costly and risky the 'minimum change' strategy in product development is less necessary for cost effectiveness
  - o the new technologies allow a profitable focus on segmented, rather than mass markets; products can be tailored to specific local conditions and needs
- a major change in labour requirements in terms of both volume and type of labour.

The current position, therefore, is that of a *diversity* of production processes and technologies (Figure 4.12), but where the relative importance of specific processes is changing. Thus, we can find a trend towards:

- an *increasingly fine degree of specialization* in many production processes, enabling their fragmentation into a number of individual operations

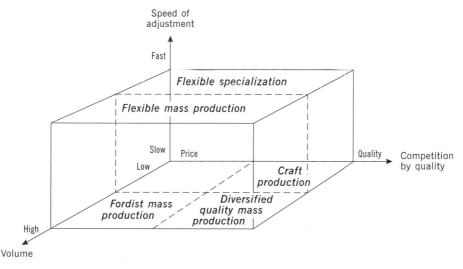

Figure 4.12 Types of production system

Source: Hollingsworth and Boyer, 1997: Figure 1.3

- an *increasing standardization and routinization* of these individual operations, enabling the use of semi-skilled and unskilled labour (this is especially apparent during the mature stage of a product's life cycle)
- an *increasing flexibility* in the production process, that is altering the relationship between the scale and the cost of production, permitting smaller production runs, increasing product variety, and changing the way production and the labour process are organized.

What does this all mean in terms of the broader question of what comes 'after Fordism'? There are strongly opposed interpretations of the nature of Fordism itself (for example, the extent to which it really constituted an all-embracing system of production, even in its heyday) and of what it is being replaced by.[35] Is it a variant on Fordism – 'neo-Fordism' – in which automated control systems are applied within a Fordist structure? Or is it a totally new 'post-Fordism', in which the new technologies create quite different forms of production organization? It is a debate that stretches way beyond the bounds of technology and technological change into the realms of the social organization of production, of the ways in which the state regulates economic activity and the nature of consumption and markets. More specifically, it reminds us that the processes involved are more than narrowly technological. As Figure 4.1 shows, there is also an organizational dimension to be considered. To repeat an earlier statement: technology is always socially embedded.

Flexible specialization: the re-birth of craft-based production?

Some assert unequivocally that flexible specialization is becoming the norm, the dominant style of production displacing Fordism. This is the 'post-Fordist' view that sees the hegemony of Fordism as being replaced not only by a new regime of flexible production but also by smaller organizational units.[36] As we have seen, Fordism was associated overwhelmingly with very large, vertically integrated firms producing standardized goods at very large volumes to benefit from economies of scale in production and selling to mass consumer markets. The development of flexible production technologies, it is claimed, leads to the resurgence of *small, independent entrepreneurial firms* emancipated from the tyrannies of mass production by the new flexibility that permits small-scale operations to serve small (perhaps local) markets.

This craft-based, 'flex-spec' interpretation of the changes in the production system also sees it as heralding a process of *re-skilling* of labour, as opposed to the relatively low skills characteristic of Fordism. The de-integration of the production system also goes hand-in-hand with a *de-integrated organizational structure* that then develops as horizontal networks of interrelated, specialist firms. The tendency, it is argued, will be for such networks to display a very strong *geographical localization* in a series of 'new industrial districts'.[37]

However, not all agree with this diagnosis.[38] There are many different types of 'flexibility': flexibility in volume of output, product flexibility, employment flexibility, flexible working practices, flexible machinery, flexible forms of organization.

The argument is not that flexible technologies do not exist – clearly they do – but that they are not as straightforward, as universal, or as pervasive as is often claimed. Equally, the Fordist system was never as ubiquitous as is sometimes claimed. Less rigid and smaller-scale production has always co-existed with mass production methods.

Japanese-inspired versions of 'after-Fordism'

Rather different organizational successors to Fordism have been suggested by writers impressed by the demonstrable economic success of the Japanese economy during the 1970s and 1980s. Although the Japanese production system has undoubtedly built heavily upon the new IT-based technologies, including the intensive use of robotic equipment, it is the 'organizational' technologies, developed within the Japanese system, that were far more important to its post-war economic success. Many of these were in place before the onset of the break-up of Fordism in the West and reflect, in particular, the conditions within which the Japanese economy was rebuilt after 1945.

There are several aspects of the contemporary Japanese production system that, taken together, may be regarded as an alternative both to classic Fordism and to the craft-based/flexible specialization model described above. Table 4.2 summarizes some of the contrasts between the three systems.

In particular,

- The organization of production is *flexible*, both in the use of facilities and, especially, in the way in which the *labour force* is organized:

  In Japan, work teams, job rotation, learning-by-doing, and flexibility have been used to replace the functional specialization, task fragmentation and rigid assembly-line production of US Fordism … there are few job classifications, work rules overlap, and production is organized on the basis of teams. Since tasks are allocated by team, workers can cover for each other and experiment with new allocations and machine configurations … Shopfloor learning is a basic characteristic of post-Fordist production in Japan … Management in post-Fordist Japan can be characterized as comprised of many 'little brains' sharing information, as opposed to the one 'big brain' directing many 'appendages' of Fordism … Learning by doing at many levels makes the Japanese firm an information-laden enterprise with problem-solving capabilities which far exceed its Fordist counterparts.[39]

- There is an obsessive preoccupation with *quality control* as an intrinsic element at all stages of the production process. The concept of Total Quality Management (TQM) involves building in quality from the beginning rather than checking for faults at the end. There is a continuous drive towards 'zero tolerance' of faults. This requires the development of a particular set of attitudes within the workforce at all levels.

- Customer–supplier relationships take on a particularly close form as firms attempt to reduce inventory to a minimum through the use of *just-in-time* (JIT), rather than the *just-in-case* (JIC) systems characteristic of Fordism.[40]

Table 4.3 summarizes the differences between the two systems.

Table 4.2 The major characteristics of craft production, Fordist mass production, and Japanese flexible production

Characteristic	Craft production	Fordist mass production	Japanese flexible production
Technology	Simple, but flexible tools and equipment using unstandardized components	Complex, but rigid single-purpose machinery using standardized components. Heavy time and cost penalties involved in switching to new products	Highly flexible methods of production using modular component systems. Relatively easy to switch to new products
Labour force	Highly skilled workers in most aspects of professional production	Very narrowly skilled workers design products but production itself performed by unskilled/semi-skilled 'interchangeable' workers. Each performs a very simple task repetitively and in a predefined time and sequence	Multi-skilled, polyvalent workers operate in teams. Responsibilities include several manufacturing operations plus respon-sibility for simple main-tenance and repair
Supplier relationships	Very close contact between customer and supplier. Most suppliers located within a single city	Distant relationships with suppliers, both functionally and geographically. Large inventories held at assembly plant 'just in case' of disruption of supply	Very close relationships with a functionally tiered system of suppliers. Use of 'just-in-time' delivery systems encourages geographical proximity between customers and suppliers
Production volume	Relatively low	Extremely high	Extremely high
Product variety	Extremely wide – each product customized to specific requirements	A narrow range of standardized designs with only minor product modifications	Increasingly wide range of differentiated products

Source: Based, in part, on material in Womack et al., 1990

Table 4.3 The characteristics of 'just-in-case' and 'just-in-time' systems

'Just-in-case' system	*'Just-in-time' system*
Characteristics	
Components delivered in large, but infrequent, batches	Components delivered in small, very frequent, batches
Very large 'buffer' stocks held to protect against disruption in supply or discovery of faulty batches	Minimal stocks held – only sufficient to meet the immediate need
Quality control based on sample check after supplies received	Quality control 'built in' at all stages
Large warehousing spaces and staff required to hold and administer the stocks	Minimal warehousing space and staff required
Use of large number of suppliers selected primarily on the basis of price	Use of small number of preferred suppliers within a tiered supply system
Remote relationships between customer and suppliers	Very close relationships between customer and suppliers
No incentive for suppliers to locate close to customers	Strong incentive for suppliers to locate close to customers
Disadvantages	
Lack of flexibility – difficult to balance flows and usage of different components	Must be applied throughout the entire supply chain
Very high cost of holding large stocks	Reliance on small number of preferred suppliers increases risk of interruption in supply
Remote relationships with suppliers prevents sharing of developmental tasks	
Requires a deep vertical hierarchy of control to coordinate different tasks	

Source: Based on material in Sayer, 1986

A particular interpretation of the Japanese 'Toyotist' system sees the entire process as being *lean*:

> lean production ... is 'lean' because it uses less of everything compared with mass production – half the human effort in the factory, half the manufacturing space,

half the investment in tools, half the engineering hours to develop a new product in half the time. Also it requires keeping far less than half of the needed inventory on site, results in many fewer defects, and produces a greater and ever growing variety of products.[41]

The use of the term 'Toyotism' to describe the Japanese alternative to Fordism reflects the leading role played by the automobile industry in general, and of Toyota in particular (see Chapter 11).

These various technological/organizational innovations developed within the post-war Japanese economy are generally accepted as contributing towards the country's spectacular economic success during the 1970s and 1980s. Where interpretations differ is, first, on how far the Japanese system is transferable to other national and local contexts and, second, on the labour implications of the system. On the first point, it is worth noting that the source of many of the practices of 'just-in-time'/'total quality control' production was, in fact, Western engineers, such as the American William Deming. Some of the ideas were imported into Japan during the 1950s and 1960s (having been largely ignored in the United States and Europe) as part of the process of technological learning and adapted to local conditions. In that sense, therefore, they form part of a broader – but specific – set of institutional practices: the distinctive Japanese 'business system' which is unlikely to be totally transferable to other contexts.

The diffusion of these practices outside Japan has occurred in two ways:

- Through the overseas expansion of Japanese firms themselves and their attempts to transfer their domestic practices to different contexts. The evidence suggests great variability in the extent to which firms are able to implement such transfers without a considerable degree of adaptation to local conditions, producing 'hybrid factories'.[42]
- Through the 'demonstration effect' – often provided by Japanese firms operating overseas but also through the acceptance and promotion of these ideas as 'best practice' by influential writers in management. The evidence suggests that adoption has been uneven, both geographically and between different industries.

The second controversial issue concerns the implications of the Japanese system for labour. Here views are strongly polarized.[43] The positive interpretation claims that the Japanese flexible production system is more humane than the Fordist mass production system, citing the following.

- It involves the re-skilling of workers, notably through the requirements of multi-task operations within work teams, job rotation and 'learning by doing'.
- It provides workers with a substantial degree of control and involvement.
- Team working and self-management reduce alienation.
- Individualized payments systems create significant incentives for workers.
- It is based upon long-term (lifetime) employment contracts.

On the other hand, critics of the Japanese system argue that, far from being more humane, it is actually highly exploitative of the labour force:

- The multi-skilled, team-based workforce is subject to strict managerial control. Work teams are used as the means to extend such control.
- The emphasis on continuous improvement places great pressure on workers.
- The individual payments system is used as part of a managerial strategy to 'divide and rule' the workforce.
- Long-term contracts apply only to 'core' workers in large companies; the remaining workforce is peripheralized with no job security.

The conclusion to be drawn from this discussion of the various 'after-Fordism' alternatives is that it is unwise to seek a single alternative at all. It follows from this that the prediction of much of the post-Fordist literature (especially in its flexible specialization form) that the new flexible technologies herald the rebirth of the small firm at the expense of the large is an exaggeration. The flexible technologies change the possibilities for firms of all sizes but in different ways. It is more satisfactory, therefore, to see current trends as consisting of a mix of the several systems described in this section. In any case, precisely how production systems are actually devised and implemented will vary greatly between specific national and, indeed, local contexts. It is to this issue that we now turn.

Geographies of innovation[44]

Innovation – the heart of technological change – is fundamentally a *learning* process. Such learning – by 'doing', by 'using', by observing from, and sharing with, others – depends upon the accumulation and development of relevant knowledge. Without doubt, the development of highly sophisticated communications systems facilitates the diffusion of knowledge at unprecedented speed and over unprecedented distances. As we have seen in this chapter, information can flow virtually instantaneously and globally both within and between organizations and between individuals (provided, of course, that they are 'connected' into the global communications systems). Nevertheless, 'conditions of knowledge accumulation are highly localized'.[45] Knowledge is *produced in specific places* and often used, and enhanced most intensively, in those same places. Hence,

> to understand technological change, it is crucial to identify the economic, social, political and geographical context in which innovation is generated and disseminated. This space may be local, national or global. Or, more likely, it will involve a complex and evolving integration, at different levels, of local, national and global factors.[46]

One reason for the continuing significance of 'localness' in the creation and diffusion of knowledge lies in a basic distinction in the nature of knowledge itself, which is broadly of two kinds:[47]

- *codified (or explicit) knowledge* – the kinds of knowledge that can be expressed formally in documents, blueprints, software, hardware, etc.

- *tacit knowledge* – the deeply personalized knowledge possessed by individuals that is virtually impossible to make explicit and to communicate to others through formal mechanisms.

This distinction is fundamentally important to understanding the role of space and place in technological diffusion. Codified knowledge can be transmitted easily across distance. It is through such means that, throughout history, political, religious and economic organizations, for example, have been able to 'act at a distance'; to exert control over geographically dispersed activities.[48] Developments in transportation and communications technologies have enabled such 'acting' or 'controlling' to take place over greater and greater distances. Tacit knowledge, on the other hand, has a very steep 'distance-decay' curve. It requires direct experience and interaction; it depends to a considerable extent – though not completely by any means – on geographical proximity. It is much more 'sticky'.

A key consideration, therefore, is the specific socio-technological context within which innovative activity is embedded; what is sometimes called the *innovative milieu*. This consists of a mixture of both tangible and intangible elements:

- the economic, social and political institutions themselves
- the knowledge and know-how which evolves over time in a specific context (the 'something in the air' notion identified many decades ago by Alfred Marshall)
- the 'conventions, which are taken-for-granted rules and routines between the partners in different kinds of relations defined by uncertainty'.[49]

The geographical scale of such innovative milieux may vary from the national down to the regional/local.

A great deal of attention has been given in the technology literature to the notion of *national innovation systems*[50] – the idea that the specific combination of social, cultural, political, legal, educational and economic institutions and practices varies systematically between national contexts. Such nationally differentiated characteristics help to influence the kind of technology system that develops and its subsequent trajectory. These underlying forces help to explain the gradual shifts in national technological leadership evident in successive K-waves. Despite the claims of the hyperglobalists that national distinctiveness is declining, the evidence strongly suggests that national variations in technology systems – and, therefore in technological competence – persist.

At the heart of such national systems we invariably find geographically localized innovative milieux. National systems of innovation, in other words, consist of *regional/local innovation clusters*. The basis of such localized clusters lies in several characteristics of the innovation process that are highly sensitive to geographical distance and proximity:[51]

- *Localized patterns of communication*. Geographical distance greatly influences the likelihood of individuals within and between organizations sharing knowledge and information links.

- *Localized innovation search and scanning patterns.* Geographical proximity influences the nature of a firm's search process for technological inputs or possible collaborators. Small firms, in particular, often have a geographically narrower 'scanning field' than larger firms.

- *Localized invention and learning patterns.* Innovation often occurs in response to specific local problems. Processes of 'learning by doing' and 'learning by using' tend to be closely related to physical proximity in the production process.

- *Localized knowledge sharing.* Because the acquisition and communication of tacit knowledge is strongly localized geographically there is a tendency for localized 'knowledge pools' to develop around specific activities.

- *Localized patterns of innovation capabilities and performance.* Geographical proximity, in enriching the depth of particular knowledge and its use, can reduce the risk and uncertainty of innovation. In other words,

> geography plays a fundamental role in the process of innovation and learning, since innovations are in most cases less the product of individual firms than of the assembled resources, knowledge and other inputs and capabilities that are localized in specific places. The clustering of inputs such as industrial and university R&D, agglomerations of manufacturing firms in related industries, and networks of business-service providers may create scale economies, facilitate knowledge-sharing and cross-fertilization of ideas and promote face-to-face interactions of the sort that enhance effective technology transfer … Two main features help explain the advantage of spatial agglomeration in this context: the involvement of inputs of knowledge and information which are essentially 'person embodied', and a high degree of uncertainty surrounding outputs. Both require intense and frequent personal communications and rapid decision-making, which are arguably enhanced by geographic proximity between the parties taking part in the exchange. Indeed, in the present era of rapid global dissemination of codified knowledge, we may even argue that tacit, and spatially more 'sticky' forms of knowledge are becoming more important as a basis for sustaining competitive advantage.[52]

Local innovative milieux consist essentially of a *nexus of untraded interdependencies* set within a temporal context of *path-dependent* processes of technological change. We outlined the major elements of these processes in general terms in Chapter 2. The point of emphasizing the 'untraded' nature of the interdependencies within such milieux is to distinguish the 'cement' which binds this kind of localized agglomeration from that which may be associated with the minimization of transaction costs (for example, of materials and components transfers) through geographical proximity. The terms technology district[53] or *technopole*[54] may be used to describe such technology clusters.

Table 4.4 lists some of the major technology districts/technopoles identified in empirical research. Many of them are associated with major metropolitan areas although some have developed outside the metropolitan sphere in rather less urbanized areas. Most are the outcome of the historical process of cumulative, path-dependent growth processes although a few are the deliberate creations of national technology policy.[55] But whatever their specific origin – and this will vary from place to place because of historical and geographical contingencies – these technological agglomerations form one of the most significant features of the contemporary global economy.

Table 4.4   Some leading technology districts/technopoles

United States	Europe	Asia
Southern California (including Silicon Valley)	M4 Corridor, London	Tokyo
	Munich	Seoul–Inchon
	Stuttgart	Taipei–Hsinchu
Boston, MA	Paris-Sud	Singapore
Austin, TX	Grenoble	
Seattle, WA	Montpellier	
Boulder, CO	Nice/Sophia Antipolis	
Raleigh–Durham, NC	Milan	

The geographies of innovation, therefore, consist of a complex set of networks and processes operating *within and across* various spatial scales, from the global, through the national, the regional and the local. Figure 4.13 provides a highly simplified picture of this very complex process. Although such local innovative milieux are based upon strongly place-specific networks, the development of TNCs means that some of the actors may be both 'insiders' and 'outsiders' at the same time. Indeed, TNCs have a strong incentive to attempt to become insiders in order to tap into local innovative systems. This linking together of territorially based local networks with non-territorially based TNC networks is a distinctive

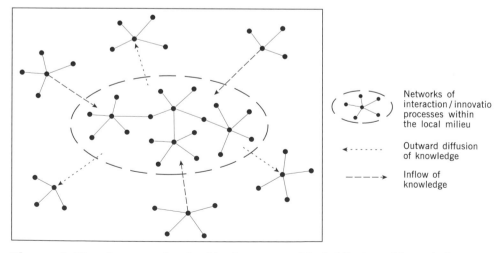

Figure 4.13   Processes involved in the geographical diffusion of knowledge

Source: Based on Malmberg et al., 1996: Figure 3

and important feature of the structure of the global economy[56] and is one to which we will return in Chapter 8. More specifically, in the context of this chapter, although the local region is undoubtedly important,

> as a source of innovation, learning, and tacit knowledge production … [it is possible that]…*relational* or *organizational proximity* (i.e. that which is achieved through communities of practice) might be more important than *geographical proximity* in constituting the 'soft architecture of learning'.[57]

Conclusion

The aim of this chapter has been to identify some of those features of technological change that are most important in the globalization of economic activity. Technological change is at the dynamic heart of economic growth and development; it is fundamental to the evolution of a global economic system. We focused on four specific aspects of technological change.

First, we explored the process of technological change as an evolutionary process in which much change is gradual and incremental, often unnoticed but none the less extremely significant. But there are periodic radical transformations of existing technologies – revolutionary developments in clusters of technologies – that dramatically alter not only products and processes in one industry but that also pervade the entire socio-economic system. These are the shifts in the techno-economic paradigm that seem to be associated with the long-waves of economic change.

Second, we concentrated on what is undoubtedly the major technological driving force today: information technology. IT is transforming both the technologies of transportation and communication and also the technologies of products and processes. IT is spreading into all sectors of the economy and to all types and sizes of organization but it is still the very large business organization, particularly the TNC, which is reaping the greatest benefits.

Third, we argued that the claim that we are shifting from one hegemonic (Fordist) system to another hegemonic (post-Fordist) system is far too sweeping and simplistic to capture the complex reality of a world based upon increased flexibility of production and organization. There are a number of alternatives to Fordism that, although they are all based upon the new flexibilities, take on rather different forms. We need to recognize the existence of such diversity.

Part of that diversity is related to the fourth focus of the chapter: the strongly localized nature of innovation and technological change. The path-dependent nature of technological change and the social conditions within which such change occurs give major importance to the *geography* of the process. In this regard, Dosi concludes that 'national (and regional) systems of innovation are there to stay, even in a more globalized world, and that they will continue, albeit in different forms, to shape the growth possibilities of different geographical areas and institutional entities.'[58]

Notes

1 Schumpeter (1943: 83).
2 Toffler (1971).
3 Freeman (1988).
4 Chesnais (1986).
5 Stewart (1997).
6 This kind of perspective on technological change is based upon the work of Dosi et al. (1988), Freeman (1982, 1987) and Perez (1985). See also Metcalfe and Dilisio (1996).
7 Borrus, quoted in Cohen and Zysman (1987: 183; emphasis added).
8 Freeman and Perez (1988).
9 Freeman (1987: 130).
10 See Freeman, Clark and Soete (1982); Freeman and Perez (1988); Hall and Preston (1988).
11 Hall and Preston (1988: 6).
12 Useful discussions of information technology can be found in Cairncross (1997), Castells (1996), Freeman (1987), Graham and Marvin (1996) and Hall and Preston (1988).
13 Rennstich (2002: 174).
14 For broad-ranging discussions of these technologies see Brunn and Leinbach (1991), Castells (1996), Graham and Marvin (1996) and Hall and Preston (1988).
15 *The Economist* (2 June 2001).
16 *The Independent* (30 August 2000).
17 Henderson and Castells (1987: 6).
18 A geo-stationary (or geo-synchronous) satellite has to be positioned above the Equator and to orbit at an altitude of 22,300 miles.
19 These data are from Baylin (1996).
20 *The Financial Times* (28 July 1998).
21 *The Financial Times* (17 September 1994).
22 *The Economist* (23 September 2000).
23 Hachten (1974: 99).
24 Castells (1996: 337, 338, 339).
25 Castells (1996: 341).
26 Batty and Barr (1994) provide an excellent discussion of the development of the Internet. See also Castells (1996); Ogden (1994); Warf (1995, 2001).
27 See, for example, the Special Report on 'Geography and the net' in *The Economist* (11 August 2001) and papers in the Special Issue of *American Behavioral Scientist* (2001), notably those by Zook, Townsend, and Brunn and Dodge.
28 *The Economist* (11 August 2001: 18).
29 Govindarajan and Gupta (2000: 279–80).
30 The concept of the product life cycle has been applied in a number of different ways. For its treatment within the marketing literature see O'Shaughnessy (1995). The product life cycle has also been employed by Hirsch (1967, 1972), Vernon (1966, 1979) and Wells (1972) to explain international trade and international production.
31 Casson (1983: 24).
32 Smith (1981).
33 Piore and Sabel (1984).
34 Perez (1985).
35 Important contributions are provided by the various chapters in Amin (1994) and by Ruigrok and van Tulder (1995) and Sayer and Walker (1992).
36 Piore and Sabel (1984).
37 The major features of this debate are captured in Amin (1994), Amin and Robins (1990), Harrison (1997), Hudson (2001), Malmberg (1996), Piore and Sabel (1984) and Storper (1995, 1997).
38 For example, Gertler (1988), Harrison (1997), Hudson (2001) and Sayer and Walker (1992) are all critical of the simplistic view that replaces Fordism with a system based on flexibility.
39 Kenney and Florida (1989: 144, 145).
40 Schonberger (1982).
41 Womack, Jones and Roos (1990: 13).

42 Abo (1994, 1996).
43 This section is based mainly on Table 1 in Peck and Miyamachi (1995).
44 Bunnell and Coe (2001) provide an excellent overview of the current literature on the geographies of innovation.
45 Metcalfe and Dilisio (1996: 58).
46 Archibugi and Michie (1997: 2).
47 This distinction derives originally from Polanyi (1962), although his treatment is more nuanced than the dichotomous classification suggests. As he points out, all knowledge is actually embedded in tacit knowledge.
48 Law (1986).
49 Storper (1995: 208).
50 Lundvall (1992); Nelson (1993). For recent reviews of the national innovation systems concept, see Archibugi and Michie (1997); Archibugi, Howells and Michie (1999); Freeman (1997); Lundvall and Maskell (2000); Patel and Pavitt (1998).
51 Howells (2000: 58–9).
52 Malmberg and Maskell (1997: 28–9).
53 Storper (1992).
54 Castells and Hall (1994).
55 This is certainly so in the case of the Hsinchu Science-based Industry Park in Taiwan (see Mathews, 1997).
56 See Amin and Thrift (1992).
57 Gertler (2001: 18), quoting Amin.
58 Dosi (1999: 41).

CHAPTER 5

'The State is Dead ... Long Live the State'

'Contested territory': the state in a globalizing economy

A dominant myth in much of the globalization literature is that we now live in a borderless world where states no longer matter. This isn't a novel view. 'The implications of international integration for domestic policy have been of concern for more than two hundred years – basically since the birth of industrial capitalism'.[1] Rather more recently – around 30 years ago – Charles Kindleberger bluntly asserted 'the nation state is just about through as an economic unit'.[2] Where Kindleberger led others have followed. While recognizing that the position of the state is being redefined, I emphatically reject the view that it is no longer a major player. While some of the state's capabilities are, indeed, being reduced, and while there may well be a process of 'hollowing out' of the state, the process is not a simple one of uniform decline on all fronts.[3] The state remains a most significant force in shaping the world economy. It has, whether explicitly or implicitly, played an extremely important role in the economic development of *all* countries. All governments intervene to varying degrees in the operation of the market and, therefore, help to shape different parts of the global economic map. I agree with Porter's claim that 'while globalization of competition might appear to make the nation less important, instead it seems to make it more so'[4] and with Wade's assertion that 'reports of the death of the national economy are greatly exaggerated'.[5]

Just as the degree of interconnection within the world economy has increased dramatically, the same can be said of the world political system. This can best be depicted, as in Figure 5.1, as a *web of interdependencies*, involving governments, international organizations (such as the United Nations, the International Monetary Fund, World Bank, World Trade Organization) and transnational organizations (non-governmental bodies, such as transnational corporations, international trades unions, environmental groups, and welfare organizations). Such a framework emphasizes both the *permeability* of state boundaries and also the *polycentric* nature of the global political system, 'with states as merely one level in a complex system of overlapping and often competing agencies of governance'.[6] States thus exist within a world system of differential power relationships operating at different geographical scales.

But what do we mean by the term 'state'? We need to clarify the meaning of three terms that are often used interchangeably:[7]

- A *state* is a portion of geographical space within which the resident population is organized (i.e. governed) by an authority structure. States have externally recognized sovereignty over their territory.

- A *nation* is a 'reasonably large group of people with a common culture, sharing one or more cultural traits, such as religion, language, political institutions, values, and historical experience. They tend to identify with one another, feel closer to one another than to outsiders, and to believe that they belong together. They are clearly distinguishable from others who do not share their culture.' A nation is, to use Benedict Anderson's term, an *imagined community*. Note that whereas a state has a recognized and defined territory, a nation may not.

- A *nation-state* is the condition where 'state' and 'nation' are coterminous. 'A nation-state is a nation with a state wrapped around it. That is, it is a nation with its own state, a state in which there is no significant group that is not part of the nation.'

Although it is often regarded as a natural institution (for all of us it has always been there), the nation-state is actually a relatively recent phenomenon. It emerged from the particular configuration of power relationships in Europe following the Treaty of Westphalia in 1648. Since then, the map of nation-states has been redrawn continuously, sometimes peacefully and incrementally, often violently through revolution. Certain key phases are evident:

- In the late 18th century, the American and the French revolutions created entirely new political structures and had an enormous influence on political change.

- In the mid-19th century, the modern state of Germany was created followed by that of Italy.

Figure 5.1

The web of interdependencies in the global political system

Source: McGrew, 1992: Figure 1.7

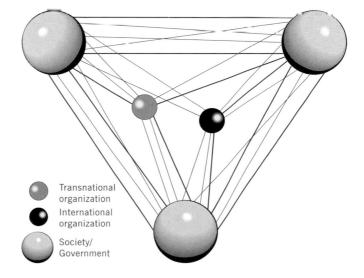

Transnational organization

International organization

Society/ Government

- In the first half of the 20th century, the world geopolitical map was drastically redrawn in 1917 (the Bolshevik revolution in Russia), in 1918 (after World War I), and in 1945 (after World War II).

- The waves of decolonization that swept through Africa and Asia in the 1960s created a whole new set of nation-states.

- More recently, the collapse of the former Soviet Union, after 1989, resulted in the creation not only of a new Russian Federation but also a number of newly independent states throughout Eastern Europe. In the 1990s, the fragmentation of the former Yugoslavia also resulted in the emergence (or re-emergence) of new states there.

As a result, the number of nation-states has grown dramatically to around 190 at the present time (Figure 5.2).

But that isn't all. An important feature of the contemporary world is the tension that exists between the triad of nation, state and nationalism. Increasingly, it seems, there are more and more 'nations without states':

> a community endowed with a stable but dynamic core containing a set of factors which have generated the emergence of a specific national identity … [but where] … the state, that is the political institution with which the nation should ideally identify, is missing … The nation-state has traditionally based its legitimacy upon the idea that it represents the nation, in spite of the fact that often the state once created had to engage in nation-building processes aiming at the forced assimilation of its citizens. It now becomes apparent that, in many cases, these processes have largely failed; the re-emergence of nationalist movements in nations without states proves it … Most so-called nation-states are not constituted by a single nation which is coextensive with the state; internal diversity is the rule. The nation-state … is being forced to respond to challenges from within … [as well as challenges from without through the processes of globalization] … The nationalism of nations without states emerges as a socio-political movement that defends the right of peoples to decide upon their own political destiny.[8]

In the early years of the third millennium, tensions in many different parts of the world constantly remind us of this problem. In some cases, there has been a degree of peaceful 'national devolution' of power (for example, for Scotland and, to a lesser extent, Wales, in the United Kingdom). But, for the most part, separatist movements are engaged in conflict with the state in which they are (wrongly in their view) embedded (obvious examples include the Quebecois in Canada, the Basques in Spain and France, the indigenous groups in Chiapas, Mexico, the East Timoreans in Indonesia, the Palestinians in Israel).

In this chapter and in Chapter 6, we explore the roles of the nation-state in the contemporary global economy. In this chapter the focus is set at a deliberately general level.

- First, we explore the nature of the nation-state as a *container* of distinctive institutions and practices.

- Second, we examine the *regulatory* role of the state through the lens of three specific policy areas of direct relevance to the reshaping of the global economy: trade, foreign investment and industry policies.

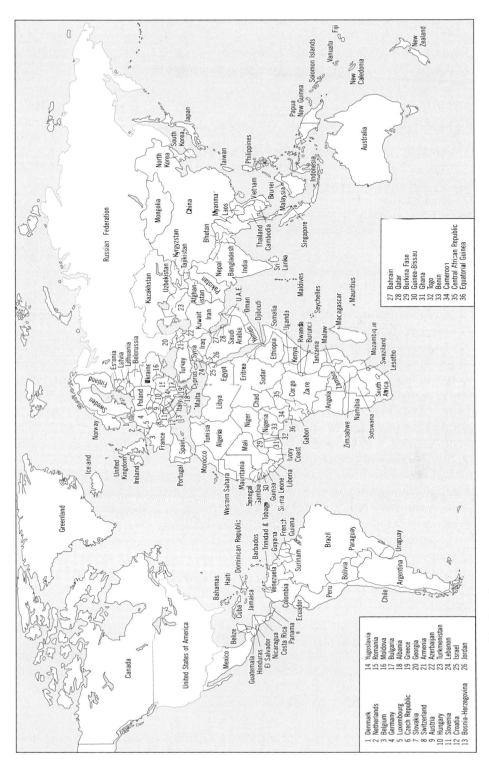

Figure 5.2 A world of nation-states

- Third, we look at the extent to which states can be regarded as *competition* states.
- Fourth, we explore the state as *collaborator,* engaging increasingly in bilateral and multilateral trading arrangements with other states.

States as *containers* of distinctive cultures, practices and institutions[9]

The cultural context

All economic activity is *embedded* in broader cultural structures and practices.[10] However, 'culture' is an extremely slippery concept to define. Here it is taken to be

> a learned, shared, compelling, interrelated set of symbols whose meanings provide a set of orientations for members of a society. These orientations, taken together, provide solutions to problems that all societies must solve if they are to remain viable.[11]

The nation-state is one of the primary *containers* of such cultural structures and practices – of distinctive 'ways of doing things'. The term 'container' should not be taken too literally. It is used here as a fairly loose metaphor to capture the idea that nation-states are *one* of the major ways (there are, of course, others) in which distinctive institutions and practices are 'bundled together'. Of course, such containers are not (except in very rare cases) hermetically sealed off from the outside world. The container is permeable or leaky to varying degrees. Most obviously, a major impact of the modern information technology and communications systems discussed in Chapter 4 is to make national containers even more permeable. But that does not mean that the container no longer exists at all. Indeed, there is a good deal of compelling evidence to show the persistence of national distinctiveness – although not necessarily uniqueness – in structures and practices which help to shape local, national and global patterns of economic activity.

From an economic perspective, there are relatively few comprehensive and robust analyses of how cultures vary between countries. A classic study, still relevant today, is Hofstede's massive survey of more than 100,000 workers employed by the American company IBM in 50 different countries.[12] The great strength of Hofstede's study is that, by focusing on a controlled population within a common organizational environment, he was able to isolate *nationality* as a variable. Hofstede identified four distinct cultural dimensions:

- *Individualism versus collectivism.* Societies vary in the extent to which people, in general, are motivated to look after their own individual interests – where ties between individuals are very loose – and those in which ties are very close and the collectivity (family, community, etc.) is the important consideration.
- *Large or small power distance.* Societies vary in how they deal with inequalities (for example, in power and wealth) between people. This is reflected in the

extent to which authority is centralized and in the degree of autocratic leadership within society.

- *Strong or weak uncertainty avoidance.* In some societies, the inherent uncertainty of the future is accepted; each day is taken as it comes, that is, the level of uncertainty avoidance is weak. In other societies, there is a strong drive to try to 'beat the future'. Efforts (and institutions) are made to try to create security and to avoid risk. These are strong uncertainty avoidance societies.

- *Masculinity versus femininity.* Societies vary in the roles assigned to males and females. Societies can be classified according to how sharply the social division between males and females is drawn. Societies with a strong emphasis on traditional masculinity allocate the more assertive and dominant roles to men. They differ substantially from societies where the social sex role division is small and where such values are less evident.

Hofstede went on to show how different countries could be characterized in terms of their positions on varying combinations of these four dimensions. Figure 5.3 summarizes the results in terms of eight country clusters. Although it is always rather dangerous to classify phenomena into statistical boxes, the categories identified by Hofstede seem intuitively reasonable. Most of us would be able to recognize our own national contexts, whilst also realizing the danger of using simple stereotypes without due care.

Group	1. Anglo	2. Germanic	3. Nordic	4. More developed Asian

Characteristics	Low power distance
	Low to medium
	   uncertainty avoidance
	High individualism
	High masculinity
		Low power distance
		High uncertainty avoidance
		Medium individualism
		High masculinity
			Low power distance
			Low to medium
			   uncertainty avoidance
			Medium individualism
			Low masculinity
				Medium power distance
				High uncertainty avoidance
				Medium individualism
				High masculinity

Countries	*Australia Britain Canada Ireland New Zealand United States*	*Austria Germany Israel Italy South Africa Switzerland*	*Denmark Finland Netherlands Norway Sweden*	*Japan*
Group	**5. Less developed Asian**	**6. Near Eastern**	**7. More developed Latin**	**8. Less developed Latin**
Characteristics	High power distance Low uncertainty avoidance Low individualism Medium masculinity	High power distance High uncertainty avoidance Low individualism Medium masculinity	High power distance High uncertainty avoidance High individualism Medium masculinity	High power distance High uncertainty avoidance Low individualism Whole range on masculinity
Countries	*India Pakistan Philippines Singapore Taiwan Thailand*	*Greece Iran Turkey*	*Argentina Belgium Brazil France Spain*	*Chile Colombia Mexico Peru Portugal Venezuela*

Figure 5.3 National variations in cultural characteristics

Source: Based on Hofstede, 1980: 336

Varieties of capitalism

Over time, and under specific historical circumstances, societies have developed distinctive ways of organizing their economies, even within the apparently universal ideology of capitalism. In fact, capitalism comes in many different varieties.[13] Of course, although in one sense all societies are unique, it is equally apparent that there are identifiable patterns so that some societies are more alike than others. In the context of the contemporary world economy, most attention has focused on the differences between three distinctive types of political-economic system:

- *neo-liberal market capitalism* – exemplified by the United States and, to a lesser extent, the United Kingdom
- *social-market capitalism* – exemplified by Germany, Scandinavia and many other European countries
- *developmental capitalism* – exemplified by Japan, South Korea, Taiwan, Singapore and most other East Asian countries.

Table 5.1 summarizes the major characteristics of each of these variants of capitalism in terms of three key characteristics: the dominant ideology of the state; the nature of the political institutions; and the nature of the economic institutions.

Table 5.1   Three major varieties of capitalism: the US, German and Japanese models

	United States	*Germany*	*Japan*
Dominant ideology	Free-enterprise liberalism	Social partnership	Technonationalism
Political institutions	Liberal democracy Divided government Interest-group liberalism	Social democracy Weak bureaucracy Corporatist legacy	Developmental democracy Strong bureaucracy Reciprocity between state and firms
Economic institutions	Decentralized, open markets Unconcentrated, fluid capital markets Antitrust tradition	Organized markets Tiers of firms Dedicated, bank-centred capital markets Certain cartelized markets	Guided, closed, bifurcated markets Bank-centred capital markets Tight business networks Cartels in sunset industries

Source: Based on Doremus et al., 1998: Table 2.1

The essence of these governance models is their differing conception of the 'proper' role of government in regulating the economy. In *neo-liberal market capitalism*, as the term suggests, market mechanisms are used to regulate all or most aspects of the economy; individualism is a dominant characteristic, short-term business goals tend to predominate, and the state does not overtly attempt to plan the economy strategically. The dominant philosophy is 'shareholder value' – facilitating maximum returns to the owners of capital. In *social-market capitalism*, in contrast, a higher premium is placed upon collaboration between different actors in the economy with a broader identification of 'stakeholders' beyond that of owners of capital. In *developmental capitalism*, the state plays a much more central role (although not usually in terms of public ownership of productive assets). This mode of governance

> has as its dominant features precisely the setting of ... substantive social and economic goals ... the government will give greatest precedence to industrial policy, that is, to a concern with the structure of domestic industry and with promoting the structure that enhances the nation's international competitiveness. The very existence of an industrial policy implies a strategic, or goal-oriented, approach to the economy.[14]

Although change undoubtedly occurs in all social systems, unless they are insulated from the outside world, such distinctive forms of capitalism tend to persist over time:

> forms of economic coordination and governance *cannot* easily be transferred from one society to another, for they are embedded in social systems of production distinctive to their particular society ... Economic performance is shaped by the entire social system of production in which firms are embedded and not simply by specific principles of management styles and work practices ... institutions are embedded in a culture in which their logic is symbolically grounded, organizationally structured, technically and materially constrained, politically defended and historically shaped by specific rules and norms ...
>
> There are inherent obstacles to convergence among social systems of production of different societies, for where a system is at any one point in time is influenced by its initial state. Systems having quite different initial states are unlikely to converge with one another's institutional practices. Existing institutional arrangements block certain institutional innovations and facilitate others ...
>
> Despite the emphasis on the logic of institutional continuity, this is not an argument that systems change along some predetermined path. There are critical turning points in the history of highly industrialized societies, but the choices are limited by the existing institutional terrain. Being path dependent, a social system of production continues along a particular logic until or unless a fundamental societal crisis intervenes.[15]

States as *regulators* of trade, foreign investment and industry

Recognizing that countries continue to differ as 'containers' of distinctive structures and practices is important in emphasizing that we do not live in a homogenized world. In this section, we focus specifically on some of the ways states *regulate* how their economies operate as they attempt to control what happens within, and across, their boundaries. Of course, states do not merely 'intervene' in markets. 'The institutions of the state apparatus are not simply involved in regulating economy and society, for *state activity is necessarily involved in constituting economy and society* and the ways in which they are structured and territorially organized.'[16] Although a high level of contingency may well be involved (no two states behave in exactly the same way), certain regularities in basic policy stance can be identified. These will reflect the kinds of cultural, social and political structures, institutions and practices in which the state is embedded. The precise policy mix adopted by a state will be influenced by:

- *The nation's political and cultural complexion and the strength of institutions and interest groups.* These are the kinds of characteristics discussed in the previous section of this chapter. In general, conservative governments are less inclined to pursue interventionist policies than liberal or socialist governments. However, much depends on the power of institutions and interest groups within the national economy: for example, business and financial interests, labour unions, environmental groups. The particular form of political structure – whether centralized or federal – will also be important. An especially significant factor, therefore, is the degree of consensus or conflict between institutions and interest groups, a situation in which the nation's history and culture will be significant influences.

- *The size of the national economy, especially that of the domestic market.* This is especially relevant for trade policies: the larger the domestic market the less important relatively is external trade likely to be and vice versa. However, much will depend upon:

- *The nation's resource endowment, both physical and human.* A weak natural resource endowment will necessitate the import of essential materials that, in turn, must be paid for by exports of other, usually manufactured, products. A strong endowment of particular types of raw material may well influence the kind of economic policy pursued. Similarly, the size, composition and skill level of a nation's potential labour force will constrain the kinds of policy that may be pursued.

- *The nation's relative position in the world economy, including its level of economic development and degree of industrialization.* A nation's 'degrees of freedom' depend very much on its relative position in the world economy and, in particular, on the extent of its dependence on external trade and investment. The kinds of industry and trade policies pursued will also tend to reflect the nation's degree of industrialization. The policy emphasis of established industrial nations will differ from nations in the process of industrializing.

At one extreme, the *macroeconomic* policies pursued by governments to control domestic demand or to manage the money supply have extremely important implications for the distribution and redistribution of economic activity. Two basic types of macroeconomic policy tend to be used by the state to manage its national economy:

- *Fiscal policies*. These consist primarily of policies to raise or lower taxes on companies and/or individual citizens and to determine appropriate levels and recipients of government expenditure. Raising taxes dampens down domestic demand, lowering taxes stimulates demand. (Although, as the Japanese experience during the 1990s showed, such automatic responses to changes in fiscal policy do not always occur.) Similarly, raising or lowering public expenditure – or targeting specific types of expenditure – can influence the level of economic activity in the economy.

- *Monetary policies*. These are policies aimed at influencing the size of the money supply within the country and at either speeding up, or slowing down, its rate of circulation (its velocity). The main mechanism employed is manipulation of the interest rate on borrowing. Lowering interest rates should stimulate economic activity through increased investment or private expenditure while, conversely, raising interest rates should dampen down activity. Again, however, rapid and automatic adjustment does not always occur. In the international context, exchange rates also have to be taken into account because their level (and volatility) affect the costs of exports and imports.

At a more tangible and material level, governments generally provide – or at least secure the provision of – those 'conditions of production that are not and cannot be obtained through the laws of the market'.[17] One example is the *physical infrastructure* of national economies – roads, railways, airports, seaports, telecommunications systems – without which private sector enterprises, whether domestic or transnational, could not operate. They are the providers, too, of the human infrastructure: in particular of an educated labour force as well as of sets of laws and regulations within which enterprises must operate. Between these two extremes of government involvement in the workings of the economy lie those policies whose explicit purpose is to influence the level, composition and distribution of production and trade. It is these policies – concerned with trade, foreign investment and industry – which form the substance of this section.

National governments have at their disposal an extensive kit of regulatory tools with which to control and to stimulate economic activity and investment within their own boundaries and to shape the composition and flow of trade and investment at the international scale. Although often viewed separately, trade policies, policies towards foreign investment and industry policies overlap to a very considerable degree. The boundaries between them may be blurred. Each may reinforce – or counteract – the other. They may be employed as part of a deliberate, cohesive, all-embracing national economic strategy as in the *developmental-capitalist* state or, alternatively, individual policy measures may be implemented in an ad hoc fashion with little attempt at coordination as in the *neo-liberal market capitalist* state model.

Trade policies

Of all the measures used by nation-states to regulate their international economic position, policies towards trade have the longest history. The shape of the emerging world economy of the 17th and 18th centuries was greatly influenced by the mercantilist policies of the leading European nations. (Mercantilism is based upon the idea that a nation's wealth and security depend upon its ability to regulate and control its external trade at the expense of others.) The successful challenge to Britain's industrial supremacy in the late 19th century by the United States and Germany was based on their strongly mercantilist (i.e. protectionist) trade policies. The deep world recession of the 1930s was also characterized by national retreat behind trade barriers. Since the late 1970s a new wave of protectionism – *neo-mercantilism* – has swept the world economy as governments have attempted to 'manage' trade in a variety of ways. Indeed, a whole new area of strategic trade policy has emerged in recent years, particularly in the United States, as we shall see in Chapter 6.

Trade policy is unique in that, since the late 1940s, it has been set within an *international* institutional framework. Until 1995 this was the GATT (the General Agreement on Tariffs and Trade); since 1995, international trade has been regulated through the WTO (World Trade Organization). We will discuss this international regulatory framework in some detail in Chapter 18. Here, we just need to note that national policies have to operate within an international system.

Figure 5.4 summarizes the major types of trade policy pursued by national governments. In general, policies towards imports are restrictive whereas policies towards exports, with one or two exceptions, are stimulatory.

Policies towards imports	**Policies towards exports**
1. Tariffs	Financial and fiscal incentives to export producers
2. Non-tariff barriers	
Import quotas (e.g. 'voluntary export restraint', 'orderly marketing agreements')	Export credits and guarantees
Import licences	Setting of export targets
Import deposit schemes	
Import surcharges	Operation of overseas export promotion agencies
Rules of origin	
Anti-dumping measures	
Special labelling and packaging regulations	Establishment of Export Processing Zones and/or Free Trade Zones
Health and safety regulations	
Customs procedures and documentation requirements	'Voluntary export restraint'
Subsidies to domestic producers of import-competing goods	Embargo on strategic exports
Countervailing duties on subsidized imports	Exchange rate manipulation
Local content requirements	
Government contracts awarded only to domestic producers	
Exchange rate manipulation	

Figure 5.4 Major types of trade policy

Policies on *imports* fall into two distinct categories:

- *Tariffs* are, essentially, taxes levied on the value of imports that increase the price to the domestic consumer and make imported goods less competitive (in price terms) than otherwise they would be.

In general, the tariff level rises with the stage of processing. Tariffs tend to be lowest on basic raw materials and highest on finished goods, as Table 5.2 shows. The purpose of such 'tariff escalation' is to protect domestic manufacturing industry while allowing for the import of industrial raw materials. Thus, although tariffs may be regarded simply as one means of raising revenue, their major use has been to protect domestic industries: either 'infant' industries in their early delicate stages of development, or 'geriatric' industries struggling to survive in the face of external competition.

Table 5.2 Tariff escalation on developing country products

	Tariff (%)	
	Mid-1980s	Post-1995[a]
All industrial products (excluding oil)		
Raw materials	2.1	0.8
Semi-manufactured products	5.3	2.8
Finished products	9.1	6.2
Natural-resource-based products		
Raw materials	3.1	2.0
Semi-manufactured products	3.5	2.0
Finished products	7.9	5.9

[a]Based on the commitments given in the Uruguay Round.

Source: Based on material in GATT, *Annual Report, 1994*

Largely through the successive rounds of international negotiation in the GATT, the general level of tariffs in the world economy has declined very substantially. Figure 5.5 shows that in 1940 the average tariff on manufactured products was around 40 per cent; today the average tariff is around 4 per cent.

- *Non-tariff barriers (NTBs).* While tariffs are based on the value of imported products, non-tariff barriers are more varied: some are quantitative, some are technical.

Although, in general, tariffs have continued to decline, the period since the mid 1970s witnessed a marked increase in the use of non-tariff barriers. Indeed, today NTBs are probably more important than tariffs in influencing the level and composition of trade between nation-states. It has been estimated that NTBs affect more than a quarter of all industrialized country imports and are even more extensively used by developing countries. Certainly much of what has been termed the 'new protectionism' consists of the increased use of NTBs.

As Figure 5.4 shows, such non-tariff barriers can take a variety of forms. The most important is the import quota, a limit on the quantity of a particular product that may be imported. In some cases, quotas are established as part of so-called 'orderly marketing agreements' or 'voluntary export restraints'. Such devices have become extremely important in several of the sector cases described in Part III of this book. In addition to quotas, importers may be required to seek import licences, to pay deposits or be subject to import surcharges. 'Rules of origin' requirements may be imposed in order to restrict imports from specific countries and to encourage local production. Companies suspected of setting their prices in overseas markets at levels below those in their domestic markets may be subjected to anti-dumping penalties.

A second group of NTBs consists of various regulatory and bureaucratic devices such as special labelling or packaging regulations, health and safety requirements, customs procedures and documentation. A third category overlaps closely with some of the industry and foreign investment policies discussed later. For example, domestic producers may be subsidized to compete more effectively against imports; firms may have to comply with specific local content requirements; government contracts may be confined to domestic producers. There is no doubt that the greatly increased operation of NTBs has led to intensified tensions and stresses in the world trading system.

Figure 5.4 also lists the major *export* policies that may be pursued. Again, some of these may overlap with industry and foreign investment policies. In particular, financial and fiscal advantages may be granted to exporting firms. In addition, specific tax and tariff concessions may be applied: export earnings may be tax-free or taxed at a lower rate; tariffs may be waived or reduced on those imports that are essential for export activities. Governments may operate an export credit guarantee scheme, fix export targets and operate export promotion offices overseas. Particular geographical areas may be set aside as export processing or free trade zones. As we shall see in Chapter 6, these have proliferated rapidly in many developing countries in recent years.

Figure 5.5

Reduction in average tariffs on goods, 1940–1995

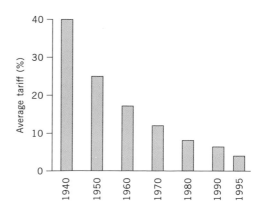

Most export policies are, of course, aimed at stimulating a nation's exports but there are some instances where policies are adopted to restrict exports. One example is the kind of 'orderly marketing arrangement' mentioned earlier, whereby the exporting nation 'voluntarily' restricts its exports of a good to the other nation (often the term 'voluntary' is hardly appropriate). Another example of export restriction occurs in the case of strategically or militarily sensitive items. Many countries operate such selective export embargoes. Finally, one measure common to both import and export policy is the manipulation of the nation's currency exchange rate. Devaluation of the currency makes exports cheaper and imports more expensive and vice versa. However, the ability of a government to manipulate its exchange rate in a controlled way in a world of floating and volatile currencies is very limited.

Foreign direct investment policies

In a world of transnational corporations and of complex flows of investment at the international scale, national governments have a clear vested interest in the effects of FDI, whether positive or negative. From a national viewpoint, such investment is of two types: *outward* investment by domestic enterprises and *inward* investment by foreign enterprises. Few national governments operate a totally closed policy towards FDI, although the degree of openness varies considerably. In general, the developed market economies tend to be less restrictive in their policies towards FDI than the developing market economies. One obvious reason is the fact that the developed economies are the major sources of, as well as the dominant destinations for, the world's FDI. There are, of course, some exceptions to this general pattern.

Figure 5.6 summarizes the major types of national policy towards foreign direct investment. Most national policies are concerned with inward investment although governments may well place restrictions on the export of capital for investment (for example, through the operation of exchange control regulations) or insist that proposed overseas investments be approved before they can take place. A far more extensive battery of policies exists in the case of inward investment. The inward investment policies listed in Figure 5.6 fall into four broad categories.

- The first category relates to the *entry* of foreign firms into a national economy. Governments may operate a 'screening' mechanism to attempt to filter out those investments that do not meet national objectives, either economic or political. Foreign firms may in fact be excluded entirely from particularly sensitive sectors of the economy or the degree of foreign penetration in a sector may be limited to a certain percentage share. More generally, there may be a restriction on the extent to which individual firms may be owned or controlled by a foreign enterprise. Government may insist that only joint ventures involving indigenous capital may be permitted, possibly on a 51:49 per cent basis in favour of domestic firms. Another possibility is for government to require foreign firms to employ local personnel in managerial positions. Generally speaking, of course, foreign firms must comply with prevailing national codes of business conduct – to be good corporate citizens – including those relating to the disclosure of information about the firm's activities. This latter point is frequently a major bone of contention between TNCs and national governments.

- The second category of policies shown in Figure 5.6 relates to the *operations* of foreign firms. A particularly common requirement is for such firms to meet a certain level of 'local content' in their manufacturing activities. Such a requirement is designed to increase the positive effects of a foreign investment on indigenous suppliers and to reduce the level of imported materials and components. Conversely, a government may insist on a foreign firm exporting a specified proportion of its output. One of the major elements in the foreign investment 'package' is that of technology. As we shall see in later chapters, there is much dispute about the extent to which TNCs do, in fact, transfer technology beyond their own corporate boundaries. Governments may wish to stimulate such technological diffusion through restrictive or stimulative measures.

- The third set of policies relates to government attitudes towards *corporate profits and the transfer of capital.* All governments are concerned to minimize the outflow of capital; on the other hand, TNCs invariably wish to remit at least part of their profits and capital abroad. Similarly, TNCs aim to minimize their liability to taxation; national governments wish to maximize their tax yield. Hence, variations in the restriction on the remittance of capital and profits, together with variations in the level and methods of taxing TNC profits, are extremely important.

Policies relating to inward investment by foreign firms

Government screening of investment proposals
Exclusion of foreign firms from certain sectors or restriction on the extent of foreign involvement permitted
Restriction on the degree of foreign ownership of domestic enterprises
Insistence on involvement of local personnel in managerial positions
Compliance with national codes of business conduct (including information disclosure)

Insistence on a certain level of local content in the firm's activities
Insistence on a minimum level of exports
Requirements relating to the transfer of technology
Locational restrictions on foreign investment

Restrictions on the remittance of profits and/or capital abroad
Level and methods of taxing profits of foreign firms

Direct encouragement of foreign investment: competitive bidding via overseas promotional agencies and investment incentives

---

Policies relating to outward investment by domestic firms

Restrictions on the export of capital (e.g. exchange control regulations)
Necessity for government approval of overseas investment projects

---

Figure 5.6   Major types of FDI policy

- The final category of policies towards foreign investment shown in Figure 5.6 aims to *stimulate* inward investment. Indeed, an increasingly common feature of today's world economy – of developed as well as of developing economies – is the scramble to attract foreign investment. Competitive bidding via overseas promotional agencies and investment incentives has become endemic throughout the world (see Chapter 9). The important point about such international (and inter-regional) competition is that for certain types of investment it is truly global in extent. For the cost-oriented transnational investments, for example, countries and localities in Europe or North America may well be in direct competition with locations in Asia and Latin America.

Historically, there were very large differences in the policy positions adopted by countries towards inward direct investment. At the broadest level, developed countries tended to adopt a more liberal attitude towards inward investment than developing countries, although there were exceptions within each broad group. For example, among developed countries France had a much more restrictive stance than most other European countries. Among developing countries, Singapore had a particularly open policy, far more so than most other Asian countries. In the past two decades, however, national FDI policies have tended to converge in the direction of liberalization.

In many cases, this liberalization of foreign investment policies has been part of broader, market-oriented reforms of economic policy and has proceeded in parallel with trade liberalization, deregulation and privatization.

The recent trend to more open investment policies has been particularly evident in the removal or relaxation of regulatory barriers to the entry of FDI … There has also been a shift away from the imposition of performance requirements and a liberalization of regulations concerning the transfer of funds. In addition, there has been increasing acceptance of standards of non-discriminatory treatment of foreign

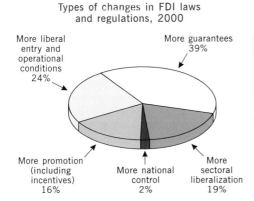

Types of changes in FDI laws and regulations, 2000

National regulatory changes, 1991 – 2000

	1991	1994	1997	2000
Number of countries making changes in their regulatory regimes	35	49	76	69
Number of changes more favourable to FDI	80	108	135	147
Number of changes less favourable to FDI	2	2	16	3

More liberal entry and operational conditions 24%

More guarantees 39%

More promotion (including incentives) 16%

More national control 2%

More sectoral liberalization 19%

Figure 5.7 Changes in national regulation of FDI

Source: Based on UNCTAD, 2000b: Box I.1; Table I.3

investors and of international standards on matters such as compensation in case of expropriation …

At the same time, however, there are several qualifications to this liberalization trend. First, the trend has not been homogeneous and significant differences between foreign investment regimes persist. Second, virtually all countries maintain some restrictions, often of a sectoral nature, on the entry of foreign investment. In this connection, an issue that has attracted attention is the existence of reciprocity requirements with regard to the entry and treatment of foreign investment.[18]

Although national differences still exist, therefore, they are now rather less stark than in the past. Figure 5.7 summarizes the major regulatory changes towards FDI between 1991 and 2000.

Industry policies

National policies towards trade or foreign direct investment are explicitly concerned with international or cross-border issues and, therefore, are most obviously relevant to our interest in the globalization of economic activity. But there is a third policy area – industry policy – that, although essentially concerned with internal issues, also has broader international implications. Indeed, it is becoming increasingly apparent that the boundaries between trade, foreign direct investment and industry policies are often extremely blurred. Figure 5.8 lists the major types of regulatory industry policies that may be used by national governments.

The most obvious stimulatory measures are the various financial and fiscal incentives governments may offer to private sector firms. As Figure 5.8 shows, the financial measures most commonly used fall into two categories. On the one hand, governments may provide capital grants or loans to firms to supply part or all of the investment required for a particular productive venture. The other major financial, or rather fiscal, incentive employed by governments is that of tax concessions. Under this banner a whole variety of measures may be employed. For example, firms may be permitted to depreciate or write down their capital investment against tax at an accelerated rate, they may be granted tax reductions or even tax exemptions. Such concessions may be for a specified period: the so-called 'tax holiday'.

Governments may also use various types of employment and labour policy to help shape and encourage industrial activity within their boundaries: in other words, to regulate the workings of the labour market. Most governments are concerned to stimulate employment and, therefore, to reduce unemployment. In pursuit of such aims, firms may be encouraged to increase their labour force by direct subsidy. Training may be paid for – or even provided in government establishments – to provide a labour force with appropriate skills. Labour unions may be encouraged, discouraged, or even prohibited altogether.

National governments tend to be the largest individual customers in any economy for an enormous variety of goods and services (including, of course, defence). Thus government policies of procurement are extremely important. The award of large government contracts or, conversely, their withdrawal or cancellation, may make or break a private sector enterprise and have enormous employment repercussions.

The rapid and far-reaching technological developments (discussed in Chapter 4) have led most governments to try to stimulate research and development in key sectors and to encourage technological collaboration between firms. The perceived need to stimulate entrepreneurial activity has produced a whole battery of policies to encourage small and medium-sized enterprises. Governments may also attempt to restructure firms – and even entire industries – to improve their international competitiveness.

Regulation of national industrial activity can also take a variety of other forms, as the right hand box of Figure 5.8 suggests. State ownership of productive assets is present in many countries although a current trend in many market economies is towards increased *privatization*. Entry into particular sectors may be discouraged through the operation of merger and competition policies although, again, there is a current trend towards the *deregulation* of certain industries such as telecommunications and financial services. Company legislation in general is designed to regulate the ways in which companies can be formed and operate and may be reinforced by specific taxation policies. Regulation of the labour market

Investment incentives:
  Capital-related
  Tax-related
 	
Labour policies:
  Subsidies
  Training
 	
State procurement policies
 	
Technology policies
 	
Small firm policies
 	
Policies to encourage industrial restructuring
 	
Policies to promote investment
	Merger and competition policies
	
	Company legislation
	
	Taxation policies
	
	Labour market regulation.
	  Labour union legislation
	  Immigration policies
	
	National technical and product standards
	
	State ownership of production assets
	
	Environmental regulations
	
	Health and safety regulations

Some or all of these policies may be applied either generally or, more commonly, selectively. Selectivity may be based on several criteria:

1 Particular sectors of industry, e.g.
(a) to bolster declining industries
(b) to stimulate new industries
(c) to preserve key strategic industries

2 Particular types of firm, e.g.
(a) to encourage entrepreneurship and new firm formation
(b) to attract foreign firms
(c) to help domestic firms against foreign competition
(d) to encourage firms in import-substituting or export activities

3 Particular geographical areas, e.g.
(a) economically depressed areas
(b) areas of 'growth potential' or new settlement

Figure 5.8 Major types of industry policy

may be pursued through the encouragement or discouragement of labour union activity or through immigration policies. Technical and product standards are usually defined in specific national terms as are health and safety regulations and environmental regulations.

As Figure 5.8 suggests, the various stimulatory and regulatory policies may be applied *generally* across the whole of a nation's industries or they may be applied *selectively*. Such selectivity may take a number of forms: particular sectors of industry, particular types of firms (including, for example, the efforts to attract foreign firms), particular geographical locations.

States as *competitors*

Do states compete? Is it correct to think of nations as being in competition with each other just as firms compete with other firms? The generally accepted view among both policy-makers and academics has been strongly in the affirmative. Books, government reports, newspaper articles, television programmes in virtually all countries resound with the language and imagery of the competitive struggle between states for a bigger slice of the global economic pie. Prior to the East Asian crisis of the late 1990s, much of the concern focused on the perceived loss of economic standing by the United States and European countries *vis-à-vis* Japan and the East and South East Asian NIEs. Today, although the focus may be a little different, the rhetoric remains the same. Indeed, the Swiss business school, IMD, publishes an annual *World Competitiveness Yearbook* with a 'competitiveness scoreboard' (or 'league table') of 49 countries based on no fewer than 286 individual criteria!

Expressed in the simplest terms, if the goals of business organizations are to achieve maximum (or at least satisfactory) profits, one of the goals of nation-states is to maximize the material welfare of their societies. In an increasingly integrated and interdependent global economy, it is argued, nations are forced to *compete* with one another in a struggle to attain such goals. States compete to enhance their international trading position and to capture as large a share as possible of the gains from trade. They compete to attract productive investment to build up their national production base that, in turn, enhances their international competitive position. States, then, can be regarded as *competition states*, in which they take on some of the characteristics of firms as they strive to develop their strategies to create competitive advantage.[19]

In stark contrast to this conventional view, Paul Krugman has described the claim that states are competition states as 'a dangerous obsession'. In a highly polemical and wide-ranging critique of both the concept and some of its proponents, notably in the United States, Krugman attacks both the empirical evidence and the policies based upon the competitiveness concept.[20] The main points of his argument are as follows:

- *Nations are not like firms*. 'The bottom line for a corporation is literally its bottom line: if a corporation cannot afford to pay its workers, suppliers, and bondholders, it will go out of business ... Countries, on the other hand, do not go out of business. They may be happy or unhappy with their economic performance, but they have no well-defined bottom line. As a result, the concept of national competitiveness is elusive' (p. 31).

- *International trade is not a zero-sum game.* 'Coke and Pepsi are almost purely rivals … if Pepsi is successful, it tends to be at Coke's expense. But the major industrial countries, while they sell products that compete with each other, are also each other's main export markets and each other's main suppliers of useful imports. If the European economy does well, it need not be at US expense' (p. 34).

- *Empirical evidence does not support the concept.* 'As a practical, empirical matter the major nations of the world are not to any significant degree in economic competition with each other. Of course, there is always a rivalry for status and power – countries that grow faster will see their political rank rise. So it is always interesting to *compare* countries. But asserting that Japanese growth diminishes US status is very different from saying that it reduces the US standard of living' (p. 35).

Krugman takes a highly cynical view of the motives of the perpetrators of 'the competitive obsession' (he claims it sells books, shifts responsibility for domestic failures on to external forces, and is a useful device for political leaders). But there is more to his argument than this.

> Thinking and speaking in terms of competitiveness poses three real dangers. First, it could result in the wasteful spending of government money supposedly to enhance … [national] … competitiveness. Second, it could lead to protectionism and trade wars. Finally, and most important, it could result in bad public policy on a spectrum of important issues … To make a harsh but not entirely unjustified analogy, a government wedded to the ideology of competitiveness is as unlikely to make good economic policy as a government committed to creationism is to make good science policy … competitiveness is a meaningless word when applied to national economies. And the obsession with competitiveness is both wrong and dangerous.[21]

Not surprisingly, Krugman's claims have generated considerable controversy.[22] But whether Krugman is right or wrong in his analysis, there is little likelihood of policy-makers actually heeding his warnings and refraining from both the rhetoric and the reality of competitive policy measures. As long as the concept of national (and local) competitiveness remains in currency then no single state is likely to opt out. This is the classic 'Prisoner's Dilemma' of game theory, where a choice that seems rational to the *individual* in his/her search for maximum advantage produces an outcome that is *collectively* less beneficial. It involves 'second guessing' what others are likely to do. In the context of international economic policy-making, although all states might benefit from cooperation a powerful incentive exists to try to gain at the expense of other states.

In contrast to Krugman, Michael Porter not only asserts that nations compete but also that their individual competitive advantages are fundamentally created through processes internal to the country:

> Competitive advantage is created and sustained through a highly localized process. Differences in national economic structures, values, cultures, institutions, and histories

contribute profoundly to competitive success. The role of the home nation seems to be as strong as or stronger than ever. While globalization of competition might appear to make the nation less important, instead it seems to make it more so. With fewer impediments to trade to shelter uncompetitive domestic firms and industries, the home nation takes on growing significance because it is the source of the skills and technology that underpin competitive advantage.[23]

In other words, in Porter's view, the particular combination of conditions within nation-states has an enormous influence on the competitive strengths of the firms located there.

Porter conceives of this situation as a 'diamond': an interconnected system of four major determinants (Figure 5.9). Connecting all of the components (other than 'chance') by double-headed arrows emphasizes the fact that the 'diamond' is a mutually reinforcing system.

As Figure 5.9 shows, each of the determinants can be broken down into several subcomponents:

● *Factor conditions* in a particular country are a combination of 'given' and 'created' factors. 'The factors most important to competitive advantage in most industries, especially the industries most vital to productivity growth in advanced economies, are not inherited but are *created within a nation, through processes that differ widely across nations and among industries … Thus, nations will be competitive where they possess unusually high quality institutional mechanisms for specialized factor creation'* (pp. 74, 80, emphasis added).

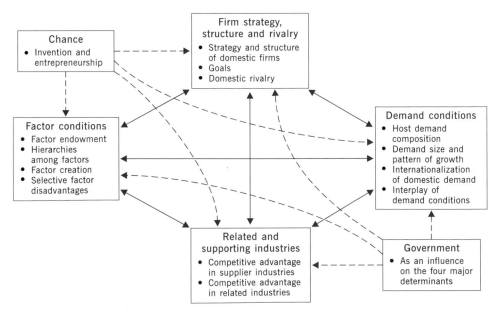

Figure 5.9 National competitive advantage: the Porter 'diamond'

Source: Based on material in Porter, 1990: ch. 3

The most important include the levels of skills and knowledge of the country's population and the provision of sophisticated physical infrastructure, including transport and communications.

- *Demand conditions*, especially those in the home market, are regarded as being particularly important. 'The home market usually has a disproportionate impact on a firm's ability to perceive and interpret buyer needs ... proximity to the right type of buyers ... [is] ... of decisive importance in national competitive advantage' (pp. 86, 87). However, the extent to which a nation's firms are connected into international markets also increases national competitiveness.

- *Related and supporting industries* that are internationally competitive constitute a third major determinant of national competitive advantage. 'Perhaps the most important benefit of home-based suppliers ... is in the *process of innovation and upgrading*. Competitive advantage emerges from close working relationships between world-class suppliers and the industry' (p. 103). However, Porter stresses that 'the benefits of both home-based suppliers and related industries ... depend on the rest of the "diamond". Without access to advanced factors, home demand conditions that signal appropriate directions of product change, or active rivalry, for example, proximity to world-class domestic suppliers, may provide few advantages' (p. 107).

- *Firm strategy, structure and rivalry*. 'The way in which firms are managed and choose to compete is affected by national circumstances ... Some of the most important are attitudes toward authority, norms of interpersonal interaction, attitudes of workers toward management and vice versa, social norms of individualistic or group behaviour, and professional standards. These, in turn, grow out of the education system, social and religious history, family structures, and many other often intangible but unique national conditions' (p. 109). Porter lays great emphasis on the importance of intense rivalry between domestic firms, arguing that this creates strong pressures on firms to innovate in both products and processes, to become more efficient and to become high quality suppliers of goods and services. 'Vigorous local competition not only sharpens advantages at home but pressures domestic firms to sell abroad in order to grow ... Toughened by domestic rivalry, the stronger domestic firms are equipped to succeed abroad. It is rare that a company can meet tough foreign rivals when it has faced no significant competition at home' (p. 119).

In addition to the four primary competitive determinants that, in combination, form his 'diamond', Porter attributes secondary importance to two other components:

- *The role of chance* – for example, the occasionally random occurrence of innovations or the 'historical accidents' that may create new entrepreneurs – are seen to be important

- *The role of government*. Porter explicitly refuses to regard government as a competitive determinant of the same order as the four primary determinants of his 'diamond'. While describing the various policies that governments might implement, he sees government as merely an 'influence' on his four determinants, a contingent, rather than a central, factor; a contributor to the environment in which, as in the biological realm, the selective survival of species occurs.

One of Porter's distinctive contributions to the discussion of the sources of national competitiveness is the importance he attributes to *localized geographical clustering*. This is a topic we have already discussed at various points in previous chapters. But Porter was the first writer within the business management field to give it a central position as a competitive influence. His treatment is conventional. It expresses in his own, rather distinctive, terminology, the kinds of interdependencies outlined in Chapter 2.

> Competitors in many internationally successful industries, and often entire clusters of industries, are often located in a single town or region within a nation … Geographic concentration of firms in internationally successful industries often occurs because the influence of the individual determinants in the 'diamond' and their mutual reinforcement are heightened by close geographical proximity within a nation … *Proximity, then, elevates the separate influences in the 'diamond' into a true system.*[24]

Porter's analysis of the competitive advantage of nations has attracted much attention. On the one hand its 'recipes' have been adopted by many governments, both national and local, in their attempts to improve their competitive position. On the other hand, it has attracted considerable criticism:

- It is highly reductionist in compressing immense complexity into a simple four-pointed 'diamond'.
- Its underplaying of the role of the state in pursuit of national competitiveness is a significant omission. *All* states perform a key role in the ways in which their economies operate, although they differ substantially in the specific measures they employ and in the precise ways in which such measures are combined.
- It neglects the influence of the transnationalization of business activity on national 'diamonds': 'there is ample evidence to suggest that the technological and organizational assets of TNCs may be influenced by the configuration of the diamonds of the foreign countries in which they produce and that this, in turn, may impinge upon the competitiveness of the resources and capabilities in their home countries'.[25]

States as *collaborators*: the proliferation of regional economic blocs[26]

While there is controversy over whether states do, or should, see themselves as competitive states, there is no doubt that states *collaborate* with other states to achieve specific economic and welfare goals. Such collaborations can take many forms. Here we focus on one dimension: the tendency for states to develop political-economic relationships at the *regional* scale. Indeed, regionalism has become one of the dominant features of the contemporary global economy. But it is not a new phenomenon by any means. In this section, we look, first, at the general characteristics of regional bloc formation before focusing specifically on regional arrangements within the three elements of the 'global triad'.

General processes and patterns of regionalism

Preferential trading arrangements (PTAs) are the fundamental basis of regional economic blocs. PTAs simply involve states agreeing to provide preferential access to their markets to other members of the regional group – primarily through tariff reductions, at least initially. Thus, 'preferential trading arrangements have a two-sided quality, liberalizing commerce among members while discriminating against third parties'.[27] In that respect, regional trading blocs are essentially *discriminatory* in nature and, as such, they contravene the general principle of non-discrimination established by the GATT (see Chapter 18). However, provision for such integration was incorporated in Article XXIV of the GATT, which allowed for the creation of free trade areas and customs unions, subject to certain provisos. Most regional blocs have a strongly defensive character; they represent an attempt to gain advantages of size in trade by creating large markets for their producers and protecting them, at least in part, from outside competition. Consequently, the most important of the regional blocs – particularly the European Union and the NAFTA – have a very considerable influence on patterns of world trade.

The classic analysis of the trade effects of regional blocs identifies two opposing outcomes:

- *Trade diversion* occurs where, as the result of regional bloc formation, trade with a former trading partner (now outside the bloc) is replaced by trade with a partner inside the bloc.
- *Trade creation* occurs where, as the result of regional bloc formation, trade replaces home production or where there is increased trade associated with economic growth in the bloc.

In addition, regional trading blocs also have a major influence on flows of *investment* by transnational corporations. The effects of regional integration on direct investment, like that on trade, can also be conceptualized in terms of 'creation' and 'diversion'. In the latter case, the removal of internal trade (and other) barriers may lead firms to realign their organizational structures and value-adding activities to reflect a regional rather than a strictly national, market. This, by definition, 'diverts' investment from some locations in favour of others. These are issues we will discuss in more detail in subsequent chapters.

Despite a widespread view that regional blocs are a relatively new phenomenon, they have, in fact, been an important feature of the global economic landscape since the middle of the 19th century. But their basis and their nature have changed over time. Four 'waves of regionalism' can be identified:[28]

- During the second half of the 19th century there were a number of trade agreements in place, especially in Europe: for example, the German *Zollverein*, the customs unions between the Austrian states, and between several of the Nordic countries. 'As of the first decade of the twentieth century, Great Britain had concluded bilateral arrangements with forty-six states, Germany had done so with thirty countries, and France had done so with more than twenty states' (p. 596).

- After the disruptive effects of World War I (1914–1918) a new wave of regional arrangements occurred but, this time, in a more discriminatory form. 'Some were created to consolidate the empires of major powers, including the customs union France formed with members of its empire in 1928 and the Commonwealth system of preferences established by Great Britain in 1932. Most, however, were formed among sovereign states … The Rome Agreement of 1934 led to the establishment of a PTA involving Italy, Austria and Hungary. Belgium, Denmark, Finland, Luxembourg, the Netherlands, Norway and Sweden concluded a series of economic agreements throughout the 1930s … Outside of Europe, the United States forged almost two dozen bilateral commercial agreements during the mid-1930s, many of which involved Latin American countries' (p. 597).

- Since the end of World War II (1939–1945) there have been two distinct waves of regionalism. 'The first took place from the late 1950s through the 1970s and was marked by the establishment of the EEC, EFTA, the CMEA, and a plethora of regional trade blocs formed by developing countries. These arrangements were initiated against the backdrop of the Cold War, the rash of decolonisation following World War II, and a multilateral commercial framework, all of which colored their economic and political effects' (p. 600).

- The fourth wave of economic regionalism – from the late 1980s onwards – occurred in the drastically changed geopolitical circumstances of the collapse of the Soviet-led system and the increased uncertainties of a more fragmented political and economic situation. 'Furthermore, the leading actor in the international system (the United States) is actively promoting and participating in the process. PTAs also have been used with increasing regularity to help prompt and consolidate economic and political reforms in prospective members, a rarity during prior eras. And unlike the interwar period, the most recent wave of regionalism has been accompanied by high levels of economic interdependence, a willingness by the major economic actors to mediate trade disputes, and a multilateral (that is, the GATT/WTO) framework …' (p. 601).

Types of regional economic integration

All of the regional collaborative arrangements that have been established over the years have been based on the principle of preferential trading arrangements. However, there are, in fact, several different types of politically negotiated regional economic arrangements involving different degrees of economic and political integration. The following four types of regional arrangement are especially important; they are listed in order of increasing economic and political integration:

- the *free trade area*, in which trade restrictions between member states are removed by agreement but where member states retain their individual trade policies towards non-member states

- the *customs union*, in which member states operate both a free trade arrangement with each other and also establish a common external trade policy (tariffs and non-tariff barriers) towards non-members

- the *common market*, in which not only are trade barriers between member states removed and a common external trade policy adopted but also the free movement of factors of production (capital, labour, etc.) between member states is permitted

- the *economic union*, which involves the highest form of regional economic integration short of full-scale political union; in an economic union, not only are internal trade barriers removed, a common external tariff operated and free factor movements permitted but also broader economic policies are harmonized and subject to supranational control.

As Figure 5.10 shows, the progression is cumulative; each successive stage of integration incorporates elements of the previous stage, together with the additional element that defines each particular stage.

The number of regional trading arrangements has grown dramatically. Some of the major recent developments are shown in Box 5.1. By 2001, according to the WTO, there were some 170 regional trading arrangements in operation covering more than 40 per cent of total world trade. The WTO was predicting that the number could increase to 250 by 2005 and would then account for more than 50 per cent of world trade. Not surprisingly, the WTO was concerned that such regional arrangements could damage the rules-based, multi-lateral trading regime established in the GATT in the 1940s and now enshrined in the WTO.[29]

The vast majority of the regional economic groupings fall into the first two categories of the classification shown in Figure 5.10: the free trade area and the customs union. There are a small number of common market arrangements, but

Figure 5.10

Types of regional economic integration

	Free Trade Area	Customs Union	Common Market	Economic Union
Removal of trade restrictions between member states	●	●	●	●
Common external trade policy towards non-members		●	●	●
Free movement of factors of production between member states			●	●
Harmonization of economic policies under supranational control				●

Box 5.1 Acceleration of regional trading arrangements

In Europe
- In the 1980s, membership of the European Community expanded further from 9 to 12
- In 1991, an agreement was reached with all but one of the members of the European Free Trade Association (EFTA) to form a European Economic Area (EEA)
- The Treaty on European Union was signed at Maastricht in December 1991
- The Single European Community Market legislation came into being in January 1993
- In 1995, the European Union expanded its membership from 12 to 15
- In 1995, free trade agreements were implemented between the European Union (EU) and the three Baltic republics of Estonia, Latvia and Lithuania
- In 1997/1999, negotiations began with 13 countries to enlarge the EU
- In 1999, European Monetary Union was established among 11 member states

In the Americas
- In 1989, the United States and Canada implemented the Canada–United States Free Trade Agreement
- In 1994, the North American Free Trade Agreement (NAFTA) came into being, whereby Mexico joined Canada and the United States in the first example of a developing country becoming fully integrated in a free trade area with highly developed countries
- In 1995, the MERCOSUR customs union between Argentina, Brazil, Paraguay and Uruguay came into being
- Further development of the Andean Pact (involving Bolivia, Colombia, Ecuador, Peru and Venezuela) occurred
- A free trade area involving Mexico, Colombia and Venezuela came into force
- In 2000, Mexico signed a free trade agreement with the European Union
- In 2001, an agreement was concluded by 34 countries (including the US) to establish a Free Trade Area of the Americas

In the Pacific Basin, where, apart from AFTA, there is no formal regional trade bloc
- The 19 member states of the Asia–Pacific Economic Forum agreed to remove all regional trade barriers by 2020
- ASEAN/AFTA membership increased to include Cambodia, Laos, Myanmar, Vietnam
- In 2000, proposals initiated to consider a free trade arrangement between the ASEAN countries and the countries of north-east Asia (China, South Korea, Taiwan)
- In 2001, 13 East Asian countries (ASEAN, Japan, China, South Korea) agreed currency swap arrangements

only one group – the European Union – comes close to being a true economic union. In fact, not only is there enormous variation in the scale, nature and effectiveness of these regional trade groupings but also there is, in some cases, a considerable overlap of membership of different groups, especially in Latin America. Such diversity must be borne in mind when considering the likely geographical effects of regional integration both internally (on member states and communities) and externally (on the rest of the world economy). Table 5.3 shows the major regional agreements currently in force.

Table 5.3 Major regional economic blocs

Regional group	Membership	Date	Type
EU (European Union)	Austria, Belgium, Denmark, France, Finland, Germany, Greece, Ireland, Italy, Luxembourg, Netherlands, Portugal, Spain, Sweden, United Kingdom	1957 (European Common Market) 1992 (European Union)	Economic union
NAFTA (North American Free Trade Agreement)	Canada, Mexico, United States	1994	Free trade area
EFTA (European Free Trade Association)	Iceland, Norway, Liechtenstein, Switzerland	1960	Free trade area
MERCOSUR (Southern Cone Common Market)	Argentina, Brazil, Paraguay, Uruguay	1991	Common market
ANCOM (Andean Common Market)	Bolivia, Colombia, Ecuador, Peru, Venezuela	1969 (revived 1990)	Customs union
CARICOM (Caribbean Community)	Antigua & Barbuda, Bahamas, Barbados, Belize, San Cristobal, Dominica, Grenada, Guyana, Jamaica, Montserrat, St Kitts & Nevis, St Lucia, St Vincent & the Grenadines, Trinidad & Tobago	1973	Common market
AFTA (ASEAN Free Trade Agreement)	Brunei Darussalam, Cambodia, Indonesia, Laos, Malaysia, Myanmar, Philippines, Singapore, Thailand, Vietnam	1967 (ASEAN) 1992 (AFTA)	Free trade area

Integration within the global triad: Europe, North America, East Asia

In Chapter 3 we observed the strong tendency for a disproportionate share of global production, trade and investment to be concentrated in three 'mega-regions' – the so-called global triad of North America, Europe and East Asia. Such concentration reflects, first and foremost, the basic economic-geographical processes we discussed in Chapter 2: the preference for proximity to markets and suppliers and a general tendency to 'followership' in location decision-making. Two questions are posed: first, how do *political* processes of economic integration map on to the triad regions? Second, are all three regions going down the same path of closer formal integration as opposed to the 'natural' integration processes stimulated by geographical proximity?

Figure 5.11 provides an answer to the first question. It shows both the internal structure of the regional arrangements in Europe, the Americas and the Asia–Pacific, and also the connections within and between these regions. The picture is immensely complex. In both the EU and the NAFTA, in particular, a core of member states is surrounded by a periphery of states or groups of states with varying degrees of connection and association with the core bloc. At the same time, each of the three regions is connected with the others, though, again, the strength and formality of such inter-regional connections varies greatly.

To answer the second question – whether or not the three mega-regions are going down a similar path of regional integration – we need to look in more detail at each of them.

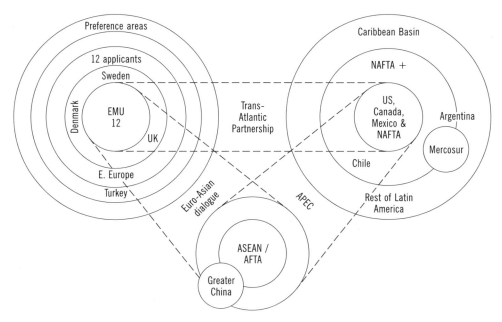

Figure 5.11   The complex relationships within and between regional economic blocs

Source: Cable, 1999: Figure 3.3

The European Union

The EU (Figure 5.12) is by far the most highly developed and structurally complex of all the world's regional economic blocs. Although initially established as a six-member European *Economic* Community by the Treaty of Rome in 1957, it was always more than simply an economic institution. The initial stimulus was the desire to bring together France and Germany in such a way that traditional enmities could no longer find their outlet in another round of European wars and also to strengthen Western Europe in the face of the perceived Soviet threat. The EEC developed out of an earlier, more modest, development: the European Coal and Steel Community (ECSC), which explicitly tied together the French and German coal and steel industries as part of the broader political agenda.

From the initiation of the EEC in 1957, we can distinguish five major stages of development:

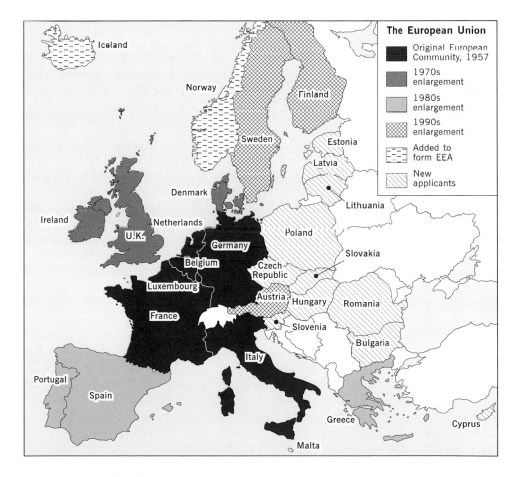

Figure 5.12 The European Union

- *1958–1968*. The elimination of customs duties between the six founder member states (Belgium, France, West Germany, Italy, Luxembourg, Netherlands) and, in 1968, the introduction of a common external tariff.

- *1973–1986*. (a) The enlargement of the Community with the accession of Denmark, Ireland and the United Kingdom in the 1970s and of Greece, Portugal, Spain in the 1980s. (b) The establishment of preferential trading agreements with EFTA countries; with countries around the Mediterranean rim; with countries in Africa, the Caribbean and the Pacific (the so-called ACP nations – all former European colonies).

- *1986–1992*. (a) A renewed attempt to complete the Single European Market. By the mid-1980s it had become apparent that a single market did not actually exist. (b) A drive by the 'core' members of the EC to move towards full economic and monetary union. The signing of the Treaty on European Union at Maastricht in December 1991 created the European Union.

- *1992–1999*. (a) The further enlargement of the EU to 15 with the accession of Austria, Finland and Sweden. (b) The attempts to implement fully both the Single European Market and the Maastricht Treaty in the context of both internal divisions and the pressures to enlarge the EU.

- *1999–present*. (a) Introduction of the European single currency in 11 (later 12) of the 15 member states. European Central Bank established with responsibility for setting EU-wide interest rates. (b) Negotiations with European countries wishing to become members of the EU. (c) Inter-governmental conference at Nice in 2001 to reform the EU's voting system.

The stimulus to complete the Single European Market in 1992 was the claim by the European Commission that the multiplicity of internal barriers to trade and factor movement that still remained were weakening the Community's ability to operate in highly competitive global markets. Despite the provisions of the Treaty of Rome, individual countries were increasingly resorting to tactics which prevented or delayed the import of certain products from other member nations: for example, the insistence on particularly stringent health and safety checks on products, the heavy bureaucracy of customs and frontier procedures, even the awarding of government contracts to domestic producers and the contravention of EC regulations on the subsidizing of domestic industries. Each individual member state was guilty of some such practices, although some were more guilty than others. The Commission argued that the costs of 'non-Europe' amounted to a significant loss of potential GDP and of jobs as well as a lessened ability to compete with the United States and Japan.[30]

The Single European Act proposed the removal of three major sets of barriers – physical, technical and fiscal – together with the liberalization of financial services, the opening of public procurement and other measures. Such internal liberalization and deregulation, it was argued, would create a virtuous circle of growth for the European Community as a whole, its member states and for those business firms which successfully take advantage of the changes. The virtuous growth scenario is shown in Figure 5.13.

Figure 5.13

The virtuous circle of growth:
predicted outcomes of the
completion of the single European
market

Source: Based, in part, on Cecchini, 1988:
Chart 10.1

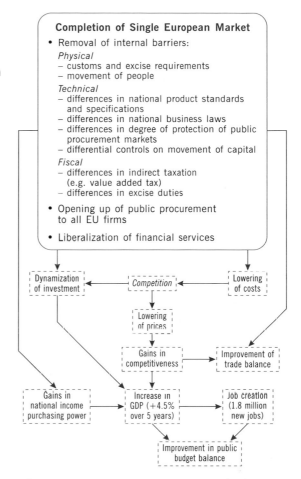

Such a scenario was based upon a host of assumptions, not all of which were especially realistic. In fact, it is virtually impossible to measure precisely the actual effects of the Single Market process. The European Commission claimed that by 1996 some 90 per cent of the legislation to complete the single market was in place. The Commission also claimed that European Union GDP in 1994 was between 1.1 and 1.5 per cent higher than it would have been without the Single Market (the Cecchini prediction was an increase of 4.5 per cent over 5 years) and that between 300,000 and 900,000 people in work would have been unemployed without the Single Market (the Cecchini prediction was the creation of 1.8 million new jobs).[31]

However, even if such an optimistic outcome were to occur for the EC as a whole, there is bound to be great *geographical unevenness* in those outcomes. There would be winners and there would be losers at both national and subnational levels. Consequently, the Single European Act (SEA) also included a *social* dimension – a reinforcement of existing EC social and regional programmes. As a result, the Structural Funds budget was doubled between 1988 and 1993.

The Treaty on European Union (TEU), signed at Maastricht in late 1991, marked a much more ambitious political agenda aimed at creating a fully fledged economic union. In particular it,

- strengthened social provisions by (a) incorporating the Social Charter; (b) the enlargement of the EC Structural Funds; (c) the creation of a new Cohesion Fund to assist poorer areas of the Union and
- set out the mechanisms for the creation of a single European currency and monetary union (EMU). This was planned for implementation by 1999, provided that member states were able to meet the strict convergence criteria stipulated in the Maastricht Treaty.

In fact, European Monetary Union did come into effect in 1999, when 11 of the 15 member states not only agreed to join the system but also met the technical criteria for doing so. The countries opting out of the EMU at that time were Denmark, Greece, Sweden and the United Kingdom (Greece joined in 2001). The issue of monetary union, and the adoption of a single European currency (the *euro*) crystallize some of the most difficult political problems within the EU, notably the sensitive issue of national sovereignty. Within the EMU, national control over monetary policy – notably the setting of interest rates – has been passed upwards to the European Central Bank based in Frankfurt. The ECB, therefore, has an immense influence over the economies of individual member states.

The pros and cons of a single European currency are finely balanced. The major benefits are the reduced costs and uncertainties associated with having to deal with many separate currencies within a single market and the overall stability this is intended to produce. Set against this is the fact that an individual state's ability to use monetary mechanisms to deal with periodic economic crises is greatly reduced. It is still too early to judge the effect and the effectiveness of the EMU. In its first two years of existence, the euro was very weak, especially against the US dollar (and against the UK pound). The fact that three EU member states – Denmark, Sweden, the United Kingdom – remain outside the EMU is a major source of uncertainty. Denmark has voted in a referendum to remain outside whilst opinion in both Sweden and the United Kingdom is strongly divided. Continued existence of 'insiders' and 'outsiders' will inevitably pose problems for the EU.

The other major issue facing the EU is its proposed enlargement. Thirteen countries have applied formally to join the EU (see Figure 5.12). The first group consists of Cyprus, Estonia, Hungary, Poland, the Czech Republic, Slovenia and Malta. It is planned that they should enter in 2004. The second group, for whom no accession date has been agreed, consists of Bulgaria, Latvia, Lithuania, Romania and Slovakia. The thirteenth applicant is Turkey, a country that has been striving to join the EU for many years but whose political circumstances have been judged to be unfavourable to EU membership at the present time.

Of course, enlargement of the Community is nothing new. As Figure 5.12 shows, it experienced several phases of enlargement between 1957 and the mid-1990s: from six, to nine, to 12 and to 15 member states. But now the circumstances are very different. First, further enlargement would be difficult under any circumstances; it will be doubly difficult given the scale and complexity of the current internal agenda. Second, a particular concern is that the nature of the applicant states would place very severe strains on the EU's budget, quite apart from making the process of decision-making even more complex. Thus, one view is that a 'variable geometry' or 'variable speed' European Union is likely to emerge with a core of states, fully integrated economically and financially, surrounded by various groups of countries with different degrees of integration.

The North American Free Trade Agreement

The NAFTA is, without doubt, a highly significant political-economic development. By integrating two highly developed countries and one large developing country into a single free trade area it changes the economic map of North America quite radically. The income gap between the United States and Mexico is very much greater than that between the richest and poorest states within the EU. The NAFTA came into force in 1994, but its origins can be traced back into the 1980s. One important building block, although this was not its intent, was the Canada–United States Free Trade Agreement (CUSFTA) signed in 1988 and implemented in 1989. Hitherto, the United States had not taken advantage of the GATT article allowing for exceptions to the principle of non-discrimination in the case of free trade areas and customs unions.

In part, this change of stance reflected the United States' desire to have a 'lever' to achieve more open trade in the world economy by showing that it was prepared to make discriminatory agreements. Other United States motives included the desire to reduce trade barriers with Canada, to improve the treatment of US direct investment in Canada, and to achieve specific deals on government procurement policies, trade in services, and other areas. To the Canadian government, a free trade agreement offered the prospect of obtaining more secure access to the US market with less risk of being inhibited by various non-tariff barriers, to secure an effective dispute settlement procedure, and to protect its cultural industries by agreement. As the CUSFTA was being signed, two other developments were also occurring. George Bush (senior) had made freer trade with Mexico a campaign issue in 1988. At the same time, President Carlos Salinas of Mexico made clear his determination to negotiate a free trade area with the United States. Within a short time of bilateral talks starting, Canada had joined in an obvious defensive response.

The arguments in favour of creating the NAFTA varied between the three parties. For the United States, it formed part of its long-term objective of ensuring stable economic and political development in the western hemisphere and also gave access to Mexican raw materials (especially oil), markets and low-cost labour. It also promised further US leverage in a world of increasing regional integration. The Canadian government was anxious to consolidate the recent CUSFTA. The motives of the Mexican government were primarily to help to lock in the economic reforms of the previous few years, to create a magnet for inward investment, not only from the United States but also from Europe and Asia, and to secure access to the United States and Canadian markets.

It is interesting to note the different kinds of debate that surrounded the NAFTA compared with that surrounding the CUSFTA.

Notably, although the Canada–US agreement sparked a heated debate in Canada, it was barely noticed in the United States. The United States has almost three times as much trade with Canada as with Mexico, yet the similarities between Canadian and

US institutions made the prospect of free trade with Canada a fairly routine issue for the general US electorate. In Canada, however, there were more concerns about system differences, namely that the United States represented a threat to the Canadian welfare system.

... The NAFTA discussion in the United States was much more politically salient, and the debate was highly charged. Its opponents argued that the NAFTA meant a fundamental threat to US domestic institutions, especially ones affecting the environment and labour standards ... NAFTA opponents in the United States also argued that NAFTA would have major consequences for US employment. Ross Perot made the memorable prediction that NAFTA would give rise to 'a giant sucking sound' as jobs left the United States for Mexico. In fact, much of the popular debate over NAFTA focused on its employment impact rather than its effects on welfare.[32]

A similar fear was expressed in Canada. One politician stated that 'this whole concept is a nightmare of US continentalists come true: Canada's resources, Mexico's labour, and US capital'.[33] In Mexico, the fear was expressed that the country would become even more dominated by the United States.

The NAFTA took effect in 1994. The Clinton administration, which took office in 1993, used the 'fast-track' route to get the agreement through the US Congress. The main provisions of the NAFTA are summarized in Box 5.2. Note that, in addition to the various trade provisions, two 'side agreements' (on the environment and on labour standards) were incorporated to meet US and Canadian concerns. However, in contrast to the EU, there are no social provisions.

The aims of the NAFTA were gradually to eliminate most trade and investment restrictions between the three countries over a 10- to 15-year period. The possibility of other countries joining the NAFTA was left open to negotiation but it is important to stress that NAFTA is not a customs union. It does not incorporate a common external trade policy. Each of the three NAFTA members is free to make free trade agreements with other states outside the NAFTA (as Mexico has recently done, for example, with the EU). In sum, the NAFTA is light years removed from the far more advanced European Union and there is not the remotest possibility that the NAFTA will evolve into anything approaching the EU. Each is a reflection of its specific circumstances. Nevertheless, there are moves under way to create a much larger Free Trade Area of the Americas. So far, such negotiations are at an early stage.

Regional economic collaboration in East Asia and the Pacific

Two of the three global triad regions – Europe and North America – now have politically based economic integration arrangements, although they differ greatly in the depth and breadth of that integration. No comparable regional arrangement exists in East and South East Asia, although there have been various attempts over the past 40 years to develop inter-nation cooperation within what was, prior to the 1997 crisis, the world's fastest growing economic region.[34] Both the nature and the scale of regional economic collaboration in the Asia–Pacific region are very different from the situation in Europe and North America. In general, regional arrangements in the Asia–Pacific are much looser, less formalized and more open than in the other two triad regions.[35] There are, in effect, two main regional economic

Box 5.2 Major provisions of the North American Free Trade Agreement

General provisions
- Tariffs reduced over a 10- to 15-year period, depending on the sector
- Investment restrictions lifted (except for oil in Mexico; cultural industries in Canada; airline and radio communications in the United States)
- Immigration is not covered, with the exception that movement of some white-collar workers to be eased
- Any member state can leave the Agreement with 6 months' notice
- The Agreement allows for the inclusion of any additional country
- Government procurement to be opened up over 10 years
- Dispute resolution panels of independent arbitrators to resolve disagreements
- Some 'snap-back' tariffs allowed if surge in imports hurts a domestic industry

Sector-specific provisions
- *Agriculture*: Most tariffs between US and Mexico removed immediately. Tariffs on 6% of products – corn, sugar, some fruits and vegetables – fully eliminated after 15 years. For Canada, existing agreement with US applies
- *Automobiles*: Tariffs removed over 10 years. Mexico's quotas on imports lifted over same period. Cars eventually to meet 62.5% local content rule in order to be free of tariffs
- *Energy*: Mexican ban on private sector exploration continues, but procurement by state oil company opened up to US and Canada
- *Financial services*: Mexico gradually to open up financial sector to US and Canadian investment. Barriers to be eliminated by 2007
- *Textiles*: Agreement eliminates Mexican, US and Canadian tariffs over 10 years. Clothes eligible for tariff breaks to be sewn with fabric woven in North America
- *Trucking*: North American trucks can be driven anywhere in the three countries by year 2000

Side agreements
- *Environment*: The three countries liable to fines, and Mexico and the US sanctions, if a panel finds repeated pattern of not enforcing environmental laws
- *Labour*: Countries liable for penalties for non-enforcement of child, minimum wage and health and safety laws

Other arrangements
- The US and Mexico to set up a North American Development Bank to help finance the clean-up of the US–Mexico border
- The US to spend roughly $90m in the first 18 months retraining workers losing their jobs because of the Agreement

Source: Based on material in *The Financial Times*, 17 November 1993

collaborations. One – *AFTA* – is confined to South East Asia. The other – *APEC* – is a much wider and looser arrangement, that involves not only East Asia but also Australia and New Zealand, the United States, Canada, Mexico, and also a part of Latin America (Chile). Let's look at each of these in turn.

AFTA – the ASEAN Free Trade Agreement   Figure 5.14 shows the current membership of AFTA. Ten countries are involved although they are extremely varied in their political and economic structures and levels of economic development. AFTA was initiated at the 1992 meeting of Heads of Government of ASEAN (the Association of South East Asian Nations). ASEAN itself had been established at a meeting in Bangkok in 1967 as a group of four, then, six South East Asian countries (Singapore, Malaysia, Thailand, Indonesia, the Philippines, Brunei).

> ASEAN as an intergovernmental institution established to promote regional cooperation, offers a striking contrast to the Western institutions such as EU and NAFTA … it is … based on a different concept of institutionalisation.
>
> ASEAN does not have a charter or a constitution, or any other legal instrument regarding the establishment, basic structure and function of the Association. The foundation of ASEAN was 'declared' by the brief Bangkok Declaration which set out only the aims, the basic principles, and the minimum machinery …
>
> Paying full respect for the sovereignty and independence of each member state is one of the fundamental principles of the Association … most of the decisions have been made by consensus through the consultation based on the ASEAN tradition, which means to negotiate and consult thoroughly till achieving an agreement …
>
> [The] mechanism for dispute settlement also reflects ASEAN's preference for an informal approach. This is a striking contrast with the Western approach to dispute settlement in which preference is clearly on the side of judicial settlement based on clear rules and binding decisions.[36]

Such a system has both its strengths and its weaknesses. One of its strengths is that it has helped what is a very diverse group of countries to maintain positive relationships. One of its weaknesses is that firm and rapid response to problems is often difficult, especially in the light of the principle of non-interference in domestic matters of member states. As Figure 5.14 shows, ASEAN's membership grew substantially in the second half of the 1990s. Vietnam became a member in 1995; Laos and Myanmar in 1997 (Myanmar's accession was much criticized by Western countries because of its human rights record). Cambodia was also due to become a member in 1997 but the coup there postponed membership until 1999.

ASEAN has had only limited success in stimulating economic activity. As a consequence, in 1992, the original six member states agreed to initiate an ASEAN Free Trade Agreement with the aim of removing all internal trade barriers by 2008. Subsequently, it was agreed to move the deadline for completion forward to 2003. However, incorporating the four very different economies of Cambodia, Laos, Myanmar and Vietnam is a difficult task. In addition, increasing competitive pressures on the ASEAN region from other East Asian countries (notably China) has forced the Association to look towards making agreements with other countries in East Asia. In 2000, talks began on establishing a free trade area with Japan,

China and South Korea. Although almost certainly a long-term project, the notion of 'ASEAN-plus-3' is a potentially significant development.

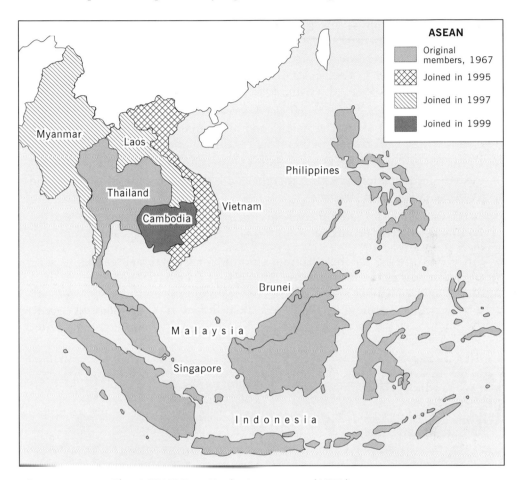

Figure 5.14   The ASEAN Free Trade Agreement (AFTA)

APEC – The Asia–Pacific Economic Cooperation Forum   The other major regional economic organization is the Asia–Pacific Economic Cooperation Forum (APEC) which was established in 1989 on the initiative of the Australian government. Figure 5.15 shows the extremely diverse composition of APEC. It includes not only the obvious East and South East Asian states themselves (including China and Taiwan) but also Australia and New Zealand on the one hand and the United States, Canada, Mexico, and Chile on the other.

   Not surprisingly, the individual agendas of particular members vary considerably.[37]

● The United States wishes both to signal its commitment to Asia and to strengthen its own economic and political interests there. It does not wish to concede Asia to Japan and to find itself confined to the Western hemisphere.

- Japan is keen to develop further its relationships with ASEAN countries but does not wish to alienate the United States (still its most important ally despite economic tensions between the two countries). But Japan wishes to avoid the danger of a protectionist North American trade bloc and thus prefers to have the NAFTA group as part of APEC.

- The smaller Asian countries are fearful of dominance by any one of the three biggest economies in Pacific Asia: Japan, the United States and China. Yet they also wish to see all three involved in the region.

- However, Malaysia is concerned to preserve 'Asianness' and has proposed the establishment of an East Asia Economic Caucus (EAEC) which would exclude the North American countries, Australia and New Zealand.

- China is keen to become more widely accepted in the international economic system. Participation in APEC, together with its desire to be admitted to the WTO, is part of that strategy.

- Australia, which initiated the whole concept of APEC, sees it as legitimizing its role as a Pacific economy.

Since 1993, APEC has held annual summits of its national leaders to set the future direction of the organization. In 1994, at the Indonesian summit, it was agreed that the member states would commit themselves to the achievement of free trade and investment by the year 2020 at the latest. The advanced countries

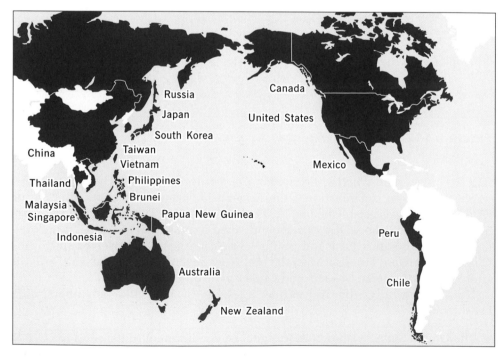

Figure 5.15   The Asia–Pacific Economic Cooperation Forum (APEC)

agreed to meet that objective by 2010. However, such liberalization is to be conducted in an open, multilateral, manner and not in an inward-looking manner. 'APEC was promoted by several nations who are deeply committed to the multilateral trading system and concerned about the possibilities of a world divided into blocks that discriminate against outsiders. Thus, while APEC is itself a regional arrangement, it has the paradoxical mission of combating (preferential) regionalism'.[38] Such 'open regionalism' creates the possibility of new members.

At the same time as addressing trade liberalization measures, APEC has also drawn up an agenda to discuss wider issues, including competition policy, environmental protection and dispute procedures. Of course, agreeing to talk about such issues is a far cry from actually implementing measures. The 1996 summit in the Philippines did not produce any significant progress, even in the already agreed area of trade liberalization. The 2000 meeting in Brunei merely asserted the need for a new round of trade liberalization in the WTO. Unlike the formal integration agreements shown in Table 5.3, (see p. 150) APEC is, at present, little more than a loose collection of states. Cynics have translated the APEC acronym as '*A Perfect Excuse to Chat*', but this is to denigrate what could be a significant initiative. It is more appropriate to see APEC, as being 'engaged in the early stages of institution building for cooperation'.[39] Nevertheless, it would seem that the creation of a formal region-wide economic bloc in the Asia–Pacific region, along the lines of the EU or even of NAFTA, is not a serious probability.

Conclusion

The aim of this chapter has been to assert the continuing significance of the state as a major influence in the global economy and to describe some of the general ways in which it operates. This does not mean, of course, that the roles and the functions of the state have not changed. On the contrary, the position of the state is being redefined in the context of a polycentric political-economic system in which national boundaries are more permeable than in the past. Nevertheless, the nation-state continues to contribute significantly towards the shaping and reshaping of the global economic map. We focused on four major aspects of the role of the state.

First, we explored the role of the state as *container* of distinctive 'ways of doing things'. The persistence of national differences in cultural, political, social and economic processes and institutions, that have evolved in a path-dependent manner over time, helps to perpetuate 'varieties of capitalism' rather than convergence to a single form. Such fundamental differences help to explain the other aspects of state behaviour discussed in this chapter.

Second, we focused on the state as *regulator* of economic activity. All states operate a broad variety of policies aimed at influencing the level and nature of economic activities within, and across, their borders. But the precise mix of policies and how they are implemented varies according to a number of variables. In particular, the policy stance varies according to the ideological complexion of the state as well as the nature of the country's position in the global economic system.

Third, we examined the state as *competitor*, a controversial issue, as the debate stimulated by Paul Krugman demonstrated. Whether or not states *should* compete, the fact is that they think they do and they behave accordingly.

Fourth, we addressed the issue of regional integration as an example of the state as *collaborator*. While there has certainly been an acceleration in the number of regional trading agreements in recent years, most are very limited in the depth and extent of their integration. Very few have proceeded beyond the first stage of a simple free trade arrangement. Only the European Union has gone very far along the road of economic integration. Ultimately, however, regional blocs of whatever degree of economic integration originate, and are given legitimacy by, nation-states, which continue to be extremely important building blocks in the global economy.

Notes

1 Garrett (1998: 793).
2 Kindleberger (1969: 207).
3 See Dicken, Peck and Tickell (1997); Garrett (1998); Gilpin (2001); Hirst and Thompson (1999); Hudson (2001); Porter (1990); Wade (1996); Weiss (1998). The concept of the 'hollowing out' of the state is discussed by Jessop (1994).
4 Porter (1990: 19).
5 Wade (1996).
6 Hirst and Thompson (1999: 268–9).
7 Glassner (1993: 35–40, from whom the following definitions are drawn).
8 Guibernau (1999: 17, 18).
9 Agnew and Corbridge (1995) and Taylor (1994) discuss the general notion of states as 'containers' and the nature and significance of territoriality and space in geopolitics.
10 See Granovetter and Swedberg (1992); Smelser and Swedberg (1994); Zukin and DiMaggio (1990).
11 Terpstra and David (1991: 6).
12 Hofstede (1980, 1983).
13 There is a substantial literature on 'varieties of capitalism'. See, for example, Berger and Dore (1996); Hall and Soskice (2001); Hollingsworth and Boyer (1997); Turner (2001); Whitley (1999).
14 Johnson (1982: 20).
15 Hollingsworth (1997: 266, 267–8).
16 Hudson (2001: 48–9; emphasis added).
17 Hudson (2001: 76).
18 WTO (1996: 61–2).
19 See Guisinger (1985).
20 Krugman (1994).
21 Krugman (1994: 41, 43, 44).
22 See, for example, the contributions in Rapkin and Avery (1995).
23 Porter (1990: 19).
24 Porter (1990: 154, 155, 157; emphasis added).
25 Dunning (1992: 142).
26 Analyses of regional economic blocs are provided by Cable and Henderson (1994); Gamble and Payne (1996); Gibb and Michalak (1994); Lawrence (1996); Mansfield and Milner (1999).
27 Mansfield and Milner (1999: 592).
28 For a detailed discussion of these four waves, see Mansfield and Milner (1999: 595–602).
29 *The Financial Times* (30 November 2001).
30 Cecchini (1988).
31 Quoted in *The Financial Times* (30 October 1996).
32 Lawrence (1996: 72–3).

33 Quoted in McConnell and MacPherson (1994: 179).
34 Haggard (1995), Table 3.4, lists the various cooperation bodies within the Asian-Pacific region.
35 See Higgott (1999).
36 Liao (1997: 150-1).
37 Lawrence (1996: 86–7).
38 Lawrence (1996: 87–8).
39 Higgott (1993: 303).

CHAPTER 6
Doing Things Differently: Variations in State Economic Policies

From the general to the specific

The basic theme of Chapter 5 was that not only do states continue to be very important agents in the global economy but also that they continue to vary in how they behave. This is because states have histories that embed elements of path-dependency in their contemporary behaviour. The result, from an economic perspective, is the existence – and persistence – of different varieties of capitalism. In this chapter, we look at some concrete examples of how states of different kinds attempt to influence both their domestic economies and also their position within the global political-economic system.

The chapter is organized into four parts:

- First, we can identify some common patterns in state behaviour in the past few decades.
- Second, we focus on some of the older industrialized countries: the neo-liberal states of the United States and the United Kingdom, and the more interventionist states of continental Europe.
- Third, we examine the case of Japan, arguably the archetypal 'developmental state'.
- Fourth, we explore some of the diversity of policy positions adopted by some of the newly industrializing economies in East Asia and Latin America, and the newly emerging market economies in Eastern Europe.

A degree of convergence

For several decades after World War II, the role of the state expanded considerably, notably through the provision of welfare benefits for particular segments of the population and the development of a considerable (though varied) degree of public ownership of productive assets. The majority of economies, outside the command economies of the state-socialist world, became *mixed economies*. Certain economic sectors, such as telecommunications, railways, energy, steel and the like, became state-owned or controlled in many countries. As a result, government spending as a percentage of GDP rose very substantially, as Figure 6.1 shows. In the OECD countries, such spending increased from less than 20 per cent of GDP in the early 1960s to 35 per cent in the early 1980s. In the developing countries, the average growth was from around 15 per cent to 27 per cent. Of course, the pattern varied across individual countries. Among the industrialized countries, for example, relative state expenditure was lowest in the United States and highest in the Scandinavian countries and in France.

However, from the mid-1980s onwards, this trend was reversed. Instead of taking a direct role in the functioning of the economy, the tendency has been for the state to withdraw from many areas of involvement. In fact, this has not reduced government expenditure as much as might have been expected; the rhetoric has often been stronger than the reality. But there has been a broadening and deepening *marketization* of the state's activities, extending the principles of market transactions into more and more aspects of public life. This is apparent not only in the older industrialized countries but also in many developing countries and, most dramatically of course, in the former state-socialist countries of Eastern Europe, the former Soviet Union, and China. Such *market liberalization* consists, primarily, of two processes: *deregulation* and *privatization*.

To varying degrees virtually all industrialized countries have jumped aboard the deregulation bandwagon. In fact, the issues are far less simple than the 'deregulationists' claim. Because no activity can exist without some form of regulation (otherwise anarchy ensues), 'deregulation cannot take place without the creation of new regulations to replace the old'.[1] In effect, what is often termed *de*-regulation is really *re*-regulation. As we saw in Chapter 5, both trade and foreign direct investment regimes have been strongly liberalized over the past two decades or so. Processes of deregulation have also spread to most economic sectors, notably in financial services and telecommunications. But they are, of course, much contested by non-business interest groups who (rightly) see major dangers in the increasing pervasiveness of a neo-liberal agenda. These are issues we will return to in the final chapter.

Figure 6.1

Growth of the state: increasing central government spending as a percentage of GDP

Source: Based on World Bank, 1997: Figure 1.2

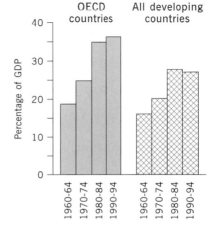

Parallel to the processes of deregulation are processes of *privatization*. The state has been pulling out of a whole range of activities in which it was formerly centrally involved and transferring them to the private sector. The selling of state-owned assets, and the greater participation of the private sector in the provision of both 'private' goods (the 'de-nationalization of state-owned economic activities) and 'public' goods (such as healthcare or education) has been a pervasive, though uneven, movement. In effect, as Figure 6.2 suggests, around the long-term secular trend shown by the middle curve in the diagram we can see cyclical variations in state involvement. But the diagram also suggests that there is a lower limit to state involvement below which the state's ability to govern by consent would be threatened. The state cannot completely wither away.

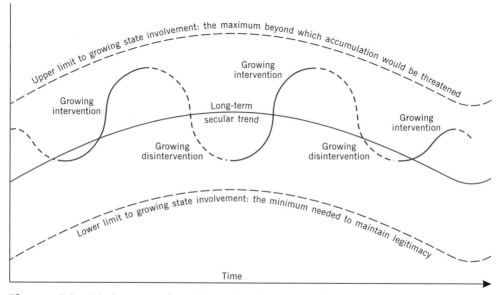

Figure 6.2   Limits to, and variations in, the extent of state involvement in the economy

Source: Hudson, 2001: Figure 3.3

The older industrialized economies of Europe and the United States

As we saw in Chapter 5, the continental European countries on the one hand, and the United States on the other, represent distinctively different types of capitalism. Historically, a major difference between most continental European countries and the United States has been the centrality of some kind of industrial policy, together with a greater degree of social accountability of business in the former, and the absence of such policy and accountability in the latter.[2] The position of the United

Kingdom is especially ambiguous. It occupies a somewhat intermediate position between the virtually pure market capitalism of the United States and the kinds of social market capitalism practised in continental Europe, but with a tendency in some areas (notably labour market policy) to be closer to the United States. One policy thread, common to all the European states, including the United Kingdom, has been that of *regional policy* – of designating specific geographical areas for special assistance. All the major industrial nations of Western Europe operate a regional economic policy to varying degrees. In contrast, direct federal involvement in regional or area economic development in the United States is relatively limited.

The United States

The policy stance of the United States reflects both the sheer scale and wealth of the domestic economy and also a basic philosophy of non-intervention by the federal government in the private economic sector. As far as industry is concerned the role of the federal government has generally been a *regulatory* one whose aim is to ensure the continuation of competition. Action at the federal level has been based primarily on macroeconomic policies of a fiscal and monetary kind. The aim has been to create an appropriate investment climate in which private sector institutions could flourish. This has not, however, prevented the federal government from rescuing specific firms – especially very large ones – from disaster. At the other end of the size spectrum, the Small Business Administration has provided aid to stimulate new and small firms. Federal procurement policies are generally non-discriminatory but the sheer size of federal government purchases, particularly in the defence and aerospace industries, has exerted an enormous influence on US industry. Entire economic sectors, regions and communities are heavily dependent on the work created by federal defence and other procurement contracts.

US policy towards international trade in the post-World War II period has been one of urging liberalization and the reduction of tariffs through multilateral negotiations through the GATT and, more recently, the WTO. As the strongest economy it has been, like Britain in the 19th century, the leading advocate of free trade. Even so, the federal government has intervened with the use of tariff and non-tariff barriers to protect particular interests. It has, for example, negotiated orderly marketing arrangements with various countries, notably Japan, in sectors such as textiles (see Chapter 10), automobiles (Chapter 11) and electronics (Chapter 12). The use of non-tariff barriers on a bilateral basis by the United States was explicitly specified in the Trade Act of 1974. This Act in effect heralded the emergence of a 'new protectionism' in the United States and the beginnings of a movement towards a *strategic trade policy* (STP).[3]

Strategic 'trade' policy is a rather misleading term; in reality it also involves industry policy and, by extension, FDI policy as well. In other words, it consists of some of the elements of a developmental system. The demand in the United States has been, increasingly, for a shift away from 'free' trade towards 'fair' trade – 'fairness' being defined by the United States itself. According to this view, because other countries – notably Japan, other East Asian NIEs, and most European countries – are themselves allegedly engaged in unfair trade practices it is only reasonable for the United States to take a similar stance.

A major new step in US trade policy came with the introduction of the Omnibus Trade and Competitiveness Act of 1988 (OTCA), stimulated by the persistence of a massive trade deficit and the difficulties being encountered by a number of particular American industries. The OTCA incorporated a strongly *unilateralist* approach to trade negotiations, rather than the multilateralist approach enshrined in the GATT, and hitherto strongly supported by the United States. The key clause – what some have called the 'crowbar' – was the so-called 'Super 301' clause, whose aim was to achieve reciprocal access to what the United States defines as unfairly restricted markets. The difference between the 1974 and 1988 Acts, in this respect, was that the Super 301 clause was directed to *entire countries*, not just individual industries. If, over a defined period of time, the named countries do not abandon what the United States defines as 'unjustifiable' or 'unreasonable' trade practices, the United States may retaliate unilaterally by restricting access to the US market for all goods from the countries in question.

The shifts towards a more strategic trade policy have been particularly evident in high-technology sectors, seen to be at the centre of a country's competitive position. The basic rationale is that, in imperfectly competitive markets, governments must intervene in favour of their domestic firms. The argument is based upon two grounds.[4]

- 'the "first mover advantage" that a country or firm captures by pre-empting foreign rivals …[that] …provides the opportunity for firms and countries to consolidate and extend their competitive advantage' and
- the externalities or spillovers that enhance the competitiveness of other parts of the domestic economy.

There is no question that, in general, the United States remains very much a neo-liberal market economy. In the eyes of the world, however, the United States is increasingly being seen as having a unilateralist tendency very much at odds with its traditional multilateral trading stance. The United States has become increasingly embroiled in a whole series of trade disputes – with Japan, with the East Asian NIEs, and with the European Union (at the time of writing the major issue was steel imports). Such disputes have focused upon US allegations of unfair trading practices by these countries in specific economic sectors. There is also concern that the United States has a tendency to introduce *extra-territorial* trade legislation to achieve its broader political objectives. One example was the Helms–Burton law that penalized foreign companies from doing business with Cuba.

US trade policy is complicated by the structure and composition of the US Congress and the ways that new trade policies have to be negotiated with domestic interest groups. In the early period of the new Republican Bush administration in 2001, for example, the attempts to renew the so-called 'fast-track' authority, in which Congress agrees not to amend trade agreements made by the US government with other countries, ran into difficulty. Congressional agreement was not reached until summer 2002.

> Labour and the environment have become the symbolic divide in the US debate over globalization. For moderate Democrats, linking trade deals to labour rights and environmental protection has become the only politically acceptable way to support further trade liberalization, while also addressing union fears that US jobs will bleed away to Mexico, Vietnam or other countries with low wages and lax environmental standards … But most Republicans still view the linkage with suspicion, seeing it either as a covert form of protectionism in which imports will be blocked on dubious grounds, or as a backhanded effort to impose a liberal regulatory agenda on the US.[5]

Particular debate surrounds trade relationships with China, seen not only as a growing competitive threat in economic terms but also as a political adversary, with unacceptable human rights practices.

A major shift in US trade policy since the 1980s has been a willingness to develop regional trading arrangements with other countries, notably within the NAFTA, APEC and the recently proposed Free Trade Area of the Americas (see Chapter 5). These are important developments in the US policy context, but the fact that the degree of actual economic and political integration in these arrangements – even in the NAFTA – is so limited means that they do not impose the kinds of constraints on US policy that EU membership imposes on European countries.

The United Kingdom

Although the United Kingdom is a member of the European Union (though not, at present, of the European Monetary Union), its policy position contrasts in a number of ways with that of the continental European countries. At the same time, the UK has been more interventionist than the United States, although since the 1980s it has taken on much more of a neo-liberal policy agenda. The UK, then, is something of a hybrid case.

During the mid-1960s, there was an attempt at national economic planning, but this was short-lived because of external pressures on the country's balance of payments and a currency crisis. Successive changes of government in 1970 and 1974 brought, first, a return to government disengagement from economic intervention and then a renewal of government involvement in the form of an explicit industrial strategy. The 1980s marked a major policy shift in the UK, as the Thatcher government developed a strongly neo-liberal agenda with, at least in intent, a much-reduced role for the state, an emphasis on individualistic competition, and a macroeconomic policy that resulted in massive decline and restructuring of the manufacturing sector. Deregulation and privatization became the dominant policy throughout the economy. This broad policy stance, although modified in some of its more extreme dimensions, has continued to prevail, even during the 'New Labour' administration from 1997 onwards. Certainly, UK policies in such areas as labour market issues are closer to those of the United States than of EU states.

In 2001, the UK government published a strategy document on 'competitiveness and the knowledge economy' designed to position the UK more strongly within the global economy. The policy focused on five specific areas:

- encouraging investment in key technologies and IT infrastructure
- reforming the regulatory and financial environment to facilitate companies' access to capital.
- building 'dynamic clusters' in the regions (the influence of Michael Porter was considerable in this respect)
- improving labour force skills, especially technical and IT skills
- ensuring a reduction in EU bureaucracy.

Of course, the UK's membership of the EU has a major impact on certain aspects of policy. For example, all international trade negotiations are carried out at the EU, not the national, level. There are limits on the size and nature of financial subsidies that can be offered to firms (for example, foreign firms) to influence the location of investment. There are rules of competition that affect mergers and acquisitions.

France

Within Europe, France has had the most explicit state industrial policy, a reflection of a tradition of strong state involvement dating back to the 17th century. France's industrial policy in the post-war period has been an integral part of the series of national economic 'indicative' plans. A major component of French industrial policy has been the promotion of 'national champions' in key industrial sectors, often through state ownership of large-scale enterprises. In its desire for technological independence, especially from the United States, France invested massively, but selectively, in certain sectors. In the late 1960s and early 1970s, for example, four *grands programmes* were launched in the nuclear, aerospace, space technology and electronics industries. Subsequently, the focus was narrowed to high-growth sectors in energy conservation equipment, office information systems, robotics, biotechnology and electronics. The French government has also exerted considerable influence on its industry through its purchasing policies and through its control of the major financial institutions. Both powers were used extensively.

France's current policy position retains many of these traditional qualities (and an especially strong antagonism to the Anglo-American neo-liberal economic model). Indeed, in some respects (as in the new labour laws reducing working hours introduced by the Jospin government in 2001), this stance continues to be reinforced. In the late 1990s, for example, total tax and social charges on business as a share of GDP were almost 25 per cent in France compared with 20 per cent in Italy, 16 per cent in Germany and 13 per cent in the UK.[6] Although there has been considerable privatization in France, the state retains a very considerable direct involvement in the economy. In fact, 'the state's holdings remain enormous. It owns all of Electricité de France (EDF) and Gaz de France (GDF), the post office, the rail system, the Paris airports, the nuclear industry, the aero-engine maker SNECMA; most of Air France, France Télécom and Thomson Multimedia; significant bits of Renault, the Bull computer company and – still – Crédit Lyonnais. In all, the state has a majority stake in some 1,500 companies and a minority in 1,300 others.'[7]

Germany

The major exception among the continental European nations to a centralized approach to industrial policy has been Germany. In part, at least, this reflects the fact that Germany is a federal political unit with power divided between the federal government and the provinces (*Länder*). But although often described as 'light', the federal government's role has been far from insubstantial. It has pursued policies of active intervention in industrial matters, including a substantial programme of financial subsidy. Such involvement has to be seen within the German model of a social-market economy.

The German economy is characterized both by a considerable degree of competition between domestic firms and also by a high level of consensus between various interest groups, including labour unions, the major banks and industry. The provincial governments have played an important role in the economy including, in recent times, the aiding of large manufacturing companies in distress. Like the French, the German government has also intervened to stimulate technological development in key sectors. During the 1970s, in particular, government financial support for research and development became focused on the development of advanced technologies – especially computer technologies – with wide application.

The major challenge facing Germany since 1990, of course, has been to cope with the fundamental transformation of the economy brought about by reunification. Putting together the strongest economy in Europe (the former West Germany) with one that, for half a century, had existed in a completely different ideological system, is an immense undertaking. It has put enormous strains on the federal budget because of the huge problems of rebuilding infrastructure and dealing with problems of unemployment brought about by restructuring.[8] More than a decade after reunification, the complex system of regional transfer payments between different parts of Germany, designed to even out imbalances, is under stress. Since 1995, the regional transfer system has been supplemented by a federally financed Solidarity Pact. But richer regions are beginning to question the basic notion of 'cooperative federalism'.[9]

Japan[10]

Japan can be regarded as the archetypal *developmental* state in which the government's economic role has been very different from that in the West.

> Historically, the State assumed an active economic role in Western
> economies in order to correct what were considered to be the private
> sector's economic and social failures. Japanese historical tradition, on the
> other hand, grants to government a legitimate role in shaping and
> helping to carry out industrial policy.[11]

However, although most commentators have regarded Japan's remarkable economic success between the 1950s and the 1990s as being closely related to the role of the Japanese government, others disagree.[12]

There has long been a high level of consensus between the major interest groups in Japan on the need to create a dynamic national economy. To some extent, this consensus is a cultural characteristic of Japanese society, with its deep roots in familism. But it also reflects the poor physical endowment of Japan and the limited number of options facing the country when, in the 1860s, it suddenly emerged from its feudal isolation. In other words, consensus is also a pragmatic stance built up over more than a hundred years. Given virtually no natural resources and a poor agricultural base, Japan's only hope of economic growth lay in building a strong manufacturing base, both domestically and internationally through trade. In this process, the state has played a central role not through direct state ownership but rather by *guiding* the operation of a highly competitive domestic market economy. Indeed, there has been relatively little state-owned

---

1 *Constructed medium-term econometric forecasts of the development of, and needed changes in, the Japanese industrial structure*
   - Established indicative plans or 'visions' of the desirable goals for the private sector
   - Made specific comparisons of cost structures of Japanese and foreign competitors

2 *Arranged for preferential allocation of capital to selected strategic industries*
   - Involved governmental and semi-governmental banks
   - Ministry of Finance guided commercial banks to coordinate their lending policies with MITI's industrial strategy
   - Financial support of an industry implied guidance (though not control) by MITI

3 *Targeted key industries for the future and put together a package of policy measures to promote their development*
   - Pre early 1980s the major measure was protection against foreign competition in the Japanese domestic market
   - Protectionism abandoned in early 1980s. Emphasis then on financial assistance, tax breaks, incentives given through administrative guidance, anti-trust relief (to facilitate 'research cartels')

4 *Formulated industrial policies for 'structurally recessed industries'*
   - MITI designated a specific industry as 'structurally recessed'
   - The ministry responsible for that sector formulated a stabilization plan specifying how the capacity to be scrapped should be shared between enterprises. The plans were drawn up in consultation with the Industrial Structure Council of MITI
   - Costs of scrapping production facilities were shared between the government and the private sector

---

Figure 6.3   The role of MITI in Japan's industrial development

Source: Based on Johnson, 1985: 66–7

enterprise in Japan and a generally much smaller public sector than in most Western economies.

Since World War II, the key government institution concerned with both industry policy and trade policy – the two are seen to be inextricably related in Japan – has been *MITI* – the *Ministry of International Trade and Industry* (recently renamed *METI* – the *Ministry for Economy, Trade and Industry*). After its establishment in 1949, MITI became the real 'guiding hand' in Japan's economic resurgence, although its role has often been misunderstood and exaggerated in the West. Figure 6.3 identifies the major roles played by MITI in the Japanese economy. Until the 1960s Japan operated a strongly protected economy and it was not until 1980 that full internationalization of the Japanese economy was reached. During the 1950s and early 1960s MITI, together with the Ministry of Finance, exerted very stringent controls on all foreign exchange, foreign investment and over the import of technology.

In fact, imported technology played a most significant part in the rebuilding of the Japanese economy. Technology was imported largely through licensing from foreign suppliers and not via the direct investment of foreign firms in Japan itself. The technologies were chosen to meet the needs of particular industries – those regarded by MITI as being the ones necessary to achieve national objectives. The selected industries were further aided by preferential financing and tax concessions and were also protected from foreign competition. Within Japan, however, intense competition and rivalry was encouraged between Japanese firms with the result that domestic production costs were kept down and efficiency increased. Within the selected industries, MITI encouraged mergers to create large-scale enterprises, although such moves were not always successful. For example, MITI failed in its attempt radically to restructure the automobile industry.

Initially, MITI focused its energies on the basic industries of steel, electric power, shipbuilding and chemical fertilizers but then progressively encouraged the development of petrochemicals, synthetic textiles, plastics, automobiles and electronics. Japan was transformed from a low-value, low-skill economy to a high-value, capital-intensive economy. The foundation of this transformation was the clearly targeted, selective nature of Japanese industry policy together with a strongly protected domestic economy. In 1971 a new industrial policy emerged designed to meet the problems of environmental pollution, urban congestion, and rural depopulation by shifting the focus of Japanese industry further towards high-technology, knowledge-intensive industries. In 1974 MITI published the first of its *long-term visions* of how the Japanese industrial structure ought to evolve to meet changed circumstances, both domestically and internationally. Appropriately in the Land of the Rising Sun, 'sunset' industries were to be scaled down and 'sunrise' industries encouraged. In effect, MITI used an industrial life-cycle model as the basis for deciding its strategic priorities (Figure 6.4).

Japanese economic policy has been strongly mercantilist. Within that framework, a key element has been the specific treatment of foreign direct investment which, for much of the post-war period, was extremely tightly regulated. The technological rebuilding of the Japanese economy was based on the purchase and licensing of foreign technology and *not* on the entry of foreign branches or subsidiaries. Although the inward investment laws have now been liberalized and

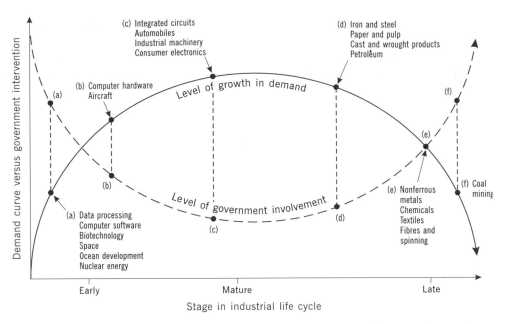

Figure 6.4 The relationship between Japanese government intervention and stage in the industry life cycle

Source: Okimoto, 1989: Figure 1.5

foreign firms do indeed operate within Japan, their relative importance remains quite small (see Chapter 3).

As far as outward investment by Japanese firms is concerned, the situation changed dramatically during the 1970s and 1980s, when Japanese overseas investment grew at a spectacular rate (Chapter 3). This was consistent with MITI's policy of internationalizing the economy and reflected the economically strategic role that Japanese overseas investment came to play. Two major developments in the external environment were especially important. The first was the upsurge in protectionist measures in North America and Europe in such industries as automobiles and electronics from the mid-1970s (see Chapters 11 and 12). The 'laser-like' targeting of Japanese industrial policy resulted in a major backlash from various Western economies. As a direct result of the erection of non-tariff barriers Japanese firms began to invest heavily in production facilities overseas. The second external stimulus to increased overseas direct investment was *endaka*; the major rise in the value of the Japanese yen which resulted from the 1985 Plaza Agreement among the Group of Five international finance ministers. This political decision stimulated an upsurge in overseas investment by Japanese firms to take advantage of lower production costs, particularly in East Asia.

Since the early 1990s, Japanese policy has been especially exercised by the problem of a high-value currency, with contentious trading relationships with the United States and Europe, and especially with the deep domestic recession which accompanied the collapse of the so-called 'bubble economy' at the end of the

1980s. Currently, there is much debate about whether Japan can recover its economic dynamism and what kind of role the government should play. Certainly attempts throughout the 1990s by successive Japanese governments to stimulate the domestic economy through fiscal mechanisms were not successful. Despite these seemingly intractable problems, however, we should not forget the fact that Japan is still the second largest economy in the world.

The newly industrializing and emerging market economies

Although they are frequently grouped together, the world's newly industrializing and emerging market economies are a highly heterogeneous collection of countries. They vary enormously in size (both geographically and in terms of population); in their natural resource endowments; in their cultural, social and political complexions. But they all tend to have one feature in common: the central role of the national state in their economic development (in the case of the former state-socialist economies, of course, the central role of the state was their defining characteristic).

Despite many popular misconceptions, none of today's NIEs is a free-wheeling market economy in which market forces have been allowed to run their unfettered course. They are, virtually without exception, *developmental* states; market economies in which the state performs a highly interventionist role.[13] Having said that, the precise role of the state – the degree and nature of its involvement – varies greatly from one NIE to another. In some cases, state ownership of production is very substantial; in others it is insignificant. In some cases, the major policy emphasis is upon attracting foreign direct investment; in others such investment is tightly regulated and the policy emphasis is upon nurturing domestic firms. Thus, although the recurring central theme which runs through the current economic behaviour of all NIEs is the role of the state, each individual NIE performs a specific variation on that general theme; a reflection of its specific historical, cultural, social, political and economic complexion.[14]

Types of industrialization strategy

Broadly speaking, a developing country may pursue one or more of three basic types of strategy:

- exports of indigenous commodities
- import-substituting industrialization (ISI) – the manufacture of products that would otherwise be imported, based upon protection against such imports
- export-oriented industrialization (EOI).

Which of these strategies is, or can be, pursued depends upon a number of factors: the economy's resource endowment (both physical and human); its size (particularly of its domestic market); its international context (especially the rate of growth of world trade and the policies of TNCs); and the attitude of the national government. For example, not all developing countries possess a natural resource endowment that could form the basis of a local processing industry. Even those that have such an asset may experience difficulty in setting up a local industry. Both developed country tariffs and also international freight rates tend to be higher on processed than on unprocessed materials. In addition, where TNCs are involved it may be corporate policy to locate processing operations elsewhere. Of course, neither Singapore nor Hong Kong, for example, had the material base to support such a strategy anyway.

The general pattern of industrialization beyond the commodity export phase has, with few exceptions, been one of an initial emphasis on import substitution followed eventually by a shift to export-oriented policies. The aim of import substitution is to protect a nation's infant industries so that the overall industrial structure can be developed and diversified and dependence on foreign technology and capital reduced. To this end, many of the policies listed in Figures 5.4, 5.6 and 5.8 have been employed. In particular, very high tariffs are imposed on those sectors chosen for protection. Import quotas and other devices, including licences, deposits and multiple exchange rates, are also used, as are incentives to encourage domestic production.

The import-substitution strategy, in theory, is a long-term *sequential* process involving the progressive domestic development of industrial sectors through a combination of protection and incentives:

- *Stage 1*: domestic production of consumer goods
- *Stage 2*: domestic production of intermediate goods
- *Stage 3*: domestic production of capital goods.

Invariably, the process begins with the heavy protection of domestic consumer goods industries to stimulate local production. As Figure 6.5 shows, tariffs on

Figure 6.5

Import-substitution policies: tariffs on consumer goods, intermediate goods, and capital goods in the early 1960s

Source: United Nations (1964) *Economic Bulletin for Latin America*, 9, 1: Table 5

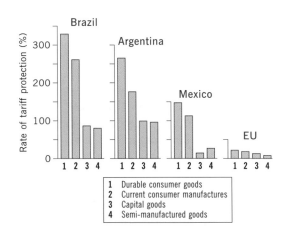

consumer goods in the major Latin American countries in the early 1960s were many times greater than the tariffs on intermediate and capital goods (and very much higher than EC tariffs). As a result, domestic consumer goods' production in at least some of the countries pursuing an import-substitution policy grew considerably, although much depended on the size of the domestic market. Although dependence on imported consumer goods certainly declined, however, dependence on the import of intermediate and capital goods – and, therefore, on foreign technology – increased. In most cases progression beyond the production of consumer goods did not occur to the extent anticipated.

Hence, various critics describe import-substituting industrialization as 'halfway' industrialization. The hoped-for domestic multiplier effects and the stimulus of a broader industrial structure do not necessarily occur. Where the domestic market is small, local production of consumer goods cannot achieve appropriate economies of scale so that domestic prices remain high. The necessarily high level of imports of intermediate and capital goods imposes balance of payments constraints. Yet there tend to be strong pressures from domestic vested interests – especially the protected consumer-goods manufacturers – against reducing the protection afforded to that sector in favour of other manufacturing sectors.

The realization that an import-substituting strategy cannot, on its own, lead to the desired level of industrialization began to dawn in a growing number of countries; some during the 1950s, rather more during the 1960s. Generally it was the smaller industrializing countries that first began to shift towards a greater emphasis on *export-orientation* because of the constraints imposed upon such a policy by their small domestic market. Increasingly, an export-oriented, outward-looking industrialization strategy became the conventional wisdom among such international agencies as the Asian Development Bank and the World Bank.

A shift towards export-based industrialization was made possible by such developments as:

- the rapid liberalization and growth of world trade during the 1960s
- the 'shrinkage' of geographical distance through the enabling technologies of transportation and communications
- the global spread of the transnational corporation and its increasing interest in seeking out low-cost production locations for its export platform activities.

Invariably, such export orientation was based upon a high level of government involvement in the economy. The usual starting point was a major devaluation of the country's currency to make its exports more competitive in world markets. The whole battery of export trade policy measures shown in Figure 5.4 was invariably employed by the newly industrializing economies. In effect, these amounted to a subsidy on exports that greatly increased their price competitiveness. Of course, the major domestic resource on which this export-oriented industrialization rests has been that of the labour supply – not only its relative cheapness but also its adaptability and, very often, its relative docility. Indeed, in many cases, the activities of labour unions have been very closely regulated.

In fact, the 'paths of industrialization' followed by individual NIEs have been rather more complex than this simple sequence suggests. Figure 6.6 sets out a five-phase sequence of industrialization based upon the experiences of the Latin American and East Asian NIEs. A number of important points can be made:[15]

- The distinction commonly drawn between inward-oriented Latin American industrialization strategies and outward-oriented East Asian industrialization strategies is misleading. 'While this distinction is appropriate for some periods, a historical perspective shows that each of these NICs has pursued *both* inward- and outward-oriented approaches ... rather than being mutually exclusive alternatives, the ISI and EOI development paths in fact have been complementary and interactive.'[16]
- The initial stages of industrialization were common to NIEs in both regions; 'the subsequent divergence in the regional sequences stems from the ways in which each country responded to the basic problems associated with the continuation of primary ISI.'[17]

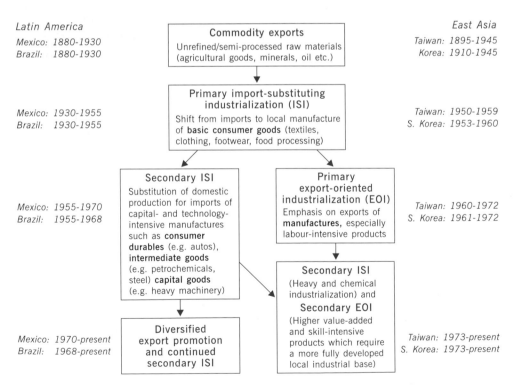

Figure 6.6 Paths of industrialization in Latin America and East Asia: common and divergent features

Source: Based on material in Gereffi, 1990: Figure 1.1 and p. 17

- 'The duration and timing of these development patterns vary by region. Primary ISI began earlier, lasted longer, and was more populist in Latin America than in East Asia ... The East Asian NICs began their accelerated export of manufactured products during a period of extraordinary dynamism in the world economy ... [after 1973] ... the developing countries began to encounter stiffer protectionist measures in the industrialized markets. These new trends were among the factors that led the East Asian NIEs to modify their EOI approach in the 1970s.'[18]
- Some degree of convergence in the strategies of the Latin American and East Asian NIEs began to occur in the 1970s and 1980s. Each 'coupled their previous strategies from the 1960s (secondary ISI and primary EOI respectively) with elements of the alternate strategy in order to enhance the synergistic benefits of simultaneously pursuing inward- and outward-oriented approaches.'[19]

The attraction of *foreign direct investment* has been an integral part of both import-substituting and export-oriented industrialization in many developing countries, although the extent to which particular countries have pursued this strategy varies considerably. In general, the Latin American NIEs have been more restrictive in their attitudes to foreign direct investment than the Asian NIEs, although there have been recent shifts towards more liberal investment policies in Latin America. In general, ownership requirements in Latin America have tended to be stricter than in most East Asian countries and the number of sectors in which foreign involvement is prohibited rather greater. Despite such differences in national attitude, most developing countries now compete fiercely to attract foreign firms using *labour* – its quantity, quality and relative cheapness – as their major bargaining chip. Those industrializing countries located close to major world markets such as North America or the EU, or which have special trading relationships with them, also emphasize their key locational attribute of proximity.

Export processing zones

Amongst all the measures used by developing countries to stimulate their export industries and to attract foreign investment, one device in particular – the export processing zone (EPZ) – has received particular attention.[20] An EPZ can be defined as:

> a relatively small, geographically separated area within a country, the purpose of which is to attract export-oriented industries, by offering them especially favourable investment and trade conditions as compared with the remainder of the host country. In particular, the EPZs provide for the importation of goods to be used in the production of exports on a bonded duty free basis.[21]

EPZs are, in effect, *export enclaves* within which special conditions apply:

- special investment incentives and trade concessions
- exemption from certain kinds of legislation

- provision of all physical infrastructure and services necessary for manufacturing activity – roads, power supplies, transportation facilities, low-cost / rent buildings
- waiver of the restrictions on foreign ownership, allowing 100 per cent ownership of export-processing ventures.

Within developing countries, EPZs have been located in a variety of environments. Some have been incorporated into airports, seaports or commercial free zones or located next to large cities. Others have been set up in relatively undeveloped areas as part of a regional development strategy. Figure 6.7 shows the geographical distribution of the major EPZs although, in fact, there are now several hundred EPZs in operation, with more being constructed.[22] Almost all of the EPZs in operation were established after 1971. Before the mid-1960s there were only two EPZs in the developing countries – in India and in Puerto Rico.

Numerically, some 90 per cent of all EPZs in the developing countries are located in Latin America, the Caribbean, Mexico and Asia. The Mexican EPZs are in the specific form of the *maquiladoras* related to the arrangement with the United States (see Chapter 3). In the Caribbean region, there are 35 EPZs in the Dominican Republic, 15 in Honduras and nine in Costa Rica, where around 30 per cent of the country's manufacturing employment is located in such zones. EPZs themselves vary enormously in size, ranging from geographically extensive developments to a few small factories; from employment of more than 30,000 to little more than 100 workers. In particular cases, like Penang, Malaysia, they can be very large indeed.

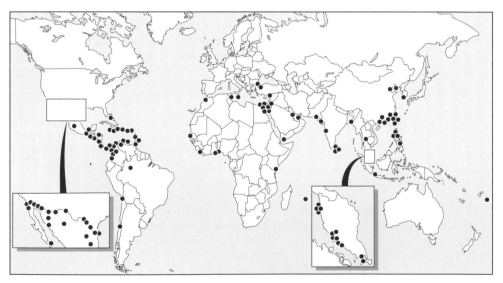

Figure 6.7 Export processing zones in developing countries

Source: Based on ILO, 1988: Table 20; UNCTAD, 1994: Table IV.2; press reports

Here, the number of plants in the Penang EPZ grew from 31 in 1970 to 743 in 1997 whilst employment in those EPZ plants grew from 3,000 to almost 200,000.[23] Total employment in developing country EPZs is well over 5 million.

Although EPZs in developing countries come in a great variety of sizes and bear the stamp of their specific national context, they also share many features in common. The overall pattern of incentives to investors is broadly similar, as is the type of industry most commonly found within the zones. The production of textiles and clothing and the assembly of electronics dominate. Almost half the total labour force in the Asian EPZs is engaged in the electronics industry. In the Mexican *maquiladoras*, 60 per cent of the workforce is employed in electrical assembly and a further 30 per cent in textiles and clothing. The characteristics of the labour force itself are similarly uniform, with a dominance of young female workers.

Having outlined some of the major features of NIE policies in general terms we need to acknowledge the substantial diversity that exists between individual countries. In order to do this, the following sections present a sketch of five NIEs: South Korea, Taiwan, Singapore, China and Mexico, together with a brief outline of developments in the emerging market economies of Eastern Europe.

South Korea[24]

South Korea (officially the Republic of Korea) came into being in 1948, following the partition of Korea into two parts. From 1910 to 1945, Korea had been a Japanese colony, very tightly integrated into the imperial economy. Between 1948 and 1988, when political liberalization occurred, South Korea was governed by a succession of authoritarian, military-backed and strongly nationalistic governments. These governments, particularly that led by Park Chung Hee (1961–79), operated a strong state-directed economic policy articulated through a series of five-year plans. As Figure 6.6 showed, the emphasis changed over time from primary ISI, through primary EOI, secondary ISI and secondary EOI. Two important developments during the 1950s helped to provide the basis for these strategies: the land reform of 1948–50, which removed the old landlord class and created a more equitable class structure, and the redistribution of Japanese-owned and state properties to well-connected individuals which helped to create a new Korean capitalist class.[25]

A powerful economic bureaucracy was created, with a key role played by a new Economic Planning Board (EPB). At the same time, the financial system was placed firmly in the hands of the state; the banks were nationalized, and the Bank of Korea brought under the control of the Ministry of Finance. This highly centralized 'state-corporatist' bureaucracy, in effect, 'aggressively orchestrated the activities of "private" firms'.[26] In particular, the state made possible – and actively encouraged – the development of a small number of extremely large and highly diversified firms – the *chaebol* – that continue to dominate the Korean economy.

By controlling the financial system, particularly the availability of credit, the Korean government was able to operate a strongly interventionist economic policy. The *chaebol* were consistently favoured through their access to finance (including the preferential allocation of subsidized loans) and very strong, long-term relationships were developed between them and the state. From the 1960s Korean policy had a strong sectoral emphasis as the state decided which particular industries should be supported through a battery of measures, including financial subsidy and protection against external competition. The precise sectoral focus changed over time, as Table 6.1 indicates. From an emphasis on consumer products during the primary import-substituting industrialization phase, the emphasis shifted, first, to chemicals, petroleum and steel and, subsequently, to automobiles, shipbuilding and electronics.

Access to modern technologies was a major need of the developing Korean economy and these, for the most part, had to be acquired from abroad. Like Japan at a similar stage in its development, Korea, for the most part, eschewed the channel of inward foreign investment for such technology transfer. Indeed, South Korea adopted the most restrictive policy towards inward foreign investment of all the four leading Asian NIEs. Until 1983 it operated strict rules on foreign direct investment that restricted the permitted level of foreign ownership, specified a minimum export performance and local content level. Korean government policy has been to build a very strong domestic sector. As a consequence, the share of FDI in the Korean economy has been extremely low (see Table 3.10).

Table 6.1 Changing sectoral focus in Korea's developmental strategy

Developmental phase	Major industries
Primary import-substituting industrialization	Food, Beverages, Tobacco, Textiles, Clothing, Footwear, Cement, Light manufacturing (e.g. wood, leather, rubber, paper products)
Primary export-oriented industrialization	Textiles and apparel, Electronics, Plywood, Wigs Intermediate goods (chemicals, petroleum, paper, steel products)
Secondary import-substituting industrialization and secondary export-oriented industrialization	Automobiles Shipbuilding Steel and metal products Petrochemicals Textiles and apparel Electronics Videocassette recorders Machinery

Source: Based on material in Gereffi, 1990: Table 1.6

Starting in the early 1980s, however, the emphasis of Korean economic policy shifted towards a greater degree of (restricted) liberalization. State control of the financial system was eased in 1983. The domestic market was opened up to a greater degree of imports. Inward foreign direct investment began to be encouraged following the 1984 Foreign Capital Inducement Law that greatly increased the number of manufacturing industries open to foreign investment. Some relaxation of the country's extremely stringent labour laws occurred. South Korea has had the most restrictive – often repressive – labour laws and practices of all the East Asian NIEs.[27]

In 1988, the military regime was replaced by a democratically elected government, although one which had not entirely lost the authoritarian habit. Some attempts were made to persuade the *chaebol* to change some of their practices, but with only limited success. Most significantly, in the mid-1990s Korea applied to join the OECD, membership of which exerted increased pressure on Korea to make major changes to its financial and economic system, especially to make its markets more open and to adopt a US-style neo-liberal system.

During the 1990s, in effect, much of Korea's traditional industry policy was diluted.[28] Major changes were made in policies of financial regulation, exchange rate management, and investment coordination. The formerly tightly controlled financial sector was significantly liberalized and the policy of exchange rate management virtually abandoned. The central pillar of South Korean industrial policy for 40 years – the coordination of investment – began to be dismantled.

When the East Asian financial crisis of 1997 hit Korea, the country's problems – as for the other affected East Asian economies – were attributed by the IMF, and by the Western financial community in general, to the existence of an over-regulated, state-dominated economy with excessively close (even corrupt) relationships between government and business. Yet, in the case of Korea, that was no longer entirely the case. It could be argued, in fact, that the Korean government had already gone too far in abandoning the principles on which its spectacular economic growth had been based. Clearly, certain reforms were needed as both the Korean economy itself and the broader global environment were changing. Not least was the need to reform the *chaebol*, which had come to distort the economy and which were, themselves, in great financial difficulty. That battle is still being fought. But 'while it is unwise to suggest that a return to the traditional model is possible and desirable, the country's headlong dash towards the Anglo-American institutional model, half voluntary and half under IMF pressure, does not seem particularly desirable, especially given the rather poor record that such reform programmes have produced in many developing and transition economies in the last two decades.'[29]

Taiwan[30]

Taiwan shares a number of common features with Korea. First, like Korea, Taiwan was a Japanese colony (from 1895 to 1945) and was tightly integrated into the Japanese economic system. A substantial industrial base and physical infrastructure was established by the Japanese to utilize local labour and materials; land reform was instituted. Second, Taiwan also has a difficult external political situation to face: the claim by the People's Republic of China (PRC) over Taiwan as an integral part of the mainland. Present-day Taiwan was established by the Kuomintang (KMT) regime fleeing from the mainland in 1949. Third, Taiwan has followed a broadly similar developmental path to that of Korea (compare Tables 6.1 and 6.2, for example). Both can be described as 'authoritarian corporatist' states. But there are also some significant differences between the Taiwanese and the Korean experiences. In fact, Taiwan shares some common features with Hong Kong, notably the massive influx of Chinese population (including many actual or potential entrepreneurs) from mainland China at the time of the communist revolution in 1949 and the greater importance of small entrepreneurial firms in the domestic economy. At the same time, Taiwan – like Korea – has operated strict labour laws (though in a less repressive manner).

The state-led Taiwanese development experience can be divided into three phases:[31]

- from 1945 to 1960
- from 1960 to the early 1970s
- the 1970s to the present.

Figure 6.8 shows the changing pattern of state leadership of different industries in Taiwan over these periods.

The period 1945 to 1960 was the 'Initial Nationalist Period' during which the incoming government, retreating from the communist forces on the mainland, acquired all the former Japanese-owned assets. 'Given the country's lack of raw

Table 6.2 Changing sectoral focus in Taiwan's developmental strategy

Developmental phase	Major industries
Primary import-substituting industrialization	Food, Beverages ,Tobacco, Textiles, Clothing, Footwear, Cement, Light manufacturing (e.g. wood, leather, rubber, paper products)
Primary export-oriented industrialization	Textiles and apparel, Electronics, Plywood, Plastics Intermediate goods (chemicals, petroleum, paper, steel products)
Secondary import-substituting industrialization and secondary export-oriented industrialization	Steel Petrochemicals Computers Telecommunications Textiles and apparel

Source: Based on material in Gereffi, 1990. Table 1.6

Figure 6.8

Changing levels of state control and leadership in Taiwan's industrial development

Source: Based on Wade, 1990b: Figure 9.1

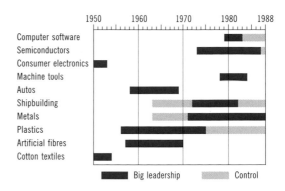

materials and a population then growing at over 3 per cent a year, raising living standards required labour-intensive manufacturing. Recapturing the mainland – which remained a central preoccupation of the government during the 1950s – required the development of some upstream industries. Over the 1950s the basis was laid for production of plastics, artificial fibres, cement, glass, fertilizer, plywood and many other industries, but above all, textiles.'[32]

The characteristic approach was for the state itself to set up such new upstream industries and then either to continue to operate them under state ownership or to transfer them to private entrepreneurs. A distinctive feature of Taiwan's development, compared with Korea, was a heavy direct involvement in production through state ownership. Textiles (building on the large number of immigrants with expertise in the industry), plastics and synthetic fibres formed the dominant focus of Taiwan's industrialization strategy in this initial period. In each case, the state played the initiating role. An explicit statement in the Second Four-Year Plan (1958–61) showed the government's determination to steer the direction of investment: 'the Government should positively undertake to guide and help private investments so that they do not flow into enterprises which have a surplus production and a stagnant market'.[33] There was, in this phase, a very strong emphasis on import-substitution.

From 1960 until the early 1970s, Taiwan concentrated on a dual developmental strategy of establishing new export sectors and continuing import-substitution. Much greater emphasis was placed on the development of heavy, capital goods industries. During this period, the level of state ownership of productive activities declined: from 56 per cent of total manufacturing output in 1952 to 21 per cent in 1970. Even so, the level of public ownership was considerably higher than in other East Asian NIEs.

The third period of Taiwanese policy (from the 1970s onward) was marked by several changes of emphasis. The world recession of the 1970s, together with the thawing of the relationship between the United States and China, necessitated an intensified emphasis on export orientation which, it was believed, required a high degree of state involvement. Sectoral priorities included, in particular, petrochemicals, electrical machinery, electronics, precision machine tools, computer terminals and peripherals. The government also addressed how such sectors should best be organized. 'Some subsectors were identified as suitable for development by local firms, others as requiring joint ventures with foreign companies and public enterprises (especially petrochemicals), and still others as suitable for a mix of foreign and local private firms (electronics).'[34] Following the oil price shock of the mid-1970s, the Taiwanese government initiated major investments in the heavy and chemical industries in order to reduce the country's exposure to external supply shocks.

Although an emphasis on new technology had existed since the 1950s, there was now an intensification of the drive to upgrade the educational and technological levels in the economy and to move the balance of the economy towards 'non-energy intensive, non-polluting, and technology-intensive activities like machine tools, semiconductors, computers, telecommunications, robotics, and biotechnology'.[35] In 1973, the government set up the Industrial Technology Research Institute (ITRI); in 1980 it established the Hsinchu Science-based Industry Park 'where foreign and domestic high-technology firms operate in close proximity to ITRI laboratories and where the government is willing to take up to 49 per cent equity in each venture'.[36]

Like Korea, Taiwan has controlled the inflow of foreign direct investment into the domestic economy, although rather less strictly. The foreign sector is important to the Taiwan economy (see Table 3.10) but far less so than for some of the other East Asian NIEs. Initially, Taiwan seems to have been less restrictive than Korea, but it became rather more selective after the 1970s. In particular, foreign firms have been discouraged from an involvement in labour-intensive industries but encouraged to invest in higher-level activities. Taiwan does, indeed, attempt to lure foreign investment into selected sectors but also operates several of the performance requirements shown in Figure 5.6.

Taiwan was far less seriously affected by the 1997 East Asian financial crisis than Korea. But Taiwan's contested political status casts a long shadow over its future. On the one hand, the 50-year rule of the authoritarian KMT ended in March 2000, when the Democratic Progressive Party (DPP) was elected into government. On the other hand, the DPP is a long-established advocate of independence from the PRC and this creates further potential tensions. In November 2001, Taiwan was admitted to the WTO as Chinese Taipei, with the status of a 'separate customs territory'.

Singapore[37]

Singapore is by far the smallest of all the East Asian NIEs. Like both Korea and Taiwan, Singapore had a very long history as a colony (within the British system). Unlike the two larger Asian countries, however, it was less tightly integrated into its imperial system, although it played a highly significant role as a commercial *entrepôt*, a reflection of its strategic geographical position. Singapore became fully independent in 1965 when it separated from Malaysia. Since then, although Singapore is a parliamentary democracy, it has been governed by one political party (the People's Action Party). For the first 30 years of its existence it was led by one powerful individual, Lee Kuan Yew.[38]

After separation from Malaysia the option of pursuing an import-substituting industrialization strategy was clearly a non-starter. Consequently, from the very beginning, the Singapore government pursued a very aggressive policy of export-oriented, labour-intensive manufacturing development. Concentration on manufacturing – especially labour-intensive manufacturing – was adopted because of the need to reduce a very high unemployment rate in a society that, at the time, had one of the fastest population growth rates in the world. The twin pillars of the policy were those of complementary economic and social planning, the latter being much more overt than in other East Asian NIEs.

The particular ways in which Singapore has operated its export-oriented policy have been substantially different from those of Korea and Taiwan. Most significantly, the central pillar was a strategy of attracting foreign direct investment. As a result, the Singaporean economy is overwhelmingly dominated by foreign firms, notably from Japan and the United States. As Table 3.10 showed, FDI accounts for over 90 per cent of the country's GDP. The most explicit industrialization measures, therefore, were those of incentives to inward investors, using a sectorally selective process, with particular attention being devoted initially to electronics, petroleum and shipbuilding. The government agency responsible was the Economic Development Board (EDB), which still plays an extremely influential role in the Singapore economy. With a few exceptions, Singapore operated a free port policy, with little use of trade protectionist measures. The second set of direct measures used to promote industrial development was the establishment of a high-quality physical infrastructure.

At the same time, a series of social policy measures was introduced aimed at creating an amenable environment for foreign investment. Major housing programmes, partly funded through the state's compulsory savings scheme (the Central Provident Fund) were undertaken. More specifically, the government effectively incorporated the labour unions into the governance system by establishing a National Trades Union Council. 'Strikes and other industrial action were declared illegal unless approved through secret ballot by a majority of a union's members. In essential services, strikes were banned altogether … These labour market regulations resulted in the creation of a highly disciplined and depoliticised labour force in Singapore.'[39] Thus, through a whole battery of interlocking policies, the Singapore government has created a very high growth, increasingly affluent, industrialized society in which foreign firms have played the dominant economic role in production but within a highly regulated political and social system.

Mainly in response to the unexpected (but, in fact, short-lived) economic shock of 1985 when, for the first time since independence, Singapore experienced a decline in GDP growth, the government initiated a new policy thrust. A major feature of the 'New Direction' for the Singapore an economy was to be a reduced dependence on the manufacturing sector and a shift towards the business services sectors. Incentive packages were introduced to attract foreign firms prepared to set up service operations in Singapore. A specific part of that policy was the Operational Headquarters Scheme that aimed to encourage TNCs to locate their high-end R&D, administrative and management activities there.

Singapore sets out to market itself as a global business centre on the basis of the very high quality of its physical and human infrastructure and its strategic geographical location. Government policies are geared towards this goal that also includes an explicit strategy to 'regionalize' the Singaporean economy by encouraging domestic firms to set up operations in Asia while Singapore develops as the 'control centre' of a regional division of labour. One element in this is the promotion of the Singapore–Johor Bahru (Malaysia)–Batam (Indonesia) Growth Triangle. More ambitiously, the government is driving a series of initiatives using government-linked corporations to develop major infrastructural projects in Asia and, more broadly, to develop international networks.[40] At the same time, the emphasis on research and development and technological upgrading continues.

As Singapore entered the third millennium, its government faced a number of major challenges. One set of challenges arises from the country's position in south-east Asia. What has so far been a major source of strength looks less so in the aftermath of the Asian financial crisis and in the light of the potential shift in the regional centre of gravity towards the north-east, notably China. Although Singapore was far less directly affected by the 1997 crisis than its neighbours it has suffered indirectly from the decline in their economies and from the political uncertainties in Indonesia. More problematical in the longer run is the emergence of China as East Asia's dominant economy (other than Japan). The relative shift in patterns of foreign direct investment towards China (see Chapter 3) poses a major threat to Singapore, whose economy has been based on inward FDI above all else.

Two policy initiatives have been developed by the Singaporean government to meet these challenges (in addition to those already in place). First, Singapore's financial system has been significantly reformed and liberalized. Second, Singapore has taken the lead in pushing for greater 'Asian regionalism'. The other set of challenges are internal to Singapore itself and concern the extent to which this highly paternalistic state is able to loosen its grip on the country's political and social life without damaging its economic influence.

China[41]

In discussing the development of South Korea, Taiwan and Singapore (as well as the other emerging NIEs in Asia) it is always necessary to emphasize the key role played by the state. In the case of China, of course, such an assertion is unnecessary. As a centrally controlled command economy there is no doubt about the state's centrality. The point about China is that, after several decades of self-imposed separation from the world economy, it has become an immensely significant regional – and increasingly a global – player. What makes China so significant, in the long run, is its sheer size – some 1.2 billion people – and its massive economic potential. Whether that potential will be realized, and how far China will come to dominate the Asian regional economy, is a matter for speculation. What is important, here, is to outline the dramatic changes in Chinese policy towards the rest of the world since 1979.

The People's Republic of China (PRC) came into being in 1949 with the replacement of the nationalist government by a communist government led by Mao Zedong. For the next 30 years, China followed a policy of economic self-reliance. This policy was pursued through a series of major – often extreme – initiatives.

Initially, the new government followed the example of the Soviet Union in establishing a Five-Year Plan (1953–57). This relatively successful policy was jettisoned in 1958 when Mao announced the 'Great Leap Forward' – a total transformation of economic planning with the emphasis on small-scale and rural development. Although this initiated the notion of rural industrialization, the GLF was disastrous in its consequences, with mass famine one of the results. In 1966, policy changed again with the introduction of the 'Cultural Revolution', a phase that lasted for some ten years with, again, disastrous human and social implications.

The period after Mao's death in 1976 was one of political hiatus that was eventually resolved by the emergence of Deng Xiaoping as leader. It was under Deng's leadership that China began to jettison the self-reliance policy of the previous 30 years and to make links with the world market economies. This has been done, however, without substantial political change. In the words of the new Party Constitution of 1997, it is 'Socialism with Chinese characteristics'. The draconian response to the Tiananmen Square demonstrations in May–June 1989 was one reflection of what that can mean.

The pivotal year was 1979, when China began its 'open policy' based upon a carefully controlled trade and inward investment strategy. This was set within the so-called 'Four Modernizations' (concerned with agriculture, industry, education, and science and defence). A central element of the new policy has been the opening up of the Chinese economy to foreign direct investors. As we saw in Chapter 3, FDI has grown very rapidly indeed in China since the early 1980s and now accounts for 30 per cent of GDP (Table 3.10), although a large proportion of this FDI emanates from Hong Kong (since 1997 part of China under the 'one country, two systems' arrangement) and Taiwan. The organizational form of these investments varies from wholly owned foreign subsidiaries to equity joint ventures with Chinese partners and other partnership arrangements.

The most distinctive feature of the open policy, however, is the explicit use of geography in its implementation. Partly in order to control the spread of capitalist market ideas and methods within Chinese society, and partly to make the policy more effective through external visibility and agglomeration economies, FDI has been steered to specific locations. Initially, the foci were the four Special Economic Zones (SEZ)[42] established in 1979 at Shenzhen, Zhuhai, Shantou and Xiamen (Figure 6.9). Apart from the desire to attract United States, Japanese and European investment,

> the four original zones were located with a view to maximizing their attraction to investment from ethnic Chinese living outside China. Shenzhen, the largest, was located in Guangdong province immediately adjacent to Hong Kong. Zhuhai was set up beside the Portuguese enclave of Macau, also in Guangdong. Shantou, in the north-east of Guangdong province was established in an area with many links with South-east Asian Chinese communities and Xiamen SEZ in Fujian province was intended to attract Taiwanese investors.[43]

The Chinese SEZs share some features in common with EPZs, although they are generally much larger. They offer a package of incentives, including tax concessions, duty-free import arrangements and serviced infrastructure.

Certainly, the focus on southern Guangdong province was also related to the incorporation of Hong Kong and Macau into the PRC in the late 1990s. Well before the political control of Hong Kong reverted to China, Hong Kong businesses had set up thousands of factories in the municipalities just across the border. The original SEZs were also located in areas well away from the major urban and industrial areas in order to control the extent of their influence. However, as Figure 6.9 shows, there has been considerable development of other kinds of externally oriented locations in the form of 'open coastal cities', 'priority development areas' and export-processing zones (a list of 15 new EPZs was announced on a trial basis in June 2000). The most spectacular case is the explosive growth of Shanghai (notably the Pudong district) during the 1990s.

Despite these inflows of foreign capital and technologies, China remains a centrally controlled command economy in which state-owned enterprises (SOEs) predominate, at least in employment terms although not in terms of their share of industrial output. The SOE share of China's industrial output fell from 75 per cent in the late 1970s to around 28 per cent in 2000. But SOEs still account for about 44 per cent of China's urban employment and 70 per cent of government revenues.[44] Reform of the SOEs is an immense task and one surrounded by massive controversy. A major problem for a country trying economically to 'modernize' is the sheer inefficiency (by Western standards) and high levels of corruption of the

Figure 6.9

The geography of China's 'open policy'

Source: Based on Phillips and Yeh, 1990: Figure 9.4; Chinese government sources

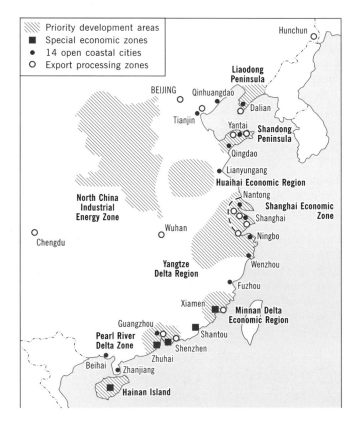

SOEs. SOEs are embedded within the communist party system and this fact pervades their operations.[45] A survey carried out by the Shanghai Stock Exchange in 2000 found that 'in listed companies 99 per cent of the main business and staffing decisions, including board appointments and salaries, are made with the approval of internal party committees'.[46]

The problems posed by the SOEs will be intensified with China's accession to the WTO (approved 1 November 2001). Although this, arguably, greatly enhances China's economic potential it also imposes severe stresses on the domestic economy and institutions. Not only will tariff levels fall from their currently high levels, thus exposing Chinese enterprises to intense competition, but also non-tariff barriers, matters relating to intellectual property rights, safety regulations, financial and telecommunications regulations will all be affected. Some 170 domestic laws are to be reformed by the Chinese government. 'However, a mass of local legislation, much of which contradicts national laws, remains in place.'[47]

Finally, we must not forget the problems posed for the Chinese state by the sheer geographical immensity of the country. It is no coincidence that China's economic policies have had a strong element of geographical concentration (Figure 6.9). The bulk of investment spending has gone, so far, to a relatively small part of the country, primarily on, or near, the coast. Despite two decades of economic reform in China, immense regional inequalities in economic well-being remain. As a consequence, in May 2000, the Chinese government announced that it was to divert most of its investment spending to the deprived western provinces in a new 'Go West Strategy'.[48]

Overall, it is clear that the institutional – and geographical – structure of the Chinese economy is in a state of flux, with much increased variety of forms. As in the past, however, the key lies in the internal political power struggles between the 'modernizers', who wish to sustain and develop the open policies of the recent past, and those who wish to retain a degree of isolation. Following the death of the initiator of the open policy, Deng Xiaoping, in 1997, his successor Jiang Zemin continued the broad open policy direction. But the key test of the survival of such a policy is its continued success in delivering economic growth and raising incomes. As China becomes more fully integrated into the global economy its freedom of manoeuvre will be reduced.

Mexico[49]

Our next example of national industrialization policies is drawn from a different part of the world: the Americas. Following the financial collapse of 1982 – when Mexico's economic turmoil precipitated the international debt crisis – Mexico has undergone dramatic political, social and economic change as the state has attempted to integrate the national economy more strongly into the global economy. But the path of transition from a strongly inward-oriented industrialized policy position to an export-oriented position has been far from smooth.

Two basic characteristics are important to an understanding of the Mexican case. The first is its location next door to the world's dominant political and economic power, the United States. The second is the fact that Mexico was governed, from 1929 to 2000, by a single party (the PRI). However, reformist pressures both inside and outside the governing party intensified, particularly from the 1980s. Thus,

> what was once seen as a predictable and stable political system became the arena for new tensions and conflicts. The onset of economic crisis coincided with, and in part, led to the emergence of a new governing elite of young technocrats, the independent mobilization of a new business class ... the growth of grassroots popular movements and the unification of the left ... All these developments implied a serious erosion of corporatist ties between state and society which had underpinned Mexican stability since the 1930s.[50]

For almost 60 years from 1929, Mexico pursued a predominantly import-substituting industrialization policy. This policy was, for the most, part quite successful. The large domestic market and a strong physical and human resource base, together with proximity to the US market, allowed Mexico to grow at high rates, especially during the period of the 1950s to 1970s. The high returns from the country's oil reserves, in particular, had a major effect on the economy. It was the collapse of oil prices in 1981 that precipitated the country's financial crisis. One component of Mexico's export policy during the import-substitution phases was the *maquiladora* programme that created a very specific form of industrial growth along the Mexico–United States border (see Figure 3.29).

Between 1982 and 1985, government policy emphasized stabilization rather than structural adjustment but external pressures from the IMF resulted in a pronounced shift in policy emphasis. 'This round of crisis and reconciliation led to important shifts within the Mexican administration. Economic difficulties had already strengthened the hand of the technocrats within the administration ... the stabilization efforts of 1985–86 were accompanied by the initiation of trade reform and negotiations that led to accession to GATT. In the Solidarity Pact of 1988 ... the government used the much-needed stabilization programme to counter strong resistance from the private sector to further trade policy reform. The government cut the maximum tariff level from 100 per cent to 20 per cent, the average tariff level dropped to just over 10 per cent.'[51]

During the second half of the 1980s, the Mexican government pursued a wide-ranging programme of liberalization involving both the deregulation of some areas of the economy and the privatization of state-owned enterprises. By 1993, some 90 per cent of Mexico's state-owned enterprises had been privatized, either wholly or partially, although the government retained control of some key companies, including PEMEX, the state oil company. The single most important act of the Salinas administration, which came to power in 1988, was to take Mexico into the NAFTA. This symbolized the government's headlong attempt to tie the domestic economy firmly into the global system and, in particular, to gain greater benefits from a formal association with the United States. One result of the major domestic reforms and the opening up of the economy was a very rapid build-up of speculative foreign capital ('hot money') that drove up the value of the peso to unsustainable levels. In December 1994, the new Zedillo administration was forced into a massive currency devaluation that heralded a new economic crisis for the country.

The most significant recent change for Mexico's economy has been political: the election in July 2000 of Vicente Fox of the PAN as President of Mexico, thus ending 71 years of control by the PRI. Fox, a former executive of Coca-Cola, set out an ambitious agenda of internal reform (to begin to remove the decades of corrupt practices) and of further opening Mexico to the global economy. Fox put forward proposals to improve Mexico's position within the NAFTA and to address the perennial problem of illegal migration to the United States (both of which will require considerable changes of attitude in the United States itself). He also entered into a free trade agreement with the European Union. For the first time in many years, Fox's accession to the presidency was not accompanied by the usual financial crisis. But the presidential term of office in Mexico is six years with no second term. It remains to be seen whether the extremely high expectations created by the defeat of the PRI can be delivered, especially as global – and US – economic growth began to slow down significantly in 2001 (see Chapter 3).

One of the major problems that undoubtedly needs to be addressed by the government is Mexico's relatively weak technological and skills base.

> In general there was a tradition of excessive technological dependence in Mexican industry in comparison to countries like the Republic of Korea or Taiwan Province of China. There was a high level of reliance on imported capital goods for a country of its industrial size and sophistication. This was accompanied by a similar and widespread reliance on inflows of foreign know-how, licences and expertise through much of industry. Despite the nationalistic stance of the government, there was a relatively large presence of foreign multinational enterprises in the advanced sectors of Mexican industry … It restricted the ability of Mexican industry to move into technologically more dynamic or sophisticated industries. While the Republic of Korea used import substitution to foster industrialization, it also pursued a strategy of independent industrial technological development. Mexico was never as export-oriented, nationalistic, State-led and technologically ambitious as the Republic of Korea.[52]

The emerging market economies of Eastern Europe[53]

One of the biggest – and most unexpected – political events of the late 1980s was the sudden unravelling of the formerly monolithic state-socialist systems of the Soviet Union and Eastern Europe. For more than 70 years in the case of the Soviet Union, and for more than 40 years in Eastern Europe, these economies had been organized and controlled by the central state. The rigid and all-embracing hierarchical system of centralized economic planning, that evolved in the 1920s and 1930s in the Soviet Union, formed the basis of the post-war industrial development both of the Soviet Union itself and also of the Eastern European countries that came under Soviet domination. Figure 6.10 shows how the centralized economic planning system worked. Regionally, the economies were integrated within the framework of the Council for Mutual Economic Assistance (CMEA) established in 1949.

The system was organized on the basis of five-year plans and the allocative mechanism of materials balances. Initially, each of the satellite countries' economic systems was organized as an individual microcosm of the Soviet system. The principles of national self-sufficiency and the development of heavy industry were extended throughout the CMEA countries. One result was a considerable degree of duplication. For example, each of the CMEA countries built up large-scale iron and steel production. Little attempt was made to develop intra-regional specialization on the basis of individual countries' comparative advantage.

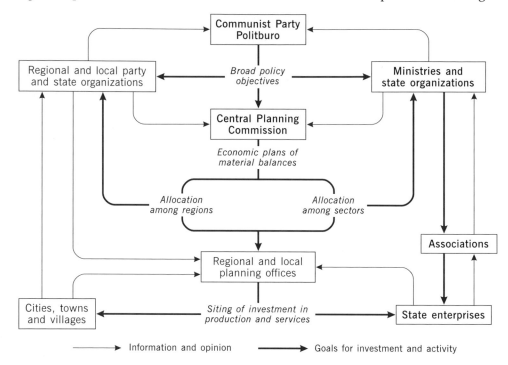

Figure 6.10 The centralized economic planning system in the state-socialist economies of the Soviet Union and Eastern Europe before 1989

Source: Based on Hamilton, 1979: Figure 5

During the mid-1960s, therefore, a number of economic reforms were carried through, although the process and extent of reform was extremely uneven. Apart from Yugoslavia (which had earlier opted out of the Soviet-dominated system), the most substantial reforms occurred in Hungary under its New Economic Mechanism introduced in 1968. In the Hungarian case, much emphasis was placed on export-oriented industrial development; foreign trade both within and, increasingly, outside the CMEA became a key state objective. By the 1980s, the rigid centralized planning system that originated in the Soviet Union, and was imposed on the Eastern European countries after 1945, had become far less rigid, although to a varying degree from one country to another. Nevertheless, the system remained, in essence, a centrally planned system in which the state set the overall goals and objectives and the bounds to what was allowed in terms of enterprise autonomy and foreign involvement.

In 1989, following Mikhail Gorbachev's 'revolution from the top' in the Soviet Union,[54] the entire state-socialist system in Eastern Europe began to unravel. Of course, China had already undertaken major steps to open its economy to the global market in 1979 (see above) but there the communist state apparatus remained in place. In the Soviet Union *both* the political *and* the economic system collapsed with repercussions that spread rapidly throughout the other CMEA countries.

> With the zeal of religious converts, most of the countries that had languished behind the Iron Curtain set about transforming themselves into market economies. They were not short of Western advisers, most of whom saw the collapse of the command economies as vindication of the Washington Consensus. Unlike Latin America or Asia, most of the new 'transition economies', as they became known, were not dragging themselves up from impoverished and overwhelmingly rural roots. Most already had some sort of industrial framework, however rusty, and has assured at least the basic needs of the majority of their citizens. *But the challenges they faced were breathtaking – it was not just a question of changing a few policies but of creating entirely new financial institutions and a completely fresh system for generating and allocating economic resources.*[55]

Among the many 'breathtaking' problems were those relating to property rights (the ownership of productive assets), how – and how quickly – former state-owned enterprises were to be transformed into privately owned enterprises (the creation of a previously non-existent entrepreneurial class, the extent and nature of the involvement of foreign capital, the position of labour, the nature of the legal system, the reconciliation of market forces with social goals).

Although the problems facing the so-called transition economics of Eastern Europe were essentially the same in each case, the precise way they are being approached varies, depending on their individual histories before and during the era of centralized control.[56] So, for example, whereas Hungary privatized much of its economy fairly early in the transition process (including the banking and energy sectors), the process of privatization of such sectors and of telecommunications has been slower in Poland. On the other hand, the liberalization of markets in Poland created an incredible growth of new companies (some 3 million). Of course, the processes of economic reform and 'marketization' have been stimulated further by the desire of Eastern European countries to become members of the European Union (see Chapter 5).

Conclusion

The aim of this chapter has been to outline the diversity of ways that states occupying different positions within the global economy, and having different political-ideological stances, have attempted to govern their national economies. Despite the undoubted shift towards a greater degree of 'marketization' of virtually all national systems since the 1980s (through processes of deregulation and privatization), very significant variety of state involvement in the economy persists. Convergence is limited by the interaction between the specific histories of individual states (their 'path-dependency') and economic-political processes operating at the global scale. The result is a mosaic of specific national systems displaying elements of both convergence and diversity.

Thus, there is a clear and continuing difference between the United States on the one hand and the countries of continental Europe on the other. The United

Kingdom occupies a rather uneasy position between the two, being closer to the United States in some respects but closer to continental Europe in others. Even though the existence of the EU provides a common regulatory framework for its member states, significant variety still exists in the precise manner of national-state involvement in individual EU economies.

The same point of continuing diversity, rather than homogeneity, can be made of both the East Asian NIEs and the emerging market economies of Eastern Europe. The East Asian NIEs, in particular, have too often been grouped together under the umbrella of an 'East Asian development model' whereas, as we have seen, the mode of state involvement has differed widely between them. A similar point can be made about the newly emerging market economies of Eastern Europe. Despite the homogenizing pressures of the former Soviet-dominated, state-socialist system that was in place from the late 1940s to the early 1990s, the legacy of earlier national histories is influencing the particular form of marketization in these countries.

Notes

1 Cerny (1991: 174).
2 See Gilpin (2001: ch. 7).
3 See Krugman (1986, 1990); Ostry (1990); Richardson (1990); Tyson (1993); Yoffie (1993).
4 Ostry (1990: 60).
5 *The Financial Times* (14 February 2001).
6 *The Economist* (5 June 1999).
7 *The Economist* (26 May 2001).
8 Gretschmann (1994: 471).
9 *The Financial Times* (2 May 2001).
10 Accounts of Japanese economic policy are provided by Dore (1986), Johnson (1982, 1985) and Okimoto (1989). Porter, Takeuchi and Sakakibara (2000) provide a strong critique of Japanese government policy.
11 Magaziner and Hout (1980: 29).
12 Porter, Takeuchi and Sakakibara (2000).
13 The chapters in Gereffi and Wyman (1990) analyse the diverse 'paths of industrialization' in Latin America and East Asia. See also Brohman (1996); Stallings (1995, Part II).
14 Douglass (1994: 543).
15 Gereffi (1990: 18).
16 Gereffi (1990: 18).
17 Gereffi (1990: 21).
18 Gereffi (1990: 21).
19 Gereffi (1990: 22).
20 See Chant and McIlwaine (1995); ILO (1988, 1998); Sklair (1989).
21 UNIDO (1980: 6).
22 ILO (1998).
23 ILO (1998).
24 Especially useful accounts of South Korean industrialization policy are by Amsden (1989), Koo and Kim (1992), Wade (1990a,b). Chang (1998a) discusses the 1997 financial crisis.
25 Koo and Kim and (1992).
26 Wade (1990a: 320). See also Amsden (1989).
27 Deyo (1992).
28 See the detailed analysis provided by Chang (1998a).
29 Chang (1998a: 230).

30 This account of Taiwan's developmental strategies is based primarily on Wade (1990a,b). See also Lall (1994); Whitley (1999).
31 Wade (1990a).
32 Wade (1990a: 77).
33 Quoted in Wade (1990a: 81–2).
34 Wade (1990a: 96).
35 Wade (1990a: 98).
36 Wade (1990a: 98).
37 Singapore's developmental policies are discussed by Lall (1994), Lim (1988), Lim and Pang (1986), Ramesh (1995), Rodan (1991) and Yeung (1998).
38 Lee Kuan Yew's autobiography (Lee, 2000) provides a unique (though certainly not unbiased) account of Singapore's rapid economic development.
39 Yeung (1998: 392).
40 The 'regionalization' strategy of Singapore is discussed in detail by Yeung (1998, 1999b).
41 Useful accounts of Chinese economic development policy are provided in Benewick and Wingrove (1995), Crane (1990), Nolan (2001) and Wong, Lau and Li (1988).
42 The Special Economic Zones are analysed by Crane (1990), Phillips and Yeh (1990), Thoburn and Howell (1995) and Wong and Chu (1995).
43 Thoburn and Howell (1995: 173).
44 *The Economist* (30 September 2000).
45 Nolan (2001).
46 *The Financial Times* (2 July 2001).
47 *The Financial Times* (7 May 2000).
48 *The Financial Times* (8 May 2000).
49 Mexican industrialization policies are discussed in Haggard (1995), Harvey (1993), Lall (1994), Sklair (1989) and Villareal (1990).
50 Harvey (1993: 4).
51 Haggard (1995: 81).
52 Lall (1994: 81).
53 Useful discussions of the transition from state-socialism to market-capitalism in that region are provided by Czaban and Henderson (1998), Grabher (1995), Myant et al. (1996), Offe (1996), Tudor (2000) and Whitley (1999).
54 Offe (1996: 30).
55 Tudor (2000: 39; emphasis added).
56 Whitley (1999: 209).

CHAPTER 7

Transnational Corporations: The Primary 'Movers and Shapers' of the Global Economy

The significance of the TNC

More than any other single institution, the transnational corporation (TNC) has come to be regarded as the primary shaper of the contemporary global economy. It has been the rise of the TNC – especially of the massive 'global' corporation – that is seen to pose the major threat to the autonomy of the nation-state. As shown in detail in Chapter 3, there has not only been a massive growth of foreign direct investment (FDI), but also the sources and destinations of that investment have become increasingly diverse. But FDI is only one measure of TNC activity. Because the FDI data are based on ownership of assets they do not capture the increasingly intricate ways in which firms engage in transnational operations through various kinds of collaborative ventures and through the different modes through which they coordinate and control transactions within geographically dispersed production networks. It is for this reason that we adopt a much broader definition of the TNC than that normally used in the conventional literature:

> A transnational corporation is a firm that has the power to coordinate and control operations in more than one country, even if it does not own them.

The significance of the TNC lies in three basic characteristics:

- its ability to coordinate and control various processes and transactions within production networks, both within and between different countries
- its potential ability to take advantage of geographical differences in the distribution of factors of production (for example, natural resources, capital, labour) and in state policies (for example, taxes, trade barriers, subsidies, etc.)
- its potential geographical flexibility – an ability to switch and to re-switch its resources and operations between locations on an international, or even a global, scale.

Hence, much of the changing geography of the global economy, is shaped by the TNC through its decisions to invest, or not to invest, in particular geographical locations. It is shaped, too, by the resulting flows – of materials, components, finished products, technological and organizational expertise – between its geographically dispersed operations. Although the relative importance of TNCs varies considerably – from sector to sector, from country to country, and between different parts of the same country – there are now very few parts of the world in which TNC influence, whether direct or indirect, is not important. In some cases, indeed, TNC influence on an area's economic fortunes can be overwhelming.

However, we should not assume that TNCs are all of a kind: identical economic monsters that stamp an identical footprint on the landscape. On the contrary, TNCs are highly differentiated not only in size and geographical extent but also in the ways in which they operate. Far from being the 'placeless' organizations often claimed, TNCs continue to reflect many of the basic characteristics of the home country environments in which they remain strongly embedded, despite the growing extent of their transnational operations.

The aim of this, and the following, chapter is to explore the activities of TNCs, particularly in the context of the increasingly complex networks of interrelationships within which all TNCs are embedded. Indeed, the concept of the *network* underlies the whole of the discussion in these two chapters. TNCs, like all social organizations, are complex networks embedded within broader network structures. The chapter is organized into three parts:

- First, we outline some of the broad theoretical explanations of *why* firms should attempt to transnationalize their activities beyond exporting their products to foreign markets.
- Second, we address the question of *how* firms transnationalize their activities in an organizational sense: the kinds of organizational architectures they have evolved to implement their competitive strategies.
- Third, we confront the commonly held view that TNCs are becoming 'placeless' – the argument that firms are converging towards a universal business model.

Why (*not*) transnationalize? Some general explanations

TNCs are essentially capitalist enterprises. As such they must behave according to the basic 'rules' of capitalism, the most fundamental of which is the drive for *profit*. Of course, business firms may well have a variety of motives other than profit, such as increasing their share of a market, becoming the industry leader, or simply making the firm bigger. But, in the long run, none of these is more important than the pursuit of profit itself. A firm's profitability is the key barometer to its business 'health'; any firm that fails to make a profit at all over a period of time is likely to go out of business (unless rescued by government or acquired by another firm). At best, therefore, firms must attempt to increase their profits; at worst, they must defend them.

The drive is, increasingly, for *global profits*. Expressed in the simplest terms, profit (P) is the difference between the revenue (R) which a firm receives from selling its products and the cost (C) of producing and distributing the firm's goods or services: P = R – C. Obviously, therefore, profit can be increased either by raising R or reducing C, or by a combination of the two. The transnationalization of a firm's operations may be motivated by either or both of them. The process is a totally logical extension of the firm's 'normal' mode of expansion: from local, to regional, to national and then to global scales of operation. This does not mean, of course, that all firms will inevitably transnationalize. Some will find themselves unable to do so because of either internal or external constraints; others may choose not to do so and to remain within a limited geographical niche.

Of course, a capitalist market economy is an intensely *competitive* economy. One firm's profit may be another firm's loss unless the whole system is growing sufficiently strongly to permit all firms to make a profit. Even so, some will make a larger profit than others. Two key features of today's world are: first, competition is increasingly global in its extent and, second, such competition is extremely volatile. 'This creates an environment of hypercompetition – an environment in which advantages are rapidly created and eroded.'[1] Firms are no longer competing largely with national rivals but with firms from across the world.

In this section, we outline two broad theoretical approaches. The first is a *macro-level* approach based upon the nature of the capitalist system itself. The second approach is set at the *micro-level* of the firm.

A macro-level approach: internationalization of the circuits of capital[2]

One of the most useful attempts to explain the internationalization of economic activity at the general level is based upon the concept of the circuits of capital. This, in turn, is embedded within Marx's conceptualization of the capitalist system as a whole. From this perspective, the internationalization of economic activity and its major vehicle, the TNC, can be regarded simply as being part of the normal expansive process of capitalist development. The 'laws of motion' of capitalism emerge from the drive to enhance profits and to accumulate capital through increasing appropriation of surplus value from the process of production. To Marxist writers, the basis of the extraction of surplus value, or profit, is the exploitation of human labour power by capitalist firms that own the means of production. The internationalization of production, from this perspective, is the extension of the system of labour exploitation and class struggle to a global scale.

The capitalist economic process can be envisaged as a continuous circuit, basically a very simple idea, as Figure 7.1(a) shows. Its essence is that 'the money at the end of the process is greater than that at the beginning and the value of the commodity produced is greater than the value of the commodities used as inputs.'[3] Thus, money (M) is used to purchase 'commodities' (C) in the form of raw materials and labour. These inputs are transformed in the process of production (P) and acquire increased value (C'). When exchanged for money (M') this increased value can be used to purchase a further round of inputs for the production process, and so the circuit continues.

The basic circuit of capital shown in Figure 7.1(a) can be expanded into three distinct, but completely interconnected, circuits: money capital, productive capital and commodity capital. The important point is that each of these three circuits of capital has become progressively *internationalized*.

- The *circuit of commodity capital* was the first of the three circuits to become internationalized, in the form of world trade.
- The *circuit of money capital* was the second to become internationalized, in the form of the flow of portfolio investment capital into overseas ventures.
- The *circuit of productive capital* was the most recent to become internationalized, in the form of the massive growth of transnational corporations and of international production.

A major merit of this 'circuits of capital' approach to the internationalization of economic activity is that it emphasizes the totally *interconnected* nature of finance,

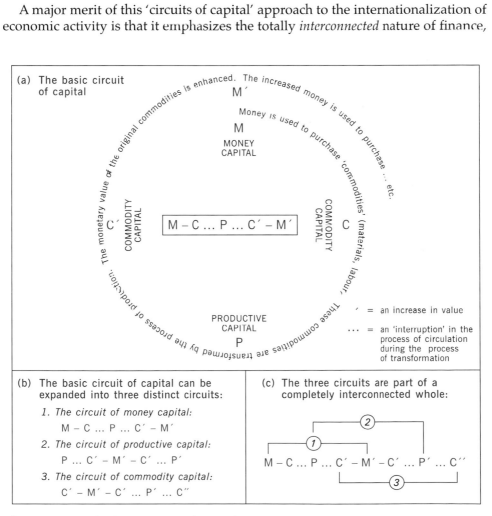

Figure 7.1 The basic circuit of capital

production and commodity trade. But it is insufficiently specific to deal with such questions as the precise form (geographical, organizational, sectoral) of transnational corporate activity. This is certainly not to deny the value of the circuits of capital approach even to non-Marxists, who can readily accept the broad framework it provides even if they do not necessarily subscribe to the entire explanatory package, particularly the overwhelming emphasis on class struggle. It serves to remind us that, in trying to explain the globalization of economic activity, we are dealing with the workings of a dynamic capitalist market *system*, and not just individual agents within it.

Micro-level approaches: the search for an integrative framework

An alternative approach to understanding the internationalization of economic activity through the TNC is to take a *firm-specific*, rather than a general-system, view. Before the early 1960s, there was no adequate theory of the TNC at the firm-specific level. Since then numerous writers have contributed to this mode of explanation.[4] However, there was a strong tendency in the past to base such explanations on a stereotypical TNC – the very large, mostly American enterprise – and theoretical conclusions were extrapolated to apply to all TNCs, whatever their geographical origin. But TNCs from different source nations and of different sizes differ considerably in their characteristics and their behaviour (see the third section of this chapter). Any satisfactory explanation, therefore, must encompass the diversity of the world's TNC population. However, there is no single accepted micro-level theoretical explanation. Here, we outline three of the more important contributions: by Stephen Hymer, Raymond Vernon and John Dunning.

Stephen Hymer: the undisputed pioneer[5]

Before Hymer's pioneering study in 1960 there was no specific theory of why firms engage in transnational production: FDI was treated as just another variant of international capital theory. Hymer struck out in a new direction, drawing his inspiration from a completely different source: industrial organization theory and, especially, that part which dealt specifically with barriers to entry. Hymer started from the assumption that, in serving a particular market, domestic firms would have an intrinsic advantage over foreign firms: for example, a better understanding of the local business environment, including the nature of the market, business customs and legislation, and the like. Given such domestic-firm advantage, a foreign firm wishing to produce in that market would have to possess some kind of firm-specific asset that would offset the advantages held by domestic firms. Such assets are primarily those of firm size and economies of scale, market power and marketing skills (for example, brand names, advertising strength), technological expertise (either product, process, or both), or access to cheaper sources of finance. On these bases, then, a foreign firm would be able to out-compete domestic firms in their own home territory.

Hymer's contribution was truly seminal. It was the first time that the firm had been taken as the specific focus of explanation and that international production (rather than international trade) had been the explicit object of analysis. He emphasized the importance of market imperfections in stimulating the internationalization of production. He showed how, once established, the control of overseas productive assets itself became a source of competitive advantage.

Of course, Hymer's theory had its limitations. It was much better at explaining why and how firms might *begin* to become internationalized as producers and less good at explaining their subsequent development from an established international position. But, like all theories, it needs to be evaluated in terms of the level of understanding prevailing at the time he was writing. By that criterion, his contribution was immense. His particular focus on the internal ownership-specific characteristics of firms has become an accepted part of the theoretical literature.

Raymond Vernon and the product life cycle[6]

The concept of the product life cycle, already encountered in Chapter 4 (see Figure 4.11), was specifically adopted and adapted as an explanation of the evolution of *international production* by Raymond Vernon in 1966. Vernon's major contribution was to introduce an explicitly *locational* dimension into the product cycle concept that, in its original form, had no spatial connotation at all. Figure 7.2 shows Vernon's PLC model based upon the US experience (it should be read in conjunction with Figure 4.11).

Vernon's starting point was the assumption that producers are more likely to be aware of the possibility of introducing new products in their home market than producers located elsewhere (the parallel with Hymer is close in this respect). The kinds of new products introduced, therefore, would reflect the specific characteristics of the domestic market. In the United States case, the high average-income level and high labour costs tended to encourage the development of new products that catered to high-income consumers and were labour-saving (both for consumer and producer goods).

In this first phase of the product cycle, as Figure 7.2 shows, all production would be located in the United States and overseas demand served by exports.

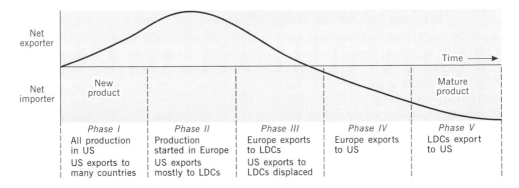

Figure 7.2 The product life cycle and its effects on the location of US production and trade

Source: Wells, 1972: Figure 15

But this situation would be unlikely to last indefinitely. US firms would eventually set up production facilities in the overseas market either because they saw an opportunity to reduce production and distribution costs or because of a threat to their market position. Such a threat might come from local competitors or from government attempts to reduce imports through tariff and other trade barriers.

It follows from the nature of the product cycle model that the first overseas production of the product would occur in other high-income markets. In the specific case of US investment this tended to be Western Europe and Canada. The newly established foreign plants would come to serve these former export markets and thus displace exports from the United States. These would be redirected to other areas where production had not yet begun (phase II in Figure 7.2). Eventually, the production cost advantages of the newer overseas plants would lead the firm to export from them to other, third-country, markets (phase III) and even back to the United States itself (phase IV). Finally, as the product becomes completely standardized, production would be shifted to low-cost locations in developing countries (phase V). It is interesting to note that when Vernon first suggested this possibility he regarded it as a 'bold projection'. At that time (the mid-1960s) there was still little evidence of developing country export platforms serving European and US markets.

How valid is the product life cycle as an explanation of the locational evolution of TNCs? There is no doubt that a good deal of the *initial* overseas investment by US firms did fit the product cycle sequence quite well. But it can no longer explain the majority of international investment by TNCs. As these firms have become more complex globally it is unrealistic to assume a simple evolutionary sequence from the home country outwards.[7] Even within strongly innovative TNCs, the initial source of the innovation and of its production may be from any point in the firm's global network. In addition, as we saw in Chapter 3, much of the world's FDI is reciprocal, or cross, investment between the industrialized countries. Such investment cannot easily be explained in product cycle terms. However, although its current relevance is limited, Vernon's adaptation of the PLC model represented an extremely important step in the development of a body of theory. Like Hymer's contribution, it should be evaluated in those terms.

John Dunning's 'eclectic' paradigm[8]

Most of the micro-level explanations of FDI and of international production are partial. In contrast, John Dunning has attempted to integrate various strands of explanation, derived from a variety of theoretical approaches – the theory of the firm, organization theory, trade theory and location theory – to create what he calls his *eclectic* theory of international production.

According to Dunning, a firm will engage in international production when all of the following three conditions are present.

- A firm must possess certain *ownership-specific advantages* not possessed by competing firms of other nationalities.
- Such advantages must be most suitably exploited by the firm itself rather than by selling or leasing them to other firms. In other words, the firm will *internalize* the use of its ownership-specific advantages.

- There must be *location-specific factors* that make it more profitable for the firm to exploit its assets in foreign, rather than in domestic, locations.

Ownership-specific advantages are assets internal to a firm. They are those 'which an enterprise may create for itself (e.g. certain types of knowledge, organization and human skills) or can purchase from other institutions, but over which, in so doing, it acquires some proprietary right of use. Such ownership-specific assets may take the form of a legally protected right, or of a commercial monopoly, or they may arise from the size, diversity or technical characteristics of firms.'[9] Stephen Hymer's influence is very clear. The most obvious ownership advantages relate to size and market power. Large firms, in general, are in a better position to obtain their production inputs at more favourable rates than smaller firms. They generally have better access to finance either from their own retained earnings or because of a better credit rating on the financial markets. Technology – of production, of marketing (including brand-names) and of organization in general – is a particularly important source of advantage. Technology, in the broadest sense of 'know-how', is an intangible asset easily transferable from one location to another.

Why should a firm *internalize* the use of its ownership-specific advantages by investing in foreign production? It could simply export its products at 'arm's length' through the usual trade channels. Alternatively, its technology could be licensed to domestic firms in foreign countries on payment of fees or royalties. Both of these alternatives are, in fact, used extensively. The main reason such alternatives may not be followed lies in the nature of the markets for materials, for intermediate goods and for finished products. In the miraculous world of neo-classical economics, markets are assumed to operate perfectly. If this were to be so then there would be no advantage to a firm in attempting to bypass the market. The global economic system would consist of a whole series of discrete transactions between independent buyers and sellers.

But the world is not like this. *Markets are imperfect.* The greater the imperfection, the greater will be the incentive for a firm to perform the function of the market itself by *internalizing* market transactions.[10] The most obvious example is vertical integration where a firm decides to control either its own sources of supply or the destination of its outputs. In both cases, the functions of independent material suppliers or of wholesale and retail merchants are absorbed – internalized – within the firm.

Uncertainty is the major incentive for a firm to internalize factor or product markets. The greater the degree of uncertainty – whether over the availability, price or quality of supplies or of the price obtainable for the firm's product – the greater is the advantage for the firm to control these transactions itself. Internalization is especially likely to occur in the case of knowledge. Many firms, especially large firms, but also all those in high-technology industries, spend huge sums of money on research and development. To ensure a satisfactory return on such investment, and to protect against predators, firms have a strong incentive to retain the technology for use within their own organizational boundaries. Rather than sell or lease the technology to another firm abroad the firm sets up its own production facilities and exploits its technological advantage directly.

The third element in the international production question is that of *location-specific factors*, 'those which are available, on the same terms, to all firms whatever their size and nationality, but which are specific in origin to particular locations and *have to be used in those locations*'.[11] Thus, in the absence of more favourable locational conditions abroad, a firm would serve foreign markets by exports from a domestic base. Several major types of location-specific factor are especially important in the context of international production, although their precise significance will vary according to the type of activity involved:

- markets
- resources
- production costs
- political conditions (including degrees of political risk)
- cultural/linguistic affinities.

Dunning's self-styled eclectic theory has been criticized as being merely 'a list of factors likely to be important in the explanation of the modern ... [TNC] ... rather than the explanation itself. Theoretical relations between the different factors too often remain untheorized.'[12] But this is to devalue Dunning's contribution rather too heavily. Seen as a pragmatic framework that attempts to integrate significant elements of other bodies of explanation, Dunning's approach is extremely useful as a conceptual structure within which specific cases can be examined. It has certainly been very widely adopted within the international production literature.

TNC development as a sequential process

At the level of the individual firm, we can explain why international production occurs and why TNCs exist at all in terms of specific combinations of Dunning's three conditions. But is there an identifiable *evolutionary sequence* of TNC development? Does the transition from a firm producing entirely for its domestic market to one engaged in foreign production follow a systematic development path? The answer to these questions is both 'yes' and 'no'. Yes, in the sense that certain common patterns of development are evident. No, in that we should not expect all firms to follow the same sequence, or in the sense that all firms will inevitably become TNCs. It is useful to consider the broad path of TNC development, however, because it helps to give some sense of the dynamics of the processes involved.

Figure 7.3 illustrates the kind of ideal-type sequence most commonly identified. It begins with the assumption that, initially, the firm is purely a domestic firm in terms both of production and of markets. In all national economies the majority of firms are of this type. However, the limits of the firm's domestic market may be reached and overseas markets may need to be penetrated to maintain growth and profitability. It is generally assumed that this is done initially through exports, using the services of overseas sales agents. Such agents are independent of the exporting firm. However, the benefits of internalization (discussed earlier) may eventually stimulate the firm to exert closer control over its foreign markets by setting up overseas sales outlets of its own. This may be achieved in one of two ways: by setting up an entirely new facility or by acquiring a local firm (possibly the previously used sales agency itself). Acquisition is, in fact, one of the most common methods of entry both to new product markets and to new geographical markets (see Chapter 3). It offers the attraction of an already functioning business compared with the more difficult, and possibly more risky, method of starting from scratch in an unfamiliar environment. Eventually the time may come – although not inevitably – when the need is felt for an actual foreign production facility. Again, this may be achieved through either acquisition or 'greenfield' investment.

There is a good deal of anecdotal evidence to support this sequence of development among firms that actually became TNCs. For example, in the case of Japanese firms investing in Europe, actual manufacturing operations came rather late following a long period of development of Japanese service investments in the form of the general trading companies, banks and other financial institutions and the sales and distribution functions of the manufacturing firms themselves.[13] However, there is nothing inevitable about the progression through each of the stages. Figure 7.3 shows a number of possible variations on the main theme and also suggests that the general sequence may be 'short-circuited' at various points. A firm may bypass the intermediate stages and set up overseas production facilities as a first step. One way in which this may occur is through the acquisition of another domestic firm that already has foreign operations. Thus, a firm may become transnational almost incidentally.

The sequential model of Figure 7.3 assumes that the process is market-driven. It fits the conventional view of the TNC as developing progressively, from serving markets through exports to direct investment in foreign production facilities. The broader definition of the TNC adopted in this book allows for other possibilities.

Figure 7.3

A sequential model of
TNC development

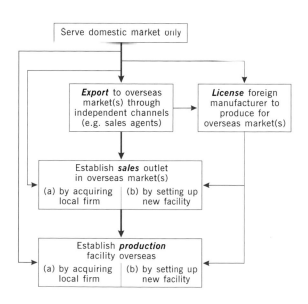

Recall that a key element of this broader definition is the *coordination* of international production. It is quite possible for a firm to begin its transnational activities by coordinating production in foreign locations in order to serve the firm's own domestic market rather than to serve overseas markets. This might subsequently evolve into a more elaborate transnational network of operation. We shall return to some of these alternative forms of transnational organization in the next chapter.

Types of transnational production

Although, at first sight, there seems to be a bewildering variety of types of transnational production being pursued by TNCs, in fact, we can boil them down to two broad categories, although the boundary between them is by no means as sharp as this dichotomy suggests:

- market-oriented production
- asset-oriented production.

Market-oriented production

Most foreign direct investment in both manufacturing and service industries is designed to serve a specific geographical market by locating inside that market. The good or service produced abroad may be virtually identical to that being produced in the firm's home country, although there may well be modifications to suit the specific tastes or requirements of the local market. In effect, such specifically market-oriented investment is a form of horizontal expansion across national boundaries. Two attributes of markets are especially important:

- The most obvious attraction of a specific market is its *size*, measured, for example, in terms of per capita income. Figure 7.4 shows the enormous variation in income levels (represented by gross national product [GNP] per capita) on a global scale. Per capita GNP in the high-income countries as a whole averages more than $26,000; in the lowest-income group of developing countries, the average per capita income is not much greater than $500. In many cases, of course, income levels are very much lower than this. The largest geographical markets in terms of incomes, although not in terms of population, are obviously the United States and Western Europe. Such variations in per capita GNP provide a crude indication of how the *level* of demand will vary from place to place across the world.
- Countries with different income levels will tend to have a different *structure* of demand. As incomes rise, so does the aggregate demand for goods and services. But such increased demand does not affect all products equally. In the economist's terminology, different products have different income elasticities of demand. We would expect populations in countries with low income levels to spend a larger proportion of their income on primary products (basic necessities) and, conversely, countries with high income levels to spend a higher proportion of their income on 'higher-order' manufactured

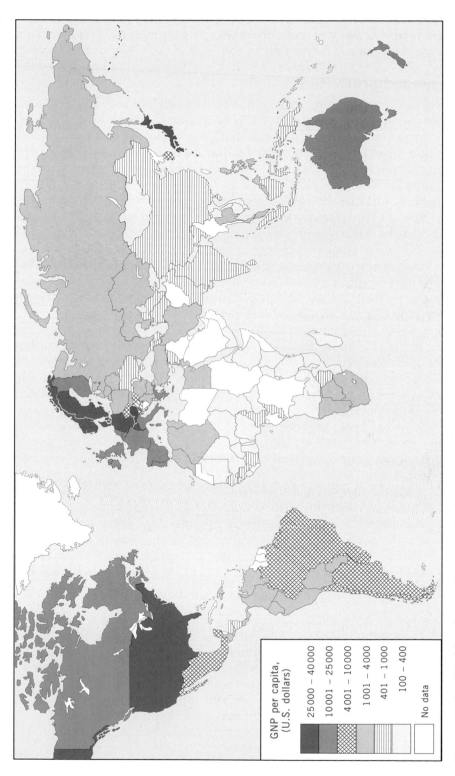

Figure 7.4 World variations in per capita income

Source: Based on World Bank, 2001: Table 1

GNP per capita,
(U.S. dollars)

25000 – 40000
10001 – 25000
4001 – 10000
1001 – 4000
401 – 1000
100 – 400

No data

goods and services. Thus, countries with different per capita income levels will tend to differ greatly in both the magnitude and the nature of their consumption patterns.

Asset-oriented production

Very few of the various assets needed by a firm to produce and sell its specific products and services are available in the same quantity and quality everywhere. They are almost always unevenly distributed geographically. This is most obviously the case in the natural resource industries, but it is also true in the case of labour and, as we saw in Chapter 4, for what has become one of the major drivers of change in today's economy: knowledge.

Firms in the natural resource industries must, of necessity, locate at the sources of supply. Often, such investments form the first element in a sequence of vertically integrated operations whose later stages (processing) may be located quite separately from the source of supply itself. In many cases, final processing of natural resources occurs close to the final market. Natural resource-oriented foreign investments have a very long history. However, other asset-oriented foreign investments by manufacturing firms are a relatively recent phenomenon and are very closely related to the kinds of technological developments discussed in Chapter 4.

For example, in discussing the production process in Chapter 4, we noted that the relative importance of the various production factors tends to vary according to the stage in the product's life cycle and, especially, with the maturity of the technology. More generally, the precise mix of production factors varies from one industry to another. In one sense, therefore, the key consideration is the relative importance of the individual factors in the firm's cost structure. But there is more to it than this. A particular factor of production may well be the most important element in a firm's total costs yet it may exert a negligible locational influence if its cost does not vary over space. If a factor costs the same everywhere it has a zero locational cost.[14]

Technological changes in production processes and in transportation have evened out the significance of location for some of the traditionally important factors of production (for example, natural resources). Many now hold the view that, at least at the global scale, the single most important location-specific factor is *labour*, particularly in terms of the skills and knowledge embodied in labour. The locational significance of labour as a production factor[15] is reflected in a number of ways:

- Geographical variations in labour *knowledge and skills*. Knowledge and skills depend on such conditions as the breadth and depth of education and on the particular history of an area's development. As a result, there are wide geographical variations in the availability of different types of labour. One very approximate indicator at the global scale is the variation in educational levels (for example, extent of literacy, enrolment in various stages of education, public expenditure on education, etc.). As might be expected, there is a very high correlation between these measures and the distribution of per capita income shown in Figure 7.4.

- Geographical variations in *wage costs*. International differences in wage levels can be staggeringly wide, as Figure 7.5 shows. These figures should be treated with some caution; they are averages across the whole of manufacturing industry and are therefore affected by the specific industry mix. Some industries have much higher wage levels than others. Even so, the contrasts are striking.

- Geographical differences in *labour productivity*. Spatial variations in wage costs are only a partial indication of the locational importance of labour as a production factor. The 'performance capacity' of labour varies enormously from place to place, a reflection of a number of influences including education, training, skill, motivation, as well as the kind of machinery and equipment in use.

- Geographical variations in the extent of *labour controllability*. Largely because of historical circumstances, there are considerable geographical differences in the degree of labour 'militancy' and in the extent to which labour is organized through labour unions. The proportion of the workers who are members of labour unions has declined markedly in some countries, notably the United States (from 35 per cent in 1955 to under 15 per cent in 1996[16]), the United Kingdom (from 48 per cent in 1970 to 30 per cent in 1997), and France (from 21 per cent in 1970 to 9 per cent in 1997[17]). The fact that most firms are very wary of 'highly organized' labour regions is demonstrated by their tendency to relocate from such regions or to make new investments in places where labour is regarded as being more malleable.

- A further significant dimension of labour is that it tends to be far *less mobile geographically* than capital, particularly over great distances. In general, labour is strongly *place-bound*, although the strength of the tie varies a great deal between different types of labour. On average, male workers are more mobile than female workers; skilled workers are more mobile than unskilled workers; professional white-collar workers are more mobile than blue-collar workers. But there are exceptions to such generalizations, as shown by the substantial waves of labour migration at different periods of history and towards particular kinds of geographical location. Such flows do not, however, contradict the basic point that labour is strongly differentiated spatially and deeply embedded in local communities in distinctive ways.[18]

Figure 7.5

Hourly earnings in selected countries

Source: Based on material in International Monetary Fund, *Yearbook of Financial Statistics* 2000; United Nations, *Statistical Yearbook* 2000

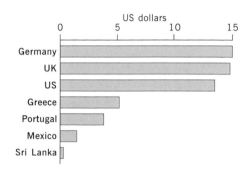

Global variations in production costs are a significant element in the transnational investment-location decision. This is obviously the case for asset-oriented investments but it is also a critical consideration for market-oriented investments. In that case there is always a trade-off to be calculated between the benefits of market proximity on the one hand and locational variations in production costs on the other. But the problem is not merely one of variations in production costs at a single moment in time or even the obvious point that such costs change over time. A particularly important consideration is the *uncertainty* of the level of future production costs in different locations. One way of dealing with such uncertainty is for the TNC to locate similar plants in a variety of different locations and then to adopt a flexible system of production allocation between plants. However, this strategy may well be complicated further by the volatility of currency exchange rates between different countries. What appears to be a least-cost source with one set of exchange rates may look very different if there is a major change in these rates.

A diversity of organizational architectures: how transnational operations are coordinated

One of the basic 'laws' of growth of any organism or organization is that as growth occurs its internal structure has to change. In particular, the *functional role* of its component parts tends to become more *specialized* and the links between the parts become more *complex*. As the size, organizational complexity and geographical spread of TNCs have increased the internal interrelationships between their geographically separated parts have become a highly significant element in the global economy. However, the precise manner in which TNCs organize and configure their production chains arises from a number of interrelated influences,[19] notably:

- the *firm's specific history*, including
  - characteristics derived from its *home-country embeddedness*
  - its *culture and administrative heritage* in the form of accepted practices built up over a period of time, producing a particular 'strategic predisposition'[20]
- the nature and complexity of the *industry environment(s)* in which the firm operates, including the nature of competition, technology, regulatory structures, etc.

Coping with complexity: a diversity of organizational structures

The traditional approach to changing organizational structures – based primarily on the hierarchical Western (that is, US) model – has depicted it as a sequential process whereby firms transformed their organizational structures from a *functional* form, in which the firm is subdivided into major functional units (production, marketing, finance, etc.), into a *divisional* form (usually product-based). In such a divisional structure, each product division is responsible separately for its own functions, particularly of production and marketing, although some functions (especially finance) tend to be performed centrally for the entire corporation. Each product division also usually acts as a separate profit centre. The main advantage of the divisional structure is usually seen to be one of a greater ability to cope with product diversity. Thus, as large US firms became increasingly diversified during the 1950s and 1960s they also tended to adopt a divisional structure.

Adoption of a divisional structure gave firms greater control over their increasingly diverse product environment. However, operating across national boundaries, rather than within a single nation, poses additional problems of coordination and control. Largely through trial and error, TNCs groped their way towards more appropriate organizational structures. Figure 7.6 shows four commonly used structures. Which one is actually adopted depends upon a number of factors, including the age and experience of the enterprise, the nature of its operations and its degree of product and geographical diversity.

The form most commonly adopted in the early stages of TNC development – at least when there are several overseas subsidiaries – is simply to add on an *international division* to the existing divisional structure (Figure 7.6a). This has tended to be a short-lived solution to the organizational problem if the firm continues to expand its international operations. In such a hybrid structure problems of coordination inevitably arise. Tensions develop between the parts of the organization operated on product lines (the firm's domestic activities) and those organized on

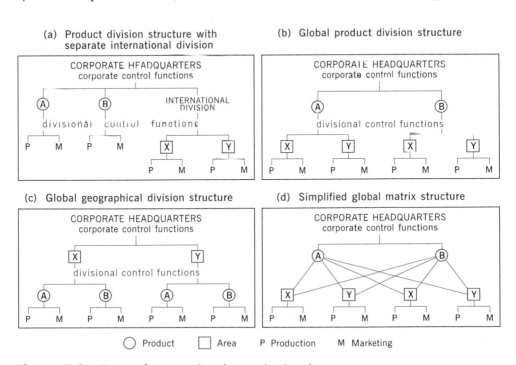

Figure 7.6   Types of transnational organizational structure

an area basis (the international operations). The need arises for an organizational form that can integrate both the domestic and international operations of the firm.

There are two obvious possibilities. One is to organize the firm on a *global product* basis: in other words, to apply the product-division form throughout the world and to remove the international division (Figure 7.6b). The other possibility is to organize the firm's activities on a *worldwide geographical* basis (Figure 7.6c). But neither of these structures resolves the basic tension between product- and area-based systems. For such reasons some of the largest TNCs adopted sophisticated *global grid* or *global matrix* structures (Figure 7.6d), containing elements of both product and area structures and involving dual reporting links.

There is plenty of evidence to support such a sequence, especially amongst US and some other Western firms. Equally, however, there is also plenty of evidence to demonstrate far greater organizational diversity. Bartlett and Ghoshal have suggested a useful typology of three major ideal-type TNCs, together with a fourth type which, they argue, is in the process of emerging.[21] Table 7.1 summarizes the major features of each type.

1   The *'multinational organization' model* emerged particularly during the interwar period. Firms were stimulated by a combination of economic, political and social forces to decentralize their operations in response to national market differences. This ideal-type model is characterized by a decentralized federation of overseas units and simple financial control systems overlaid on informal personal coordination (Figure 7.7). The company's worldwide operations are organized as a portfolio of national businesses. This was the kind of transnational organizational form used extensively by expanding European companies. Each of the firms' national units has a very considerable degree of autonomy and a predominantly 'local' orientation. It is able, therefore, to respond to local needs but its fragmented structure lessens scale efficiencies and reduces the internal flow of knowledge.

Figure 7.7

Multinational organization model

Source: Based on Bartlett and Ghoshal, 1998: Figure 3.1

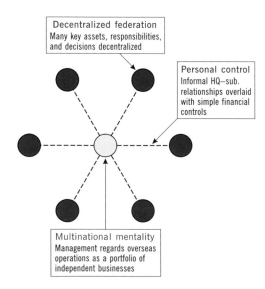

Decentralized federation
Many key assets, responsibilities, and decisions decentralized

Personal control
Informal HQ–sub. relationships overlaid with simple financial controls

Multinational mentality
Management regards overseas operations as a portfolio of independent businesses

Table 7.1   Some ideal-types of TNC organization: basic characteristics

Characteristics	'Multinational'	'International'	'Global'	'Integrated network'
Structural configuration	Decentralized federation. Many key assets, responsibilities, decisions decentralized	Coordinated federation. Many assets, responsibilities, resources, decisions decentralized but controlled by HQ	Centralized hub. Most strategic assets, resources, responsibilities and decisions centralized	Distributed network of specialized resources and capabilities
Administrative control	Informal HQ–subsidiary relationship; simple financial control	Formal management planning and control systems allow tighter HQ–subsidiary linkage	Tight central control of decisions, resources and information	Complex process of coordination and cooperation in an environment of shared decision-making
Management attitude towards overseas operations	Overseas operations seen as portfolio of independent businesses	Overseas operations seen as appendages to a central domestic corporation	Overseas operations treated as 'delivery pipelines' to a unified global market	Overseas operations seen an integral part of complex network of flows of components, products, resources, people, information among interdependent units
Role of overseas operations	Sensing and exploiting local opportunities	Adapting and leveraging parent company competencies	Implementing parent company strategies	Differentiated contributions by national units to integrated worldwide operations
Development and diffusion of knowledge	Knowledge developed and retained within each unit	Knowledge developed at the centre and transferred to overseas units	Knowledge developed and retained at the centre	Knowledge developed jointly and shared worldwide

Source: Based on material in Bartlett and Ghoshal, 1998

2   The *'international organization' model* came to prominence in the 1950s and 1960s through the large US corporations which expanded overseas to capitalize on their firm-specific assets of technological leadership or marketing power. This ideal-type involves far more formal coordination and control by the corporate headquarters over the overseas subsidiaries (Figure 7.8). Whereas 'multinational' organizations are, in effect, portfolios of quasi-independent businesses, 'international' organizations see their overseas operations as appendages to the controlling domestic corporation. Thus, the international subsidiaries are more dependent on the centre for the transfer of knowledge and the parent company makes greater use of formal systems of control. The 'international' TNC is better equipped to leverage the knowledge and capabilities of its parent company but its particular configuration and operating systems tend to make it less responsive than the 'multinational' model. It is also rather less efficient than the next ideal-type.

Figure 7.8

International organization model

Source: Based on Bartlett and Ghoshal, 1998: Figure 3.2

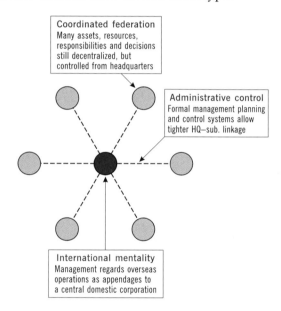

Coordinated federation
Many assets, resources, responsibilities and decisions still decentralized, but controlled from headquarters

Administrative control
Formal management planning and control systems allow tighter HQ–sub. linkage

International mentality
Management regards overseas operations as appendages to a central domestic corporation

3   The *'global organization' model* was one of the earliest forms of international business (used, for example, by Ford and by Rockefeller in the early 1900s, as well as by Japanese firms in their much later internationalization drive of the 1970s and 1980s). It is based upon a tight centralization of assets and responsibilities in which the role of the local units is to assemble and sell products and to implement plans and policies developed at the centre (Figure 7.9). In this ideal-type model, overseas subsidiaries have far less freedom to create new products or strategies or to modify existing ones. Thus, the 'global' TNC capitalizes on scale economies and on centralized knowledge and expertise. But this implies that local market conditions tend to be ignored and the possibility of local learning is precluded.

Although each of these three ideal-type models developed initially during specific historical periods, there is no suggestion that each one was sequentially replaced by the next. Each form has tended to persist, in either a pure or hybrid form, helping to produce a diverse population of TNCs. There is some correlation

Figure 7.9

Global organization model

Source: Based on Bartlett and Ghoshal, 1998: Figure 3.3

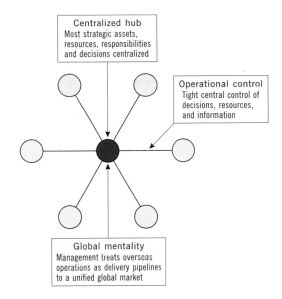

Centralized hub
Most strategic assets, resources, responsibilities and decisions centralized

Operational control
Tight central control of decisions, resources, and information

Global mentality
Management treats overseas operations as delivery pipelines to a unified global market

between organizational type and nationality of parent company but it is by no means perfect; it is better to regard firms of different national origins as having a predisposition to one or other form of organization.[22] These ideal-types, and variations upon them, are still apparent.[23] However, new forms of transnational organization are also emerging that may – although not inevitably – replace some of the existing forms. As outlined above, each of the three ideal-types possesses specific strengths, but each also has specific weaknesses.

In Bartlett and Ghoshal's view, the dilemma facing TNCs in today's turbulent competitive environment is that they need the best features of each one of the three ideal-types: to be globally efficient, multinationally flexible, and capable of capturing the benefits of worldwide learning all at the same time. Hence, they argue, we are now seeing the emergence of a fourth ideal-type TNC:

4　The *'integrated network organization' model*. characterized by a distributed network configuration and a capacity to develop flexible coordinating processes (Figure 7.10). Such capabilities apply both inside the firm, displacing hierarchical governance relationships with what Hedlund[24] terms a *heterarchical* structure, and also outside the firm through a complex network of inter-firm relationships. In other words, it implies a blurring of traditional organizational boundaries. The identification of this fourth organizational type does not imply an inevitable sequential development but merely that some firms are beginning to develop such a complex networked structure. We will explore this issue further in Chapter 8 in the broader context of networks of externalized relationships.

The point to be emphasized here is the continuing diversity of TNC organizational architectures. Firms organized on hierarchical principles not only still exist but also they may still be in a majority. The empirical evidence suggests that the newer, 'flatter', organizational forms tend to be confined to a limited number of firms in certain sectors. TNCs come in all shapes, sizes and forms of governance. Their internal architecture reflects not only the external constraints and

Figure 7.10

Integrated network organization model

Source: Based on Bartlett and Ghoshal, 1998: Figure 5.1

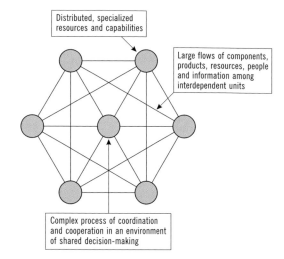

Distributed, specialized resources and capabilities

Large flows of components, products, resources, people and information among interdependent units

Complex process of coordination and cooperation in an environment of shared decision-making

opportunities they have to face – including the structures made possible by new communications technologies[25] – but also a strong element of path dependency.

Headquarter–subsidiary relationships

The various TNC structures discussed in the preceding section, and especially those shown in Figures 7.7 to 7.9, assume a clear distinction between a TNC's organizational centre – its headquarters – and its subsidiary operations. In a pure hierarchical model, the relationship is intrinsically top-down: the TNC subsidiaries simply perform the role allocated to them. In contrast, in a *heterarchical* organization (Figure 7.10) the position is far more complex. The role played by a subsidiary varies between different organizational structures and in terms of a TNC's specific strategy. Three broad types of subsidiary role can be identified:[26]

- *The local implementer*. This is a subsidiary with limited geographical scope and functions, sometimes termed a 'miniature replica'. Its primary purpose is to adapt the TNC's products for the local market.

- *The specialized contributor*. This kind of subsidiary has specific expertise that is tightly integrated into the activities of other subsidiaries in the TNC. It has a narrow range of functions and a high level of interdependence with other parts of the firm.

- *The world mandate*. Here, the subsidiary has worldwide (or possibly regional) responsibility for a particular product or type of business. In that sense, its activities are managed from the subsidiary itself, not the headquarters.

These different subsidiary roles have important implications for the impact of TNC activities on national and local economies (see Chapter 9). How these roles develop, and possibly change – for example from local implementer to world mandate – depends upon the nature of the bargaining relationships within the

TNC. Relationships within firms are highly contested processes. They are the manifestation of internal power structures and bargaining relationships. In a similar way, the individual affiliates of a firm (its subsidiaries, branches, etc.) are continuously engaged in competition to improve their relative position within the organization by, for example, winning additional investment or autonomy from the corporate centre. At the same time, the performance of each affiliate is continuously monitored against the relevant others (internal benchmarking) and this is used as an integral part of the internal bargaining processes within the firm.

In a transnational firm, operating in a diversity of national/cultural environments, the very nature of these environments – the *places* in which parts of the firm are embedded – will exert an influence on these internal bargaining processes:

> It is probably wrong to conceive of the multi-locational firm, however organizationally centralized, as a unitary agent with a singular and coherent identity. Rather, we should expect that different 'places' within the firm, organizationally and geographically, develop their own identities, ways of doing things and ways of thinking over time ... In effect, the large firm is internally regionalized ... The firm's dominant culture, created by and expressed through the activities and understandings of top management at headquarters, necessarily contains multiple subcultures. Some of these may revolve around functions and cut across places (engineers versus sales people, for example), but some will have real geographical locations – they will have grown up in specific plants in particular places.[27]

The relationships between TNCs and 'place' are the subject of the final section of this chapter.

Strategic tensions: global integration – local responsiveness

Intensification of global competition in a world that retains a high degree of local differentiation creates, for all TNCs, an internal tension between globalizing forces

Advantages	Costs and risks
The firm's oligopoly power is increased through the exploitation of scale and experience effects beyond the size of individual national markets	The TNC may be vulnerable to disruption of its entire operations (or part of them) because of labour unrest or government policy changes affecting a particular unit
The TNC is placed in a better position to exploit the growing discrepancy between a relatively efficient market for goods (created by freer trade) and very inefficient markets for production factors	Fluctuations in currency exchange rates may disrupt integration strategies, drastically altering the economies of intrafirm transactions of intermediate or final goods
The possibility of exploiting differences in tax rates and structures between countries is increased and so, therefore, is the possibility of engaging in transfer pricing	Governments may impose performance requirements or other restrictions which impede the optimal operation of the firm's integrated production chain
The specialized and integrated function of individual country operations makes hostile government action less rewarding and less likely	The task of managing a globally integrated operation is more complex and demanding than that of managing separate national subsidiaries

Figure 7.11   The advantages and disadvantages of a globally integrated strategy

Source: Based on material in Doz, 1986b

on the one hand and localizing forces on the other. As Figure 7.11 shows, there are considerable potential advantages for a firm pursuing a globally integrated strategy. But there are also considerable disadvantages. Figure 7.12 captures this basic tension within a 'global integration–local responsiveness' framework. The vertical axis shows the major pressures on firms to strive for global strategic coordination of their activities; the horizontal axis shows the countervailing pressures on firms to develop locally responsive strategies. Hence, TNCs

> must balance pressures for integrating globally with those for responding idiosyncratically to national environments … [however] there may not be a single optimal point on the fragmentation–unification continuum but a range of tenable positions.[28]

Figure 7.12

The global integration–local responsiveness framework

Source: Based on material in Prahalad and Doz, 1987: Figure 2.2; pp. 18–21

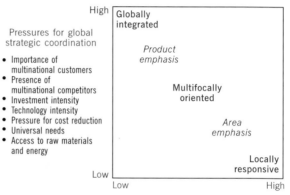

There is now quite a lot of evidence to suggest that TNCs are being forced to take the 'local' more seriously.[29] Indeed, some archetypal 'global' firms have been forced to rethink their strategy. A good example is Coca-Cola. In the late 1990s, Coca-Cola experienced a series of problems in its international operations, leading it to announce a new 'think local, act local' strategy and to reduce the excessive degree of centralization that had been the hallmark of the company for decades. In the words of Coca-Cola's new chairman, Douglas Daft,

> For a couple of years, the world was moving in one direction, and we were moving in another. We were heading in a direction that had served us very well for several decades, generally moving towards consolidation and centralized control. That direction was particularly important during the 1970s and 1980s when we were 'going global' … The world, on the other hand, began moving in the 1990s in a different direction … the very forces that were making the world more connected and homogeneous were simultaneously triggering a powerful desire for local autonomy and preservation of unique cultural identity … we had a lesson to learn. And what we learned was … that the next big evolutionary step of 'going global' now has to be 'going local'. In other words, we had to rediscover our own multi-local heritage.[30]

Significantly, many TNCs make much of the 'global–local tension' in their corporate literature, although how much of this is mere rhetoric rather than actual reality is difficult to determine. For example,

- The Swedish transnational corporation ABB claims to have perfected 'the art of being local worldwide'.
- The American financial services TNC, J.P. Morgan asserts that 'the key to global performance is understanding local markets'.
- The Anglo-Dutch firm Unilever describes itself as a 'multi-local multinational'.
- The Hong Kong Bank (HSBC) boasts of its 'local insight, global outloook'.
- The Japanese electronics firm Sony was one of the first to use the term 'glocalization' to describe its international corporate strategy.

The myth of the 'placeless' TNC[31]

> Before national identity, before local affiliation … before any of this comes the commitment to a single, unified global mission … Country of origin does not matter. Location of headquarters does not matter. The products for which you are responsible and the company you serve have become denationalized.[32]

One of the central claims of 'hyperglobalists' is that international firms are inexorably, and inevitably, abandoning their ties to their country of origin and, by implication, converging towards a universal *global* organizational form. The argument is, essentially, that technological and regulatory developments in the world economy have created a 'global surface' on which a dominant organizational form will develop and inexorably wipe out less efficient competitors who are no longer protected by national or local barriers. Such an organization is, it is argued 'placeless' and 'boundary-less'.

In this section we challenge this scenario of placeless global corporations using two kinds of empirical evidence. The first is essentially *quantitative* and uses statistical data on the operations of the world's leading TNCs. If the 'global corporation' hypothesis is valid then we would expect to find that at least the majority of these largest TNCs have the overwhelming majority of their assets and employment outside their home country. In Table 7.2, the world's 100 largest TNCs are ranked by their *transnationality index* (TNI). The two right-hand columns of Table 7.2 also show the percentage of each firm's assets and employment that are located outside the firm's home country.

The results are quite striking. The average TNI is only 52.6. Only 57 of the 100 companies have an index of greater than 50; a mere 16 have an index greater than 75. Significantly, the 14 most transnational firms (in terms of the TNI) originate from small countries (Switzerland, the UK, the Netherlands, Belgium, Canada). Conversely, the biggest TNCs in terms of total foreign assets all have relatively low TNI scores. For example, the largest TNC (in terms of foreign assets), General Electric, ranks 75th on the index of transnationality; GM (4th in terms of foreign

Table 7.2 How 'global' are the world's top 100 transnational corporations?

Rank	TNI Index	Company	Country	Industry	Foreign share of Assets (%)	Foreign share of Empl. (%)
1	95.4	Thomson Corp.	Canada	Publ'g and printing	98.6	92.5
2	95.2	Nestlé	Switzerland	Food/beverages	89.9	97.2
3	94.1	ABB	Switzerland	Electrical equipt	88.2	96.3
4	93.2	Electrolux	Sweden	Electrical	92.9	90.4
5	91.8	Holcim	Switzerland	Constr. materials	91.9	93.4
6	91.5	Roche Group	Switzerland	Pharmaceuticals.	90.4	85.6
7	90.7	BAT	UK	Food/tobacco	84.0	96.8
8	89.3	Unilever	UK/Neths	Food/beverages	90.4	90.5
9	88.6	Seagram Company	Canada	Beverages/media	73.1	..
10	82.6	Akzo Nobel	Netherlands	Chemicals	85.0	81.0
11	82.4	Nippon Oil Co.	Japan	Petroleum	88.7	74.5
12	81.9	Cadbury–Schweppes	UK	Food/beverages	88.8	79.7
13	79.4	Diageo	UK	Beverages	69.3	82.6
14	78.3	News Corporation	Australia	Media	61.2	72.5
15	76.9	L'Air Liquide Grp	France	Chemicals	..	..
16	76.6	Glaxo Wellcome	UK	Pharmaceuticals	70.2	74.1
17	73.8	Michelin	France	Rubber/tyres	..	..
18	73.7	BP	UK	Petroleum	74.7	77.3
19	72.5	Stora Enso OYS	Finland	Paper	..	..
20	71.6	AstraZeneca	UK/Sweden	Pharmaceuticals	37.3	83.3
21	70.3	TotalFina	France	Petroleum	..	67.9
22	68.0	ExxonMobil	USA	Petroleum	68.8	63.6
23	67.8	Danone Groupe	France	Food/beverages	62.9	..
24	67.1	McDonald's Corpn	USA	Eating places	57.6	82.8
25	65.6	Alcatel	France	Electronics	52.1	74.1
26	65.2	Coca-Cola	USA	Beverages	83.3	..
27	64.7	Honda	Japan	Automobiles	58.4	..
28	63.8	Compart Spa	Italy	Food	..	68.2
29	62.2	Montedison Group	Italy	Chemicals	..	71.7
30	61.4	Volvo	Sweden	Automobiles	..	53.4
31	60.9	Ericsson	Sweden	Electronics	44.5	57.4
32	60.9	BMW	Germany	Automobiles	69.1	40.1
33	60.2	Bayer	Germany	Chemicals	58.0	53.2
34	60.2	RTZ	Australia/UK	Mining	61.2	62.5
35	59.9	Philips	Netherlands	Electronics	76.2	..
36	59.2	BASF	Germany	Chemicals	57.0	44.4
37	58.9	Bridgestone	Japan	Rubber/tyres	44.6	69.0

(Continued)

Table 7.2 (*Continued*)

Rank	Index	Company	Country	Industry	Assets (%)	Empl. (%)
38	58.2	Renault	France	Automobiles	..	..
39	57.5	Crown Cork & Seal	USA	Packaging	62.6	..
40	57.1	Canon	Japan	Electronics	48.4	52.8
41	56.8	Siemens	Germany	Electronics	..	56.7
42	56.7	Sony	Japan	Electronics	..	61.0
43	56.3	Royal Dutch Shell	Neths/UK	Petroleum	60.3	57.8
44	56.2	Motorola	USA	Electronics	58.0	55.3
45	55.7	Volkswagen Group	Germany	Automobiles	..	48.3
46	55.3	Robert Bosch	Germany	Auto components	..	49.8
47	54.6	Cemex	Mexico	Const. materials	58.8	..
48	54.2	Johnson & Johnson	USA	Pharmaceuticals	67.1	50.7
49	54.0	Aventis	France	Pharmaceuticals	..	..
50	53.7	IBM	USA	Computers	51.1	52.6
51	53.7	Daimler–Chrysler	Germany	Automobiles	31.7	48.3
52	53.1	Hewlett–Packard	USA	Electronics	..	49.1
53	52.7	Royal Ahold	Netherlands	Retailing	69.9	19.2
54	51.7	Elf Aquitaine	France	Petroleum	43.5	..
55	51.6	Repsol	Spain	Petroleum	70.3	..
56	51.2	Texaco	USA	Petroleum	..	..
57	51.2	Mitsubishi Motors	Japan	Automobiles	27.6	..
58	49.1	Suez Lyonnaise des Eaux	France	Diversified/utility	..	68.2
59	48.9	Mannesmann	Germany	Telecomms/ engineering	..	44.9
60	46.2	Dow Chemical	USA	Chemicals	39.7	42.8
61	45.5	AES Corporation	USA	Utility	48.8	..
62	44.7	Peugeot	France	Automobiles	39.2	30.3
63	43.5	Usinor	France	Steel	47.4	34.9
64	43.3	Viag	Germany	Diversified	..	51.0
65	42.4	Veba Group	Germany	Diversified	27.1	37.7
66	41.3	Du Pont	USA	Chemicals	36.3	38.3
67	40.9	ENI Group	Italy	Petroleum	47.2	..
68	40.7	Nissan	Japan	Automobiles	..	..
69	40.4	Texas Utilities Co.	USA	Utility	42.5	39.2
70	40.3	Procter & Gamble	USA	Soaps/cosmetics	33.3	..
71	39.3	Matsushita	Japan	Electronics	19.2	49.5
72	38.4	Fujitsu	Japan	Electronics	36.2	38.6

(*Continued*)

Table 7.2 (*Continued*)

TNI Rank	TNI Index	Company	Country	Industry	Foreign share of Assets (%)	Foreign share of Empl. (%)
73	38.0	Telefonica	Spain	Telecomms	37.8	..
74	38.0	Hutchison Whampoa	Hong Kong	Diversified	..	50.9
75	36.7	General Electric	USA	Electronics	34.8	46.1
76	36.4	Metro	Germany	Retailing	37.7	32.4
77	36.1	Ford	USA	Automobiles	..	52.5
78	34.7	Carrefour	France	Retailing	36.5	..
79	34.2	Chevron	USA	Petroleum	49.4	25.8
80	34.0	Vivendi	France	Utility/media	..	..
81	33.4	Fiat	Italy	Automobiles	18.9	44.6
82	30.9	Toyota	Japan	Automobiles	36.3	6.3
83	30.7	General Motors	USA	Automobiles	24.9	40.8
84	29.8	Petroléos de Venezuela	Venezuela	Petroleum	16.9	..
85	29.7	Mitsubishi Corpn	Japan	Diversified	31.3	45.5
86	29.1	Mitsui & Co.	Japan	Diversified	30.6	..
87	28.4	Merck & Co.	USA	Pharmaceuticals	25.6	38.2
88	26.0	Marubeni Corpn	Japan	Trading	19.9	..
89	25.9	Lucent Technologies	USA	Electronics	22.4	23.5
90	25.8	Wal-Mart Stores	USA	Retailing	60.4	..
91	24.3	Edison International	USA	Electronics	23.1	..
92	23.3	Toshiba	Japan	Electronics	13.2	24.4
93	23.3	Atlantic Richfield	USA	Petroleum	..	..
94	22.9	RWE Group	Germany	Utility/diversified	19.0	..
95	19.8	Southern Company	USA	Utility	25.0	21.0
96	17.9	Hitachi	Japan	Electronics	16.0	..
97	16.1	Sumitomo Corpn	Japan	Trading	31.5	..
98	15.8	Nissho Iwai	Japan	Trading	23.6	..
99	13.7	Itochu	Japan	Trading	22.2	..
100	12.9	SBC Communications	USA	Telecomms	..	..

TNI-Index of transnationality. Represents the average of foreign assets to total assets, foreign sales to total sales and foreign employment to total employment.

Source: Calculated from UNCTAD, 2001: Table III.1

assets) ranks 83rd; IBM ranks 50th; VW ranks 45th; Toyota, the 6th-ranking TNC in foreign assets ranks 82nd on the broader index. On this measure, therefore, there is little evidence of TNCs having the share of their activities outside their home countries that might be expected if they were global firms.[33]

Hu provides a rather broader approach to this question. He suggests the following criteria for evaluating the extent to which TNCs are 'global':[34]

- In which nation or nations is the bulk of the corporation's assets and people located?
- By whom are the local subsidiaries owned and controlled, and in which nation is the parent company owned and controlled?
- What is the nationality of the senior positions (executive and board posts) at the parent company, and what is the nationality of the most important decision-makers at the subsidiaries in host nations?
- What is the legal nationality of the parent company?
- To whom would the group as a whole turn for diplomatic protection and political support in case of need?
- Which is the nation where the tax authorities can, if they choose to do so, tax the group on its worldwide earnings rather than merely its local earnings?

On the basis of his empirical analysis of a sample of TNCs, Hu concludes that 'these criteria usually produce an unambiguous answer: that it … [the TNC] … is a national corporation with international operations (i.e. foreign subsidiaries)'. Thus, despite many decades of international operations, TNCs – at least in quantitative terms – remain distinctively connected with their home base.

However, such quantitative analysis provides only a partial answer to the questions posed in this section. It tells us something about the relative *geographical extent* of TNC activities outside the home country and, to that degree, demonstrates the continuing emphasis on the home base. But it distinguishes only between home and foreign. A firm might have a TNI of, say, 80 (meaning that 80 per cent of its activities were outside its home country) but all of those activities might be located in just one foreign country. An example would be the large number of US firms that operate only in Canada. Neither does it help us to establish whether or not TNCs of different national origins are becoming similar in their modes of operation. It is at least possible that TNCs may retain more of their assets and employment in their home country but still be converging organizationally and behaviourally towards a universal, global form. To address this issue we need a different type of empirical evidence from that which merely measures the geographical dispersion of a firm's activities. We need evidence that explicitly compares TNCs from different countries of origin in a systematic and qualitative manner.

The geographical embeddedness of transnational corporations

The theoretical basis for claiming that TNCs 'produced' in different places will continue to display a significant degree of organizational differentiation lies in the notion that all business firms are embedded within specific social, cultural, political and institutional contexts that help to influence the ways in which they develop.[35]

At least in origin, TNCs are 'locally grown'; they develop their roots in the soil in which they were planted. The deeper the roots the stronger will be the degree of local embeddedness, such that they should be expected to bear at least some traces of the economic, social and cultural characteristics of their home country. In other words, they continue to contain elements of the local within their modes of operation ... the local social-cultural milieu is a major influence on how firms evolve and behave even when their operations are geographically very extensive. This is not to argue a case for cultural determinism or even to argue that all firms of a given nationality are identical. Clearly they are not. But they do tend to share some common features.[36]

Although such embeddedness may occur at a variety of interrelated geographical scales the most significant scale would appear to be that of the national state, the major 'container' within which distinctive practices develop (see Chapter 5). Such 'national containers' of distinctive assemblages of institutions and practices help to 'produce' particular kinds of firms. In the TNC literature, Dunning was one of the first to make the explicit connection between what he terms the 'ownership-specific' advantages of firms and the 'location-specific' characteristics of national states.

Table 7.3 Links between selected ownership-specific advantages and country-specific characteristics

Ownership-specific advantage	*Country characteristics favouring such advantages*
Size of firm	Large, standardized markets Liberal regime towards mergers and concentration
Managerial expertise	Pool of managerial talent Educational and training facilities
Technology-based advantages	Good R&D facilities Government support of innovation Pool of scientific and technical labour
Labour and/or mature, small-scale intensive technologies	Large pool of labour (including technical labour) Appropriate consultancy services
Production differentiation	High-income national markets High-income elasticity of demand
Marketing economies	Highly developed marketing/advertising system Consumer-oriented society
Access to (domestic) markets	Large national market No restrictions on imports
Capital availability and financial expertise	Well-developed, reliable capital markets Appropriate professional advice

Source: Based on Dunning, 1979: Table 6

Table 7.3 outlines some of the links that exist between the ownership-specific advantages of firms and the location-specific characteristics of the firm's *home* country. It is this link that helps to explain the different characteristics of TNCs from different source nations. For example, the large domestic market and high level of technological sophistication of the US domestic eeonomy have helped to produce the distinctive characteristics of US TNCs. The lack of natural resources and the strong involvement of government and other institutions in technological and industrial affairs helps to explain the particular attributes of Japanese TNCs.

Although the conditions in which firms develop in their home country environments exert an extremely powerful influence on their behaviour, the impact of the host (foreign) environments in which they operate may also be influential. For a whole variety of reasons – political, cultural, social – non-local firms invariably have to adapt some of their domestic practices to local conditions. It is virtually impossible to transfer the whole package of firm advantages and practices to a different national environment. For example, Japanese overseas manufacturing plants tend to be 'hybrid' forms rather than the pure organizations found in Japan itself.[37] The same argument applies to US firms operating abroad. Even in the United Kingdom, where the apparent 'cultural distance' between the US and the UK is less than in many other cases, there is a very long history of American firms having to adapt some of their business practices to local conditions. What results, therefore, is a varying mix of home-country and host-country influences.

A comparison between US, German and Japanese companies

Empirical evidence of the relationships between firms and the national contexts in which they are embedded is now considerable. A particularly powerful set of arguments is provided by a detailed comparative study of US, German and Japanese TNCs using a series of structural and behavioural dimensions, including their modes of corporate governance, corporate financing systems and their strategic behaviour (notably in relation to R&D, direct investment and intra-firm trade).[38] The results are summarized in Table 7.4.

The experiences of US, German and Japanese firms provide strong arguments to counter the convergence hypothesis. There appears to be

> little blurring or convergence at the cores of firms based in Germany, Japan, or the United States … Durable national institutions and distinctive ideological traditions still seem to shape and channel crucial corporate decisions … the domestic structures within which a firm initially develops leave a permanent imprint on its strategic behavior … At a time when many observers emphasize the importance of cross-border strategic alliances, regional business networks, and stock offerings on foreign exchanges – all suggestive of a blurring of corporate nationalities – our findings underline, for example, the durability of German financial control systems, the historical drive behind Japanese technology development through tight corporate networks, and the very different time horizons that lie behind American, German, and Japanese corporate planning.[39]

Table 7.4 Differences between US, German and Japanese TNCs

	US TNCs	German TNCs	Japanese TNCs
Corporate governance and corporate financing	Constrained by volatile capital markets; short-termist perspectives. Finance-centred strategies	Relatively high degree of operational autonomy except during crises. Long-term perspectives. Conservative strategies	Bound by complex but reliable networks of domestic relationships. Long-term perspectives. Market-share-centred strategies
	High risk of takeover. 90% of firm shares held mainly by individuals, pension funds, mutual funds. Less than 1% held by banks	Low risk of takeover. Firm shares held mainly by non-financial institutions (40%). Significant role of regional bodies	Very low risk of takeover – mainly confined to within network
	Banks provide mainly secondary financing, cash management, selective advisory role	Banks play a lead role. Supervisory boards of companies are strongly bank-influenced	High degree of cross-shareholdings within group. Lead bank performs a steering function
	Ratio of bank loan/corporate financial liabilities = 25–35%	Ratio of bank loan/corporate liabilities = 60–70%	Ratio of bank loan/corporate liabilities = 60–70%
Research and development	Corporate R&D expenditure peaked in 1985 at 2.1% of GDP. Declining	Corporate R&D expenditure declined steeply in late 1980s/early 1990s. At 1.7% of GDP lower than US and Japan	Corporate R&D grew very rapidly in 1980s. Overtook US in 1989. Peaked at 2.2% of GDP. Real cuts made only as last resort
	Diversified pattern; innovation-oriented. Some propensity to perform R&D abroad	Narrow focus. Some propensity to perform R&D abroad	High tech and process orientation. Very limited propensity to perform R&D abroad
Direct investment and intra-firm trade	Extensive outward investment. Substantial competition from inward investment	Selective outward investment. Moderate competition from inward investment	Extensive outward investment. Very limited competition from inward investment.
	Moderate intra-firm trade; high propensity to outsource	High level of intra-firm trade	Very high level of intra-firm and intra-group trade

Source: Based primarily on material in Pauly and Reich, 1997

East Asian business organizations[40]

With the rise of East Asia as a major dynamic growth point in the global economy, it became commonplace to contrast the 'East Asian way' of doing business with that of firms in the West. There is some basis in such a comparison because we do find certain common features of East Asian business organizations that, in combination, tend to differentiate them from Western firms:[41]

- formation of intra- and inter-firm business relationships
- reliance on personal relationships
- strong relationships between business and the state.

However, it is a mistake to think in terms of *one* East Asian business model because, in fact, there is considerable *diversity* between firms from different East Asian countries. This can be illustrated by looking briefly at the cases of Japanese, Korean, Taiwanese and Overseas Chinese business organizations.

The Japanese *keiretsu*[42]

> Intercorporate alliances in the contemporary Japanese economy are marked by an elaborate structure of institutional arrangements that enmesh its primary decision-making units in complex networks of cooperation and competition … There is a strong predilection for firms in Japan to cluster themselves into coherent groupings of affiliated companies extending across a broad spectrum of markets.[43]

The precise composition and structure of such business groupings is immensely varied. Here we focus on the industrial groupings generally known as *keiretsu*. Five diagnostic characteristics of these groups can be identified:[44]

- Transactions are conducted through alliances of *affiliated* companies. This creates a form of organization intermediate between vertically integrated firms and arm's-length markets.
- Inter-firm relationships tend to be *long-term* and stable, based upon mutual obligations.
- These inter-firm relationships are *multiplex* in form, expressed through cross-shareholdings and personal relationships as well as through financial and commercial transactions.
- Bilateral relationships between firms are embedded within a broader 'family' of *related companies*.
- Inter-corporate relationships are imbued with *symbolic significance* which helps to sustain links even where there are no formal contracts.

Figure 7.13 shows the simplified structure of the two basic types of *keiretsu*:

- *Horizontal keiretsu* are highly diversified industrial groups organized around two key institutions: a core bank and a general trading company (*sogo shosha*). Three of the horizontal *keiretsu* groups (Mitsubishi, Mitsui, Sumitomo) are the successors of the pre-war family-led *zaibatsu* groups that were abolished after 1945. The others are primarily bank-centred.

- *Vertical keiretsu* are organized around a large parent company in a specific industry (for example, Toshiba, Toyota, Sony).

Figure 7.13

The basic elements of the Japanese *keiretsu*

Source: Based, in part, on Gerlach, 1992: Figure 1.1

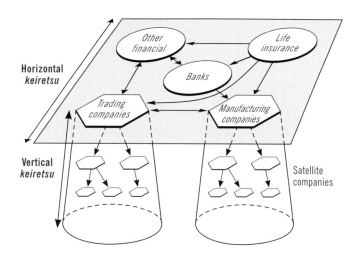

Although distinctive, these two types are not mutually exclusive. For example, Toshiba, is itself a parent company controlling hierarchically substantial numbers of satellite companies including parts suppliers, in a vertically integrated Toshiba group. At the same time, Toshiba is also a member of the horizontally integrated Mitsui industrial group. In fact, the webs of interrelationships are extremely complex, as Figure 7.14 shows. The organizational scale of the leading *keiretsu* is immense. For example, the eight leading horizontal *keiretsu* shown in Figure 7.14 consist of around 900 separate companies but, in effect, they control, in total, some 12,000 companies. In the early 1990s, the 163 leading companies of the six major *keiretsu* effectively controlled more than 40 per cent of all Japanese non-financial enterprises and some 32 per cent of total assets.[45]

A comparison of Korean and Taiwanese business groups[46]

In the previous chapter we noted the strong similarities in the history and developmental experiences of Korea and Taiwan. Their business firms also share common features of Confucian-based familism. 'In all these background variables – economic, social, and cultural – Taiwan and South Korea are as nearly the same as could be imagined between any two countries in the world today. Yet the economies of these two countries are organized in radically different ways.'[47] Although business organizations in both countries are organized as networks of family-owned firms their mode of organization differs considerably.

In Korea, the dominant type of business group is the *chaebol*. The model for such firms was the Japanese *zaibatsu*, the giant family-owned firms which had been so important in the pre-World War II development of the Japanese economy.

The *chaebol* 'are highly centralized, most being owned and controlled by the founding patriarch and his heirs through a central holding company. A single person in a single position at the top exercises authority through all the firms in the group. Different groups tend to specialize in a vertically integrated set of

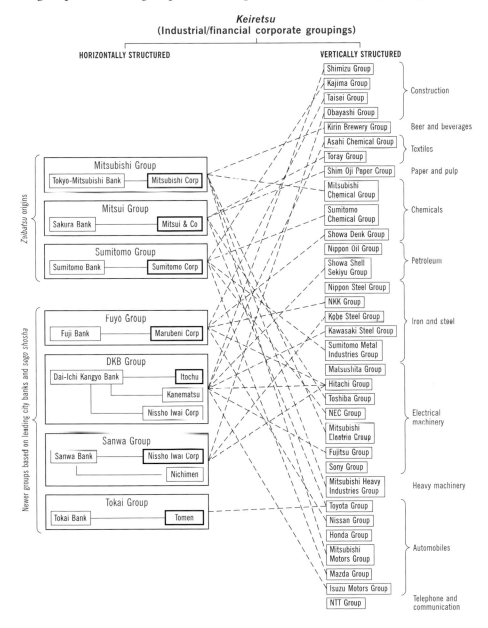

Figure 7.14   Major relationships between horizontal and vertical *keiretsu*

Source: Based on material in Dodwell Marketing Consultants (1992) *Industrial Groupings in Japan*. Tokyo: Dodwell

economic activities.'[48] As a result, the Korean economy became highly concentrated and oligopolistic, while the small- and medium-sized firm sector is relatively underdeveloped. Not only this, but many of these smaller firms are tightly tied into the production networks of the *chaebol*.

Indeed, the *chaebol* have developed as some of the most highly vertically integrated business networks in the world: 'the firms in the *chaebol* are the principal upstream suppliers for the big downstream *chaebol* assembly firms ... in Samsung electronics, most of the main component parts for the consumer electronics division are manufactured and assembled in the same compound by Samsung firms.'[49] In the terminology of Figure 2.7, the *chaebol* are 'producer-driven' networks. Taiwanese business networks, in contrast, have low levels of vertical integration. The more horizontal Taiwanese networks consist of two main types: 'family enterprise' networks and 'satellite assembly' networks (independently owned firms that come together to manufacture specific products primarily for export. The Taiwanese business groups, therefore, are closer to the 'buyer-driven' networks of Figure 2.7.

These contrasts between the Korean and Taiwanese business groups – despite the strong similarities between the two countries – have been explained as arising from

> differences in social structures growing out of the transmission and control of family property. In South Korea, the kinship system supports a clearly demarcated, hierarchically ranked class structure in which core segments of lineages acquire elite rankings and privileges. These are the 'great families' ... In Taiwan, however, the Confucian family was situated in a very different social order ... Unlike Korea (and in the early Chinese dynasties), where the eldest son inherited the lion's share of the estate and all the lineage's communal holdings, in late imperial China the Chinese practiced partible inheritance, in which all sons equally split the father's estate ... This set of practices preserved the household and made it the key unit of action, rather than the lineage itself ... In summary, although based on similar kinship principles, the Korean and the Chinese kinship systems operate in very different ways.[50]

Such differences in socio-cultural practices largely explain the contrasts between the ways that business firms are organized in the two neighbouring countries.

The Overseas Chinese family business network[51]

> In many Western economies, the main efficiencies in coordination derive from large-scale organization. In the case of the Overseas Chinese, the equivalent efficiencies derive from networking.[52]

A very different kind of business network is to be found within the Overseas Chinese entrepreneurial system that underpins much of the dynamic economic development not only of Hong Kong and Taiwan but also throughout much of South East Asia. Their essence has been described as 'weak organizations and strong linkages'.[53] Personal relationships based on reciprocity (*guanxi*) play a central role in contrast to the situation in Western firms where formal contractual arrangements are the norm.[54] The basis of the Overseas Chinese business system is to be found in the specifics of cultural and historical experience and in the set of norms and values derived from a common historical experience and implemented in particular contexts. Three influences have been especially strong in influencing how Overseas Chinese businesses are managed and operated: Confucian value systems of familism and respect for authority; experience as refugees; experience of oppression.

On this basis, five basic characteristics of Overseas Chinese business behaviour can be identified:[55]

- Control of the firm must be retained in the long-term interest of family prosperity.
- Family assets must be protected by hedging of risks.
- Key decision-making should be confined to an inner circle.
- Dependence on outsiders ('non-belongers') for key resources must be limited.
- Inter-firm transactions should be based on 'networks of interpersonal obligation. Personalizing trust-bonds substitutes for a system of legal contracts so normal in Western contexts.'

By its very nature, the Overseas Chinese business is less visible than other, more formally structured businesses. Because of its extended network form, any approach based solely on the legal definition of the firm will fail to capture its full extent. It will also underestimate its significance as a form of transnational business activity. Yet the significance throughout Asia and, increasingly, in other parts of the world, of Overseas Chinese family business networks – what has been termed 'the worldwide web'[56] or the 'bamboo network'[57] – is immense and growing. Their entrepreneurial drive has enabled them to occupy dominant positions in what are otherwise non-Chinese societies, such as Indonesia and Malaysia.

The precise form of Overseas Chinese business varies between different social and institutional contexts in different parts of Asia. In the case of Hong Kong and Singapore, for example, the different political and institutional contexts have produced distinctive entrepreneurial characteristics even though, in both cases, their ethnicity is comparable.

> The spirit and ethos of capitalism in Hong Kong have produced socially, culturally, and politically specific business systems … Ethnic Chinese industrialists in Hong Kong are known for their entrepreneurship and higher propensity to engage in risky business and overseas ventures … The peculiar neoliberal political economy in Hong Kong had several consequences for transnational entrepreneurship in Hong Kong. First, the private sector assumed a leading role in Hong Kong's economic development … Second, the lack of direct state intervention in Hong Kong's industrialization and economic development processes has contributed to the growth of domestic companies in both large firm sectors and small firm networks … Third, the financial system in Hong Kong … is highly favourable for the development of the service sector.[58]

In contrast, although the Chinese mode of business is clearly present in Singapore, it differs in some important respects from that in Hong Kong. The main reason is the very different institutional structure in Singapore:

Notably, a large proportion of local investments, particularly in the manufacturing sector, came from foreign firms and GLCs ... [government-linked companies] ... and their various subsidiaries. The role of indigenous private enterprises in Singapore's industrialization is rather limited ... The majority of Singaporeans have become contented with their job security and are less willing to take specific kinds of risks to launch new business ventures.[59]

Continuity and change?

Despite many decades of international operations, TNCs remain distinctively connected with their home base. This does not mean that TNCs necessarily retain a 'loyalty' to the states in which they originate, although there are strong pragmatic forces at play.

> However great the global reach of their operations, the national firm does, psychologically and sociologically, 'belong' to its home base. In the last resort, its directors will always heed the wishes and commands of the government which has issued their passports and those of their families.[60]

But the nature of the embeddedness process is far more complex than this. The basic point is that TNCs are 'produced' through an intricate process of embedding in which the cognitive, cultural, social, political and economic characteristics of the national home base play a dominant part. TNCs, therefore, are 'bearers' of such characteristics, which then interact with the place-specific characteristics of the countries in which they operate to produce a set of distinctive outcomes. But the point is that the home-base characteristics invariably remain dominant. This is not to claim that TNCs from a particular national origin are identical. This is self-evidently not the case; within any national situation there will be distinctive corporate cultures, arising from the firm's own specific corporate history, which predispose it to behave strategically in particular ways.

> The global corporation, adrift from its national political moorings and roaming an increasingly borderless world market is a myth ... The empirical evidence ... suggests that distinctive national histories have left legacies that continue to affect the behavior of leading ... [TNCs] ... The scope for corporate interdependencies across national markets has unquestionably expanded in recent decades. But history and culture continue to shape both the internal structures of ... [TNCs] and the core strategies articulated through them.[61]

But this does not imply that nationally embedded business organizations are unchanging. On the contrary, the very interconnectedness of the contemporary global economy means that influences are rapidly transmitted across boundaries. This will, inevitably, affect the way business organizations are configured and behave. There 'is essentially a process of co-evolution through which different business systems may converge in certain dimensions and diverge in other attributes'.[62]

For example, the *keiretsu* have been at the centre of Japanese economic development during the post-war period. But the financial crisis in Japan that has persisted since the bursting of the 'bubble economy' at the end of the 1980s has put them under considerable pressure to change at least some of their practices. In particular, the recent influx of foreign capital to acquire significant, sometimes controlling, shares in some of these companies has had a catalytic effect. The most notable example was the acquisition by the French automobile company Renault of 36.8 per cent of the equity of Nissan (see Chapter 11). There are strong pressures, particularly from Western (notably US) finance capital for Japanese business groups to open up to outsiders, to reduce or eliminate the intricate cross-shareholding arrangements, and to become more like Western (that is to say, US) firms with their emphasis on 'shareholder value' rather than the broader socially based 'stakeholder' interests intrinsic to Japanese companies. While, without doubt, some changes are occurring, it would be a mistake to assume that Japanese firms will suddenly be transformed into US-clones. The Japanese have a very long history of adapting to external influences by building structures and practices that remain distinctively Japanese.

Similarly, Korean and other East Asian firms have come under enormous pressure to change some of their business practices in the aftermath of the region's financial crisis of the late 1990s. In Korea, the *chaebol* are being drastically restructured and the relationships with the state reduced. Among Chinese businesses, the strong basis in family ownership and control is being challenged both by internal and external forces. Greater involvement in the global economy is forcing these firms to modify some of their practices.[63] And yet it would be extremely surprising if the distinctive nature of nationally based TNCs were to be replaced by a standardized, homogeneous form despite the dreams of the hyperglobalists.

Conclusion

The aim of this chapter has been to explore three specific aspects of transnational corporations. First, we outlined some of the major theoretical explanations of why firms become 'transnational' by looking at both the macro-structural theories based upon the internationalization of the circuits of capital and the micro-scale theories cast at the level of the individual firm. In fact, both levels of explanation provide valuable insights into the internationalization process; it is important to develop explanatory frameworks that combine both general elements of the way in which the system as a whole operates with specific elements of how firms behave.

Second, we focused on TNCs as networks of internalized relationships. We identified a variety of organizational structures designed to enable firms to implement their chosen competitive strategies. The particular type employed appears to be influenced both by the firm's specific history and by the nature and complexity of the industry environment(s) in which it operates. Strategically, all TNCs have to resolve the basic tension between globalizing pressures on the one hand and localizing pressures on the other.

Third, we challenged the popular conception of the 'placeless' transnational corporation. All TNCs have an identifiable home base and the characteristics of that base continue to exert an influence on how firms behave as they develop international networks of operations. We showed clear differences between TNCs of different nationalities; there is little evidence of TNCs converging towards a single model. Of course, as TNCs move into new environments they have to adapt, to a greater or lesser degree, to local circumstances. On both counts, however, it is very apparent that 'geography matters' to how TNCs coordinate and configure their production chains and, by extension, to the kind of impact they have on the states and communities within which they operate. This is an issue which we will look at specifically in Chapter 9.

Notes

1 D'Aveni (1994: 2).
2 For a detailed treatment of international production as the process of the internationalization of the various circuits of capital see Palloix (1975, 1977).
3 Harvey (1982: 69).
4 General reviews of the theoretical literature are provided by Dunning (1993) and Pitelis and Sugden (1991).
5 Hymer's pioneering contribution was contained in his 1960 PhD thesis which was not published until after his tragic death in the 1970s (Hymer, 1976).
6 Vernon's thinking on the product life cycle evolved through a series of contributions: 1966, 1971, 1974, 1979. Hirsch (1967) and L.T. Wells (1972) both made significant contributions to the product cycle interpretation of international production and trade. Cantwell (1997) evaluates the current applicability of the PLC model.
7 Vernon (1979) recognized some of these limitations.
8 Dunning first introduced his eclectic approach to international production in the 1977. Through many subsequent publications he has developed it into what he now calls the *eclectic paradigm* (see Dunning, 1993).
9 Dunning (1980: 9).
10 The initial concept of 'internalization' (although not the term itself) is generally attributed to Coase (1937). Its specific application to TNCs dates from the 1970s through the work of Buckley and Casson (1976) and Rugman (1981) amongst others.
11 Dunning (1980: 9; emphasis added).
12 Taylor and Thrift (1986: 11).
13 Dicken and Miyamachi (1998); Dicken, Tickell and Yeung (1997); Mason (1994).
14 See Smith (1981) for a discussion of the distinction between basic costs and locational costs.
15 But, as Herod (1997: 2) argues, working people are much more than 'crude abstractions in which labour is reduced to the categories of wages, skill levels, location, gender, union membership and the like, the relative importance of which is weighed by firms in their locational decision-making'.
16 Walker (1999: 264).
17 *The Financial Times* (9 March 2001).
18 Peck (1996) provides an excellent discussion of the 'place' of labour within the capitalist market system. See also Herod (2001); Hudson (2001).
19 For a discussion of some of these influences and how they affect firm structure and behaviour see Bartlett and Ghoshal (1998), Hedlund (1986), Heenan and Perlmutter (1979) and Schoenberger (1997).
20 Heenan and Perlmutter (1979).
21 Bartlett and Ghoshal (1998).
22 Heenan and Perlmutter (1979).
23 Harzing (2000) provides empirical support for the Bartlett and Ghoshal typology. Malnight (1996) provides an interesting case study of Citibank's transition to a network-based organizational structure.
24 Hedlund (1986: 218–30).
25 Roche and Blaine (2000).

26 Birkinshaw and Morrison (1995: 732–5). See also Bartlett and Ghoshal (1998).
27 Schoenberger (1999: 210–11).
28 Kobrin (1987: 104, 107).
29 Precisely what is meant by 'local' is very variable. For most TNCs, it seems to refer either to individual countries or even to entire regions such as the EU, North America, or Asia.
30 Douglas Daft, chairman of Coca-Cola, in *The Financial Times* (27 March 2000).
31 There is a considerable literature on this topic. See, for example, Dicken (2000); Dicken and Malmberg (2001); Dicken and Thrift (1992); Doremus, Keller, Pauly and Reich (1998); Hu (1992, 1995); Kogut (1999); Pauly and Reich (1997); Reich (1996); Ruigrok and van Tulder (1995); Schoenberger (1999); Whitley (1999).
32 Ohmae (1995: 94). Reich (1991) adopts a similar viewpoint.
33 Using slightly different data, Ruigrok and van Tulder (1995: ch. 7) reach similar conclusions.
34 Hu (1992).
35 The concept of embeddedness was introduced in economic sociology (see Chapter 5: n. 10).
36 Dicken, Forsgren and Malmberg (1994: 34).
37 Abo (1994, 1995). See also Beechler and Bird (1999).
38 Doremus, Keller, Pauly and Reich (1998). See also Pauly and Reich (1997).
39 Pauly and Reich (1997: 1, 4, 5, 24).
40 There is large, and growing, literature on this topic. General treatments can be found in Mirza (2000), Orrù, Biggart and Hamilton (1997), Whitley (1999) and Yeung (2000).
41 Yeung (2000: 408).
42 The Japanese *keiretsu* are discussed by Aoki (1984), Fruin (1992), Gerlach (1992) and Helou (1991).
43 Gerlach (1992: xiii).
44 Gerlach (1992: 4).
45 Helou (1991).
46 This section is based primarily on Hamilton and Feenstra (1998).
47 Hamilton and Feenstra (1998: 124).
48 Wade (1990a: 324).
49 Hamilton and Feenstra (1998: 128–9).
50 Hamilton and Feenstra (1998: 134, 135).
51 See Kao (1993); Redding (1991); Whitley (1992); Yeung and Olds (2000).
52 Redding (1991: 45).
53 Redding (1991: 30).
54 Yeung and Olds (2000).
55 Redding (1991: 36).
56 Kao (1993)
57 Weidenbaum and Hughes (1996).
58 Yeung (2002: 98).
59 Yeung (2002: 99).
60 Stopford and Strange (1991: 233).
61 Doremus, Keller, Pauly and Reich (1998: 3, 9).
62 Yeung (2000: 425).
63 See Yeung (2000: 411–24).

CHAPTER 8

'Webs of Enterprise': The Geography of Transnational Production Networks

In Chapter 7 we outlined some of the explanations of why TNCs exist and how they have devised appropriate organizational architectures to implement their competitive strategies. In this chapter, we focus specifically on the *geography of the networks of relationships* that exist both within and between firms as they attempt to coordinate and configure their production activities at different geographical scales. Three sets of relationships are explored:

- First, we examine TNCs as changing geographical networks of internalized relationships.
- Second, we explore the complex networks of externalized relationships that exist between independent and quasi-independent firms.
- Third, we make the connections between the organizational and geographical dimensions of TNC networks to see how they map on to geographical spaces and territories and, in so doing, create the multi-scalar, multi-dimensional structures that form the framework within which the global economy is organized.

The geography of the TNC: configuring the firm's internal production network

> Every business is a package of functions and within limits these functions can be separated out and located at different places.[1]

This observation, made more than 70 years ago, is even more valid today. Over time, the 'limits' have become less and less restrictive, mainly because of developments in the enabling technologies of transportation and communications. As a result, the *geographical extent* to which a TNC can separate out its component units, and their precise *geographical configuration*, have been transformed. Different parts of a TNC have different locational needs and, because these needs can be satisfied in various types of geographical location, each part tends to develop rather distinctive spatial patterns. Some TNC functions tend to be geographically dispersed; others geographically concentrated.

We can illustrate these processes by looking at three of the most important functions of the TNC:

- corporate and regional headquarters offices
- research and development facilities
- production units.

Other functions in the production network, such as marketing, sales, distribution and servicing, tend to be distributed very widely in accordance with the firm's geographical markets. Indeed, a local market presence is becoming essential in most economic sectors.

Corporate and regional headquarters[2]

The *corporate headquarters* is the locus of overall control of the entire TNC, responsible for all the major strategic investment and disinvestment decisions that shape and direct the enterprise: which products and markets to enter or to leave, whether to expand or contract particular parts of the enterprise, whether to acquire other firms or to sell off existing parts. One of its key roles is *financial*. The corporate headquarters generally holds the purse strings and decides on the allocation of the corporate budget between its component units. Headquarters offices are, above all, handlers, processors and transmitters of *information* to and from other parts of the enterprise and also between similarly high-level organizations outside. The most important of these are the major business services on which the corporation depends (financial, legal, advertising) and also, very often, major departments of government, both foreign and domestic. A recent study of headquarters operations in different countries showed that the size and complexity of corporate headquarters varied substantially between firms from different home countries.[3]

Regional headquarters offices constitute an intermediate level in the corporate organizational structure, having a geographical sphere of influence encompassing several countries. Regional headquarters perform several distinctive roles,[4] depending on the stage and nature of development of the TNC's activities. Most commonly, their primary responsibility is to *integrate* the parent company's activities within a region, that is, to coordinate and control the activities of the firm's affiliates (manufacturing units, sales offices, etc.) and to act as the intermediary between the corporate headquarters and its affiliates within its particular region. Regional headquarters, therefore, are both coordinating mechanisms within the TNC and also an important part of the TNC's 'intelligence-gathering' system. A regional headquarters' major role may also be *entrepreneurial*: to act as a base to initiate new regional ventures or to demonstrate to governments that the company has a commitment to the region. In either case, regional headquarters act as 'strategic windows' on regional developments and opportunities.[5]

These characteristic functions of corporate and regional headquarters define their particular *locational* requirements.

- Both require a *strategic location* on the global transportation and communications network in order to keep in close contact with other, geographically dispersed, parts of the organization.
- Both require access to *high-quality external services* and a particular range of *labour market skills*, especially people skilled in information processing.

● Since much corporate headquarters activity involves interaction with the head offices of other organizations, there are *strong agglomerative forces* involved. Face-to-face contacts with the top executives of other high-level organizations are facilitated by close geographical proximity. Such high-powered executives invariably prefer a location that is rich in social and cultural amenities.

On the global scale only a relatively small number of cities contain a large proportion of both corporate and regional headquarters offices of TNCs. Such *global cities* are sometimes described as the geographical 'control points' of the global economy. Figure 8.1 presents a simplified map of these centres (the links shown are diagrammatic only; they are intended solely to give an impression of a connected network of cities). Three global cities – New York, Tokyo and London – stand head and shoulders above all the others. Below them is a tier of other key cities in each of the three major economic regions of the world – Western Europe, North America and Asia – with other representation in Australia and Latin America. One of the striking features of the geography of corporate headquarters is that very few, if any, major TNCs have moved their ultimate decision-making operations outside their home country. This is a further indicator of the continuing significance of the home base for corporate behaviour.

Both the further integration of the European Union and the rapid growth of the East Asian economies have stimulated the need for regional headquarters in those areas. For example, many US TNCs that have had a presence in Europe for a very long time have established European headquarters to coordinate their regional operations. A number of Japanese TNCs have set up regional headquarters in Europe as the scale and extent of Japanese operations within Europe have increased.[6] In East Asia, the Singapore government introduced an Operational Headquarters Scheme to attract foreign TNCs to establish their regional headquarters in Singapore.[7] *Within* individual countries, on the other hand, the

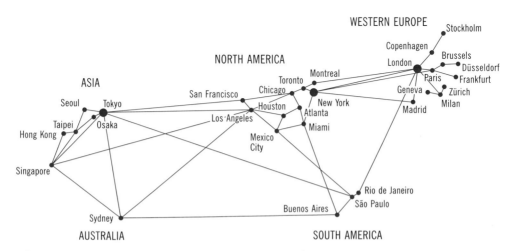

Figure 8.1 Major corporate and regional headquarters centres

Source: Based, in part, on Cohen, 1981: 307–8; Friedmann, 1986: Figure 1

locational pattern of both corporate – and especially regional – headquarters is far from static, with substantial geographical decentralization of corporate head-quarters out of the city centres of New York and London. In the case of London, most of these shifts are short distance to the less congested outer reaches of the metropolitan area. In the United States, on the other hand, there appears to be a much higher degree of locational change in headquarters functions. Detailed empirical research for the period 1974–89 shows a considerable degree of change. There have been considerable shifts away from New York towards such regional cities as Atlanta, Dallas–Fort Worth, Philadelphia and St Louis.[8] Occasionally – though rarely – truly spectacular individual locational shifts do occur, as in the case of Boeing's decision to relocate its headquarters from Seattle to Chicago in 2001.

Nevertheless, the fact that the locational needs of corporate and regional head-quarters of TNCs are satisfied most readily in very large cities means that they tend not to be spread very widely within any particular country. In the United Kingdom, for example, there are very few corporate headquarters of major firms or regional headquarters of foreign TNCs outside London and the South East; in France few locate outside Paris. In Italy the most important centre is Milan, in the highly industrialized north, which is more important than Rome as a location for foreign TNCs.

Research and development facilities[9]

In general, TNCs spend more on R&D than other firms, as part of their drive to remain competitive and profitable on a world scale. Innovation – of new products or new processes – is critically important for such firms in an increasingly com-petitive global economy. The R&D function is, therefore, highly significant for the TNC. Indeed, it has become even more important with the intensified pace and changing nature of technology.

The process of R&D is a complex sequence of operations (Figure 8.2), in which three major phases can be identified. Each phase tends to have rather different locational requirements, although, in each case, the TNC has to reconcile several factors. One of these is the advantage of scale economies derived from concen-trating R&D against the need to locate R&D closer to other corporate functions or to markets.

- *Phase I* is largely concerned with applied scientific and marketing research. The primary need is for access to the basic sources of science and marketing information – universities, research institutes, trade associations, and the like.
- *Phase II* is concerned with product design and development. It tends to require large-scale teamwork, that is, access to a sufficiently large supply of highly qualified scientists, engineers and technicians.
- *Phase III* is the 'de-bugging' phase and the adaptation of the new product to local circumstances. Its locational requirements are for quick two-way contact with the users of the innovation: the production or marketing units themselves.

These phases may be carried out either inside the TNC itself or through external networks: a *double network* system.[10] In this section we are concerned with the first of these – internalized R&D.

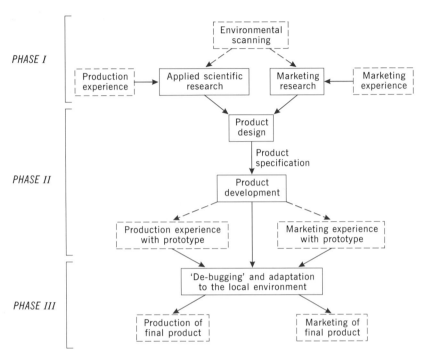

Figure 8.2 Major phases in the R&D process

Source: Based, in part, on Buckley and Casson, 1976: Figure 2.7

The type of R&D undertaken by TNCs in their own foreign locations can be classified into three major categories (Figure 8.3). The lowest level of R&D activity is the *support laboratory*, whose primary purpose is to adapt parent company technology to the local market and to provide technical back-up. It is the equivalent of phase III in Figure 8.2 and is by far the most common form of overseas R&D facility. The *locally integrated R&D laboratory* is a much more substantial unit, in which product innovation and development are carried out for the market in which it is located. It is the equivalent of phase II in Figure 8.2. The *international interdependent R&D laboratory* is of a quite different order. Its orientation is to the integrated global enterprise as a whole rather than to any individual national or regional market. Indeed, there may be few, if any, direct links with the firm's other affiliates in the same country. Only a small number of technologically intensive global corporations operate international interdependent laboratories.

The kind of R&D activity, and its locational pattern, varies according to the specific market orientation of the TNC:[11]

> **Support laboratory**
>
> *Function:* Technical service centre; translator of foreign manufacturing technology
>
> *Reason for establishment:* Response to market growth; differing market conditions; expectation of continuing stream of technical service projects

> **Locally integrated R&D laboratory**
>
> *Function:* Local product innovation and development; transfer of technology
>
> *Reason for establishment:* Improved status of subsidiaries; concept of overseas operations as fully developed business entities; identification of new business opportunities outside home country. Frequently develop out of support laboratories

> **International interdependent R&D laboratory**
>
> *Function:* Basic research centre; close links with international research programme; may or may not interact with the firm's foreign manufacturing affiliates
>
> *Reason for establishment:* Operation of coordinated world R&D programmes as part of global product strategies involving the manufacture of a single product line for world markets. Units tend to be created by direct placement

Figure 8.3   Three major types of corporate R&D facility

Source: Based on material in Hood and Young, 1982

- TNCs with a strong *home market* orientation tend to carry out little foreign R&D other than of the support laboratory type. Such firms tend to regard their foreign sales as not requiring any further R&D beyond that carried out for their domestic market.

- *Host-market* TNCs – those oriented towards the national (or regional) market in which their foreign operations are located – operate both support laboratories and also higher-level locally integrated laboratories. The most important locational criteria are proximity to the firm's foreign markets and the fact that the firm's foreign operations are sufficiently substantial to justify separate R&D activities. Such activities tend to be located in the firm's biggest and most important foreign markets.

- *Global-market* firms are the globally integrated corporations whose orientation is to global, rather than to national, markets. Their R&D activities include both support and locally integrated laboratories but, in addition, their adoption of a globally integrated production strategy leads them to establish specially designed international interdependent research laboratories. The major locational criteria for these global-market R&D activities are the availability of highly skilled scientists and engineers, access to sources of basic scientific and technical developments – especially of high-quality universities – and an appropriate infrastructure.

Precisely how these R&D activities are organized by TNCs varies considerably, particularly according to the type of organizational structure in use. TNCs operating a more hierarchical structure continue to organize their R&D in a highly centralized, top-down, fashion. In contrast, TNCs with a more heterarchical, multi-centred, structure tend to have a more decentralized R&D system. However, there is considerable disagreement over the extent to which TNCs continue to concentrate their R&D in their home countries or to disperse it more widely to their foreign subsidiaries.

> Foreign R&D investment tends to follow production abroad. The more production is located abroad, the more likely research and development will be too. Research and development by … [TNCs] complements their manufacturing and sales activities in major markets. For most countries, there is a strong correlation between foreign affiliate shares of R&D expenditures and their domestic sales. But most … [TNCs] have kept their core technologies or strategic projects in their home economies, and do design and development abroad to adapt products to local markets … Much evidence points to growing international dispersion of [trans]national corporations' research activities. Key know-how is increasingly scattered internationally, although predominantly within the Triad of Europe, Japan, and North America.[12]

Hence, there seem to be two trends in the internationalization of R&D activities by TNCs. On the one hand, TNCs continue to show a very strong preference for keeping their high-level R&D in their home countries. For example, the United States Office of Technology Assessment calculated that only around 13 per cent of the total R&D performed by US manufacturing TNCs was located abroad.[13] Detailed empirical analyses of patent data for almost 600 firms[14] produced the following conclusions:

- Only 43 firms in the sample (7.6%) located more than half of their technological activities outside their home country.
- More than 40 per cent of the sample performed less than 1 per cent of their technological activity abroad.
- More than 70 per cent of the sample performed less than 10 per cent of this activity abroad.
- Very little of the overseas R&D activity of firms from the United States, Japan, Germany, France and Italy is located outside the 'global triad'.
- Most of the apparent increases in overseas R&D came about through merger and acquisition rather than through internal growth and geographical expansion.

On the other hand, there is undoubtedly evidence of increasing geographical dispersal of R&D activities within TNC networks, although the dominant share of these activities continues to remain in the TNC's home country. For example, by the late 1990s, there was 'an increasing share of company-financed R&D performed abroad by US firms *as compared to domestically financed* industrial R&D … US firms' investment in overseas R&D increased three times faster than did company-financed R&D performed domestically.'[15]

Why should home-country bias in R&D persist? Why do TNCs show such a strong preference for keeping their major R&D activities close to their home base?

The answer lies in the importance of the kinds of *untraded interdependencies* discussed earlier (see Chapters 2 and 4):

> Two key features related to the launching of major innovations may help explain the advantages of geographic concentration: the involvement of inputs of knowledge and information that are essentially 'person-embodied', and a high degree of uncertainty surrounding outputs. Both of these are best handled through geographic concentration. Thus it may be most efficient for firms to concentrate the core of their technological activities in the home base with international 'listening posts' and small foreign laboratories for adaptive R&D.[16]

Not only are corporate R&D activities strongly concentrated in TNCs' home countries, but also the spatial pattern *within* nations is very uneven. The support laboratories are the most widely spread in so far as they generally locate close to the production units, although not every production unit has an associated support laboratory. But the larger-scale R&D activities tend to be confined to particular kinds of location. The need for a large supply of highly trained scientists, engineers and technicians, together with proximity to universities and other research institutions, confines such facilities to large urban complexes. These are often those that are also the location of the firm's corporate headquarters. A secondary locational influence is that of 'quality of living' for the highly educated and highly paid research staff: an amenity-rich setting including a good climate and potential for leisure activities as well as a stimulating intellectual environment.

Spatial patterns of corporate R&D in both the United States and the United Kingdom illustrate both of these locational influences. In the United States, corporate R&D is still predominantly a big-city activity despite recent growth in smaller urban areas. The pull of the amenity-rich environment is illustrated by the considerable concentration of R&D activities in locations such as Los Angeles, San Francisco and San Diego in California, Denver–Boulder in Colorado and the 'Research Triangle' in North Carolina. In the United Kingdom, corporate R&D, like corporate headquarters and regional offices, is disproportionately concentrated in South East England. Almost two-thirds (61 per cent) of the research undertaken by foreign affiliates in the UK is located in the south-eastern region of the country (compared with only 40 per cent of domestic firms' research).[17]

Production units

There are clearly some identifiable geographical regularities in the patterns of both corporate headquarters and R&D functions. This is because the locational needs of corporate offices and R&D laboratories are broadly similar for all firms, regardless of the particular industries in which they are involved. This is not so for production units, whose locational requirements vary considerably according to the specific organizational and technological role they perform within the enterprise and the geographical distribution of the relevant location-specific factors. It is certainly true that, compared with corporate headquarters and R&D facilities, production units of TNCs have become more and more dispersed geographically. But there is no single and simple trend or pattern of dispersal common to all activities, whether at the global scale or within individual nations. The pattern varies greatly from one industry to another. Figure 8.4 illustrates diagrammatically four types of geographical orientation which a TNC might adopt for its production units.

(a) Globally concentrated production

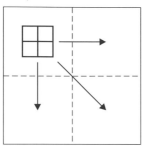

(b) Host-market production

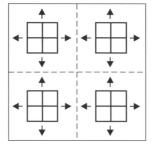

(c) Product-specialization for a global or regional market

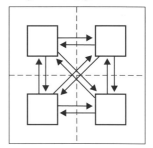

All production occurs at a single location. Products are exported to world markets

Each production unit produces a range of products and serves the national market in which it is located. No sales across national boundaries. Individual plant size limited by the size of the national market

Each production unit produces only one product for sale throughout a regional market of several countries. Individual plant size very large because of scale economies offered by the large regional market

(d) Transnational vertical integration

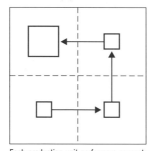

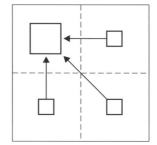

Each production unit performs a separate part of a production sequence. Units are linked across national boundaries in a 'chain-like' sequence – the output of one plant is the input of the next plant

Each production unit performs a separate operation in a production process and ships its output to a final assembly plant in another country

Figure 8.4 Some major ways of organizing the geography of TNC production units

Globally concentrated production

Figure 8.4(a) presents the simplest case. All production is concentrated at a single geographical location (or, at least, within a single country) and exported to world markets through the TNC's marketing and sales networks. This is a procedure consistent with the classic global strategy shown in Table 7.1 – the kind of strategy followed by many Japanese companies until their relatively recent move towards more dispersed global production.

Host-market production

In Figure 8.4(b), production is located in, and oriented directly towards, a specific host market. Where that market is similar to the firm's home market the product is likely to be identical to that produced at home. The specific locational criteria for the setting up of host-market plants are:

- size and sophistication of the market as reflected in income levels
- structure of demand and consumer tastes
- cost-related advantages of locating directly in the market
- government-imposed barriers to market entry.

In effect, this kind of production is import substituting. Most of the manufacturing plants established by US TNCs in Europe in the post-1945 period were of this kind, in many cases following a product life cycle sequence (see Chapter 7). The more recent surge of European manufacturing investment in the United States was also directly host-market related. Similarly, the large markets of some developing countries, such as Brazil, have attracted considerable numbers of TNC manufacturing affiliates whose role is to serve that market directly.

With changes in the 'space-shrinking' technologies, the establishment of a production unit in a specific geographical market becomes less necessary in purely cost terms. There are, however, two reasons for the continued development of host-market production:

- The need to be close to the market in order to be sensitive to variations in customer demands, tastes and preferences, or to be able to provide a rapid after-sales service. As we have seen, sensitivity to local geographical differences continues to be an important issue even where TNCs pursue broadly global strategies.
- The existence of tariff and, particularly, non-tariff barriers to trade. Tariff barriers have been a significant locational factor from the very early days of transnational investment. Today it is the various kinds of non-tariff barrier that have become most prevalent (see Chapter 5). Both types of barrier have stimulated TNCs to jump over them and to establish direct production units to serve the local market. The recent growth of Japanese and other Asian manufacturing investment in Europe and in North America was substantially a response to the actual, or threatened, existence of non-tariff trade restrictions.

Product specialization for a global or regional market

During the past four decades or so a radically different form of production organization has become increasingly prominent. Figure 8.4(c) shows production being organized geographically as part of a rationalized product or process strategy to serve a global or a large regional market (such as the European Union or the NAFTA). The existence of a huge internal market, together with differences in factor endowments between member nations, facilitates the establishment of very large, specialized units of TNCs to serve the entire regional market rather than single national markets. The key locational consideration, therefore, involves the 'trade-off' between

- economies of large-scale production at one or a small number of large plants and
- additional movement costs involved in assembling the necessary inputs and in shipping the final product to a geographically extensive regional market.

Transnational vertically integrated production

A rather different kind of rationalized production strategy involves geographical specialization by process or by semi-finished product. As we saw in Chapter 4, technological innovations in the production process permit a number of processes to be fragmented into separate parts and have led to a greater degree of standardization in some manufacturing operations. Parallel developments in the technologies of transportation and communications have introduced a much enhanced flexibility into the geographical location of the production process. Hence, TNCs can locate production units to take advantage of geographical variations in production costs at a global scale. In other words, transnational vertical integration becomes feasible, in which different parts of the firm's production system are located in different parts of the world. Materials, semi-finished products, components and finished products are transported between geographically dispersed production units in a highly complex web of flows.

In such circumstances, the traditional connection between production and market is broken. The output of a manufacturing plant in one country may become the input for a plant belonging to the same firm located in another country or countries. Alternatively, the finished product may be exported to a third-country market or to the home market of the parent firm. In such cases, the host country performs the role of an 'export platform'. The plants themselves are sometimes termed 'workshop affiliates'; their role is to act as international sourcing points for the TNC as a whole. Figure 8.4(d) shows two ways in which such international process specialization might be organized as part of a vertically integrated set of operations across national boundaries.

Such offshore sourcing and the development of vertically integrated production networks on a global scale were virtually unknown before the early 1960s. The pioneers were US electronics firms that set up offshore assembly operations in East Asia as well as in Mexico. Since then, the growth of such international production networks has been extremely rapid, although it is far more important in some types of activity than in others, as we shall see in the case study chapters of Part Three.

These practices of international intra-firm sourcing have become an increasingly important mechanism of global integration of production processes in which the more mobile factors, such as technology, management and equipment, are moved to the site of the least mobile. The broad trend, as Figure 8.5 suggests, is from 'stand-alone' strategies to those of more complex integration in which different parts of a firm's operations are reconfigured and relocated according to the relative advantages of alternative locations.

However, the choice of location for a production unit on the global scale is by no means as simple as it is often made out to be. It is not just a matter of looking at differences in labour costs between one country and another or at the incentives offered as part of an export-oriented government policy. Despite the enormous shrinkage of geographical distance that has occurred, the relative geographical location of parent company and overseas production unit may still be significant. The sheer organizational convenience of geographical proximity may encourage TNCs to locate offshore production in locations close to their home country even when labour costs there are higher than elsewhere. A clear example of this is Mexico, in the case of US firms, and parts of Southern and Eastern Europe in the case of European firms.

Of course, just as geographical proximity may override differentials in labour costs so, too, other locational influences may dominate in any particular case. For the largest TNCs the world is indeed their oyster. Their production units are

Figure 8.5

From stand-alone to complex integration strategies in TNCs
Source: UNCTAD, 1994: Figure III.2

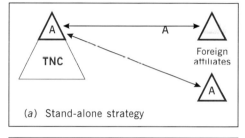

(a) Stand-alone strategy

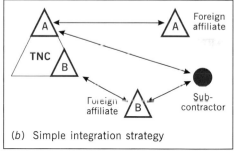

(b) Simple integration strategy

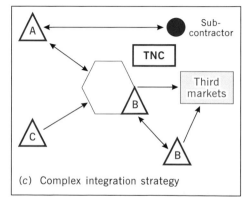

(c) Complex integration strategy

spread globally, often as part of a strategy of *dual or multiple sourcing* of components or products. This is one way of avoiding the risk of over-reliance on a single source whose operations may be disrupted for a whole variety of reasons. In a vertically integrated production sequence, in which individual production units are tightly interconnected, an interruption in supply can seriously affect the other units, perhaps those located at the other side of the world. In an extreme case, a whole segment of the TNC's operations may be halted.

Restructuring and reorganization within the TNC

Transnational corporate networks, and their resulting spatial patterns, are always in a continuous state of flux. At any one time, some parts may be growing rapidly, others may be stagnating, others may be in steep decline. The functions performed by the component parts and the relationships between them may alter. Change itself may be the result of a planned strategy of adjustment to changing internal and external circumstances or the 'knee jerk' response to a sudden crisis. Whatever its origin, however, corporate change will have a specific spatial expression. The changes that occur within the TNC itself will be projected into particular kinds of impact on the localities in which the component parts are located, relocated, expanded or contracted.

Forces underlying reorganization and restructuring

The forces that may lead to corporate reorganization and restructuring can be divided broadly into two categories:

- *External conditions*. These may be negative pressures such as declining demand, increased competition in domestic or foreign markets, changes in the cost or availability of production inputs, militancy and resistance of labour forces in particular places, the pressure of national governments to modify their activities or even to cede control. For example, the shock waves generated by the spectacular collapse of parts of the IT industry in 2001 generated huge restructuring measures among firms in those industries. Conversely, changes in external conditions may be positive rather than negative; for example, the growth of new geographical markets or the availability of new production opportunities. A good illustration is the formation of regional economic groupings, such as the European Union or the NAFTA, that dramatically alter the pattern of investment opportunities. The creation of a large regional market made up of separate nation-states, each with its own resource endowment and production cost attributes but with free movement of materials and products across national boundaries, provides an unprecedented opportunity for TNCs to restructure their production activities to serve the regional market. Investments that had made sense in the context of an individual national context are no longer necessarily rational in the wider context (see Figure 8.4c)

- *Internal pressures*. These may relate to the enterprise as a whole or to one or other individual parts: for example, sales may be too low in relation to the firm's target, production costs may be too high. In a TNC the performance of individual plants in widely separate locations can be continuously monitored and compared to assess their efficiency. A key influence is often the 'new broom factor' – a new chief executive who undertakes a sweeping evaluation of the enterprise's activities and makes changes that stamp his/her authority on the firm.

In reality, external and internal pressures may be so closely interrelated that it is often difficult to disentangle one from the other. More than this, precisely how firms both identify and respond to changes in their circumstances is very much conditioned by the firm's *culture*.[18]

Complex corporate restructuring is occurring at all geographical scales, from the global to the local, as strategic decisions are made about the organizational coordination and geographical configuration of the TNC's production network. Decisions to centralize or to decentralize decision-making powers or to cluster or disperse some or all of the firm's functions in particular ways are, however, contested decisions. They are the outcome of power struggles within firms, both within their headquarters and between headquarters and affiliates. How they are resolved depends very much on the nature and the location of the dominant coalition. Such processes also have to be seen within the context of the fundamental tension facing TNCs: whether to globalize fully or whether to respond to local differentiation (see Figure 7.12). Specifically, corporate restructuring may occur in a variety of ways (Figure 8.6), involving in some cases technological change, changes in work practices, rationalization of corporate activities, changes in the extent to which different functions are internalized, and increased international investment.

Figure 8.6

Major forms of corporate restructuring

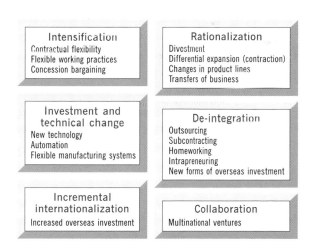

The geography of reorganization and restructuring

Whether corporate reorganization is the result of a consciously planned strategy for 'rational' change or simply a reaction to a crisis (internal or external), its spatial outcome may take several different forms. In Figure 8.7 two broad categories of spatial-organizational change are identified:

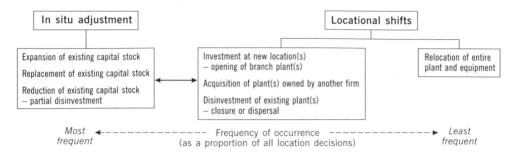

Figure 8.7 Reorganization, restructuring and spatial change: major types of investment-location decision

- *In situ adjustment* to the existing network of production units is by far the most common form of adjustment. A major advantage of the multi-locational firm is that it can make substantial adjustments without necessarily engaging in relocation. The capacity of an existing plant can be increased to achieve economies of scale or reduced (partial disinvestment) to shed surplus capacity; an existing plant's capital stock can be replaced with new technology. In such ways, the importance and even the actual function of production units can be altered as the TNC reallocates tasks among its existing geographically dispersed operations. Change at an existing plant may be either a gradual process of incremental adjustment or a more sudden change to its scale or function.

- *Locational shifts* explicitly involve abrupt change because they consist of either an increase or decrease in the number and location of plants operated by the enterprise or even, in rare cases, the physical relocation of an entire plant. The most common locational shifts within a TNC's production network are: disinvestment at an existing plant; greenfield investment at a new location; acquisition of plants belonging to another firm. Acquisition and merger are particularly important mechanisms of corporate adjustment and growth.

Thus, reorganization, restructuring and the resulting spatial changes are an inevitable aspect of the evolution of transnational corporations. The actual form such change takes depends upon forces both internal and external to the firm. The very large global corporations are developing into *global scanners*. They use their immense resources to evaluate potential production locations in all parts of the world. The performance of existing corporate units can be monitored and evaluated against competitors, against the rest of the corporate network, and also against potential alternative locations. Those existing plants that fall short of expectations created by such *benchmarking* procedures[19] may be disposed of. As plants become obsolete in one location they are closed down; whether or not new investment occurs in the same locality depends upon its suitability for the TNC's prevailing strategy. The chances are, in many cases, that the new investment will be made at a different location – quite possibly in a different country altogether.

However, we should beware of over-exaggerating the speed and ease with which TNCs can, and do, radically restructure their operations. There are 'barriers to exit' – in many cases production units represent huge capital investments which cannot be written off lightly. But there are other kinds of sunk costs that impose additional exit barriers.[20] 'These are costs that cannot be recovered (for example by selling off surplus assets) or closed out in the short run (as with the case of pension liabilities) even when an operation is terminated'.[21] Political pressures may also inhibit firms from closing plants, especially in areas of economic and social stress. On the other hand, TNCs do have a highly tuned capacity to *switch* and *re-switch* operations within their existing corporate network. They also have the resources to alter the shape of their spatial network through locational shifts.

Within these broad restructuring processes by TNCs, four general tendencies are especially apparent:

- a redefinition of the firm's core activities through stripping away activities that no longer 'fit' the firm's strategy
- a repositioning of the firm's focus along the production chain to place a greater emphasis on downstream, service functions
- a geographical reconfiguration of the firm's production network internationally – and in some cases globally – to redefine the roles and functions of individual corporate units
- an organizational reconfiguration of the firm's production network that involves redefining the boundary between internalized and externalized transactions.

It is this latter point that forms the focus of the next section of this chapter.

TNCs within networks of externalized relationships

TNCs are locked into *external* networks of relationships with a myriad of other firms: transnational and domestic, large and small, public and private (see Figure 2.6). TNCs are best understood as 'a dense network at the centre of a web of relationships'.[22] In other words, the *boundary* between what is contained inside and outside the firm has become far more blurred.[23] It is through such interconnections that a very small firm in one country may be directly linked into a global production network, whereas most small firms normally serve only a very restricted geographical area. Such interrelationships between firms of different sizes and types increasingly span national boundaries to create a set of *geographically nested relationships from local to global scales*. These inter-firm relationships are the threads from which the fabric of the global economy is woven. It is through such links that changes are transmitted between organizations and, therefore, between different parts of the global economy.

Connecting the production network: relationships between customers and suppliers

In procuring inputs of materials, intermediate products and services for its operations a firm has three broad choices:

- It can produce these inputs for itself in one of its own units.
- It can purchase the inputs from independent suppliers through conventional, arm's-length, market transactions.
- It can procure the inputs through a longer-term collaborative relationship with suppliers.

In fact, these 'make', 'buy', or 'network' alternatives generally exist all at the same time 'In practice, companies blend in-house supply, arm's-length transactions based on price and quality, and "embedded" relations grounded in networks of loyalty, reciprocity and trust in constructing their competitive strategies'.[24] Possibly between 50 and 70 per cent of manufacturing costs are spent on purchased inputs and this proportion appears to be increasing. The general trend is for a greater proportion of inputs to be outsourced to supplier firms. Some of these purchases will be 'off-the-shelf' sourcing from independent suppliers at the arm's-length market price. However, an increasingly significant proportion of such purchases is made on the basis of longer-term relationships whereby a customer firm *subcontracts* certain tasks to independent firms.

The nature of the subcontracting relationship

Subcontracting is a kind of half-way house between complete internalization of procurement on the one hand and arm's-length transactions on the open market on the other. Figure 8.8 sets out the basic elements of the subcontracting

Technical aspects of production
- Subcontracting processes ⎫
  ⎬ *Industrial*
- Subcontracting ⎭ *subcontracting*
  components
- Subcontracting *Commercial*
  entire products *subcontracting*

Nature of the principal firm
- Producer firm (both industrial and commercial subcontracting)
- Retailing/wholesaling firm (commercial subcontracting)

Type of subcontracting (motivation of the principal firm)
- Speciality subcontracting
- Cost-saving subcontracting
- Complementary or intermittent subcontracting

Types of relationship between principal firm and subcontractor
- Time period may be long-term, short-term, single-batch
- Principal may provide some or all materials or components
- Principal may provide detailed design or specification
- Principal may provide finance, e.g. loan capital
- Principal may provide machinery and equipment
- Principal may provide technical and/or general assistance and advice
- Principal is invariably responsible for all marketing arrangements

Geographical scale
- Within-border, i.e. *domestic subcontracting*
- Cross-border, i.e. *international subcontracting*

Figure 8.8 Major elements of the subcontracting system

relationship. Generally, the firm placing the order or contract is known as the 'principal firm'; the firm carrying out the order is known as the 'subcontractor'. There are two major types of subcontracting:

- *Commercial subcontracting* involves the manufacture of a finished product by a subcontractor to the principal's specifications. The subcontractor plays no part in marketing the product, which is generally sold under the principal's brand name and through its distribution channels. The principal firm may be either a producer firm, that is, one that is also involved in manufacturing, or a retailing or wholesaling firm whose sole business is distribution. This distinction is closely related to the notion of producer-driven and buyer-driven production networks discussed in Chapter 2. Whereas a producer firm may engage in both industrial and commercial subcontracting, retailers/wholesalers are confined to commercial subcontracting.

- *Industrial subcontracting* can be subdivided into three types according to the motivation of the principal firm. *Speciality* subcontracting involves the carrying out, often on a long-term or even a permanent basis, of specialized functions which the principal chooses not to perform for itself but for which the subcontractor has special skills and equipment. *Cost-saving* subcontracting is self-explanatory: it is based upon differentials in production costs between principal and subcontractor for certain processes or products. *Complementary or intermittent* subcontracting is a means adopted by principal firms to cope with occasional surges in demand without expanding their own production capacity. In effect, the subcontractor is used as extra capacity, often for a limited period or for a single operation.

The actual relationship between principal and subcontractor can also take a variety of forms, as Figure 8.8 indicates. The length of time involved may be long or short. The principal's involvement may vary in terms of finance, technology, design, the provision of materials and equipment. Invariably, however, the principal is solely responsible for marketing the finished products or for arranging further assembly or processing. This kind of subcontracting is often termed *original equipment manufacture* (OEM).

Costs and benefits of subcontracting to the participants

The precise advantage of subcontracting to the *principal firm* depends very much on the type of subcontracting involved. In general, however, it offers a number of benefits:

- The firm avoids having to invest in new or expanded plant.
- It offers a degree of flexibility: it is easier to change subcontractors than to close down or reduce the firm's own fixed capacity.
- By entering into a contractual agreement the principal firm gains a certain amount of control over the operation.
- It is one way of externalizing some of the risks and costs of certain operations.

From the viewpoint of the *subcontractor* there are both costs and benefits. The major *costs* include:

- Bearing some of the risks involved. Small subcontracting firms, in effect, perform a 'shock-absorbing' role for large firms. Subcontractors tend to be both expandable and expendable, particularly where they are small firms in an unequal power relationship with large firms.
- Devoting an excessive proportion of its total production to subcontract work for a particular customer. In effect, the subcontractor becomes part of a vertically integrated operation, but without the full benefits of such involvement. As such, its freedom to move into new products or new markets may be limited. The problem is greatest for the small subcontractor where the principal firm specifies the product in detail and where the subcontractor depends upon the principal for product and process development.

On the other hand, small firms may well *benefit* substantially from their subcontracting role:

- Access is gained to particular markets, via brand names, that would otherwise be unattainable.
- Continuity of orders (in some cases over a long period of time) is assured.
- Capital is injected in the form of equipment and access to technology.

In many respects, therefore, the subcontracting relationship is *symbiotic* – a division of labour between independent firms – in which each partner contributes to the support of the other. Subcontracting operations are especially important for small firms. Many firms start out as subcontractors to larger firms, and it is certainly an important channel through which small entrepreneurial firms can operate.

The geographical dimension of subcontracting

As a process, subcontracting is as old as industrialization itself, if not older. The 'putting-out' system was a key element of most industries from their earliest stages. It depended, essentially, on close geographical proximity between firms and their subcontractors. The very fine and intricate network of subcontracting relationships, based on the externalization of transactions in the production chain, often led to the development of highly localized industrial districts. Such tight, functionally and transactionally based, geographical agglomerations of linked economic activities declined in most Western industrial countries with increasing speed between the 1960s and early 1980s, although they persisted in Japan.

Indeed, Japan still has one of the most highly developed domestic subcontracting networks. Each large Japanese firm is surrounded by a constellation of small and medium-sized subcontracting firms which act as suppliers of components or perform specialist processes to the specification, and the timetable, laid down by the controlling large firm. The Japanese subcontracting system, with its sharp distinction between the two major segments, contributed a great deal to the international competitiveness of the Japanese economy, as discussed in previous chapters.

However, conditions in the myriad of small subcontracting firms are very different from those in the major firms. The subcontracting segment has none of the much-lauded qualities of lifetime employment and corporate paternalism that exist in the major corporations. Competitiveness within the subcontracting segment is fierce; the small firms are very heavily subservient to the stringent demands of the principal companies.

One of the most significant developments of the past 40 years has been the extension of subcontracting across national boundaries – that is, the emergence of *international subcontracting* as an important global activity. Relatively low transportation costs, plus the ability to control and coordinate the operation of a long-distance subcontracting system, allow firms to take advantage of very low labour costs in developing countries. Various kinds of international subcontracting relationship are possible. Figure 8.9 distinguishes between direct international subcontracting, that is, between independent firms located in separate countries (Figure 8.9a), or indirect, where the principal is an overseas affiliate of a transnational corporation and the subcontractor is either a local firm or perhaps an affiliate of another TNC (Figure 8.9b).

The basic driving force, for the most part, was cost-minimization, especially of labour costs. However, the adoption of *just-in-time* systems, described in Chapter 4, is possibly changing this simple cost equation. Certainly the adoption of such systems, with their need for very frequent delivery of components, would suggest a possible geographical re-concentration of supplier firms and customers. The evidence is conflicting in this regard, with a mixture of arrangements involving both long- and short-distance linkages, depending upon the particular circumstances in specific industries and firms.

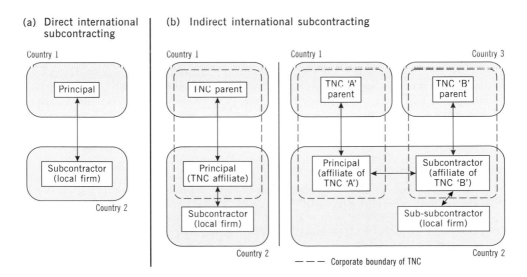

Figure 8.9 Types of international subcontracting relationship

Source: Based on Michalet, 1980: 51–2

One undoubted development is the tendency for many firms to move towards *closer functional relationships* with their suppliers. Rather than merely seeking out the lowest-cost supplier and little else, there is a strong move towards the nomination of *preferred suppliers* with whom very close relationships are developed. Such suppliers are increasingly being given greater responsibility for the quality of their outputs and, indeed, are playing a more direct role in the design of products. The result, in many cases, is the development of a system of *tiered suppliers*:

> In tiered production, the buying company deals primarily with the top-tier suppliers, while lower-tier suppliers are managed by those above them in the pyramid ... A tiered production system reduces the need for a single company to manage the entire supply chain. Close working relationships are developed between the tiers above and below ... Tighter integration with the suppliers is achieved using the tiered production model ... [25]

International strategic alliances[26]

The development of preferred supplier relationships as described in the previous section takes us towards one of the most significant developments in the global economy in recent years: the growth and spread of *strategic alliances* between firms at an international scale. In fact, collaborative ventures between firms across national boundaries are nothing new. What is new is their current scale, their proliferation and the fact that they have become *central* to the global strategies of many firms rather than peripheral to them. Most strikingly, the overwhelming majority of strategic alliances are *between competitors*. One study found that almost three-quarters of alliances were between two companies in the same market.[27] In other words, they reflect a new form of business relationship, a 'new rivalry ... in the way collaboration and competition interact'.[28]

There has been a dramatic increase in the number of strategic alliances during the past two decades:

> The number of new strategic alliances (both domestic and international) increased more than six-fold during the period 1989–1999, from just over 1,000 in 1989 (of which around 860 are cross-border deals) to 7,000 in 1999 (cross-border deals: 4,400) ... there are indications that recent alliances, particularly joint ventures, are far larger in scale and value terms than earlier partnerships. In each year of the 1990s, international partnerships linking firms from different national economies are always the majority of these alliances. International strategic alliances accounted for 68% of all alliances (numbering 62,000) between 1990–99. On average, there are about two international strategic alliances for every domestic partnership, illustrating that globalization is a primary motivation for alliances.[29]

This proliferation of strategic alliances has a strong geographical bias in three respects. First, North American firms were involved in two-thirds of the international strategic alliances formed in the 1990s. In comparison, European firms were involved in one-third, and Asian firms in just over one-quarter, of these alliances. Second, the pattern of alliances is strongly regionalized. In particular, the majority of the strategic alliances involving North American firms were located within

North America.[30] Third, specific circumstances in certain parts of the world tend to stimulate an especially high level of joint ventures. The two clearest examples of this are China and Eastern Europe. In the case of China, joint ventures with domestic enterprises have been dictated primarily by national regulation. In Eastern Europe, the preference for joint ventures reflected the perceived uncertainties and risks of doing business in the transition economies during the 1990s.

Many companies are forming not just single alliances but *networks of alliances*, in which relationships are increasingly multilateral rather than bilateral, polygamous rather than monogamous. In effect, they create new *constellations* of economic power: as a result, a new component – 'collective competition' – has been added to the economic landscape. Figure 8.10 shows four alliance-based constellations in the computer industry in the 1990s.

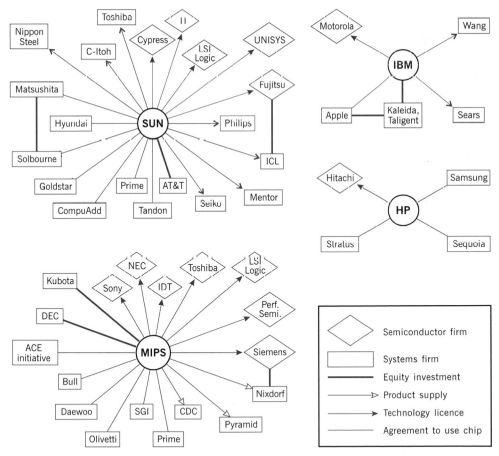

Figure 8.10   Competition between constellations of firms in the computer industry

Source: Based on Gomes-Casseres, 1996: Figure 9

Strategic alliances are formal agreements between firms to pursue a *specific* strategic objective, to enable firms to achieve a specific goal that they believe cannot be achieved on their own. In particular, an alliance involves the sharing of risks as well as rewards through joint decision-making responsibility for a specific venture. Strategic alliances are not the same as mergers, in which the identities of the merging companies are completely subsumed. In a strategic alliance only *some* of the participants' business activities are involved; in every other respect the firms remain not only separate but also usually competitors. Figure 8.11 sets such collaborative agreements in their broader organizational context.

As shown in Figure 8.11, there are three major modes of collaboration involved in strategic alliances: research-oriented, technology-oriented and market-oriented. These three categories capture the major motivations for alliances. More specifically, alliances offer the following (potential) kinds of advantage to the participants:[31]

- overcoming problems of gaining access to markets
- facilitating entry into new/unfamiliar markets
- sharing the increasing costs, uncertainties and risks of R&D and new product development
- gaining access to technologies
- achieving economies of synergy, for example, by pooling resources and capabilities, and rationalizing production.

Very often, the motivations for strategic alliances are highly specific. In the case of R&D ventures, for example, cooperation is limited to research into new products and technologies while manufacturing and marketing usually remain

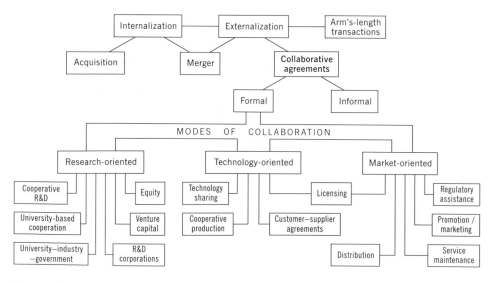

Figure 8.11 Types of inter-firm collaboration

Source: Anderson, 1995: Figure 1

the responsibility of the individual partners. Cross-distribution agreements offer firms ways of widening their product range by marketing another firm's products in a specific market area. Cross-licensing agreements are rather similar but they also offer the possibility of establishing a global standard for a particular technology, as happened, for example, in the case of compact disc players. Joint manufacturing agreements are used both to attain economies of scale and also to cope with excess or deficient production capacity. Joint bidding consortia are especially important in very large-scale projects in industries such as aerospace or telecommunications, where the sheer scale of the venture or, perhaps, the specific regulatory requirements of national governments put the projects out of reach of individual companies.

Not surprisingly, therefore, strategic alliances are more common in some industries than in others:

> Pharmaceuticals, chemicals, electronic equipment, computers, telecommunications, and financial and business services are examples of industries characterized by a large number of strategic alliances ... Although a large number of alliances are still formed in manufacturing industries, more and more strategic alliances are taking place in the services ... As the world economy becomes more service-based, strategic alliances are playing a more important role in cross-border restructuring in service sectors.[32]

As Table 8.1 shows in more detail, the majority of strategic alliances are in sectors 'typified by high entry costs, globalization, scale economies, rapidly changing technologies, and/or substantial operating risks'.[33]

Table 8.1 Strategic alliances by sector during the 1980s

Sector	Alliances		Major reasons for alliance
	No.	%	
Information technology-based	1660	39.7	Market access/structure
Microelectronics	383		Market access/structure
Telecommunications	366		Market access/structure
Software	344		Technology complementarity
Industrial automation	278		Technology complementarity
Computers	198		Market access/structure
Other	91		Market access/structure
Biotechnology	847	20.3	Technology complementarity
New materials technology	430	10.3	Technology complementarity
Chemicals	410	9.8	Market access/structure
Aviation/defence	228	5.5	Technology complementarity
Automotive	205	4.9	Market access/structure
Heavy electric/power	141	3.4	High cost risks
Instruments/medical technology	95	2.3	Reduction of innovation timespan
Consumer electronics	58	1.4	Market access/structure
Food and beverages	42	1.0	Market access/structure
Other	66	1.6	High cost risks
Total	**4182**	**100.0**	

Source: Based on material in Hagedoorn and Schakenraad, 1990

Advocates of strategic alliances claim that by cooperating, companies can combine their capabilities in ways that will benefit each partner. But not everybody shares this rosy view. Many fear that entering into such alliances will result in the loss of key technologies or expertise by one or other of the partners. More broadly, strategic alliances are clearly more difficult to manage and coordinate than single ventures; the potential for misunderstanding and disagreement, particularly between partners from different cultures, is great. Certainly many such alliances have relatively short lives. Nevertheless, the obvious attractions of international strategic alliances in today's volatile and competitive global economy are likely to guarantee their continued growth as a major organizational form.

Flexible business networks[34]

As we have seen in the two previous sections of this chapter, firms are increasingly engaging in close, longer-term, relationships with other firms, whether as part of arrangements to subcontract inputs of goods and services, or within strategic alliances. Now, however, some are suggesting that a new organizational form is emerging – the *flexible business network* – in which almost all functions in the production chain, other than those of central coordination and control, are contracted to independent firms, but in which the final product is marketed under the lead company's brand name. These dynamic and flexible business networks involve complex relationships between independent firms, each of which performs a specialist role within a coordinated network. Figure 8.12 shows, in a highly simplified form, the major elements of such a flexible network.

Such a network goes a long way beyond conventional subcontracting, strategic alliances and the 'integrated network' structure discussed in Chapter 7 (see Figure 7.10). In a flexible network, *all* aspects of the particular production and distribution sequence – the entire business system – are involved, not just some parts of that system.[35] Organizationally, the entire structure is relatively 'flat' and non-hierarchical. The essence of such a flexible network is that the participants are all separate firms with no common ownership. They are cooperative, *relational* structures between independent and quasi-independent firms that are based upon a high degree of trust, something that takes time to develop. However, this does not mean that there are not *power* differentials within the network. There certainly are.

Two examples[36] of flexible business networks can be given to provide some idea of their actual, as opposed to their theoretical, characteristics, although neither meets all the criteria discussed above.

- The first example is US toy manufacturer Lewis Galoob Toys, Inc., a multimillion-dollar company, yet, in the words of a *Business Week* report, 'hardly a company at all':

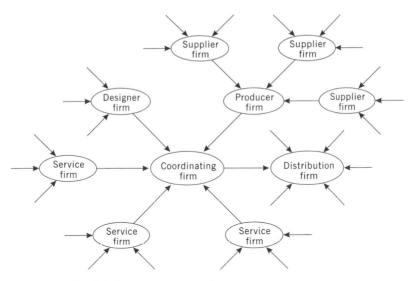

Figure 8.12   A flexible business network

Source: Based, in part, on Miles and Snow, 1986: Figure 1

A mere 115 employees run the entire operation. Independent inventors and entertainment companies dream up most of Galoob's products, while outside specialists do most of the design and engineering. Galoob farms out manufacturing and packaging to a dozen or so contractors in Hong Kong, and they, in turn, pass on the most labour-intensive work to factories in China. When the toys land in the US, they're distributed by commissioned manufacturers' representatives. Galoob doesn't even collect its accounts. It sells its receivables to Commercial Credit Corp., a factoring company that also sets Galoob's credit policy. 'In short,' says Executive Vice President Robert Galoob, 'our business is one of relationships.' Galoob and his brother, David, the company's president, spend their time making all the pieces of the toy company fit together, with their phones, facsimile machines, and telexes working overtime.[37]

- The second example, one that has become a primary target for the global protest movements, is the Nike athletic footwear company.[38] Like other athletic footwear firms, Nike does not wholly own any integrated production facilities but is characterized by 'the large-scale vertical disintegration of functions and a high level of subcontracting activity'.[39] Its development displays great flexibility in adapting to changing competitive circumstances. As Figure 8.13 shows, Nike consists of a complex tiered network of subcontractors that perform specialist roles.

  Nike's 'in-house' production may be thought of as production from its exclusive partners. Nike develops and produces all high-end products with exclusive partners, while volume producers manufacture more standardized footwear that experience larger fluctuations in demand ... Nike acts as the production co-ordinator and three categories of primary production alliance form the first tier of subcontractors. A second tier of material and component subcontractors supports production in the first tier ... [all] ... production takes place in South East Asia while the headquarters in Beaverton, Oregon, houses Nike's research facilities.[40]

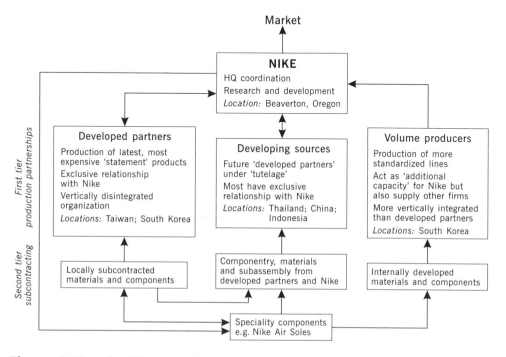

Figure 8.13 The Nike network

Source: Based on Donaghu and Barff, 1990: Figure 4; 542–4

These two examples are consistent with Gereffi's concept of a *buyer-driven commodity chain*[41] in which:

- businesses do not own any production facilities – they are not 'manufacturers' because they have no factories

- the companies are 'merchandisers' that design and/or market, but do not make, the branded products they sell

- these firms rely on complex tiered networks of contractors that perform almost all their specialized tasks

- the main job of the core company is to manage these production and trade networks and make sure all the pieces of the business come together as an integrated whole

- profits derive from unique combinations of high-value research, design, sales, marketing and financial services that allow the buyers and branded merchandisers to act as strategic brokers in linking overseas factories and traders with evolving product niches in their main consumer markets.

In these cases, we seem to be getting closer to the 'virtual firm' or, less dramatically, to the 'hollow corporation'. But they may be mere stepping-stones to an even more radical form of business organization: the *cellular-network* organization,[42] especially in such 'knowledge businesses' as advanced electronics, computer software design, biotechnology, design and engineering services, health-care, and the like. As might be expected, we can find very few examples of such cellular organizations as yet. However, the Taiwanese computer company Acer is suggested as at least moving along that path in its '21 in 21' vision: a planned federation of at least 21 self-managing, independent firms, located around the world by the 21st century, and held together by mutual interest rather than hierarchical control.

The notion of a flexible business network is a seductive one and may well prove to be the future. However, we should beware of extrapolating from individual cases to make universal generalizations. Most likely, the future will be rather like the present in that a diversity of organizational forms will probably co-exist, although the precise mix may well vary and some forms may become obsolete.

Synthesis: connecting the organizational and geographical dimensions of transnational production networks

Figure 8.14 summarizes the primary modes of TNC operations and relates them to the major contributory influences that help to determine them. As we have shown in both this chapter and in Chapter 7, these modes of operation have become both increasingly diverse and increasingly complex. The idea that TNC activity can be captured effectively through the single lens of foreign direct investment is clearly totally inadequate. Strategic alliances, international subcontracting and sourcing, and the embryonic development, at least, of dynamic and flexible network forms, add enormously to the organizational complexity of the system. The question we pose now is: how are these diverse organizational forms expressed 'on the ground'?

The situation is complicated enough even if we take a static picture. It is, of course, far more complex than this because all these networks of relationships are in a continuous state of flux. The ways in which the transnational production networks are organized, and the boundary between which functions are internalized within a firm and which are externalized and performed as a division of labour between firms, is extremely fluid. As we have seen, there are signs of increasingly

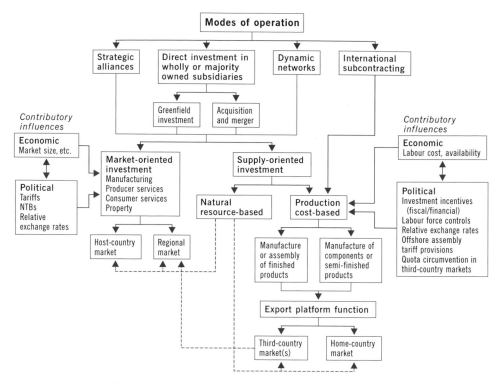

Figure 8.14 Different modes of TNC operation and conditions influencing them

flexible network forms of organizational relationship in the global economy. The combination of these various networks of relationships, both within TNCs and between independent and quasi-independent firms, creates a highly complex *geographical* structure.

'A world in its own image': Hymer's model of the relationship between TNCs and uneven development

In Chapter 7 we noted that Stephen Hymer was, in many ways, the pioneer in attempting to explain why firms became transnational. He did that from a conventional micro-economic perspective. Hymer was also the first to explore the link between the organizational structures of TNCs and their geographical form, although by then (1972) he had become a radical economist primarily concerned with problems of uneven development in the world. Hymer focused solely on the internal organization of TNCs, posing the question 'does the internal division of labour within the TNC correspond to an international division of labour?' He drew upon the theories of organizational hierarchies devised by

Alfred Chandler and the theory of location of Alfred Weber to argue that such an organizational-geographical correspondence did, indeed, exist:

> the [trans]national corporation tends to create a world in its own image by creating a division of labour between countries that corresponds to the division of labour between various levels of the corporate hierarchy. It will tend to centralize high-level decision-making occupations in a few key cities (surrounded by regional sub-capitals) in the advanced countries, thereby confining the rest of the world to lower levels of activity and income.[43]

There is obviously some validity in Hymer's view, as our discussion in this chapter has demonstrated. There *are* recognizable hierarchical spatial tendencies in the location of different parts of TNCs. Corporate headquarters *do* tend to concentrate in a small number of major metropolitan centres; regional offices *do* favour a slightly wider range of cities; production units *are* more extensively spread both within and between nations, in developed and developing countries. Not surprisingly, therefore, Hymer's model of a world being created in the image of the TNC has been widely accepted. But, of course, it grossly oversimplifies the complexity of the modern global organization of economic activity.

Hymer's scheme depicted a clear and distinct hierarchical arrangement in which the vertical division of labour within the TNC is reflected in an unambiguous geographical division of labour (Figure 8.15). Interpreted literally, this would imply that only those levels of corporate activity would be present at each appropriate level in the geographical hierarchy. The top levels of the metropolitan hierarchy would contain only high-level control functions and associated occupations. But this is clearly not the case. A major metropolis, such as London or New York does, indeed, contain the major corporate headquarters of TNCs. But it also contains lower-level control functions and, most significant of all in this context, many of the kinds of production unit that are supposed to be present only in peripheral areas of the world (see Chapter 16). It is more realistic to think of different types of geographical location as containing different *mixes* of corporate units in which the actual proportions vary. Hymer, quite reasonably given the circumstances prevailing at the time he was writing, conceived of the TNC as a very simple hierarchically organized structure. But he did not consider how the various levels of the firm interact with other firms, other than to note the agglomerative tendencies of TNC headquarters.

Figure 8.15

Hymer's model of the relationship between organizational and geographical hierarchies

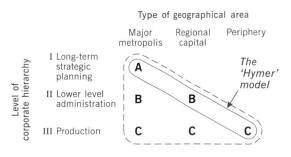

The economic landscape as networks of concentrated and dispersed activities

As we have seen in this chapter, the global economy is made up of a variety of *complex intra-organizational and inter-organizational networks* – the internal networks of TNCs, the networks of strategic alliances, of subcontracting relationships, and of other, newer, organizational forms. These intersect with *geographical networks* structured particularly around linked agglomerations or clusters of activities. As we emphasized in Chapter 2, not only is all economic activity grounded in specific places but also geographical clustering is the norm. The key issue, then, is what form do such clusters take and how are they interconnected?

Localized clusters of economic activity consist of a mixture of independent firms of various sizes and of the branch plants and affiliates of multi-plant firms, many of which are TNCs (Figure 8.16). These different types of establishment, that together constitute a place's 'organizational ecology', are connected into much larger organizational and geographical structures.[44] In the case of the branches and affiliates they are obviously part of a specific corporate structure and will be constrained in their autonomy by parent company policy. The extent to which they are functionally connected into the local economy will be enormously variable. The 'independent' firms in a local economy may, in fact, be rather less independent than they appear at first sight. Some, at least, will be integrated into the supply networks of larger firms, again including TNCs, whose decision-making functions are very distant. As we saw in our discussion of subcontracting, there are both benefits and costs associated with such a role for subcontracting firms. Other firms may be linked together through strategic alliances or they may be a part of the flexible business networks coordinated by key 'broker' firms.

In other words, it is very difficult indeed to generalize about the precise forms and relationships involved. What can be said is that the global economy is made up of intricately interconnected *localized clusters* of activity that are embedded in various ways into different forms of corporate network which, themselves, vary greatly in their geographical extent. The pattern is one of both concentration and dispersal. Some TNCs are globally extensive, others have a more restricted geographical span.

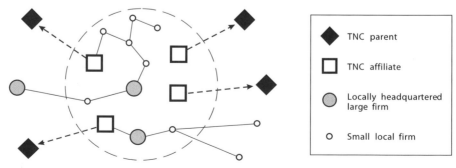

Figure 8.16 A place's 'organizational ecology'

Either way, however, firms in specific places – and, therefore, the places themselves – are increasingly connected into international and global networks.

The precise roles played by firms, and 'parts' of firms, in these networks have very significant implications for the communities in which they are based. Hymer's model captured part of this issue in terms of the uneven geographical distribution of 'high-level' and 'low-level' corporate functions. Places that consist of substantial concentrations of high-level functions are obviously substantially better-off than those that are allocated entirely low-level functions. Similarly, the kind of supply role played by firms within TNC networks will be critical to local well-being. As we have seen, customer–supplier relationships increasingly involve a greater emphasis on long-term, closer relationships based upon a high degree of mutual trust. This may encourage closer geographical relationships between customers and suppliers.

However, TNCs are increasingly operating an upper tier of preferred suppliers that are closely integrated at all stages of the production process, from design through to final production. For any one firm, such preferred suppliers will be relatively few in number and unevenly distributed geographically. TNCs still have a strong propensity to source over extensive geographical distances, including globally. Not every local economy, therefore, can hope to participate in these new supply networks. The highly complex connections between the component parts of firms, and between firms and the places in which they operate, are embedded within asymmetrical, multi-scalar *power structures*.

Four highly interconnected sets of relationships are especially relevant:

- *intra-firm relationships* – between different parts of a transnational business network, as each part strives to maintain or to enhance its position *vis-à-vis* other parts of the organization
- *inter-firm relationships* – between firms belonging to separate, but overlapping, business networks as part of customer–supplier transactions and other inter-firm interactions
- *firm–place relationships* – as firms attempt to extract the maximum benefits from the communities in which they are embedded and as communities attempt to derive the maximum benefits from the firms' local operations
- *place–place relationships* – between places, as each community attempts to capture and retain the investments (and especially the jobs) of the component parts of transnational corporations.

Each of these sets of relationships is embedded within and across *national/state* political and regulatory systems that help to determine the parameters within which firms and places interact. Notwithstanding changes in the international political economy that are reconfiguring the role and functions of nation-states, the state remains fundamentally important as both a regulator of economic transactions and as a container of distinctive institutional practices (see Chapter 5).

Even in the archetypal hierarchical business organization, in which decisions are essentially top-down, it is by no means the case that each component part simply operates as a passive recipient of decisions handed down from on high (see Chapter 7). The individual affiliates of a firm (its subsidiaries, branches, etc.) are continuously engaged in competition to improve their relative position within the organization by, for example, winning additional investment or autonomy from the corporate centre. In a transnational firm, operating in a diversity of national/cultural environments, the very nature of these environments – the *places* in which parts of the firm are embedded – will exert an influence on these internal bargaining processes. Indeed, because each affiliate is itself grounded in a specific geographical community, there is a strong stimulus for that community to try to influence the outcome of these intra-firm bargaining processes by, for example, making concessions or improving local conditions that may enhance the likelihood of the local affiliate winning out in its struggle for additional investment.

Thus, one aspect of the firm–place relationship is that between firms and those places in which they already have operations. The other aspect concerns the search by firms for new investment locations and by communities attempting to attract such investments. It is here, above all, that the firm–place and place–place bargaining processes meet through what are sometimes termed 'investment tournaments'. Indeed, one of the most striking developments of the past two or three decades has been the enormous intensification in competitive bidding between states (and between communities within the same state) for the relatively limited amount of internationally mobile investment. This is a topic we will address in Chapter 9.

Regionalizing transnational production networks

The complex organizational-geographical networks of TNCs, described in the previous section, occur at a whole spectrum of geographical scales, from the finely local at one extreme to the global at the other. An important characteristic, however, is for such networks to have a strong *regional* dimension (that is, networks organized on the multinational scale of groups of contiguous states – as in North America, Europe, or East Asia).[45] In some instances, such a tendency is reinforced by regional political structures – as in the cases of the EU or NAFTA – although this is not invariably the case. Simple geographical proximity is a stimulus for integrating operations. The rationale for such a strategy was presented in Figure 8.4.

> A regional strategy offers many of the efficiency advantages of globalization while more effectively responding to the organizational barriers it entails … From the perspective of a TNC, a regional strategy may represent an ideal solution to the competing pressures for organizational responsiveness and global integration.[46]

In particular:[47]

- Regional-scale manufacturing facilities may represent the limits of potential economies of scale.
- Regionalization allows for faster delivery, greater customization and smaller inventories than would be possible under globalization.
- Regionalization accommodates organizational concerns and exploits subsidiary strengths.

Transnational production networks organized at the regional scale are evident in most parts of the world but most especially in the three 'triad regions' of Europe, North America and East Asia, as we shall see in several of the case study chapters in Part III. In North America, the establishment of the NAFTA is leading to a re-configuration of corporate activities (especially in Mexico) to meet the opportunities and constraints of the new regional system, although it is too early yet to calculate its likely extent.[48] In Europe, the increasing integration of the European Union has led to substantial reorganization of existing corporate networks and the establishment of pan-EU systems by existing and new TNCs. 'The EU can be seen as a gigantic international production complex made up of the networks of TNCs which straddle across national boundaries and form trade networks in their own right.'[49]

There is abundant evidence of US and Japanese TNCs – as well as many European firms themselves – creating regional networks within the EU (often incorporating the transitional economies of Eastern Europe as well, especially as they are likely to be members of the EU in the medium term). However, the process is a complicated one. On the one hand, supply-side forces are stimulating a pan-EU structure of operations to take advantage of scale efficiencies. On the other hand, demand-side forces are still articulated primarily at the country-specific level, where linguistic and cultural differences play a major role in the demands for goods and services. In effect, the strategic tensions between global integration and local responsiveness, discussed in Chapter 7, are played out at the EU regional level.[50]

Although East Asia does not have the same kind of regional political framework as the EU or NAFTA, there is very strong evidence of the existence of regional production networks organized primarily by Japanese firms, although non-Asian as well as some other Asian firms (from Korea, Hong Kong and Taiwan, for example) also tend to organize their production networks regionally.[51] Within East Asia, a clear intra-regional division of labour has developed consisting of four tiers of countries: Japan; the so-called 'four tigers' of Hong Kong, Korea, Singapore and Taiwan; the South East Asian 'later industrializers' – Malaysia, Thailand, Indonesia, the Philippines; China, together with, at least potentially, countries such as Vietnam. However, the relationships between these tiers are more complex than is often suggested. The idea of a simple developmental progression starting with Japan and automatically moving on to the other Asian countries – as suggested in the so-called 'flying geese' model of development – is not tenable.[52]

Conclusion

The aim of this chapter has been to explain how economic activities are organized and reorganized through dynamic networks of relationships within and between transnational corporations and other types of firm. We have emphasized, in particular, the immense diversity of processes and outcomes, both organizational and geographical. Production networks can be articulated through many different combinations of organizational structures and geographical configurations. In this chapter we focused on three specific aspects of transnational production networks.

First, we focused on the geography of TNCs' internal networks. In pursuing their specific strategies, TNCs create not just organizational structures but also have to configure their component parts geographically. Because different parts of a firm have different locational needs, and because these may be satisfied in different types of location, each tends to take on distinctive geographical characteristics. We illustrated this by looking at three functions: corporate and regional headquarters, R&D facilities and production units. But, of course, these structures are dynamic, not static. Restructuring and reorganization are endemic within TNCs.

Second, we explored the networks of externalized inter-firm relationships within which TNCs are embedded. Again, these take on a bewildering variety of forms. There has been a strong propensity for TNCs to engage in strategic alliances and other forms of relationships with other firms, including firms that are their direct competitors. The nature and extent of the relationship between firms and their suppliers has also become increasingly complex with, in general, a tendency for major firms to develop longer-term relationships with key suppliers. We also explored the idea of the emergence of vertically disintegrated flexible network organizations, where individual functions are performed by specialist firms within a flexible network coordinated by a key firm.

Third, we connected up the organizational dimension of TNC networks (both internal and external) with the geographical dimension. Seen in this way, the global economy can be pictured as intricately interconnected localized clusters of activity embedded in various ways into different forms of corporate network that, in turn, vary greatly in their geographical extent. Some TNCs are globally extensive, others have a more restricted geographical span. Either way, however, firms in specific places – and, therefore, the places themselves – are increasingly connected into international and global networks. In particular, there is a strong tendency for TNCs to organize their transnational production networks on a regional scale.

Notes

  1 Haig (1926: 426).
  2 Recent studies of corporate and regional headquarters are by Yeung, Poon and Perry (2001) and by Young, Goold et al. (2000).
  3 Young, Goold et al. (2000).
  4 Lasserre (1996).
  5 Yeung, Poon and Perry (2001: 165).
  6 Aoki and Tachiki (1992).
  7 Yeung, Poon and Perry (2001: 169–70).
  8 Lyons and Salmon (1995: 103–4).
  9 Useful recent studies of trends in R&D activities by TNCs include Blanc and Sierra (1999), Cantwell (1997), Hotz-Hart (2000), Patel (1995) and Zanfei (2000).
 10 Zanfei (2000).
 11 Behrman and Fischer (1980).
 12 Hotz-Hart (2000: 442).
 13 Office of Technology Assessment (1994).

14 Patel (1995).
15 Blanc and Sierra (1999: 188).
16 Patel (1995: 152).
17 Cantwell and Iammarino (2000: 322).
18 Schoenberger (1997: 204).
19 See Sklair (2001: ch. 5).
20 See Clark (1994); Clark and Wrigley (1995); Schoenberger (1997).
21 Schoenberger (1997: 88).
22 Badaracco (1991: 314).
23 See Dicken and Malmberg (2001: 351).
24 Hudson (2001: 187–8).
25 Mentzer (2001: 192).
26 There is a large literature on strategic alliances. Useful contributions are by Anderson (1995), Dunning (1993: ch. 9); Gomes-Casseres (1996), Kang and Sakai (2000) and Mockler (2000).
27 Morris and Hergert (1987).
28 Gomes-Casseres (1996: 2).
29 Kang and Sakai (2000: 7).
30 Kang and Sakai (2000: 13).
31 See Hudson (2001: 206); Dunning (1993: 250).
32 Kang and Sakai (2000: 20).
33 Morris and Hergert (1987: 18).
34 Useful discussions of flexible business networks (although often using different terminology) are provided by Jarillo (1993), Miles and Snow (1986), Miles, Snow, Mathews and Miles (1999) and Nohria and Ghoshal (1997).
35 Jarillo (1993).
36 Another example, Benetton, will be discussed in Chapter 10.
37 *Business Week* (3 March 1986).
38 See Donaghu and Barff (1990); Korzeniewicz (1994).
39 Donaghu and Barff (1990: 539).
40 Donaghu and Barff (1990: 544).
41 Gereffi (1994: 99).
42 This concept is developed in some detail by Miles, Snow, Mathews and Miles (1999).
43 Hymer (1972: 59).
44 Amin and Thrift (1992); Harrison (1997).
45 See Dunning (2000); Kozul-Wright and Rowthorn (1998); Morrison and Roth (1992); Rugman (2000).
46 Morrison and Roth (1992: 45, 46).
47 Morrison and Roth (1992: 46–7).
48 Eden and Monteils (2000); Holmes (2000).
49 Amin (2000: 675).
50 *The Financial Times* (8 November 2001).
51 Abo (2000); Borrus, Ernst and Haggard (2000); Dicken and Yeung (1999); Yeung, Poon and Perry (2001).
52 Bernard and Ravenhill (1995) provide a reasoned critique of the flying geese model.

CHAPTER 9

Dynamics of Conflict and Collaboration: *Both* Transnational Corporations *and* States Matter

The ties that bind

In the preceding four chapters we have outlined the major characteristics of states and of TNCs as separate actors in shaping the global economy. However, so far, the *relationships* between TNCs and states have been merely hinted at. In this chapter we focus explicitly on these relationships because they are, in many ways, at the very centre of the processes of global shift and of global economic transformation. There is a widespread view that the major cause of the state's (alleged) demise is the counterpoised growth of the TNC. Indeed, one of the most commonly used devices is to compare the size of TNCs (measured in terms of sales revenues) to the size of national states (measured in terms of GNP).[1] Using such measures it can easily be shown that around 50 per cent of the world's one hundred largest 'economic units' are TNCs and the other 50 per cent are states: in other words, that TNCs are equivalent to, or even more important than, states. But such comparisons are superficial and, in many respects, misleading – although they are certainly eye-catching. The statistics do not measure the same thing quantitatively and they certainly do not capture the qualitative differences between the two types of institution.

The relationships between TNCs and states are far more complex and ambiguous than the popular view would have us believe:

> It is perhaps most useful ... to view the relationship between [trans]nationals and governments as both cooperative and competing, both supportive and conflictual. They operate in a fully dialectical relationship, locked into unified but contradictory roles and positions, neither the one nor the other partner clearly or completely able to dominate.[2]

This quotation captures the essence of the intricate relationships between TNCs and states: they contain elements of both rivalry and collusion.[3] On the one hand, there is no doubt that the fundamental goals of states and TNCs differ in important respects. In ideal-type terms, whereas the basic goal of business organizations is to maximize profits and 'shareholder value', the basic economic goal of the state is to maximize the material welfare of its society. Table 9.1 indicates some of the dimensions of these conflicting objectives of firms and states.

On the other hand, although their relationships may be conflictual in certain circumstances, *states and firms need each other*.

- *States need firms* to generate material wealth and provide jobs for their citizens. They might prefer such firms to be domestically bounded in their allegiance but that is not an option in a capitalist market economy. Conversely,

- *Firms need states* to provide the infrastructural basis for their continued existence: both physical infrastructure in the form of the built environment and also social infrastructures in the form of legal protection of private property, institutional mechanisms to provide a continuing supply of educated workers, and the like.

Table 9.1 Some conflicting objectives of TNCs and states

TNC objectives	State objectives
Performance	
Maximize profits and shareholder value	Maximize growth of GDP
Minimize cost base consistent with customer needs	Maximize quantity and quality of employment opportunities
Technology	
Undertake R&D at locations optimal to the needs of the firm as a whole	Stimulate the development of locally-rooted technology
Gain access to all necessary technology	
High-order functions	
Locate headquarters and other high-order functions to fit optimal pattern of the firm's overall operations	Maintain indigenous headquarters
	Attract and retain key operations of TNCs
Responsiveness	
Retain flexibility to move profits in optimal manner	Retain power to gain a fair return on local operations of TNCs through taxation policies
Retain flexibility to modify the geographical configuration of the firm's production network to meet changing conditions	Maximize the extent and benefits of local supplier linkages
	Prevent the closure/scaling down of local TNC operations
Retain flexibility to use the labour force as required	Develop a flexible, high-skill, high-earning labour force

Source: Based, in part, on Hood and Young, 2000: Table 1.1

As we showed in Chapter 5, states are both *containers* of distinctive business practices and cultures – within which firms are embedded – and also *regulators* of business activity. National boundaries, therefore, create significant differentials on the global political-economic surface; they constitute one of the most important ways in which location-specific factors are 'packaged'. They create discontinuities in the flow of economic activities that are extremely important to the ways in which TNCs can operate. In particular, states have the potential to determine two factors of fundamental importance to TNCs that may force them to modify their strategic behaviour:[4]

- the terms on which TNCs may have *access* to markets and/or resources
- the *rules of operation* with which TNCs must comply when operating within a specific national territory.

At the same time the fact that TNCs not only span national boundaries but also, in effect, incorporate *parts* of national economies within their own firm boundaries (Figure 9.1) creates major potential problems for states. The nature and the magnitude of the problem vary considerably according to the kinds of strategies pursued by TNCs. Most important is the extent to which TNCs pursue globally integrated strategies within which the roles and functions of individual units are related to that overall global strategy.[5] As we saw in Chapter 8, integrated TNCs do, indeed, geographically fragment their operations in pursuit of global profits. Some types of TNC operation create higher value for the communities in which they are located than other types of operation. More generally, states tend to be fearful about the autonomy and stability of those TNC units located within their national territory as well as concerned about the leakage of tax revenues. At the extreme, TNCs have the capability to move their operations out of specific countries.

At first sight it may seem obvious that TNCs would invariably seek the removal of all regulatory barriers that act as constraints and impede their ability to locate wherever, and to behave however, they wish. Such barriers include those relating to entry into a national market, whether through imports or a direct presence; freedom to export capital and profits from local operations; freedom to import materials, components and corporate services; freedom to operate

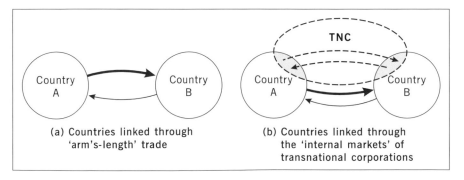

(a) Countries linked through 'arm's-length' trade

(b) Countries linked through the 'internal markets' of transnational corporations

Figure 9.1   Territorial interpenetration: the 'incorporation' of parts of a state's territory into a TNC

unhindered in local labour markets. Certainly, given the existence of differential regulatory structures in the global economy, TNCs will seek to overcome, circumvent, or subvert them. Regulatory mechanisms are, indeed, constraints on a TNC's strategic and operational behaviour.

Yet it is not quite as simple as this. TNCs may perceive the very existence of regulatory structures as an *opportunity*, enabling them to take advantage of regulatory differences between states by shifting activities between locations according to differentials in the regulatory surface – that is, to engage in *regulatory arbitrage.*[6] One aspect of this is the ability of TNCs to stimulate competitive bidding for their mobile investments by playing off one state against another as states themselves strive to outbid their rivals to capture or retain a particular TNC activity. More generally, TNCs seem to have a rather ambivalent attitude to state regulatory policies:

> TNCs have favoured minimal international coordination while strongly supporting the national state, since they can take advantage of regulatory differences and loopholes ... While TNCs have pressed for an adequate coordination of national regulation, they have generally resisted any strengthening of international state structures ... Having secured the minimalist principles of national treatment for foreign-owned capital, TNCs have been the staunchest defenders of the *national state*. It is their ability to exploit national differences, both politically and economically, that gives them their competitive advantage.[7]

More specifically, it has been argued that TNCs will tend increasingly to support a strategic trade policy in their home country, with the expectation that this will open up market access in foreign countries and enable them to benefit from large scale-economies and learning curve effects.[8]

It is clear, therefore, that the relationships between TNCs and states are exceedingly complex. Hence, it is extraordinarily difficult unequivocally to evaluate the relative costs and benefits of TNC activities to states (and their constituent interest groups). According to ideological viewpoint, TNCs either

- expand national economies or exploit them
- are a dynamic force in economic development or a distorting influence
- create jobs or destroy them
- spread new technology or pre-empt its wider use.

At one extreme, the charge is one of political interference in national affairs or of bribery of national officials. At the other extreme, the TNC is regarded as a greater force for international economic well-being than the parochially bounded nation-state. Virtually every aspect of the TNC's operations – economic, political, cultural – has been judged in diametrically opposed ways by its proponents and its opponents. The major problem in trying to evaluate the impact of TNCs is that it is a *counterfactual* problem: we cannot fully measure what would have happened if the TNC activity did not exist. In other words, we generalize about the impact of TNCs at our peril. A realistic assessment must be based on a careful evaluation of specific cases. What is true in one set of circumstances may not be true in others. We need to avoid 'knee-jerk' reactions, whether positive or negative.

In this chapter, the major ways in which TNCs may affect various aspects of national economies are outlined as a framework within which specific cases can be understood. The chapter is organized into three parts:

- First, we explore the major ways in which TNCs, through their direct presence, may impact upon host economies.
- Second, we focus on the potential effects on the home-country bases of TNCs as they transfer some of their activities to other countries.
- Third, we examine the nature of the bargaining relationships between TNCs and states.

TNCs and host economies

The establishment of an overseas facility by a TNC incorporates a package of interlocking characteristics – financial, technological, managerial – that, together, have far-reaching implications for the host economy. Figures 9.2 and 9.3 attempt to capture something of the complexity of the situation. These two, rather complex, diagrams form the basis of our discussion throughout this section and should be referred to as we examine each dimension of potential TNC impact.

Figure 9.2 sets out the broad picture, showing the major areas of potential TNC impact on host economies in the central part of the diagram. These five dimensions are analysed in detail in the following pages. Such potential impacts are contingent upon the interactions between the nature of the foreign-owned unit (this term is used to cover both manufacturing and service operations) and the nature and characteristics of the host economy.

As Figure 9.2 shows, three major aspects of the foreign unit are especially significant:

- The *mode of entry* – how the foreign unit is established (by setting up a completely new facility, by acquiring an existing indigenous enterprise or by forming a joint venture with local capital). A completely new 'greenfield' operation is generally regarded with greater favour by host countries than acquisition of existing capacity. A new unit obviously adds initially to the host country's stock of productive capacity whereas acquisition merely transfers ownership of existing capacity to a foreign firm. Of course, the outcome is never as clear-cut as this. Much depends on the effect of a new operation on other firms and, in the case of acquisition, on whether the acquired unit's performance is enhanced to the benefit of the domestic economy. The mode of entry obviously depends on the opportunities available. Foreign acquisitions have been more prevalent in developed host economies, simply because there are many more suitable candidates for takeover, but they are by no means absent in developing countries.
- The *function* of the foreign unit. Figure 9.2 reminds us that foreign branch operations tend to be established for one of three reasons: to exploit a localized resource, to serve the host market itself by substituting for imports,

or to use the host country location as a platform for exporting either finished products or components.

- The *operational attributes* of the unit, including industry type, technology employed, scale of operations and the extent to which the unit is integrated into the parent company structure.

Thus, operations with different modes of entry, different functions and different attributes will affect the host economy in different ways. Each of the major areas of impact shown in Figure 9.2 – capital and finance, technology, trade and linkages, industrial structure and employment – may be influenced in various ways according to the nature of the foreign unit involved.

Figure 9.2 identifies several characteristics of a host economy that will interact with the nature of the foreign unit itself to produce a specific outcome. As we saw in Chapter 3, most TNC activity originates from the highly industrialized and affluent developed market economies. In so far as the bulk of this activity also flows to these same developed economies, the degree of 'dissonance' between a foreign-controlled operation and the host economy is likely to be small. For less industrialized host countries, however, and for those with very different socio-cultural characteristics from those of the foreign firm's home country, the 'shock waves' may be much greater.

Figure 9.3 adopts a slightly different perspective on the question of how TNCs might impact on host economies. Rather than simply identifying the major elements of the relationship, Figure 9.3 shows in greater detail the *interconnections* between each of the individual processes. It identifies the direct and indirect ways that a foreign operation might impact on a host economy (at whatever geographical scale). It should help us to appreciate just how complex the relationship can be between the operation of a foreign-controlled unit and the host economy in which it is placed.

In this chapter we concentrate overwhelmingly on the *economic* impact of TNCs. However, they are not only extremely significant transmitters of economic change but also of *social, cultural and political* change. This is particularly the case for developing countries. A person employed in a TNC factory or office acquires not just work experience but also possibly a new set of attitudes and expectations. The effect of the employment of women is often to transform, not always favourably, family structures and practices. TNCs introduce patterns of consumption that reflect the preferences of industrialized country consumers. In this respect, the transnational advertising agencies are especially important. They 'are not simply trying to sell specific products in the Third World, but are engineering social, political and cultural change in order to ensure a level of consumption that is "the material basis for the promotion of a standardized global culture."'[9]

Having set out the broader picture, let's now look at each of the major areas of potential TNC impact in turn.[10]

Capital and finance

The inflow of capital is the most obvious impact of foreign investment on a host economy, especially for those countries suffering from a shortage of investment

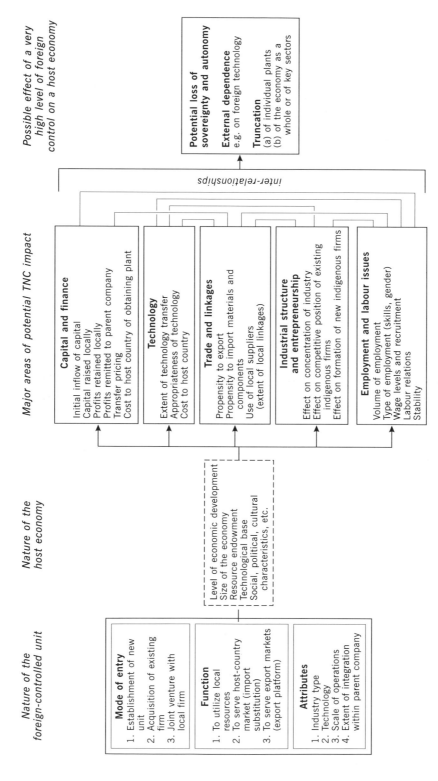

Figure 9.2 The major dimensions of the potential impact of TNCs on host economies

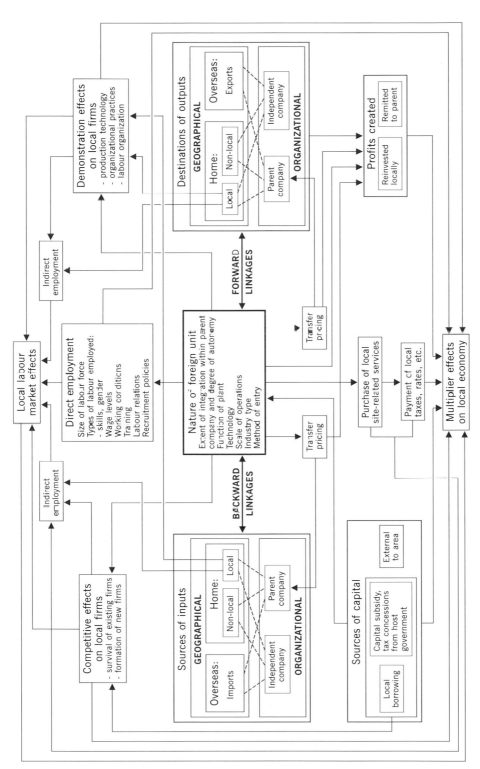

Figure 9.3   Making the connections: mapping the direct and indirect effects of a TNC on a host economy

capital. TNCs have certainly been responsible for injecting capital into host economies, both developed and developing. But not all new overseas ventures undertaken by TNCs involve the actual transfer of capital into the host economy:

> About a half of all funds obtained externally by majority-owned foreign affiliates (non-bank) of United States TNCs (non-bank) were raised in host countries in 1992 … In the case of Japan, funds raised externally financed 58 per cent of total overseas investments in 1992, with funds obtained from local (host-country) financial institutions accounting for 35 per cent of that total … Externally raised capital appears to finance a larger share of investment funds in developed than in developing countries.[11]

Thus, at least some of the capital employed by TNCs may either be borrowed on the host-country capital markets or arise from the reinvestment of retained earnings from the foreign affiliate. Local firms may be bought with local money. Local firms may even be squeezed out of local capital markets by the perceived greater attractiveness of TNCs as a use for local savings.

Even where capital inflow does occur, or where earnings are retained, there will, eventually, be a reverse flow as the foreign operation remits earnings and profits back to its parent company. This reverse flow may, in time, exceed the inflow of capital. Any net financial gain to the host country also depends on the trading practices of the TNC. A host country's balance of payments will be improved to the extent that the foreign plant exports its output and reduced by its propensity to import. A vital issue, therefore, is the extent to which financial 'leakage' occurs from host economies through the channel of the TNC. This raises the question of the ability of host-country governments to obtain a 'fair' tax yield from foreign firms, many of which are capable of manipulating the terms of their intra-corporate transactions through transfer pricing.

The problem of transfer pricing

In external markets, prices are charged on an 'arm's-length' basis between independent sellers and buyers. In the internal 'market' operated by TNCs, however, transactions are between *related* parties – units of the *same* organization. The rules of the external market do not apply. The TNC itself sets the transfer prices of its goods and services within its own organizational boundaries and, therefore, has very considerable flexibility in setting its transfer prices to help achieve its overall goals. The ability to set its own internal prices – within the limits imposed by the vigilance of the tax authorities – enables the TNC to adjust transfer prices either upwards or downwards and, therefore, to influence the amount of tax or duties payable to national governments. For example, as Figure 9.4 suggests, it would be in a TNC's interest to charge more for the goods and services supplied to its subsidiaries located in countries with high tax levels and vice versa. A similar incentive exists where governments restrict the amount of a subsidiary's profits that can be remitted out of the country.

In general, the greater the differences in levels of corporate taxes, tariffs, duties and exchange rates, the greater will be the incentive for the TNC to manipulate its internal transfer prices. TNCs, therefore, have a strong incentive to engage in

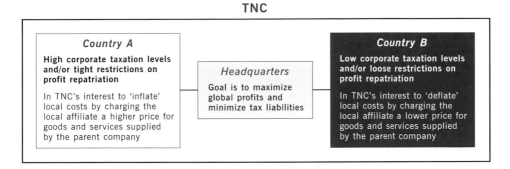

Figure 9.4 The incentives for TNCs to engage in transfer pricing

transfer pricing. The very large, highly centralized, global TNC has the greatest potential for doing so. But it has proved extremely difficult for governments (and researchers) to gather hard evidence on its actual extent:

> Picture a General Motors plant in Windsor, Ontario, producing hundreds of items for assembly in ... [autos] ... that will be sold in the Canadian market as well as for assembly by its sister plants in Michigan. No independent public market exists for many of the items, since no other firm produces these products. Nor is it obvious what the production cost may be of the items that cross the US–Canadian border – that kind of estimate will depend heavily on how the fixed costs of the Windsor plant are allocated among the many items produced, an allocation that cannot fail to be arbitrary. Without an obvious selling price or an indisputable cost price, all the ingredients exist for a pitched battle over the transfer price.
>
> When the item crossing the border is intangible, such as a right bestowed by the parent on a foreign subsidiary to use the trademark of the parent or to draw on its pool of technological know-how, the indeterminateness of a reasonable price becomes even more apparent. How much is the use of the IBM trade name worth to its subsidiary in France? How valuable is the access granted to a team of engineers in an Australian subsidiary to the databank of a parent in Los Angeles?[12]

A US House of Representatives' study claimed that more than half of almost forty foreign companies surveyed had paid virtually no taxes over a ten-year period. It was estimated that some $35 billion was being lost through the transfer pricing mechanism. In the United Kingdom, a study of 210 TNCs showed that 83 per cent had been involved in a transfer pricing dispute.[13] But even in advanced economies like the United States, or the United Kingdom, it is extremely difficult for the Internal Revenue Service to assess the actual extent of transfer pricing. It is even more difficult for developing countries to do so.

The financial gain (or loss) to host countries from foreign investment, therefore, consists primarily of the net balance of capital flows plus the net earnings from trade. But there is also the cost of actually obtaining the foreign investment to consider. Most states, national and subnational, offer very considerable financial and fiscal incentives to attract internationally mobile investment. These costs, together with the provision of the necessary physical infrastructure, may be quite substantial. We will return to this issue in the context of the bargaining process between TNCs and states later in this chapter.

Technology

Three issues are especially important in evaluating the technological impact of foreign enterprises on host economies:

- the *extent* to which technology is transferred
- the *appropriateness* of the technology transferred
- the *costs* to the host economy of acquiring the technology.

Transfer of technology

Each of the various modes of foreign involvement shown in Figure 8.14 – foreign direct investment, collaborative ventures, international subcontracting, dynamic and flexible business networks – is a potential channel for technology transfer. Simply by locating some of its operations outside its home country the TNC engages in the geographical transfer of technology. In this respect, technologies are, indeed, spread more widely. In so far as a foreign affiliate employs local labour there will be a degree of technology transfer to elements of the local population through training in specific skills and techniques. But the mere existence of a particular technology within a foreign-controlled plant does not guarantee that its benefits will be widely diffused through a host economy. The critical factor here is the extent to which the technology is made available to potential users outside the firm, either directly, through linkages with indigenous firms, or indirectly, via 'demonstration effects'.[14]

However, the very nature of the TNC may inhibit the spread of its proprietary technology beyond its own organizational boundaries. Possession and exploitation of technology are inherent features of the TNC. Such technology is not lightly handed over to other firms. Control over its use is jealously guarded: the terms under which the technology is transferred are dictated primarily by the TNC itself in the light of its own overall interests. They tend to transfer the results of innovation but not the innovative capabilities[15] – the 'know-how' rather than the 'know-why'. A major tendency, as we saw in Chapter 8, is for TNCs to locate the bulk of their technology-creating activities either in their home country or in the more advanced industrialized countries. So far, relatively little R&D has been relocated to developing countries. Nevertheless, some R&D – mostly lower-level support laboratories – has been located in a few of the newly industrializing economies. In some cases this is a direct result of host-government pressure on TNCs to establish R&D facilities in return for entry. Such leverage is probably greatest where the TNC wishes to establish a branch plant to serve the host-country market itself.

A recent comparative study[16] of TV production by TNCs in Malaysia, Mexico and Thailand found evidence of local technological development but to varying degrees in each case:

- In Malaysia, there was evidence of significant progress in automation and quality improvement, while some localization of R&D was under way.
- In Mexico, although there was also significant progress in automation and quality improvement, the technology changes were limited to minor adaptive work. There were no independent design functions.
- In Thailand, the top end of the industry was using world-class technology and quality standards but the rest of the industry was less developed. There were no local R&D and design capabilities.

Appropriateness of technology

In the case of developing countries, in particular, a major issue is the appropriateness of the technology transferred via the TNC. Do the processes and products being introduced match local conditions and local needs in a dynamic, rather than merely in a static, sense? New technologies are invariably introduced by TNCs first of all in their home country or in other industrialized countries. Since they reflect the prevailing cost and availability of factors in those countries the technologies tend to be capital- rather than labour-intensive. But in most developing countries the abundant factor tends to be low-skilled labour while capital is relatively scarce. Hence, there is much disagreement about the extent to which TNCs adapt their *process* technology for use in developing countries to make it more appropriate.

The second major issue in the appropriateness debate relates to the kinds of *products* transferred by TNCs to developing countries. This has become an especially contentious issue, largely because of some highly publicized cases. In their drive to create global markets, TNCs have sought to introduce and sell their products throughout the developing world. The creation of particular types of demand and the shaping of consumer tastes and preferences are an intrinsic part of the TNC system. The problem is that

> the use of scarce resources for the production of goods which are over-differentiated, over-packaged, over-promoted, over-specified and within the reach of only a small elite, or, if bought by the poor, at the expense of more essential products, is not conducive to 'national welfare'.[17]

Against those products that are clearly 'inappropriate' to developing countries, however, there are undoubtedly others that have brought great benefit, including those in agriculture (seeds, fertilizers) and healthcare.

A final aspect of the appropriateness issue relates to *the environmental and safety dimension* of TNC activity. Do TNCs export technologies to developing countries that are environmentally objectionable or that are less safe than they should be? There have been claims that TNCs systematically shift some of their environmentally noxious or more hazardous operations to developing countries with less stringent environmental and safety standards. Although this may indeed occur in specific cases, there is no evidence to suggest that this is the general practice. For example, the years following the introduction of stringent environmental regulations in the United States were not characterized by the widespread relocation of pollution-intensive industries to countries with lower regulatory standards.[18]

A rather different aspect of the problem relates to safety and environmental management. Industrial disasters, such as the one at the Union Carbide plant at Bhopal, India, in 1994, focused attention on the safety practices of TNCs. A frequent claim was that TNCs tend to adopt less stringent safety practices in their developing-country plants than in their home plants. The more recent conflict in Nigeria involving Shell's environmental practices in Ogoniland raised both environmental and political issues. The now-collapsed – and infamous – US company Enron clearly rode roughshod over environmental regulations in many parts of the world with some devastating effects. These cases, and others, are serious in the extreme. But it is dangerous to generalize from them to produce universal statements on the environmental behaviour of *all* TNCs. UNCTAD has carried out comprehensive surveys that give a more balanced picture.[19] While not minimizing the seriousness of specific cases, the UNCTAD findings do not support the view that TNCs in general are environmentally irresponsible.

In terms of environmental management, the UNCTAD survey found that

> Transnational corporations are increasingly taking a more strategic approach towards environmental management issues, tending to view the costs associated with environmental management as long-term investment central to successful business ventures ... TNCs are increasingly establishing targets with respect to the environmental performance of their operations ... North American TNCs, with their generally longer experience of FDI, are relatively more sensitive to or aware of the international aspects of their activities than TNCs from other regions.[20]

Questions about the appropriateness of technology tend to be less relevant in the developed economies for obvious reasons. But one issue of growing importance in such economies is the influence of foreign TNCs on *business organization and practice.* Japanese investment in Britain is a case in point. One argument for encouraging such investment during the 1980s, in particular, was that the introduction of highly efficient Japanese business methods would rub off on British industry in general and raise the level of efficiency in the economy. In other words, it is the *demonstration* of the effect of the 'social' innovations of work organization, labour relations, relationships with suppliers, and so on that are suggested to be highly appropriate to the needs of the economy.

Costs of technology transfer via the TNC

Finally, we need to consider the question of the cost to the host country involved in technology transfer through the TNC. Precisely what that cost is, and whether it is a 'reasonable' price to pay, are extremely difficult to determine. First, technology is only one part of the overall package of attributes that the TNC brings to a host country; it is difficult to separate it out. Second, assessment of the cost involved assumes that it can be measured against alternative ways of acquiring the same technology. The two major alternatives are:

- to buy or license the technology alone from its owner (the TNC) – that is, to 'unbundle' the TNC package or
- to produce the technology domestically.

The parallel usually drawn is with Japan, which rebuilt its post-war economy without the introduction of direct foreign investment, mainly by *licensing* technology from Western firms. Although a great deal of technology is licensed by developing (and developed) countries from TNCs it is not always a feasible alternative. A TNC may be unwilling to license the technology or it may charge an exorbitant price. It may be a question of the host country accepting the entire TNC package or getting nothing at all. The possibility of producing the technology domestically may be feasible for some of the more advanced industrial nations but rarely so for developing countries.

Trade and linkages

Two of the most important questions surrounding the impact of TNCs on all host economies, whether developed or developing, are:

- Their role in the host country's *trade* with the outside world. Exports and import-substituting production by TNCs contribute towards a positive trade balance, imports by TNCs contribute towards a negative trade balance, although this may be offset if the imports are essential for export-producing activities.

- The extent to which their operations are *integrated* into the local economy through *linkages with domestic firms.* This is the most significant mechanism through which technology is transferred, additional employment created and opportunities increased for the formation of new local enterprises.

Trade effects

How far TNCs actually contribute to the host economy's *trade balance* depends on a number of factors. The most obvious is the primary purpose for which the foreign operation is established (see Figure 9.2): that is, whether its role is to serve the host market itself or to serve third-country markets using the host economy as an export platform. In general, TNCs tend both to export more and to import more than local firms, but that merely reflects their greater overall size and the geographical extensiveness of their operations. Operations set up to serve the local market, by definition, will not be large exporters. How far they are large importers of intermediate products will depend upon corporate sourcing policy, local supply capabilities and host-government policy towards local content. Plants set up as export platforms, on the other hand, are bound to be major exporters – that is their entire *raison d'être.* But, again, the *net* effect on the trade balance will be influenced by the extent to which its export-producing activities depend upon imported components. Nevertheless,

> as far as export promotion is concerned, one of the main advantages that TNCs offer is the possibility of participation in their global supply networks and direct access to their sales channels and brand names ... In a large number of vertically integrated activities, TNC participation may be the only way for new entrants to access export

markets. In other activities, especially those with strong product differentiation, TNCs offer an avenue to overcome the entry barriers that can hold back exporters which lack established marketing outlets or brands.[21]

Linkages between TNCs and domestic firms[22]

Inter-firm linkages are the most important channels through which technological change is transmitted. By placing orders with indigenous suppliers for materials or components that must meet stringent specifications technical expertise is raised. The experience gained in new technologies by local firms enables them to compete more effectively in broader markets, provided, of course, that they are not tied exclusively to a specific customer. The sourcing of materials locally may lead to the emergence of new domestic firms to meet the demand created, thus increasing the pool of local entrepreneurs. The expanded activities of supplying firms, and of ancillary firms involved in such activities as transportation and distribution, will result in the creation of additional employment. But such beneficial spin-off effects will occur only *if* the foreign affiliates of TNCs *do* become linked to local firms. Where TNCs do not create such linkages they remain essentially as foreign enclaves within a host economy, contributing little other than some direct employment.

As far as local linkages are concerned the most significant are *backward* or *supply* linkages (see Figure 9.3). Here, the crucial issue is the extent to which TNCs either import materials and components or procure them from local suppliers. The actual incidence of local linkage formation by foreign-controlled plants depends upon three major influences.

1 *The particular strategy followed by the TNC and the role played by the foreign plant in that strategy*   The influence of TNC strategy on the development of local linkages within host economies has a number of aspects. One is the general corporate policy towards the sourcing of inputs, which will determine the degree of sourcing autonomy granted to individual plants. Those TNCs that are strongly vertically integrated at a global scale are less likely to develop local supply linkages than firms with a lower degree of corporate integration. But even where vertical integration is low the existence of strong links with independent suppliers in the TNC's home country or elsewhere in the corporate network may inhibit the development of local linkages in the host economy. Familiarity with existing supply relationships may well discourage the development of new ones, particularly where the latter are perceived to be potentially less reliable or of lower quality. A particularly important factor is the role of the foreign plant itself in the TNC's overall strategy, that is, whether it is oriented to the host market and is, therefore, import-substituting, or whether it is an export platform activity. Foreign plants that serve the host market are more likely to develop local supply linkages than are export platform plants.

2 *The characteristics of the host economy*   In general, we would expect to find denser and more extensive networks of linkages between TNCs and domestic enterprises in the developed economies than in the developing economies. Within developing countries such linkages are likely to be greatest in the larger

and more industrialized countries than in others. In addition, host-country governments may well play a very important role in stimulating local linkages by insisting that TNCs utilize a certain level of locally sourced materials and components. Such local content policies have become increasingly widespread in both developing and developed countries. But much depends on the relative strength of the host country's bargaining power *vis-à-vis* the TNC, and on the extent to which local supplies are of an appropriate quantity and quality. Again, it tends to be in the larger and the more industrialized developing countries that such local content policies have the greatest impact, and also in those TNC activities serving the local market. Indeed, it could be that the export-oriented industrialization strategies of developing countries actually inhibit the development of local supply linkages

3 *Time* The third influence on the establishment of TNC linkages with local suppliers is time. Particularly in view of the closer relationships between firms and their suppliers which have been emerging (see Chapter 8), it should not be expected that a foreign plant, newly established in a particular host country, would immediately develop local supplier linkages. Not only do appropriate suppliers have to be identified, but also it takes time for supplier firms to 'tune in' to a new customer's needs.

Empirical evidence of local linkage formation by TNCs is extremely variable. Overall,

> the evidence suggests that TNCs have strong, but often very uneven, effects on the development of local suppliers in host countries. As with FDI flows themselves, there appears to be growing concentration on locations that are industrially advanced and able to meet the rigours of world competition without substantial additional cost and effort. Other locations may well receive FDI but may not gain much by way of local depth and linkages.[23]

Studies within the smaller developing countries, particularly those with a short history of industrial development, tell a fairly uniform story of shallow and poorly developed supply linkages between local firms and foreign-controlled plants. A common observation is that foreign plants located in export processing zones (EPZs) are particularly unlikely to develop supplier linkages with the wider economy. In the case of the Mexican *maquiladora* plants, for example, less than 5 per cent of the inputs used are sourced from within Mexico. Additionally, most of those inputs are low-value and low-technology products whose production does little to upgrade the local technological and skill base.[24]

The cause appears to be the reluctance of corporate (rather than local) purchasing managers to buy Mexican inputs. In Indonesia, the pattern of local purchasing was more varied:

> only the firms in the automotive industry … have been increasing the local content of the final goods through the mandatory increase of purchases of locally made parts and components. In contrast, in the case of pharmaceuticals, virtually all basic raw materials had to be purchased from the [TNCs] … In the case of food processing and chemicals companies, various raw materials were procured locally, but in general these linkages did not involve an appreciable increase in the technical capabilities of local suppliers.[25]

290 Part II Processes of Global Shift

However, there are circumstances where there may be quite a high level of *local sourcing* but which does not involve *local firms*. Research in the Johor region of southern Malaysia, for example, has shown that although the new foreign manufacturing plants established there 'are sourcing a large part of their inputs in Johor … the regional effect is confined to foreign, mainly Japanese and Singaporean, suppliers. As a result, the linkages of the new manufacturing plants are only in part beneficial to the local economy.'[26] Overall, Japanese firms in the Malaysian electronics industry tend 'to rely more heavily on relocated suppliers from their home country, supporting the general belief about the effect of Japanese business ties'.[27]

The experience of the TV industry in Malaysia, Mexico and Thailand (referred to above) showed variation in the degree of local supplier upgrading:

- In Malaysia, there was limited involvement of domestic suppliers in less complex functions but considerable sourcing by foreign component suppliers.
- In Mexico, the involvement of domestic suppliers was almost zero, although some links with foreign suppliers were beginning to occur.
- In Thailand, there was limited involvement of domestic suppliers in less complex functions but considerable sourcing by foreign component suppliers.

> In the electronics industry, sourcing patterns appear to differ significantly by host country. For example, in 2001, foreign affiliates in the colour TV industry in Tijuana, Mexico, sourced about 28 per cent of their inputs locally, of which only a very small proportion (3 per cent) was supplied by Mexican-owned firms … in Malaysia, locally-procured components by foreign affiliates in the electronics and electrical industries comprised 62 per cent of exports in 1994; the corresponding figure for Thailand was 40 per cent. However, in both countries, the most strategic parts and components were supplied mainly by foreign-owned companies rather than domestic ones.[28]

A survey of 67 foreign enterprises in Shanghai, China, found evidence of weak local linkages: 45 per cent of the surveyed firms did not use domestic suppliers at all; 30 per cent used less than 50 per cent; only 25 per cent of the firms used more domestic than imported materials and parts.[29]

The situation in the older industrialized countries is rather different. Often, of course, foreign firms have been in operation for so long that they are an almost indistinguishable part of the landscape and have developed dense networks of local supplier linkages over a long period of time. On the other hand, it is by no means inevitable that TNCs in industrialized countries will develop strong local linkages. In Canada there have been persistent criticisms of the failure of many US firms to source their major inputs from Canadian suppliers. The extent of local supply links of foreign electronics plants in Scotland is also small.

Apart from the extent of linkages created by foreign firms in a host economy there is also the question of their *quality* and the degree to which they involve a beneficial transfer of technology (either production or organizational) to supplier firms. A common criticism is that many TNCs tend to procure only 'low-level' inputs from local sources, for example cleaning services and the like. This may be because of deliberate company policy to keep to established suppliers of higher-level inputs or because such inputs are simply not available locally (or are perceived not to be so). Where development of higher-level supply linkages occurs there does seem to be a positive effect on supplier firms.

Table 9.2 summarizes the differences between 'dependent' and 'developmental' linkage scenarios. Clearly, from a host-country perspective, the aim must be to achieve a linkage structure that is developmental. In this respect, much will depend upon its bargaining power (see below).

Table 9.2 Dependent and developmental linkage structures

Attribute	Dependent structure	Developmental structure
Form of local linkages	Unequal trading relationships. Conventional subcontracting. Emphasis on cost-saving	Collaborative, mutual learning. Basis in technology and trust. Emphasis on added value
Duration and nature of local linkages	Short-term contracts	Long-term partnerships
Degree of local embeddedness of inward investors	Weakly embedded. Branch plants restricted to final assembly operations	Deeply embedded. High level of investment in decentralized, multi-functional operations
Benefits to local firms	Markets for local firms to make standard, low-technology components. Subcontracting restricts independent growth	Markets for local firms to develop and produce their own products. Transfer of technology and expertise from inward investor strengthens local firms
Prospects for the local economy	Vulnerable to external forces and 'distant' corporate decision-making	Self-sustaining growth through cumulative expansion of linked firms
Quality of jobs created	Predominantly low-skilled, low-paid. May be high level of temporary and casual employment	Diverse, including high-skilled, high-income employment

Source: Based on Turok, 1993: Table 1

Industrial structure and entrepreneurship

Although not all TNCs are giant corporations, it is often the case that foreign plants are larger than their domestic competitors, especially in developing countries. Hence, the entry of a foreign plant into a host economy may have a number of repercussions on the structure of domestic industry, particularly on the competitiveness, survival and birth of domestic enterprises. But, as in other aspects of TNC impact, there is no inevitability about such structural effects. Much will depend upon the specific domestic context itself and on the relative size and market power of the TNC affiliate in that context. In general, the difference between a foreign plant and a domestic enterprise in developed market economies, especially those that are themselves sources of TNCs, will be far less than that in developing countries. The industrial structure of most developing economies tends to be much more strongly *dualistic,* with a small, technologically advanced sector (relatively speaking) which is oriented to the more modern urban market and a technologically less advanced sector characterized by traditional production and attitudes. TNCs in developing countries are, by definition, part of the technologically advanced sector of the host economy.

A major long-term effect of the entry of TNCs into a host economy – both developed and developing – is likely to be an increase in the level of *industrial concentration.* The number of firms is likely to be reduced and the dominance of very large firms increased. Thus, two important *negative* effects of TNCs on host economies are:

- the possible squeezing out of existing domestic firms and
- the suppression of new indigenous enterprises.

Both of these fears have been voiced especially strongly in the case of developing countries but there is no reason to believe that they do not also apply to particular sectors in developed economies if high TNC penetration has occurred. But, clearly, the less developed the indigenous sector the greater is the likelihood of its being swamped by foreign entry and of local entrepreneurship being suppressed.

However, we should beware of assuming that the involvement of TNCs in a particular host economy will inevitably destroy or suppress domestic enterprise. There may well be *positive* effects:

- Where substantial local linkages are forged by foreign plants, particularly on the supply side, opportunities for local businesses may well be enhanced. Existing firms may receive a boost to their fortunes or new firms be created in response to the stimulus of demand for materials or components.
- The formation of new enterprises may be stimulated through the 'spin-off' of managerial staff who set up their own businesses on the basis of experience and skills gained in employment with the foreign firm.

Employment and labour issues[30]

For most ordinary people, as well as for many governments, the most important issue in the debate over the TNC is its effect on jobs:

- Does the entry of a foreign-controlled plant create new jobs?
- What kinds of jobs are they?

- Do TNCs pay higher or lower wages than domestic firms?
- Do TNCs operate an acceptable system of labour relations?

Aggregate levels of TNC employment

TNCs employ very large numbers of workers in both developed and developing countries. In the mid-1990s, for example, over 73 million workers were employed worldwide in TNCs – although this was only around 3 per cent of the world's labour force.[31] To these we must add the *indirect* employment in linked firms that are, themselves, not TNCs. UNCTAD estimates that in developing countries between one and two jobs are created indirectly for each direct employee in a TNC. But it is extremely difficult to estimate accurately the level of indirect employment creation because, as Figure 9.5 shows, there are many ways in which such employment may be generated.

The number of *direct jobs* created in a particular TNC plant will depend upon two factors:

- the *scale* of its activities and
- the *technological* nature of the operation, particularly on whether it is capital-intensive or labour-intensive.

Employment effects	Definition or illustration
Direct employment effects	Total number of people employed within the TNC subsidiary
Indirect employment effects	All types of employment indirectly generated throughout the local economy by the TNC subsidiary
1. Macro-economic effects	Employment indirectly generated throughout the local economy as a result of spending by the TNC subsidiary's workers or shareholders
2. Horizontal effects	Employment indirectly generated among other local enterprises as a result of competition with the TNC subsidiary
(a) Narrow horizontal effects	Employment indirectly generated among local enterprises competing in the same industry as the TNC subsidiary
(b) Broad horizontal effects	Employment indirectly generated among local enterprises active in other industries than the TNC subsidiary
3. Vertical effects	Employment indirectly generated by the TNC subsidiary among its local suppliers and customers
(a) Backward effects (or linkages)	Employment indirectly generated by the TNC subsidiary among its local suppliers (of raw materials, parts, components, services, etc.)
(b) Forward effects (or linkages)	Employment indirectly generated by the TNC subsidiary among its local customers (e.g. distributors, service agents, etc.)

Note: The above employment effects, if they could be measured, should be calculated in net terms (i.e. gross employment directly or indirectly generated, minus total employment displacement).

Figure 9.5 Direct and indirect employment-generating effects of TNCs in host economies

Source: ILO, 1984: 39

The number of *indirect* jobs created will also depend upon two major factors:

- the extent of *local linkages* forged by the TNC with domestic firms, and
- the *amount of income generated* by the TNC and *retained* within the host economy. In particular, the wages and salaries of TNC employees and of those in linked firms will, if spent on locally produced goods and services, increase employment elsewhere in the domestic economy (Figure 9.3)

Against the number of jobs *created* in, and by, TNCs we need to set the number of jobs *displaced* by any possible adverse effects of TNCs on indigenous enterprises. For reasons outlined earlier, domestic enterprises may be squeezed out by the size and strength of foreign branch plants while new firm formation may be inhibited. In these respects, the effect of the TNC may be to displace existing or potential jobs in domestic enterprises. Hence, the overall employment effect of TNCs in host economies depends upon the balance between job-creating and job-displacing forces:

The *net* employment contribution of a TNC to a host economy = (DJ + IJ − JD)

> *where*: DJ = the number of direct jobs created in the TNC
> IJ = the number of indirect jobs in firms linked to the TNC
> JD = the number of jobs displaced in other firms.

Types of employment

The number of jobs created by TNCs in host economies is only part of the story. What kind of jobs are they? Do they provide employment opportunities that are appropriate for the skills and needs of the local labour force? The answer to these questions depends very much on the attributes of the foreign plant (see Figure 9.2): the type of industry, the nature of the technology used, the scale of operations and the extent to which the foreign plant is integrated into its parent organization. In particular, where the plant 'fits' into the TNC's overall structure and how much decision-making autonomy it has are key factors. In general, the fact that TNCs tend to concentrate their higher-order decision-making functions and their R&D facilities in the developed economies produces a major geographical bias in the pattern of types of employment at the global scale.

In developing countries, the overwhelming majority of jobs in TNC plants are *production* jobs. In export processing zones, of course, low-level production jobs, especially for young females, are the rule, although this partly reflects the types of industry that dominate in EPZs. More generally, the ILO suggests that the proportion of higher-skilled workers employed by TNCs in developing countries has tended to increase over time, as has the proportion of local professional and managerial staff. Such changes have progressed furthest in the more advanced industrial countries of Latin America and East Asia. Even so, the TNC labour force in developing countries remains concentrated in low-skill production and assembly occupations.

The TV example discussed earlier again shows how the experience of individual developing countries may vary in the extent of TNC-induced labour upgrading. In each case, the extent of human capital formation in the industry was very limited prior to the mid-1990s:

- In Malaysia, specialized staff were still foreign but there was significant training by leading firms and for their partners in their regional TNC networks. There was evidence of rising skill levels and increasing numbers of specialized technical and managerial staff.

- In Mexico, the first signs of upgrading were apparent as a result of significant training efforts and linking with local education institutions, rising labour skill levels, and increasing numbers of specialized technical and managerial staff.

- In Thailand, the skill levels of the labour force were low but rising with increased emphasis on labour training. But there was not much evidence of an increasing involvement of more highly educated specialist staff.

Geographical segmentation of high- and low-order occupations (white-collar and blue-collar jobs) in TNCs is not confined solely to that between developed and developing countries. As we saw in Chapter 8, a similar geographical segmentation of TNC functions exists within developed economies. This dichotomy is clearly illustrated in the United Kingdom where the higher-order occupations are disproportionately concentrated in South East England in and around London. A major criticism of the foreign branch plants located in Scotland, Wales, Northern England and Northern Ireland, therefore, is that they employ primarily lower-skilled workers and that they generally lack higher-level employment opportunities. Whether or not better types of job would be available in such areas in the absence of the foreign plants is, of course, an open question.

Wage and salary levels

In so far as TNCs take advantage of geographical differences in prevailing wage rates between countries they do, in fact, 'exploit' certain groups of workers. The exploitation of cheap labour in developing countries at the expense of workers in developed countries is one of the major charges levelled at the TNC by labour unions in Western countries (an issue we will return to in Part IV). The major problem here seems to be in relation to conditions in subcontracting companies rather than in the TNCs themselves, and there is much controversy over this issue. The general response of TNCs facing such allegations is that they do not have complete control over what goes on in independent factories, although, in the face of these criticisms, many TNCs are now implementing codes of practice to which their subcontractors must conform.

However, as far as their *directly owned affiliates* are concerned, the general consensus seems to be that TNCs generally pay either at or above the 'going rate' in the host economy. UNCTAD's conclusion is that

> Generally speaking, at the aggregate and industry levels, the workforce directly employed by foreign affiliates enjoys superior wages, conditions of work and social security benefits relative to the conditions prevailing in domestic firms. In developed countries, for instance, the average level of wages and salaries in foreign affiliates has been found without exception to be above that in domestic firms, and the gap between affiliates and domestic companies continued to grow during the 1980s ... [in developing countries] ... In Indonesia, Malaysia, Peru and Thailand, on average, TNCs pay generally higher wages than local companies.[32]

Figure 9.6 shows two aspects of the wage comparison. First, the relative height of the columns shows that TNCs pay very much more to workers in high-income countries than to those in middle- and low-income countries. This differential reflects a number of factors, including the composition of economic activity, educational and skill levels, cost of living, and so on. Second, Figure 9.6 shows that although TNCs certainly pay higher wages overall than domestic firms in the same country group (a ratio of 1.5), the pattern varies between country groups: 1.4 in high-income countries, 1.8 in middle-income countries and 2.0 in low-income countries.

Where TNCs do pay above the local rate, of course, they may well 'cream off' workers from domestic firms and possibly threaten their survival. This point relates to the kinds of *recruitment policies* used by TNCs. Invariably, TNCs tend to operate very careful screening procedures when hiring workers. This may well mean that employees for a newly established foreign plant are drawn from existing firms rather than from the ranks of the unemployed. Another aspect of recruitment, at least in some industries, is the extent to which TNCs recruit particular types of workers to keep labour costs low. In the textiles, garments and electronics industries, for example, there is a very strong tendency to prefer females to males in assembly processes and, in some cases, to employ members of minority groups as a means of holding down wage costs and for ease of dismissal. But such practices appear to be specific to particular industries and should not necessarily be regarded as universally applicable to all TNCs in all industries.

Figure 9.6

Differences in average wages paid by foreign affiliates of TNCs and domestic firms

Source: Based on Crook, 2001: 15

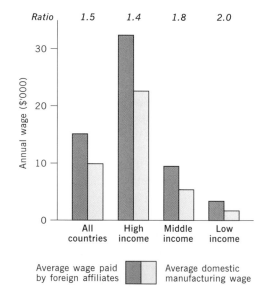

Labour relations

In many developing countries labour is either weakly organized or labour unions are strictly controlled (or even banned) by the state. Even in developed economies some major TNCs simply do not recognize labour unions in their operations while deregulation of labour markets has become widespread. But most TNCs, however reluctantly, do accept labour unions where national or local circumstances make this difficult to avoid. Whether labour unions are involved or not, the question of the nature of labour relations within TNCs focuses on whether they are 'good' or 'bad', that is, harmonious or discordant. Some studies suggest that TNCs tend to have better labour relations in their plants than domestic firms; others point to a higher incidence of strikes and internal disputes in TNCs. But it is often difficult to separate out the 'transnational' element. In the case of strikes, for example, it may be plant or firm size that is the most important influence rather than nationality of ownership.

One of the most acute concerns of organized labour is that decision-making within TNCs is too remote: that decisions affecting work practices and work conditions, pay and other labour issues are made in some far-distant corporate headquarters which has little understanding or even awareness of local circumstances. Some labour relations decisions made by TNCs are far more centralized than others in that they are either made directly at corporate headquarters or require its approval. These areas tend to relate to the operating costs of the subsidiary and reflect the parent company's concern to control financial and labour costs. However, there is considerable variation between TNCs in their degree of headquarters' involvement in labour relations.

The dispersed nature of TNC operations and the tendency towards remoteness in decision-making have made it very difficult for labour unions to organize effectively to counter such issues as plant closure or retrenchment. Two developments, although relatively limited so far, are significant.[33] One is the initiation, by global union federations (such the International Confederation of Free Trade Unions – ICFTU), of networks of workers within specific TNCs in an attempt to move industrial relations issues to the global level. The second development has occurred within the European Union:

> As part of the social protocol of the 1993 Maastricht Treaty, at least 15 million employees in some 1500 [TNCs] operating in Europe now have rights to information and consultation on all matters that affect more than one member state ... Each company that employs more than 1000 people, of whom at least 150 are located in two (or more) member states, has to meet the representation, transportation, accommodation, and translation costs of bringing employee representatives from across the European Union on an annual basis ... they provide a new opportunity for workplace representatives to meet their counterparts from every division of their companies' operations in Europe.[34]

Despite such developments, labour unions remain primarily contained within national state boundaries while TNCs are not. This structural difference creates inevitable tensions.

An especially important factor in the employment effects of TNCs on host countries is the particular role played by the foreign subsidiary in the parent company's overall strategy. Table 9.3 provides one perspective on this by relating various aspects of employment impact – level, type, job security, labour relations – to three types of subsidiary role: the miniature replica, the rationalized manufacturer and the product specialist. 'Clearly, from a host country perspective, "product

Table 9.3 Subsidiary strategies and the employment effects of TNCs on host countries

Subsidiary strategy	Spatial division of labour	Level of employment	Employment effects		Degree of job security	Labour relations
			Type of employment			
'Miniature replica'	Located in close proximity to major markets. Adds to regional imbalances	Employment creation through import-substitution and multiplier effects. Restricted long-term employment growth prospects due to limited product/ market role of subsidiary	Low-medium skill content; limited R&D		Depends on market share and market growth. In the longer term, job security is threatened by shift to global/regional strategies	Host-country industrial relations practices
'Rationalized manufacturer'	Reinforces international division of labour through international sourcing in low-cost countries and location of final assembly close to major markets	Employment creating effects limited by restricted plant role and high import content in final assembly operations	Low skill content; routine assembly work, 'screwdriver' operations; limited functional responsibility or R&D; limited workforce training		Plant status dependent on international factor cost movements; but possibly upgrading of assembly plants over time through widening product/market responsibilities and increased local content	Anti-union practices to reduce threats to coordinated and integrated global production system
'Product specialist'	Located mainly in highly developed countries or regions; access to skilled labour	Major positive effect through wide product/ market role of plant	High skill content and decentralization of R&D and other functional activities		Long-term job security due to enhanced plant status	Transfer of parent company industrial relations practices

Source: Based on Hamill, 1993: Figure 3.5

specialists" with enhanced product and market responsibilities are preferable to "miniature replicas" or "rationalized manufacturers".'[35]

Dangers of overdependence on TNCs

It should now be clear that no unambiguous, all-embracing, evaluation of TNC impact on host economies can be made. Whether a foreign plant creates net costs or net benefits will depend on the *specific context*: the interaction between the attributes and functions of the plant itself within its corporate system and the nature and characteristics of the host economy. It also depends, critically, on the alternatives realistically available. But what if a host economy – or an important sector within the economy – develops a very high level of foreign TNC involvement (the right-hand side of Figure 9.2)? Does the presence of a large foreign-controlled sector tip the balance of evaluation? In a long-term sense, the answer would appear to be 'yes'. Whereas the involvement of some foreign plants in a host economy will have beneficial effects – not only in creating employment but also in introducing new technologies and business practices – overall dominance by foreign firms is almost certainly undesirable from a host-country viewpoint. There are real dangers in acquiring the status of a *branch plant economy*.

Precisely what constitutes an undesirable level of foreign penetration is open to debate. Indeed, a country may be dominated by, and dependent upon, external forces even where there is very little direct foreign investment in the economy. This may occur, for example, where a large segment of an economy is engaged in subcontracting work for foreign customers. Here, however, our concern is with the effects of a large direct foreign presence. Until recently, most of the debate was focused upon developing countries and formed part of the broader dependency debate. But such concern is no longer confined to developing countries. Most obviously, those developed economies that have a very high level of foreign direct investment also share the same kinds of problem. For example, in 2001 two economies with historically high levels of dependence on foreign investment – Scotland and Singapore – experienced major problems associated with plant closures and contractions by foreign firms.

A high level of dependence on foreign enterprises potentially reduces the host country's sovereignty and autonomy: its ability to make its own decisions and to implement them. At the heart of this issue are the different – often conflictual – goals pursued by nation-states on the one hand and TNCs on the other. Each is concerned to maximize its own 'welfare' (in the broadest sense). Where much of a host country's economic activity is effectively controlled by foreign firms, non-national goals may well become dominant. It may be extremely difficult for the host government to pursue a particular economic policy if it has insufficient leverage over the dominant firms. The tighter the degree of control exercised by TNCs within their own corporate hierarchies and the lower the degree of autonomy of individual plants the greater this loss of host-country sovereignty is likely to be. In the *individual* case this may not matter greatly but where such firms *collectively* dominate a host economy or a key economic sector it most certainly does matter. The most significant aspect of dependence upon a high level of foreign direct investment is *technological*: the inability to generate the knowledge, inventions and innovations necessary to generate self-sustaining growth. However, this is not to argue that foreign investment should be avoided completely. On the contrary. What should be avoided by host economies is an *excessive* degree of foreign penetration.

TNCs and home economies

Most of the arguments over the possible costs and benefits of TNCs have been concerned with their effects on host economies. This is not surprising because the geographical destinations of TNC investments are far more diverse and numerous than their origins. But the issue is one that faces an increasing number of countries as more and more firms have extended their operations across national boundaries.

Questions and assumptions

What are the implications for the firm's home country of increased transnational production? Does it adversely affect the country's economic welfare by, for example, drawing away investment capital, displacing exports or destroying jobs? Or is it an inevitable feature of today's highly competitive global economy that forces firms to expand overseas in order to remain competitive? *Proponents* of overseas investment argue that the overall effects on the domestic economy will be positive, raising the level of exports and of domestic activity to a level above that which would prevail if overseas investment did not occur. *Opponents* of overseas investment, on the other hand, argue that the major effect will be to divert capital that could have been invested at home and to displace domestic exports.

The critical issue is the extent to which domestic investment could realistically be *substituted* for overseas investment:

- What would have happened if the investment had not been made abroad? Would that investment have been made at home? Or would the resources that went into the foreign investment have been used in higher levels of consumption and/or public services?
- What would have been the effect of foreign investment on domestic exports? Would the foreign sales of the product of the investment have been filled by exports from the home economy in the absence of the investment? Or would they have been taken over by foreign competitors?

Estimating the employment impact of outward investment

Establishing an overseas operation will have implications for the home country's balance of payments, through its influence on capital and financial flows and its effects on trade. But the most obvious implication for the average citizen is the effect on employment.[36] This is the issue that has received most attention in the United States, Europe and, more recently, in Japan with the acceleration of Japanese overseas investment. But the interpretations of the employment effects of outward investment are diametrically opposed because they are based on totally different assumptions.

- Labour interests are adamant that the overseas activity of TNCs dramatically increases unemployment at home because it is frequently assumed that *all* the activity undertaken overseas could realistically have been retained at home.
- Most business interests, on the other hand, are equally adamant that overseas investment increases, or at least preserves, jobs at home. The business-interest estimates tend to make the opposite assumption: that *none* of the overseas activity could realistically have been retained at home.

Quite clearly, more sophisticated approaches are needed to estimate even the approximate impact of overseas investment on domestic employment.

The possible direct employment effects of outward investment fall into four categories:[37]

- the *production-displacement effect* (DE) – employment losses arising from the diversion of production to overseas locations and the serving of foreign markets by these overseas plants rather than by home-country plants, that is, the displacement of exports
- the *export-stimulus effect* (XE) – employment gains from the production of goods for export created by the foreign investment which would not have occurred in the absence of such investment
- the *home office effect* (HE) – employment gains in non-production categories at the company's headquarters made necessary by the expansion of overseas activities
- the *supporting firm effect* (SE) – employment gains in other domestic firms supplying goods and services to the investing firm in connection with its overseas activities.

Thus, the *net employment effect* (NE) of overseas investment on the home economy is:

$$NE = -DE + XE + HE + SE$$

Unfortunately, the data needed to disaggregate employment change into these components are rarely, if ever, available so that, once again, large assumptions have to be made. That is why the estimates of the numbers of jobs either created or destroyed vary so widely – often by hundreds of thousands. Table 9.4 summarizes, in a qualitative manner, the positive and negative aspects of both direct and indirect employment effects of outward investment in terms of three attributes: quantity of jobs, quality of jobs and the location of jobs. In interpreting this table we need to bear in mind that the precise effects of outward investment on home-country employment are highly contingent on the specific circumstances involved. Unfortunately, there have been relatively few attempts to calculate actual home-country employment changes associated with outward investment.

Even though we cannot put precise numbers on jobs gained or lost through FDI, one thing is clear: *the winners and the losers are rarely the same*:

> On computer tapes, jobs may be interchangeable. In the real world they are not. A total of 250,000 new jobs gained in corporate headquarters does not, in any political or human sense, offset 250,000 old jobs lost on the production line. When Lynn, Massachusetts becomes a ghost of its former self, its jobless citizens find little

Table 9.4   The potential effects of outward investment on home-country employment

Area of impact	Direct		Indirect	
	Positive	*Negative*	*Positive*	*Negative*
Quantity	Creates or preserves jobs in home location, e.g. those serving the needs of affiliates abroad	Relocation or 'job export' if foreign affiliates substitute for production at home	Creates or preserves jobs in supplier/service industries at home that cater to foreign affiliates.	Loss of jobs in firms/ industries linked to production/ activities that are relocated
Quality	Skills are upgraded with higher-value production as industry restructures	'Give backs' or lowers wages to keep jobs at home	Boosts sophisticated industries	Downward pressure on wages and standards flows on to suppliers
Location	Some jobs may depart from the community, but may be replaced by higher-skilled positions, upgrading local labour market conditions	The 'export' of jobs can aggravate regional/local labour market conditions	The loss of 'blue-collar' jobs can be offset by greater demand in local labour markets for high-value-added jobs relating to exports or international production	Demand spiral in local labour market triggered by layoffs can lead to employment reduction in home-country plant locations

Source: UNCTAD, 1994: Table IV-1

satisfaction in reading about the new headquarters building on Park Avenue and all the secretaries it will employ. The changing composition of the workforce and its changing geographical location brought about by the globalization of US industry are affecting lives of millions of Americans in serious and largely unfortunate ways.[38]

At one time, it could be said with some accuracy that the dominant losers were production workers while the major gainers were white-collar, managerial workers. But such a simple distinction no longer holds as TNC operations have increased in complexity and as their modes of restructuring (including those of acquisition and merger) have become more diverse.

Is there a choice?

It is virtually impossible to say with any certainty that overseas investment could equally as well have been made in the firm's home country. We can make various assumptions about what might have happened, but that is all. Ultimately, the key lies with the *motivations* that underlie specific investment decisions. As profit-seeking organizations, firms invest abroad for a whole variety of reasons, for example:

- to gain access to new markets
- to defend positions in existing markets
- to circumvent trade barriers
- to diversify the firm's production base
- to reduce production costs
- to gain access to specific assets and resources.

It might be argued that foreign investment undertaken for *defensive* reasons – to protect a firm's existing markets, for example – is less open to criticism than *aggressive* overseas investment. The argument in the case of defensive investment would be that in its absence the firm would lose its markets and that domestic jobs would be lost anyway. Such investment might be made necessary by the erection of trade barriers by national governments, by their insistence on local production, or by the appearance of competitors in the firm's international markets. But, presumably, defensive investment might also include the relocation of production to low-wage countries in order to remain competitive. Here, the alternative might be the introduction of automated technology in the domestic plant that would also lead to a loss of jobs.

Although there may well be some clear-cut cases – particularly where access to markets is obviously threatened or where proximity to a localized material is mandatory – there will inevitably be many instances where there is substantial disagreement over the need to locate overseas rather than at home. The various interest groups will have different perceptions of the situation. Thus,

[Trans]national management tends to explain investments abroad as reflecting imperative requirements of markets and competition, i.e. global change factors calling for such adjustment by [trans]national enterprises in the interest of efficient production and ultimate survival. Its critics see the search for profits as the driving factor.

The available evidence is mixed as to the extent to which [trans]national enterprises do, or can, in fact exercise discretion in decisions to transfer production abroad on a large scale, especially to low-wage developing countries. Instances can be found of [trans]national enterprises apparently having set up off-shore sourcing subsidiaries, in Hong Kong, Singapore or elsewhere, before they were obliged to do so in self-defence, by competition in home or export markets, either from local firms or from [trans]national enterprises based in other home countries. On the other hand, examples can be found where such foreign subsidiaries were set up only after substantial competition, for instance from national enterprises in NICs, had been encountered in home or third markets. Still other cases exist – notably in the US television industry – where some TNCs resisted production transfers abroad until they were thrust upon the firm by imminent bankruptcy. The problem for the researcher is that, in all such cases, observable events must be compared with hypothetical, non-observable alternatives. *Thus a clear-cut general answer as to the degree to which TNCs have been wilful discretionary agents of production transfers abroad is not possible. For this purpose the actual decision-making processes of individual [trans]nationals … would need to be investigated through detailed case studies.*[39]

It is this kind of complexity that makes the adoption of national policies towards outward investment so problematic. A comprehensive restriction on such investment by domestic firms might make a national economy worse, rather than better, off. On the other hand, wholesale outflow of investment must surely be detrimental to home-country interests. If there is to be a national policy in this area, therefore, it must be both well informed and selective. It is worth recalling that Japanese policy changed from that of tight restriction on outward investment to one of carefully encouraging certain types of overseas investment as a matter of national policy.

The bargaining relationship between TNCs and states

In the final analysis, the relationship between TNCs and host countries (actual or potential) revolves around their relative bargaining power: the extent to which each can implement their own preferred strategies. The situation is especially complex when TNCs pursue a strategy of transnational integration of their activities in which individual units in a specific host country form only a part of the firm's transnational operations. In such circumstances, governments have a number of legitimate concerns:[40]

- that integrated TNCs might relocate their operations to other countries because of relative differences in factor costs
- that integrated TNCs use their integrated operations to engage in transfer pricing to reduce the taxes they pay

- that integrated TNCs retain their key competencies outside host countries (typically in the firm's home country) and locate only lower-skill, lower-technology operations in host countries
- that national decision centres no longer operate within an integrated TNC, making negotiations difficult between host-country governments and the local affiliate.

Although the degrees of freedom of TNCs to move into and out of territories at will are often exaggerated, the potential for such locational mobility is obviously there. This helps to make the TNC–state bargaining process immensely complex and highly variable from one case to another. Despite such contingency, however, we need to try to understand some of the *general* features of the bargaining process. Here, for the sake of simplicity, we assume that a host country can be regarded as a single entity in the bargaining relationship. In fact, of course, many competing interest groups are involved, including domestic business interests which may have varying attitudes towards TNCs.

Figure 9.7 is a highly simplified, hypothetical example. The vertical axis of the graph shows the rate of return a TNC may seek for a given level of investment (XA) on the horizontal axis. The bargaining range for this level of TNC investment is shown to vary between

- a lower limit (XY), which is the minimum rate of return that the TNC is prepared to accept for the amount of investment XA and
- an upper limit (XZ), which is determined by the cost to the host economy of either developing its own operation, finding an alternative investor or managing without the particular advantages provided by the TNC.

XZ is the maximum return the TNC can make for the amount of direct investment (XA) permitted by the host economy. It is in the interests of the TNC to try to raise the upper limit (XZ); conversely, it is in the interests of the host economy to try to lower that upper limit: 'The higher is the cost to the host economy of losing the proposed [investment], the greater are the possibilities for the TNC of setting the bargain near the maximum point.'[41]

Figure 9.7

A simplified model of the bargaining relationship between a TNC and a host country

Source: Based on Nixson, 1988: Figure 1

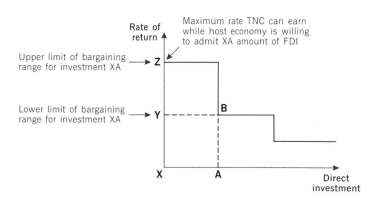

On the other hand, the more possibilities the host economy has of finding alternatives the greater are its chances of lowering that upper limit. The greater the competition between TNCs for the particular investment opportunity the greater are the opportunities for the host country to reduce both the upper and lower limits: 'In addition, the host economy has an interest in lowering the lower limit, through the creation of an advantageous "investment climate" (political stability, constitutional guarantees against appropriation, etc.) which might persuade the TNC to accept a lower rate of return.'[42]

'Locational tournaments': competitive bidding for investment

The greater the competition between potential host countries for a specific invest-ment the weaker will be any one country's bargaining position because countries will tend to bid against one another to capture the investment. Indeed, one of the most striking developments of the past two or three decades has been the devel-opment of so-called *locational tournaments*. There has been an enormous intensifi-cation in *competitive bidding* between states (and between communities within the same state) for the relatively limited amount of internationally mobile investment. Such cut-throat bidding undoubtedly allows TNCs to play off one state against another to gain the highest return for their investment. Some of the most spectac-ular examples of this process have occurred in the automobile industry because of the size and potential spin-off effects of assembly and component plants (see Chapter 11).[43]

In fact, much of the actual capital may be provided by the host government itself in the form of various kinds of financial and fiscal (tax) deals as well as in the form of physical and social infrastructure. A study by UNCTAD found that only 4 countries out of 103 did not offer some kind of fiscal incentive to inward investors during the early 1990s while financial incentives were offered in 59 out of 83 countries surveyed.[44] Table 9.5 shows just a few examples of the kinds of deals that have been made by states in their scramble to capture major investments.

Table 9.5 shows both the size of the company's investment and the amount of financial incentive provided by the state. Most interestingly, in the final column, it shows the state investment per job created. The sheer size of the numbers in the final column is staggering. It is not surprising that there was widespread conster-nation in neighbouring Southern states of the US when Alabama offered the equivalent of $167,000 per job created to attract the Mercedes–Benz plant, for which virtually all the Southern states had been competing. More recently,

> job-hungry Alabama offered $252.8m in incentives to persuade Hyundai to choose it over Kentucky, the other finalist in the contest for the Korean carmaker. The incentives include $76.7m in tax breaks, $61.8m in job training, $34m for site purchase, $10m in advertising to state employees, and assorted site and road improvements.[45]

These examples provide a graphic demonstration of the bargaining power of major TNCs to offset some of their investment costs on to the state. But it is not only new TNC investments that are involved in this process. It has become

Table 9.5 The costs of attracting foreign investment: some examples of incentive packages

Location	Company	Company investment ($m)	State investment ($m)		State's financial investment per employee ($)
Smyrna, Tenn. (USA) (1983)	Nissan (Japan)	745–848	Road access Training Total	22 7.3 33	25,384
Flat Rock, Mich. (USA) (1984)	Mazda (Japan)	745–750	Training Road improvt On-site works Economic development grant/loan Water system Total	19 5 3 21 5 48.5	13,857
Georgetown, Ky (USA) (1985)	Toyota (Japan)	823.9	Land purchase Site prepn Road improvt Training Toyota families' educn Total	12.5 20 47 65 5.2 149.7	49,900
Tuscaloosa, Ala. (USA) (1993)	Mercedes–Benz (Germany)	300	Site devt Infrastructure Private sector/goodwill Training Total	68 77 15 90 250	166,667
Spartenburg, SC (USA) (1994)	BMW (Germany)	450	Total	130	108,333
Setubal (Portugal) (1991)	Auto Europa (Ford/VW) (USA/Germany)	2603		483.5	254,451
West Midlands (UK) (1995)	Ford/Jaguar	767		128.72	128,720
North East England [1994/5]	Samsung (Korea)	690.3		89	29,675
Lorraine (France) (1995)	Mercedes–Benz (Germany)	370		111	56,923

Source: Based on UNCTAD, 1995: Table VI.3

increasingly common for TNCs to try to lever various kinds of state subsidies in order to persuade them to keep a plant in a particular location. Otherwise, it is threatened, the plant will be closed or much reduced in scale.

Like firms, therefore, states engage in *price competition* in their attempts to capture a share of the market for mobile investment. Like firms, states also engage in *product differentiation* by creating particular 'images' of themselves, such as the strategic nature of their location (it is amazing how many places project themselves as being at the 'centre' of the world), the attractiveness of the business environment, the quality of the labour force, and so on. Hence, the figures shown in Table 9.5 do not reveal the full extent of states' incentives to attract and retain TNC investment.

States undoubtedly face a major dilemma. If they do not join the bidding battle they face the probability of being left out of TNCs' investment plans, even though incentives are not the major determinant of TNC location decisions. However,

> other things being equal, incentives can induce foreign investors towards making a particular locational decision by sweetening the overall package of benefits and hence tilting the balance in investors' locational choices. Incentives can be justified if they are intended to cover the wedge between the social and private rates of return for FDI undertakings that create positive spillovers. However, incentives also have the potential to introduce economic distortions (especially when they are more than marginal) ... It is not in the public interest that the cost of incentives granted exceeds the value of the benefits to the public. But, as governments compete to attract FDI, they may be tempted to offer more and larger incentives than would be justified, sometimes under pressure from firms that demand incentives to remain in a country.[46]

Relative bargaining power

The extent to which a state feels the need to offer large incentives will depend on its relative bargaining strength in any specific case. Conversely, the extent to which a TNC is able to obtain such incentives will depend on its relative bargaining strength. The outcome will depend on a number of factors.[47]

The price that a *host country* will ultimately pay is a function of:

- the number of foreign firms independently competing for the investment opportunity
- the recognized measure of uniqueness of the foreign contribution (as against its possible provision by local entrepreneurship, public or private)
- the perceived degree of domestic need for the contribution.

The terms the *TNC* will accept are a function of:

- the firm's general need for an investment outlet
- the attractiveness of the specific investment opportunity offered by the host country, compared to similar or other opportunities in other countries
- the extent of prior commitment to the country concerned (e.g. an established market position).

Figure 9.8 sets out the major components of the bargaining relationship between TNCs and host countries. Both possess a range of 'power resources' that are their major bargaining strengths. Both operate within certain constraints that will restrict the extent to which these power resources can be exercised. The relative bargaining power of TNCs and host countries, therefore, is a function of three related elements:

● the *relative demand* by each of the two participants for resources which the other controls

● the *constraints* on each which affect the translation of potential bargaining power into control over outcomes

● the *negotiating status* of the participants.

Figure 9.8 suggests that, in general, host countries are subject to a greater variety of constraints than are TNCs, a reflection of the latter's greater potential flexibility to switch its operations between alternative locations. Nevertheless, the extent to which a TNC can implement a globally integrated strategy is constrained by nation-state behaviour. Where a company particularly needs access to a given location and where the host country does have leverage, then the bargain that is eventually struck may involve the TNC in making concessions. In general, the scarcer the resource being sought (whether by a TNC or a host country) the greater the relative bargaining power of the controller of access to that resource and vice versa.

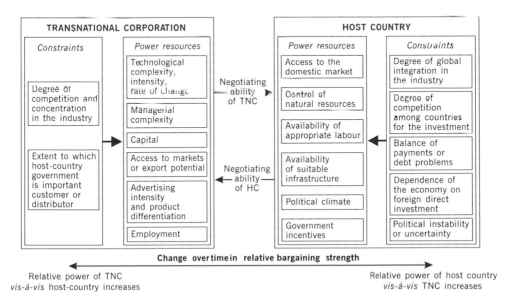

Figure 9.8   Components of the bargaining relationship between TNCs and host countries

Source: Based on material in Kobrin, 1987

For example, states that control access to large, affluent domestic markets have greater relative bargaining power over TNCs pursuing a market-oriented strategy than states whose domestic markets are small. It is in this kind of situation that the host country's ability to impose performance requirements – such as local content levels – on foreign firms is greatest. It may also give the host country sufficient leverage to persuade the inward investor to establish higher-level functions such as R&D facilities. On the other hand, the nature of the domestic market may not be a consideration for a TNC pursuing an integrated production strategy. Where, for example, the TNC's need is for access to low-cost labour that is very widely available then an individual country's bargaining power will be limited. On the whole, cheap labour is not a scarce resource on a global scale.

On the other hand, states may well have the ability to inhibit the achievement of globally integrated strategies by TNCs: 'host countries wield the power to limit the extent of, or even to dismantle, the [T]NC integrated manufacturing and trade networks with more regulations and restrictions on foreign investments and market access'.[48] At the extreme, of course, both institutions – TNCs and governments – possess sanctions which one may exercise over the other.

- The TNC's ultimate sanction is not to invest in a particular location or to pull out of an existing investment.
- A nation-state's ultimate sanction against the TNC is to exclude a particular foreign investment or to appropriate an existing investment.

The problem, of course, is that the whole process is *dynamic*. The bargaining relationship changes over time, as the bottom section of Figure 9.8 suggests. In most studies of TNC–state bargaining the conventional wisdom is that of the so-called 'obsolescing bargain', in which

> once invested, fixed capital becomes 'sunk', a hostage and a source of bargaining strength. The high risk associated with exploration and development diminishes when production begins. Technology, once arcane and proprietary, matures over time and becomes available on the open market. Through development and transfers from FDI the host country gains technical and managerial skills that reduce the value of those possessed by the foreigner.[49]

In this view, after the initial investment has been made, the balance of bargaining power is believed to shift from the TNC to the host country, in other words, to move to the right in Figure 9.8. But although this may well be the case in natural resource-based industries it is far less certain that this applies in those sectors in which technological change is frequent and/or where global integration of operations is common. In such circumstances, 'the bargain will obsolesce slowly, if at all, and the relative power of [T]NCs may even increase over time'.[50]

An example of the level and complexity of TNC–state bargaining relationship is shown by Ford's strategy to enter the Spanish market in the early 1970s.[51] This case demonstrates just how powerful a large TNC can be in persuading a host-country government to change its existing regulations. The Spanish automobile market in the early 1970s was heavily protected. Not only were tariffs on imports very high (81 per cent on cars, 30 per cent on components) but also cars built in

Spain had to have 95 per cent local content. In addition, no foreign company could own more than 50 per cent of a company operating in Spain. Such restrictions were very much in conflict with Ford's own preferences for a Spanish operation. Ford's aim was not only to penetrate the local market but also to create an export platform from which to serve the entire European market. As such, it wanted the lowest possible import tariffs on components and a minimal level of local content so that it could source components from other parts of its transnational network. Ford's preferred policy was also to have complete ownership of its foreign affiliates. On the other hand, the Spanish government was very anxious to build up its automobile industry and, especially, to increase exports.

Two years of negotiations at the highest political level ensued, a reflection of the 'new' diplomacy of the global economy in which heads of TNCs talk directly to heads of government. The eventual agreement showed just how powerful Ford's position was. The fact that virtually all other European governments were trying to entice Ford to locate in their countries gave the company substantial negotiating leverage. On the other hand, Ford regarded a Spanish location as vital to its future European operations. Under the agreement finally signed, virtually all of Ford's demands were met. In particular, for Ford the tariff on imported components was reduced from 30 per cent to 5 per cent; the local content requirement was reduced from 95 per cent to 50 per cent, provided that two-thirds of production was exported (precisely what Ford wanted to do anyway), and Ford was allowed 100 per cent ownership of its Spanish subsidiary. But, of course, the benefits were not all one way. In return, Spain gained a massive boost to its automobile industry which was subsequently enhanced by the entry of other TNCs, including General Motors. As a result, as we shall see in Chapter 11, Spain became one of the world's leading automobile producers.

However, states are not as weak as is often assumed. Or at least in certain circumstances this is the case. A prime example is China and its policy towards automobile firms.[52] The prospect of access to the world's largest and fastest-growing market led many automobile firms to try to enter China. But the Chinese government has complete control over such entry and has adopted a policy of limited access for foreign firms. Here, then, we have the obverse of the usual situation. Whereas in most cases, TNCs play off one country against another to achieve the best deal, in the Chinese case it is the state whose unique bargaining position enables it to play off one TNC against another. Of course, China is something of a special case. But although 'some developing countries have few attractive productive assets or locational advantages for which TNCs will compete with each other, and as a result may not be able to play off one TNC against another … equally there are many others who can play this game, as they have at least some "bargaining chips".'[53]

Conclusion

The aim of this chapter has been to explore the interrelationships between TNCs and states that form such an important nexus within the global economy. We have emphasized, in particular, the fact that TNC–state relationships may be both conflictual and cooperative; that TNCs and states may be rivals but, at the same time, they may collude with one another. In a real sense, states need firms to help in the process of material wealth creation while firms need states to provide the necessary supportive infrastructures, both physical and institutional, on the basis of which they can pursue their strategic objectives. Within this general framework, we focused in this chapter on three major aspects of TNC–state relationships: the effects of TNCs on host economies, the effect on home economies, and the bargaining relationships between TNCs and states.

In the cases of both host- and home-country impacts, the analytical problem is the same: it is a counterfactual problem. We cannot unambiguously establish what the situation would be like *if* the TNC were *not* involved in a particular case. Assumptions have to be made as to what the alternative situation would realistically be. The 'bottom line' is the *net effect* that takes into account the opportunities forgone by the presence or absence of the TNC. But the situation has become vastly more complex in today's highly interconnected global economy.

From a host economy's viewpoint, could the particular item of technology, the particular level of employment, and so on, be created without the involvement of the TNC? In some cases, the answer will undoubtedly be 'yes': in others, the answer will just as undoubtedly be 'no'. Similarly, from a home economy's viewpoint, what would be the effect if the country's firms were not involved in overseas activities? Can a firm opt out of international operations in today's increasingly global production environment? Would jobs at home be lost anyway even if the firm did not invest overseas? Would the economy be better or worse off? Everything will depend, as we have seen, on whether the overseas investment is obligatory or discretionary, but even here differences of viewpoint will exist. What is quite clear is that *TNCs tie national and local economies more closely into the global economy.*

Finally, we explored the bargaining processes between TNCs and states as each tries to gain maximum advantage from the other in pursuit of their own strategic objectives. States have become increasingly locked into a cut-throat competitive bidding process for investments, a process that provides TNCs with the opportunity to play off one bidder against another. But the extent to which this happens depends upon the specific *relative bargaining power* of TNCs and states. There is little doubt that there has been a shift in the relative power of TNCs and states but the position is far less straightforward than has often been supposed. Each bargaining process is different and highly contingent on the specific circumstances involved. That is why it is so difficult to make broad generalizations about the relative 'balance of power' between TNCs and states. Stopford and Strange's view is that

> governments *as a group* have indeed lost bargaining power to the [trans]nationals …
> [however] … does it follow that firms *as a group* have increased their bargaining power
> over the factors of production? Here, the argument becomes complex, for the power of
> the individual firm may be regarded as having also fallen as competition has
> intensified. New entrants have altered the rules and offer governments new
> bargaining advantage. *One needs to separate the power to influence general policy from the
> power to insist on specific bargains.*[54]

The discussion in this chapter also raises the question: does ownership matter? The usual answer is that, yes, it does; that 'home country' firms are 'better for us' than foreign firms. But is that always the case? The old aphorism was 'what's good for General Motors is good for America' (or the equivalent in other countries). However, that may not always be the case. We have to look more carefully at the actual costs and benefits of specific cases. Robert Reich posed the question to Americans: 'who is us?' His response was that 'the competitiveness of American-owned corporations is no longer the same as American competitiveness … the skills, training, and knowledge commanded by American workers – workers who are increasingly employed within the United States by foreign-owned corporations.'[55] But, like all the other issues in this debate over the relationships between TNCs and states, Reich's argument is not as straightforward as it seems. The fact remains that TNC–state relationships defy universal generalization. In Stopford and Strange's terms, it is the question of the 'specific bargain' that needs to form the basis of the much-needed research in this area.

Notes

1 For a recent use of such a comparison, see Anderson and Cavanagh (2000).
2 Gordon (1988: 61).
3 Pitelis (1991).
4 See Reich (1989).
5 Doz (1986a,b) provides an extensive discussion of these issues.
6 Leyshon (1992).
7 Picciotto (1991: 43, 46).
8 Yoffie and Milner (1989).
9 Sklair (1995: 167).
10 The *World Investment Report* published annually by UNCTAD devotes a major section each year to specific aspects of TNC impact.
11 UNCTAD (1995: 142–3).
12 Vernon (1998: 40).
13 The study was conducted by Ernst & Young and reported in *The Financial Times* (23 November 1995).
14 See UNCTAD (1995), Chapter III.
15 UNCTAD (2000a: 14).
16 UNCTAD (2000a: Table 3.11).
17 Lall and Streeten (1977: 71).
18 Leonard (1988).
19 UNCTC (1988).
20 UNCTAD (1995: 176–7).
21 UNCTAD (2000a: 12–13).
22 The UNCTAD *World Investment Report 2001* devotes particular attention to this topic. See also UNCTAD (2000a).
23 UNCTAD (2000a: 15).
24 See Fuentes et al. (1993) and Sklair (1989).
25 Hill (1993: 210).
26 Grunsven, Egeraat and Meijsen (1995: 3).
27 Linden (2000: 210).
28 UNCTAD (2001: 135).
29 Yeung and Li (2000: 628).
30 The most comprehensive reviews of the employment impact of TNCs can be found in studies by the ILO (Bailey, Parisotto and Renshaw, 1993) and by UNCTAD (1994).
31 UNCTAD (1994: 163–4).

32 UNCTAD (1994: 197, 198).
33 Wills (1998).
34 Wills (1998: 122).
35 Hamill (1993: 72).
36 See ILO (1981a,b); UNCTAD (1994).
37 Hawkins (1972).
38 Barnet and Muller (1975: 302).
39 ILO (1981b: 69–70; emphasis added).
40 Doz (1986a: 231–4).
41 Nixson (1988: 379).
42 Nixson (1988: 380).
43 See Mytelka (2000); Rodríguez-Pose and Arbix (2001).
44 UNCTAD (1995).
45 *The Financial Times* (16 April 2002).
46 UNCTAD (1995: 299).
47 Gabriel (1966).
48 Doz (1986b: 39).
49 Kobrin (1987: 611–12).
50 Kobrin (1987: 636).
51 Seidler (1976) provides an interesting narrative of the bargaining process between Ford and the Spanish government.
52 Chang (1998b: 233–4).
53 Chang (1998b: 234).
54 Stopford and Strange (1991: 215–16; emphasis added).
55 Reich (1990: 54).

PART THREE

GLOBAL SHIFT: THE PICTURE IN DIFFERENT SECTORS

CHAPTER 10

'Fabric-ating Fashion': The Textiles and Garments Industries

The textiles and garments industries exemplify many of the intractable issues facing today's global economy, particularly the trade tensions between developed and developing economies. Indeed, these industries occupy a central position in the anti-globalization debates. It is not difficult to see why. At least 20 million workers are officially employed worldwide in the textiles and garments industries, while countless unregistered workers, employed both in factories and at home, need to be added to reach a true picture, especially in the garments industry.

The textiles and garments industries were the first manufacturing industries to take on a global dimension. They are the most geographically dispersed of all industries across both developed and developing countries. Despite the changes wrought by new technologies, corporate rationalization and competition from new producers, the textiles and garments industries continue to be important sources of employment in the developed economies. In particular they employ many of the more 'sensitive' segments of the labour force: females and ethnic minorities, often in tightly localized communities. In developing countries the industries employ predominantly young female workers in conditions that recall those of the sweatshops and mills of 19th century cities in Europe and North America. Hence, the importance of textiles and garments as a basis for today's newly industrializing and less industrialized countries, together with their continued, though much diminished, importance in the older industrialized economies, have made these industries into a global political football. They are the subject of fierce political controversy between developed and developing countries and, increasingly, between the developed economies themselves. It is no accident that textiles and garments are the only industries to which special international trade regulations apply through the Multi-Fibre Arrangement.

The textiles industry was the archetypal industry of the first industrial revolution of the 18th and 19th centuries in Britain (see Figure 4.1). Its major geographical centre of production – Lancashire – became the exemplar of the 19th century industrial landscape with its oft-described 'dark satanic mills'. Its marketing capital, Manchester, became the first global industrial city – the 'Cottonopolis' – of an industrial system whose tentacles spread across the globe. All the 'newly industrializing countries' of the 19th century – the United States, Germany, France, the Netherlands – also developed large textiles industries, employing many hundreds of thousands of workers, often in strongly localized geographical clusters.

A similarly concentrated pattern occurred in the rather later development of a factory garments industry in the second half of the 19th century.

The sheer strength – both economic and political – of the British textiles industry in the 19th century effectively strangled the development of an indigenous textiles industry in the major colonies, especially India.[1] But such dominance could not last, particularly in an industry so ideally suited to the early stages of industrialization. Textiles and garments could be manufactured using relatively simple technologies and low-skill labour. The traditional craft skills of hand spinning, weaving and sewing were a ready basis for larger-scale industrial application. The capital investment required was relatively modest compared with many other types of industry. Where local supplies of the raw materials were also available there was an even more obvious case for the development of a textiles industry, although lack of indigenous supplies has not inhibited the development of highly successful textiles and garments industries in many parts of the world.

The textiles–garments production chain

The textiles and garments industries form part of a larger production chain (Figure 10.1). Each stage has its own specific technological and organizational characteristics and particular geographical configuration. Each has been changing very substantially in recent decades. The textiles industry itself consists of two major operations: the preparation of yarn and the manufacture of fabrics. Both stages are performed by firms of all sizes, from the very small domestic enterprise to the very large subsidiaries of TNCs. The general trend, however, has been for textile manufacturing to become more and more capital-intensive and for large firms to be increasingly important. The output of the textiles industry goes to three major end-users, of which the garments industry is by far the most important.

Despite some recent changes, the garments industry remains far more fragmented organizationally than the textiles industry and is less sophisticated

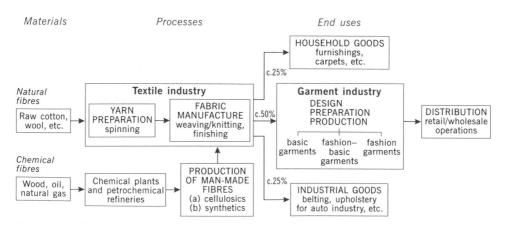

Figure 10.1   The textiles–garments production chain

technologically. It is also an industry in which subcontracting is especially prominent. Very often the design and even the cutting processes are performed quite separately from the sewing process, the latter being particularly amenable to international subcontracting. The garments industry itself produces an enormous variety of often rapidly changing products. Finally, although not part of the production sequence itself, the role of the distributors of garments – particularly the retailers – is of considerable and growing importance and has enormous implications for the organization and global geography of garments manufacture. It is in the garments industry, in fact, that *buyer-driven* production chains are especially prominent (see Figure 2.7).

Figure 10.2 suggests a generalized sequence of development through which individual producing countries appear to have passed. Six stages are shown, beginning with the embryonic stage typical of the least developed countries through to the maturity and decline of the older industrialized countries. Figure 10.2 shows the type of production likely to be characteristic of each stage and the related kinds of trade. The sequence is a very useful summary of what has happened so far but, like all such sequential models, it should not be regarded necessarily as being predictive of what will happen in the future. Although many countries have passed through some or all of these stages, the precise path of development depends upon a number of factors that, together, produce the specific geographical patterns of the textiles and garments industries.

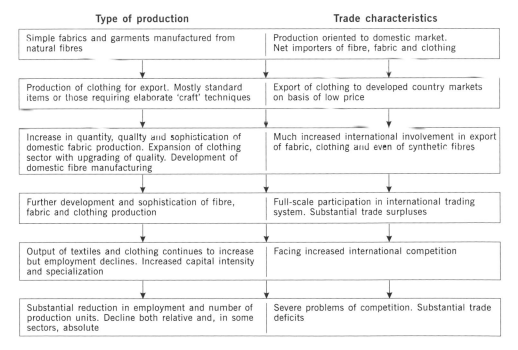

Type of production	Trade characteristics
Simple fabrics and garments manufactured from natural fibres	Production oriented to domestic market. Net importers of fibre, fabric and clothing
Production of clothing for export. Mostly standard items or those requiring elaborate 'craft' techniques	Export of clothing to developed country markets on basis of low price
Increase in quantity, quality and sophistication of domestic fabric production. Expansion of clothing sector with upgrading of quality. Development of domestic fibre manufacturing	Much increased international involvement in export of fabric, clothing and even of synthetic fibres
Further development and sophistication of fibre, fabric and clothing production	Full-scale participation in international trading system. Substantial trade surpluses
Output of textiles and clothing continues to increase but employment declines. Increased capital intensity and specialization	Facing increased international competition
Substantial reduction in employment and number of production units. Decline both relative and, in some sectors, absolute	Severe problems of competition. Substantial trade deficits

Figure 10.2 An idealized sequence of development in the textiles and garments industries

Global shifts in the textiles and garments industries

The geography of production

In Figures 10.3 and 10.4, employment is used as an indicator of the global production of textiles and garments. Although widely spread geographically, some clear concentrations stand out. In the case of textiles, China has by far the largest concentration of employment in the world, with more than 6 million employed in the industry. India, with almost 1.5 million, is a distant second. Textiles employment in the United States is around 800,000, that in Japan just under 600,000. Most of the other major textiles activity is in Europe, and in Asia. China also dominates global employment in the garments industry, with some 1.6 million workers followed, a long way behind, by the United States, the Russian Federation and then Japan. Garments manufacture also remains important in Western Europe (notably in the United Kingdom, Germany, France, and Italy) and Eastern Europe (notably in Romania and Poland). Among developing countries, the East and South East Asia region is overwhelmingly dominant. Apart from China, there are major concentrations of garments workers in Indonesia, Thailand, the Philippines, South Korea and Hong Kong.

The geography of trade

Recent geographical trends in the production of textiles and garments show a clear pattern of the continuing relative (and, in some cases, absolute) decline of developed country producers and a geographical shift of production to certain developing countries, notably in East Asia and, to a lesser extent, in Mexico, the Caribbean, Eastern Europe and some parts of the Mediterranean rim. Such shifts are more substantial in the garments industry than in the more capital-intensive parts of the textiles industry. Such shifts are even more apparent when viewed through the lens of trade, where important differences between the textiles and garments patterns can be seen.

Trade in textiles

Table 10.1 lists the 15 leading textiles exporting and importing countries in 2000. In terms of exports (Table 10.1a), only the United States and Italy, of the seven industrialized countries in the list, retained their shares of the total world exports between 1980 and 2000. All of the others lost export share. On the other hand, East Asian economies dominate the upper part of the table. China dramatically increased its share of world exports from 4.6 per cent in 1980 to 10.2 per cent in 2000. Both Korea and Taiwan also doubled their shares of the world total. In total, the Asian economies in the list increased their share of world exports from 27 per cent in 1980 to 39 per cent in 2000. The pattern of textile imports is very different, as Table 10.1(b) shows. But, again, the leading players are the United States and China. The right-hand column of Table 10.1(b) shows the textile trade balance of major countries. Note the $4.75 billion deficit of the United States, the significant $5.84 billion surplus of Italy and the very large surplus of South Korea.

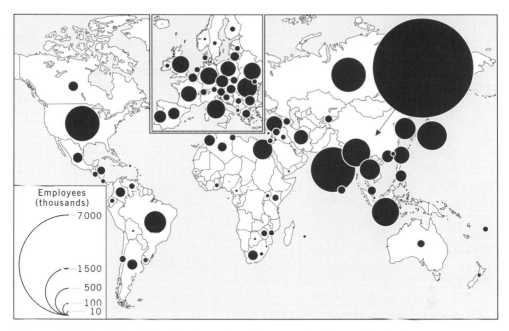

Figure 10.3 Global distribution of employment in the textiles industry
Source: Based on material in UNIDO (1996) *International Yearbook of Industrial Statistics, 1996;* ILO, 1996b

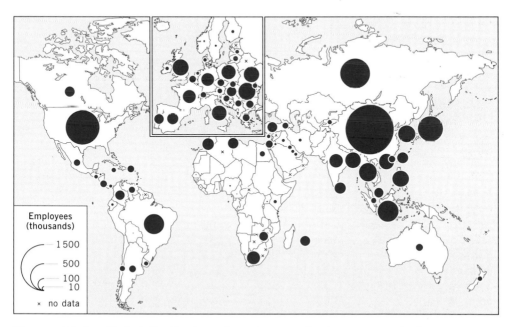

Figure 10.4 Global distribution of employment in the garments industry
Source: Based on material in UNIDO (1996) *International Yearbook of Industrial Statistics, 1996;*
ILO, 1996b

Table 10.1 The world's leading exporters and importers of textiles

Exporting country	$bn	(a) Exports		
		1980	1990	2000
China	16.14	4.6	6.9	10.2
Hong Kong, China	13.44	—	—	—
Domestic exports	1.18	1.7	2.1	0.7
Re-exports	12.27	—	—	—
South Korea	12.78	4.0	5.8	8.1
Italy	11.96	7.6	9.1	7.6
Taiwan	11.69	3.2	5.9	7.4
Germany	11.02	11.4	13.5	7.0
United States	10.96	6.8	4.8	7.0
Japan	7.02	9.3	5.6	4.5
France	6.76	6.2	5.8	4.3
Belgium	6.40	—	—	4.1
India	5.09	2.1	2.1	3.4
Pakistan	4.53	1.6	2.6	2.9
United Kingdom	4.21	5.7	4.2	2.7
Turkey	3.67	0.6	1.4	2.3
Indonesia	3.51	0.1	1.2	2.2
Above 15	116.9	71.0	76.7	74.5

Importing country	$bn	(b) Imports			Exports/Imports
		1980	1990	2000	
United States	15.71	4.5	6.2	9.4	−4.75
Hong Kong, China	13.72	—	—	—	
Retained imports	1.45	3.7	3.8	0.9	
China	12.83	1.9	4.9	7.7	+3.31
Germany	9.32	12.1	11.0	5.6	+1.70
United Kingdom	6.91	6.3	6.5	4.1	−2.70
France	6.75	7.2	7.0	4.0	+0.01
Italy	6.12	4.6	5.7	3.7	+5.84
Mexico	6.10	0.2	0.9	3.6	
Japan	4.94	2.9	3.8	2.9	+2.08
Canada	4.13	2.3	2.2	2.5	
Belgium	3.63	—	—	2.2	+2.77
Spain	3.32	0.6	1.9	2.0	
South Korea	3.00	0.7	1.8	1.9	+9.78
Netherlands	2.64	4.0	3.4	1.6	
Poland	2.43	0.5	0.2	1.5	
Above 15	89.27	55.4	62.6	53.4	+27.63

Source: Based on WTO (2001) *International Trade Statistics, 2001*: Table IV-72

Figure 10.5

Regional shares of world
trade in textiles

Source: Based on WTO (2001)
International Trade Statistics,
2001: Chart IV.13

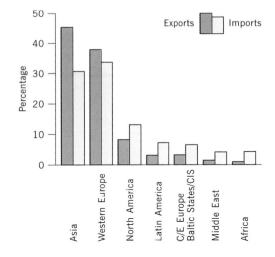

Figure 10.5 shows that world textiles exports are dominated by Asia (primarily East Asia) and Western Europe. Together these account for more than 80 per cent of the total. However, the pattern of inter-regional and intra-regional trade varies considerably between different parts of the world, as Figure 10.6 reveals:

● Around two-thirds of both Western Europe's and Asia's textile trade is intra-regional.
● For both North America (the United States and Canada) and Latin America the intra-regional share is only around one-third.
● Western Europe's main export destination is Eastern and Central Europe.
● For Asia, North America and Western Europe account for a similar share of textile exports (11 per cent each).
● North America and Latin America are very tightly connected in terms of textile exports: 40 per cent of North America textile exports go to Latin America while 60 per cent of Latin America's textile exports go to North America. In total, 77 per cent of North America's textile exports occur within the Americas and 93 per cent of Latin America's.

Trade in garments

The world's leading garments exporting and importing countries are shown in Table 10.2. Although there are similarities with textiles (see Table 10.1), the differences are significant. First, garments exports (Table 10.2a) are rather less concentrated than textile exports (63 per cent compared with 75 per cent). Second, China's dominance of garments exports is even more marked. In 1980, China generated 4 per cent of world garments exports; by 2000 its share had increased to 18 per cent (23 per cent if Hong Kong's domestic exports are included in the Chinese total). Third, Mexico has emerged as the world's fourth largest exporter of garments (4.4 per cent of the world total), a reflection of its involvement in the NAFTA.

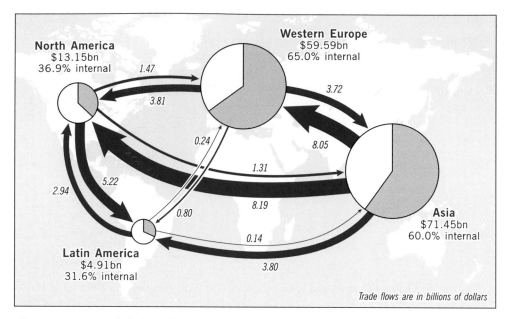

Figure 10.6 Global trade network in textiles

Source: Based on WTO (2001) *International Trade Statistics, 2001*: Table A10

Fourth, both South Korea's and Taiwan's shares declined substantially whereas Indonesia's and Thailand's shares increased. Fifth, apart from the United States (which slightly increased its share of world garments exports), all the other major developed economies lost share, although Italy's leading position was maintained.

In terms of garments imports (Table 10.2b), the most striking feature is the overwhelming dominance of the United States as the destination for almost one-third of the total. Second, Japan's share of garments imports almost tripled between 1980 and 2000 to be on a par with Germany's, at around 9 per cent of the total. The pattern of trade deficits and surpluses has long been an especially sensitive issue in the politics of international trade and, as we have noted already, in some of the anti-globalization protests. The right-hand column of Table 10.2(b) shows that, apart from Italy, all the industrialized countries have substantial trade deficits. All of these deficits pale into insignificance, however, when compared with the position of the United States which, in 2000, had a deficit in garments of almost $60 billion. The unique position of the United States in terms of garments trade is shown even more graphically in Figure 10.7 (compare this with North America's textile position in Figure 10.5). As in textiles, Western Europe and Asia dominate garments exports (with around 75 per cent of the world total). However, Asia is overwhelmingly the leading producer region.

Where the garments trade pattern differs most from that of textiles is in the extent of intra- and inter-regional trade (compare Figures 10.6 and 10.8).

- For Western Europe, intra-regional trade is even more significant in garments (77 per cent of the total compared with 65 per cent for textiles). The figure would be higher if the trade with Eastern and Central Europe were included).

Table 10.2 The world's leading exporters and importers of garments

Exporting country	$bn	1980	1990	2000
	(a) Exports			
China	36.07	4.0	9.0	18.1
Hong Kong, China	24.22	—	—	—
Domestic exports	9.94	11.5	8.6	5.0
Re-exports	14.28	—	—	—
Italy	13.22	11.3	11.0	6.6
Mexico	8.7	0.0	0.5	4.4
United States	8.65	3.1	2.4	4.3
Germany	6.84	7.1	7.3	3.4
Turkey	6.53	0.3	3.1	3.3
France	5.43	5.7	4.3	2.7
India	5.15	1.5	2.3	2.8
South Korea	5.03	7.3	7.3	2.5
Indonesia	4.73	0.2	1.5	2.4
United Kingdom	4.11	4.6	2.8	2.1
Thailand	3.95	0.7	2.6	2.0
Belgium	3.94	—	—	2.0
Taiwan	2.97	6.0	3.7	1.5
Above 15	125.4	65.6	68.2	63.1

Importing country	$bn	1980	1990	2000	Exports/imports
	(b) Imports				
United States	66.39	16.4	24.1	31.6	−57.74
Japan	19.71	3.6	7.8	9.4	
Germany	19.31	19.7	18.2	9.2	−12.47
Hong Kong, China	16.01	—	—	—	
Retained imports	1.73	0.9	0.7	0.8	
United Kingdom	12.99	6.8	6.2	6.2	−8.88
France	11.48	6.2	7.5	5.5	−6.05
Italy	6.07	1.9	2.3	2.9	+7.15
Netherlands	4.83	6.8	4.3	2.3	
Belgium	4.81	—	—	2.3	−0.87
Spain	3.77	0.4	1.5	1.8	
Canada	3.69	1.7	2.1	1.8	
Mexico	3.41	0.3	0.5	1.6	+5.29
Switzerland	3.22	3.4	3.1	1.5	
Russian Federation	2.96	—	—	1.4	
Austria	2.47	2.2	2.1	1.2	
Above 15	166.83	74.5	83.3	79.4	+27.63

Source: Based on WTO (2001) *International Trade Statistics, 2001*: Table IV-80

- For North America and Asia intra-regional trade in garments is only around one-quarter of the total.
- The Americas (North America and Latin America) constitute an even more integrated trading area in garments than in textiles. Almost two-thirds of North America's garments exports go to Latin America (giving an 'Americas' total of almost 90 per cent). For Latin America, the hemispheric tie is even stronger: almost 100 per cent of its garments exports are within the Americas but, overwhelmingly, as exports to North America (Mexico is an especially important component of this trade).
- The global pattern of Asia's garments exports is more even than the others and, indeed, than its own position in textiles. Some 27 per cent of Asia's garments are sold within Asia itself (compared with 60 per cent of its textile exports). Conversely, a much higher proportion of Asian garments exports go to North America (40 per cent) and Western Europe (23 per cent) than in the case of textiles.

Figure 10.7

Regional shares of world trade in garments

Source: Based on WTO (2001) *International Trade Statistics, 2001:* Chart IV.14

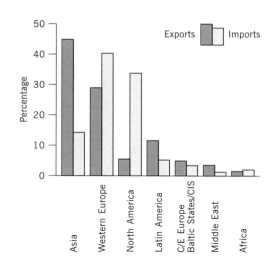

We will look more closely at the processes creating these distinctive regionalized patterns of garments exports later in this chapter.

In terms of garments-producing countries' shares of the OECD market, three major trends are apparent:[2]

- Europe is on a path of secular decline. Italy, a traditional stronghold of the garments industry, is still the number two supplier of OECD countries, but its share has been declining since 1980. The same holds for Belgium/Luxembourg, France, Germany, Greece, the Netherlands and the United Kingdom. The exception is Portugal, the country with the lowest wages in Europe, which has improved its position in the league table. However, taken together, the importance of European producers has been on the wane.

- The initial challenge to Europe was posed by Japan and the East Asian NIEs (mainly Hong Kong (China), the Republic of Korea and Taiwan Province of China), but these countries are increasingly being displaced by lower-income countries. Japan, the earliest Asian competitor, has now practically withdrawn from garments production for export altogether. The three East Asian NIEs, which were major players in 1980, remain important suppliers, but they have suffered big reductions in OECD market share.

- The largest slices of the market are being taken by large developing countries in Asia, especially China. Some ASEAN countries and India have also moved to high positions in the league table. Elsewhere, there has been strong growth in the Caribbean/Latin American area. Mexico is the largest supplier in this region. Other OECD suppliers in this region include the Dominican Republic, Honduras, Costa Rica, Guatemala, Jamaica, El Salvador, Colombia and Peru. Growth in North Africa is visible in the rising shares of Tunisia, Morocco and Egypt.

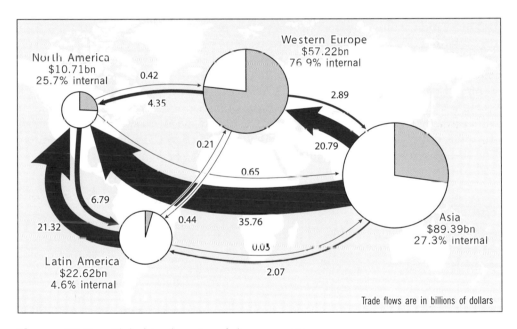

Figure 10.8 Global trade network in garments

Source: Based on WTO (2001) *International Trade Statistics, 2001*: Table A10

The dynamics of the market

Changing patterns of demand

Demand is a fundamental influence on the size, the organization and the location of the textiles and garments industries in different parts of the world. Each sector in the textiles–garments production sequence shown in Figure 10.1 relates to rather different markets. However, since some 50 per cent of all textiles production goes to the garments industry, the major influence on the demand for textiles is the demand for garments.

As Figure 10.1 also shows, garments can be divided into three major types: basic, fashion–basic and fashion garments. The major general determinant on both the *level* of demand and on the *composition* of demand (in terms of these three basic categories) is the level and distribution of personal income. Since personal incomes are so very unevenly distributed geographically at the global scale (see Figure 7.4), it is the affluent parts of the world that largely determine the level and the nature of the demand for garments. The generally low incomes in developing countries clearly restrict the size of their domestic garments markets.

The conventional economic wisdom is that, beyond the level of basic necessities, demand for garments increases less rapidly than the growth of incomes. This poses a major problem for garments manufacturers and retailers: they need to stimulate demand through *fashion change.* In other words, they need to shift consumer demand away from low-margin basic garments to higher-margin fashion-related garments. Enormous expenditure has gone into promoting fashion products and creating 'designer' labels. Such a practice covers a very broad spectrum of consumer income levels from the exceptionally expensive to the relatively cheap. Designer labelling is basically a device to *differentiate* what are often relatively similar products and to cater to – and to encourage – the segmentation of market demand for garments.

Figure 10.9 shows that 45 per cent of the dollar volume of US garments sales in the early 1990s was in the basic product category, while the remaining expenditure was split more or less evenly between fashion and fashion–basic products. The fastest growth is occurring in the fashion–basic segment.[3]

Figure 10.9

Composition of demand for different garments categories in the United States

Source: Based on Abernathy et al., 1999: Figure 1.1

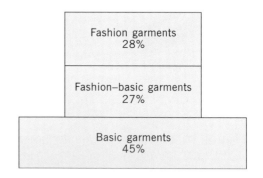

The growing power of the retailing chains[4]

Within the garments industry, in particular, demand is becoming increasingly dominated by the purchasing policies of the major multiple retailing chains. In the United States, big companies such as Wal-Mart, Sears, J.C. Penney, Dayton Hudson and K-Mart, account for a very large proportion of garments sales, as do Daiei, Mitsukoshi, Daimaru and Ito Yokado in Japan. In Germany, the leading garments retailers include Karstadt, Kaufhof, Schickendanz; in the Netherlands,

C&A; in France, Carrefour; in Britain, Marks and Spencer. However, the extent and the nature of multiple retailer dominance over the garments market vary a good deal from one country to another.

There has certainly been a 'retailing revolution' in both the United States and the United Kingdom since the 1960s. The result has been the rapid emergence of specialist garments retailers – like The Gap, The Limited, Liz Claiborne – serving niche markets. A similar process has occurred in Europe. In the United Kingdom, for example, the nature of garments retailing changed substantially during the 1980s as retailers began to focus on segmented markets, such as particular age and income groups. As in the United States, specialist chains emerged, such as Next, together with local branches of some of the US firms (notably The Gap). During the 1990s, this trend intensified further with the spread of such firms as Jigsaw, DKNY, Zara, H&M and, more recently, the Japanese company Uniqlo, catering to affluent young consumers.

These kinds of development, which are occurring to a greater or lesser degree in all developed market economies, have profound implications for textiles and garments manufacturers. The highly concentrated purchasing power of the large

(a) Traditional retailing–apparel supplier relations

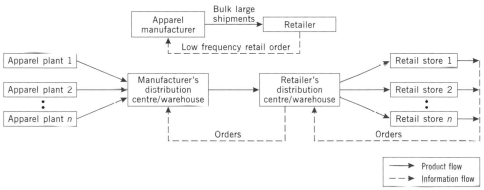

(b) 'Lean' retailing–apparel supplier relations

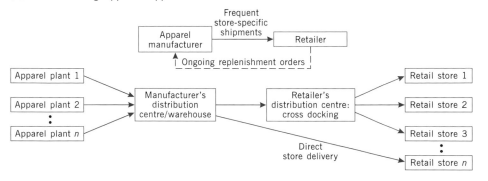

Figure 10.10   Changing relationships between garments manufacturers and retailers

Source: Abernathy et al., 1999: Figures 3.1, 4.1

retail chains gives them enormous leverage over textiles and garments manufaturers. When the market was dominated largely by the mass market retailers, demand was for long production runs of standardized garments at low cost. As the market has become more differentiated and more frequent fashion changes have become the rule, manufacturers are forced to respond far more rapidly to retailer demands and specifications. Under such circumstances, the *time* involved in meeting orders becomes as important as the cost.

> Product proliferation and shorter product cycles, reflected in ever-changing styles and product differentiation, contribute to general demand uncertainty for both retailers and manufacturers, thereby making demand forecasting and production planning harder every day. In a world where manufacturers must supply an increasing number of products with fashion elements, speed and flexibility are crucial capabilities for firms wrestling with product proliferation, whether they are retailers trying to offer a wide range of choices to consumers or manufacturers responding to retail demands for shipments.[5]

This basic shift in the structure of marketing is having repercussions throughout the textiles and garments industries and influencing both the adoption of new technologies and also corporate strategies (as we shall see in later sections of this chapter). Although relatively few retail chains are themselves manufacturers of garments, they are very heavily involved in subcontracting arrangements. The production chain in these industries is becoming transformed into a *buyer-driven* chain. Figure 10.10 shows how the modern 'lean' manufacturer–retailer system differs from the traditional system. Traditionally,

> The typical shipment between an apparel manufacturer and retail customers was large and of low frequency – usually once a season. Once delivered, the retailer held the products in central warehouses or as inventory individual stores' 'back rooms'.
> When the desired time of display and sale arrived, workers stocked the product on the selling floor and replenished from store or warehouse inventories as the selling period progressed. Inventory control relied on painstaking, manual comparisons between sales records (paper receipts) and physical counts of items on the floor, in the back room, and in warehouses.
> In contrast with the infrequent, large bulk shipments between apparel manufacturers and retailers under the traditional model, lean retailers require frequent shipments made on the basis of ongoing replenishment orders placed by the retailer. These orders are determined by real-time sales information collected at the retailer's registers via bar code scanning. SKU-level [stock-keeping units] sales data are then aggregated centrally and used to generate orders to suppliers, usually on a weekly basis.[6]

Production costs and technology

The production characteristics of each sector of the textiles–garments production sequence vary considerably (Figure 10.11), although much depends on *where* production takes place. A process that is relatively capital-intensive in one country may be relatively labour-intensive in another, depending upon relative factor endowments. Similarly, there is considerable variation in the size of production units between one country and another. In garments manufacture capital intensity is generally low, labour intensity is generally high, the average plant size is small, and the technology relatively unsophisticated.

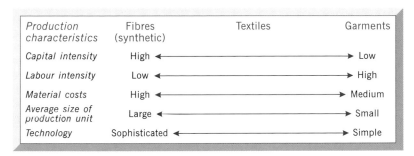

Production characteristics	Fibres (synthetic)	Textiles	Garments
Capital intensity	High ←	→	Low
Labour intensity	Low ←	→	High
Material costs	High ←	→	Medium
Average size of production unit	Large ←	→	Small
Technology	Sophisticated ←	→	Simple

Figure 10.11   Variations in production characteristics between major components of the textiles–garments production chain

Labour

Variations in labour costs

For textiles and garments manufacture there is no doubt that labour costs are the most significant production factor. Not only are textiles and garments two of the most labour-intensive industries in modern economies but also labour costs are the most geographically variable of the production costs of the industries. Figure 10.12 shows just how wide the labour cost gap can be between different countries. The spread is enormous, from over $10 per hour in the United States to 22 cents per hour in Vietnam. There is substantial labour cost differentiation within Asia between the first-generation and second-generation NIEs. Note also that the labour cost levels in Eastern Europe are comparable with those in low-cost Asia.

The difficulty of competing in labour cost terms with other producers was a major reason for the large-scale shift of the US textiles industry from New England and of the garments industry from New York to the Southern states. However, in terms of unit labour costs (which allow for levels of productivity) the United States is in a far better position *vis-à-vis* its industrialized competitors. But even allowing for productivity differences the developing countries have an enormous labour cost advantage over the developed market economies, particularly in the production of garments. The major advantage of low-labour-cost producers lies in the production of standardized items, which sell largely on the basis of price, rather than in fashion garments, in which style is more important. The difference between the two is one of *rate of product turnover*. Fashion garments have a rapid rate of turnover that tends to reflect the idiosyncrasies of particular markets. Geographical proximity to such markets is important and this helps to explain the survival of many developed country garments manufacturers. It also partly explains the relative advantage of low-cost countries located close to the major consumer markets of the United States (e.g. Mexico, the Caribbean), Europe (e.g. Central and Eastern Europe, the Mediterranean rim) and Japan (the Asian countries).

Figure 10.12

Hourly labour costs in
the garments industry
1998

Source: Based on material
in Werner International
Inc., *Hourly Labor Cost in
the Apparel Industry*

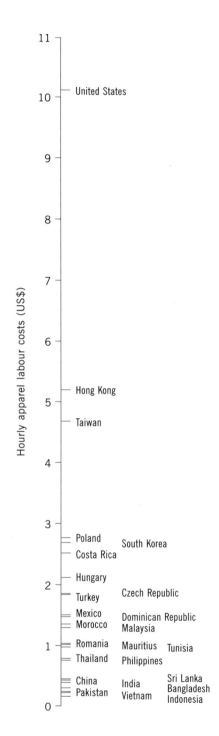

Characteristics of the labour force and conditions of work

Some 80 per cent of the workers in the garments industry and more than 50 per
cent of those in textiles manufacture are female. A substantial proportion of the

labour force is relatively unskilled or semi-skilled, with no easily transferable skills. The specific socio-cultural role of women, in particular their family and domestic responsibilities, also makes them relatively immobile geographically. A further characteristic of the textiles and garments workforce in the older industrialized countries is that a large number tend to be immigrants or members of ethnic minority groups. This is a continuation of a very long tradition, especially in the garments industry. The early garments industries of New York, London, Manchester and Leeds were a major focus for Jewish immigrants. Subsequent migrants have also seen the industry as a key point of entry into the labour market. The participation of Italians and East Europeans in both the United States and the United Kingdom has been followed more recently by the large-scale employment of blacks, Hispanics and Asians in the United States and by non-white Commonwealth immigrants in the United Kingdom.

The history of these industries – and especially of the garments industry – is one of appalling working conditions in sweatshop premises. At least in the textiles industries of the developed economies such conditions are now relatively rare; factory and employment legislation have seen to this. But the sweatshop has certainly not disappeared from the garments industries of the big cities of North America and Europe. The highly fragmented and often transitory nature of much of the industry makes the regulation of such establishments extremely difficult. The result has been a major resurgence of garments sweatshops in some big Western cities.[7]

A survey of garments establishments in San Francisco and Oakland, California, in the mid-1990s found 'more than half of them in violation of minimum wage standards. Sewing jobs for Esprit, Liz Claiborne, Izumi, and other glittering names were being done by underpaid workers'.[8] Similar problems were uncovered in the United Kingdom in a series of investigations of garments workshops in the big cities: 'Workers earn less than £2 an hour for a 50-hour week. Yet some of the UK's best-known high-street retail chains buy from these manufacturers, even though they appear to break their own guidelines.'[9]

In the rapidly growing textiles and garments industries of the developing countries the labour force is similarly distinctive. Employment in the industries tends to be spatially concentrated in the large, burgeoning cities and in the export processing zones. The labour force is overwhelmingly female and predominantly young. Many workers are first-generation factory workers employed on extremely low wages and for very long hours: a seven-day week and a 12–14-hour day are not uncommon. Employment in the garments industry in particular tends to fluctuate very markedly in response to variations in demand.

Hence, a high proportion of outworkers are employed – women working as machinists or hand-sewers at home on low, piecework, rates of pay. Such workers are easily hired and fired and have no protection over their working conditions. Many are employed in contravention of government employment regulations. Yet there is no shortage of candidates for jobs in the fast-growing garments industries of the developing countries. Factory employment is often regarded as preferable to un- or underemployment in a poverty-stricken rural environment. A factory job does provide otherwise unattainable income and some degree of individual freedom. Often the wages earned are a crucial part of the family's income and there is much family pressure on young daughters to seek work in the city factories or in the EPZs.

As a result of pressure from groups such as Oxfam, from labour unions and other anti-sweatshop groups, the major garments companies have given undertakings to monitor the operations of their suppliers and subcontractors in both developing and developed countries to remove illegal practices, especially of child labour. The major UK retailers have promised to end contracts with firms that contravene their guidelines. Similarly, in the United States in April 1997, the leading garments firms (including Nike, Liz Claiborne, Nicole Miller, L.L. Bean and Reebok) formally announced a code of conduct to eliminate domestic and overseas sweatshop conditions.

But the process of monitoring and detection is difficult. It is even more difficult to monitor the practice of home-working which, again, tends to be highly exploitative of the most disadvantaged groups who work at home for minimal rates of pay and no benefits. But in an industry as fragmented and organizationally complex as garments this is an immense task: 'Codes of conduct are awfully slippery. Unlike laws, they are not enforceable.'[10] Nevertheless, a plethora of such codes has evolved:

> Some were drafted in cooperation with human-rights groups or ethical investment specialists in the West. Others, like Bill Clinton's Apparel Industry Partnership's code, were organized according to where the [trans]nationals were headquartered ... Layered on top of these stacks of codes was one drafted by the Council on Economic Priorities, a consumer watchdog group in New York. The CEP plan would inspect factories for adherence to a set of standards covering key issues such as health, safety, overtime, child labor and the like. Under this model, brand-name [trans]nationals ... rather than trying to enforce their own codes around the world, simply place their orders with factories that have been found to be in compliance with the code. Then the factories are monitored by a private auditing company, which certifies factories that meet the code as 'SA8000' (SA stands for 'Social Accountability').[11]

Despite such confusion, and continuing evasion of such codes by some companies, there is no doubt that some progress has been made in improving conditions in these industries, although problems certainly remain. This is in issue to which we will return in Chapter 18.

Technological change

Both the cost of production and the speed of response to changes in demand are greatly influenced by the technologies used. Technological innovation can reduce the time involved in the manufacturing process and make possible an increased level of output with the same size – or even smaller – labour force. As international competition has intensified in the textiles and garments industries the search for new, labour-saving technologies has increased, especially among developed country producers. However, the potential for such innovation varies very considerably between different manufacturing operations within the production sequence.[12] Two kinds of technological change are especially important:

- those that increase the speed with which a particular process can be carried out and
- those that replace manual with mechanized and automated operation.

On both counts, technological change has been far more extensive in the textiles manufacturing sector than in the manufacture of garments.

One of the most important technological innovations in the spinning of yarn was the introduction of open-ended spinning, which combined what were formerly three separate processes into a single process using rotors instead of spindles. Spinning speeds increased at least four-fold and labour requirements reduced by approximately 40 per cent. Textile weaving has been revolutionized by the introduction of the shuttle-less loom in which the shuttle is replaced by a variety of alternative devices (for example 'rapier' looms, 'projectile' looms, 'water jet' and 'air jet' looms). Again, the major result has been a spectacular increase in the speed of the weaving process and this, together with wider loom capacity, has raised productivity by a huge factor. Parallel developments in knitting technology and in finishing – the latter now being especially highly automated – have contributed further to the increased speed of textiles operations and the consequent reduction in the number of workers needed.

In contrast, technological developments in the garments industry have been far less extensive. Indeed, there was relatively little change in garments technology between the industry's initial emergence in the late 19th century and the early 1970s. The basic sewing machine was not so very different from that in use 50 years earlier. Even today, the manufacture of garments remains a complex of related *manual* operations, especially in those items in which production runs are short. Most of the recent technological developments in the industry, including those based on micro-electronic technology, have been in the non-sewing operations: grading, laying out and cutting material in the pre-assembly stage, and warehouse management and distribution in the post-assembly stage. The application of computer-controlled technology to these operations can achieve enormous savings on materials wastage and greatly increase the speed of the process. For example, the grading process may be reduced from four days to one hour, computer-controlled cutting can reduce the time taken to cut out a suit from one hour to four minutes. But these developments do not reach the core of the problem. The sewing and assembly of garments account for 80 per cent of all labour costs in garments manufacture. So far only very limited success has been achieved in mechanizing and automating the sewing process itself.

Current technological developments in the manufacture of garments are focused on three areas:

- To increase the *flexibility* of machines: robots are being designed which can recognize oddly shaped pieces of material, pick the pieces up in a systematic manner and align the pieces on the machine correctly whilst also being able to sense the need to make adjustments during the sewing process itself.
- To address the problem of *sequential operations*, particularly the difficulty of transferring semi-finished garments from one workstation to the next while retaining the shape of the limp material.

- To develop the *unit production system* which will deliver individual pieces of work to the operator on a conveyor belt system. This will greatly reduce the large amount of (wasted) production time spent by the operator on unbundling and rebundling work pieces. The handling process has been estimated to take up to 60 per cent of the operator's total time.

The drive to introduce new technologies has been stimulated mainly by the need for developed country producers to be cost-competitive in the face of the very low labour costs in developing countries. But cost reduction is not the only benefit derived from the new technologies. At least as important, if not more so, are the *time savings* that result from automated manufacture. This has two major benefits:

- speeding up the production cycle reduces the cost of working capital by increasing the velocity of its use
- it becomes possible for the manufacturer to respond more quickly to customer demand.

Electronic point of sale (EPOS) technologies permit a direct, real-time link, between sales, reordering and production. As the production chain has become increasingly buyer-driven, these IT-based innovations have become extremely important. Not only do they permit very rapid response to sales and demand at the point of sale but also they enable the buyer-firm to pass on the costs of producing and holding inventory to the manufacturer.

The role of the state and the Multi-Fibre Arrangement

In developing economies, textiles and garments manufacture have occupied a key position in national industrialization strategies. Hence, the kinds of import-substitution and export-oriented measures outlined in Chapter 5 have been applied extensively to the encouragement of the industries in most developing countries. But it is in the older-established producing countries of Europe and North America and, more recently, Japan, faced with increasingly severe competition from low-cost or more efficient producers, that government intervention has been especially marked. The political sensitivity of these industries has forced governments to intervene. The policies adopted by governments have been of three kinds:

- those aimed at encouraging the *restructuring and rationalization* of the country's textiles and garments industries, through the use of subsidies and adjustment programmes
- those aimed at *stimulating* the industries through offshore assembly provisions (for example, by granting tariff concessions on imports of products assembled abroad using domestic materials) and through preferential trading agreements

- those aimed at *protecting* the domestic textile and garments industries from competition from low-cost producers in developing countries.

This third strategy is intimately bound up with the Multi-Fibre Arrangement.

The international regulatory framework: the Multi-Fibre Arrangement[13]

Individual national policies to stimulate domestic textiles and garments industries or to facilitate their rationalization and restructuring have been extremely important in helping to reshape the industry globally. However, the *international regulatory framework* within which the industries have operated for the past three decades – the *Multi-Fibre Arrangement* (MFA) – has been far more significant. Today, most of the world trade in textiles and garments is covered by this agreement. Its provisions and their implementation – and avoidance – have been a major factor in the changing global pattern of production and trade. Although protectionism in these industries is not solely a post-war phenomenon, the origins of its modern variant, the MFA, are to be found in the problems that faced developed country producers, particularly the United States and the United Kingdom, in the 1950s.

Faced with a massive inflow of low-price imports from Japan, Hong Kong and some other Asian producers, both the United States and the United Kingdom negotiated separate 'voluntary' agreements with the Asian exporters to restrict imports for a limited period. By 1962 such arrangements had become broadened into the Long-Term Arrangement (LTA) which regulated international trade in cotton textiles. The aim of the LTA was to encourage orderly development of the international cotton textiles market to allow the developed countries to restructure their cotton textiles industry. It allowed an importing country to limit shipments from any source in any cotton textile category which would 'cause or threaten to cause disruption in the market of the importing country'.

The LTA allowed for a gradual increase in imports from developing country signatories of 5 per cent a year. It remained in force for 11 years. During that time, however, the world picture became far more complex. First, there was the massive growth of man-made fibres not covered by the LTA. Second, an increasing number of developing countries emerged as important exporters of textiles and, especially, of garments. The precipitous decline of the industries in the developed economies continued. In 1973 a much broader trade agreement, which included the European countries and also covered man-made and other non-cotton fibres, was negotiated. This was the first Multi-Fibre Arrangement. By Article 1(2), its principal aim was:

> to achieve the expansion of trade, the reduction of barriers to such a trade and the progressive liberalization of world trade in textiles products, while at the same time ensuring the orderly and equitable development of this trade and avoidance of disruptive effects in individual markets and on individual lines of production in both importing and exporting countries.

Like the LTA, the MFA was initially negotiated for a limited period: four years from January 1974. Also like the LTA, its aim was to create 'orderly' development of trade in textiles and garments that would benefit *both* developed and developing countries. Access to developed country markets was to increase at an annual average rate of 6 per cent, although this was far below the 15 per cent sought by the developing countries. At the same time, the developed countries were to have safeguards to protect the 'disruption' of their domestic markets. Within the MFA, individual quotas were negotiated which set precise limits on the quantity of textiles and garments that could be exported from one country to another. For every single product, a quota was specified. When that quota was reached no further imports were permitted.

In practice it has been the disruptive, rather than the liberalizing, aspect which has been at the forefront of trading relationships in these industries. The MFA was renegotiated or extended four times (in 1977, 1982, 1986, 1991). Over time, the MFA became more, rather than less, restrictive. Both the EU and the United States negotiated much tighter import quotas on a bilateral basis with most of the leading developing country exporters, and in several cases also invoked anti-dumping procedures.

The effects of the MFA on world trade in textiles and garments have been immense. Without doubt, it has greatly restricted the rate of growth of exports from developing countries. These have been far lower than would have been the case in the absence of the MFA. A major initial beneficiary of this dampening of the relative growth of developing country exports was the United States, which greatly increased its penetration of European textiles and garments markets during the 1970s. During the early 1980s, however, it was the European producers who greatly increased their penetration of the US market.

An inevitable consequence of the increased restrictiveness of developing country exports of textiles and garments was a parallel increase in efforts to circumvent the restrictions. Such evasive action has taken a variety of forms. Examples include:

- for a producing country which has reached its quota ceiling in one product to switch to another item
- for false labelling to be used to change the apparent country of origin (an illegal act)
- for firms to relocate some of their production to countries which either are not signatories to the MFA or whose quota is not fully used by indigenous producers.

A major task of the Uruguay Round of the GATT negotiations in the late 1980s/early 1990s was to integrate the MFA into the GATT. The basic principle at the heart of GATT rules on international trade is non-discrimination between all trading partners. The MFA clearly contravenes this principle. From the standpoint of the industrialized countries, 'the MFA by its structure discriminates among supplier countries, most notably between industrial countries, which enjoy free access, and developing countries (and Japan), which face controls'.[14] On the other hand, the developing countries are indicted by the industrialized countries for restricting access to their domestic markets through both very high tariffs and non-tariff barriers.

As a result of the Uruguay Agreement of the GATT, trade in textiles and garments was incorporated into the WTO, with the MFA to be phased out over a ten-year period (1995–2004) but in three stages:

- Stage 1 (1995–1998): 16 per cent of tariff lines to be integrated into the GATT
- Stage 2 (1998–2002): 17 per cent of tariff lines to be integrated
- Stage 3 (2002–2004): 18 per cent to be integrated after 2002 followed by the final 49 per cent by the end of 2004.

Thus, 'integration is … heavily backloaded, putting most of the difficult liberalization off to the future'.[15] In fact, both the US and the EU 'integrated' first those products which already enter their markets freely – hardly a major concession. The United States' 10-year liberalization schedule, in effect, left the integration of 70 per cent of imports by value to the very end of the transition period.[16]

Not surprisingly, developing countries are extremely unhappy with what they regard as a deliberate dragging of feet by the world's two largest textiles and garments markets. In response, European and US producers argue that developing countries need to be more positive in increasing access to imports into their own domestic markets. Quite clearly, full liberalization of the textiles and garments industries is far from assured. Meanwhile, bilateral deals have continued to be the basis of trade with the world's leading garments producer, China, prior to its entry into the WTO. In fact, particularly acrimonious trade relationships have occurred between the United States and China, although a new agreement on textiles and garments was signed by the two countries in 1997.

Corporate strategies in the textiles and garments industries

As we have seen in this chapter, the manufacture of textiles and garments is geographically very widespread. They are relatively rare instances of globally significant industries that are extensively present in many developing countries, rather than in just a few. Vast numbers of, mostly small, developing country firms are involved in textiles and garments production. Nevertheless, the globalization of these industries has been driven primarily by developed country firms. Indeed, it is paradoxical that a significant proportion of the textiles, and especially the garments, imports that are the focus of such concern in developed countries are, in fact, organized by the international activities of those very countries' own firms. But the processes and strategies involved are both complex and dynamic.

In examining the development of these strategies two basic points need to be made.

- The globalization of the textiles and garments industries cannot be explained simply as a relocation of production from developed to developing countries in search of low labour costs. Other factors are involved, including, in particular, orientation to specific markets.

● Where firms have internationalized their production operations they have used a variety of methods, notably international subcontracting and licensing, which do not necessarily involve equity participation. We discussed such strategies in general terms in Chapter 8.

Major factors in how and where the various modes of international involvement have been used are, first, the existence of the Multi-Fibre Arrangement, with its complex system of national quotas, and, second, the volatility of currency exchange rates. Technological innovations in production and distribution processes are also part of the strategic equation in so far as they influence both the costs of producing textiles and garments and also the speed of response to changing customer demands. These, in turn, are heavily affected by the dominance of the major retail chains. Within this multiplicity of influences on the strategic behaviour of textiles and garments firms it is the need for *flexibility* that is increasingly the dominant consideration.

Corporate strategies in the textiles industry

Although textiles firms of all sizes continue to exist, the textiles industry is increasingly an industry of large firms. In the world as a whole, some thirty or so textiles corporations are especially important. They include such companies as Burlington in the United States, Toray in Japan, Coats in the United Kingdom and the Marzotto Group in Italy.

Textile firms have tended to pursue one of three major strategies in attempting to remain competitive:[17]

● produce standardized goods for large markets using economies of scale to reduce costs and to enable them to compete on price

● supply large markets on the basis of utilizing low-cost labour in offshore locations

● produce small quantities of specialized goods for specific market niches. This presupposes high-quality products which can be sold at a premium price to offset the additional costs of switching specifications.

Although the British and European textiles companies have a lengthy history of international involvement in textiles production, the first really major wave of such activity occurred in the early 1960s and was led by the Japanese textiles firms and the general trading companies (the *sogo shosha*). The Japanese textiles firms were already strongly vertically integrated in Japan itself and operated a dualistic-hierarchical network of domestic subcontractors. The *sogo shosha* were responsible for organizing a huge proportion of Japanese imports and exports and already had an intricate international distribution system. When the United States introduced the LTA in 1962 to protect its domestic market from Japanese cotton textiles imports, it triggered the first surge of overseas involvement by Japanese firms. To avoid the problem of quotas, both the textiles firms and the *sogo shosha* set up international subcontracting links in other East Asian countries. Very often, as in Japan itself, the principal firms took a small equity share in the subcontractors. Relatively quickly, therefore, the Japanese textile industry became an international, vertically integrated operation incorporating an extensive network of local Asian producers. Some of this involved direct investment but most consisted of international subcontracting arrangements.

Compared with the Japanese companies, US and European textile firms are less internationalized. In the United States the tendency has been to increase the degree of domestic concentration in the industry through acquisition and merger and to upgrade domestic productivity through heavy investment in new technology. In Europe, too, these strategies have been pursued, although some of the very large textiles companies – such as Coats – have also become increasingly internationalized. Such companies have engaged in massive rationalization and restructuring programmes since the 1970s, involving a varying mixture of: product rationalization and focus; technological innovation to reduce costs and increase flexibility; reduction of production capacity, particularly domestically but also overseas; and the use of international subcontracting, licensing and other forms of relationship with local firms in developed and developing countries.

For example, Coats was created through a series of acquisitions of British textile and garments companies during the 1970s and 1980s.[18] It gradually transformed itself from a 'production-driven firm' with an immense range of products to a 'market-driven firm' with five core businesses. Locationally, Coats adopted a two-pronged strategy:

- shifting low-value-added activities, such as undyed thread manufacture and zip production, to low-cost countries, in the eastern and southern European periphery and in Asia and
- locating dyeing factories, which are less labour-intensive, closer to its main markets.

However, Coats faces the major problem common to most textile firms: its customers themselves are continuously shifting their locations to remain competitive. This is a particular problem for its main business – threads. Hence, Coats

> is being forced to follow the textiles industry round the world … its main preoccupation is the relocation of other companies' stitching operations. Threads account for nearly half the group's sales. And it is threads that absorbs 60 per cent of its reorganization costs, as it expands in eastern Europe, closes down elsewhere on the continent and refocuses in Asia. These shifts will only end when the world's garments producers stop moving – and that may be another 10 years, as the implications of phasing out the Multi-Fibre Arrangement are worked through.[19]

Corporate strategies in the garments industry

Of all the major parts of the textiles–garments production chain, the manufacture of garments is by far the most fragmented and least dominated by large firms. In part this is explained by the relatively low technological sophistication of the garments-manufacturing process and the low barriers to entry to the industry. In part it reflects the vagaries of the market for garments, which limits long production runs and high-volume production to basic items. Even in this archetypal small-firm industry, however, large firms are becoming increasingly important. Only they can afford to invest in the new technologies and to build a worldwide brand image based on mass advertising. Thus, although the garments industry of most countries is made up of a myriad of very small firms, many of which operate as subcontractors, there is an undoubted trend towards increased concentration.

As in the case of textile firms, three broad strategies can be identified among garments producers:[20]

- production of basic goods for large markets utilizing economies of scale to lower costs and to be price competitive (however, the garments segment is far less amenable to automation than the textiles segment and there are lower limits to efficient scale)

- operation of small workshops – often in the form of 'sweatshops' in large cities – using immigrant and sometimes undocumented labour; such firms generally work as short-term subcontractors producing lower-quality garments

- production of short orders to fill manufacturers' production gaps, often for very specific segments of the market.

However, in the case of garments there is an additional organizational component of great significance: 'factory-less' firms that organize entire systems of garments production.[21] These major international *retail chains* and *buying groups* exert enormous purchasing power and leverage over garments manufacturers. Although the production and retailing of garments may be fragmented in individual markets, international buying operations are highly concentrated. In the garments industry in general the boundary between production and retailing is becoming increasingly blurred as the power within the production chain shifts further towards the buyers (including the department stores, mass merchandisers, discount chains and fashion-oriented firms). Precisely how they organize the sourcing of their garments varies according to the position they occupy within the market.

International subcontracting, licensing and other forms of non-equity international investment have been even more pervasive and influential in the garments sector than in textiles. In some cases, garments production is part of a firm's vertically integrated activities. Again, as in the case of textiles, it was Japanese firms that initiated the extensive use of international subcontracting in garments. During the 1960s and 1970s Japanese companies established subcontracting arrangements in Hong Kong, Taiwan, South Korea and Singapore. Their production was mostly exported to the United States and not to Japan's own domestic market. Again, the Japanese *sogo shosha* were at the leading edge of these international subcontracting developments, often using minority investments in local firms. Probably 90 per cent of Japanese overseas garments operations are still located in East and South East Asia and most were set up in the 1970s. Overall, however, the Japanese garments industry is far less internationalized than its textiles industry. The opposite is true of the US and European garments industries, which are far more internationalized than their textiles industries.

The strategies developed by US garments firms to cope with the intensified competition of the 1970s and 1980s were either to focus on the leading edge of the fashion market or to cut costs and raise productivity. The larger US garments firms increased their level of offshore processing using suppliers in developing countries. Some of the most interesting strategic mixes are used by those garments firms that compete in mass markets but on the basis of brand names supported by extensive advertising. A good example is the jeans manufacturer Levi Strauss. Even by the late 1970s, Levi Strauss was spending $50 million a year on worldwide advertising. But, increasingly it faced the problem of how to adapt to the demographic change which has altered the size of its traditional market segment of 15- to 24-year-olds without moving too far from its core product, the denim jean.

Levi Strauss's international production strategy has been to develop its own branch factories in both Western and Eastern Europe, in Latin America and in Asia, together with licensing agreements. At its peak, Levi Strauss employed some 40,000 workers worldwide, of which 28,000 were in North America, 7,000 in Europe and 2,000 in Asia. But in the late 1980s, the company began to make massive cuts in its operations in the United States and Europe and to shift more of its operations to lower-cost locations. In 1998, the company closed 13 of its plants in the United States and four in Europe, shedding 7,400 jobs. The following year (1999) Levi Strauss closed half of its remaining 22 US factories and eliminated 30 per cent of its US labour force (almost 6,000 jobs). At the same time, the company reduced the proportion of its production manufactured in-house to 30 per cent (in 1980 Levi Strauss had manufactured 90 per cent of its own production).[22] In 2002, a further six manufacturing plants were closed in the United States, with the loss of more than 3,000 jobs. According to the company:

> 'This is a painful but necessary decision … There is no question that we must move away from owned and operated plants in the US to remain competitive in our industry.' The company plans to outsource much more production to its overseas contractors.[23]

The adoption of offshore production strategies by European garments firms has been most pronounced among German and British companies. German garments firms have been especially heavily involved in international subcontracting arrangements. Already by the 1970s around 70 per cent of all (the then West) German garments firms, including some quite small ones, were involved in some kind of offshore production. Roughly 45 per cent of the arrangements involved international subcontracting and a further 40 per cent involved varying degrees of equity involvement by West German firms in local partners. West German firms tended to concentrate their offshore activities in Eastern Europe but with a growing presence in countries of North Africa and the Mediterranean (notably Greece, Malta and Tunisia). Overall, more than 80 per cent of all German garments imports manufactured under subcontracting arrangements came from the former East Germany, Poland, Hungary, Romania, Bulgaria and the former Yugoslavia.[24]

The case of the German fashion company Hugo Boss provides a good example of current trends. Faced with high domestic production costs, Hugo Boss has long used offshore subcontractors, primarily in Romania, the Ukraine, Poland and the Czech Republic, where wages are 50 per cent lower than in Germany. In 1989, Boss acquired an American garments producer, Joseph & Feiss of Cleveland, Ohio. In 1991, Hugo Boss itself was acquired by Marzotto, the Italian textiles and garments group. In addition to sourcing an increasing proportion of its garments overseas, the company has also moved strongly into retailing through franchising its brand-name in around 200 stores worldwide. According to its (German) chairman, 'we are no longer a production-oriented company. Today, we are a company with a strong emphasis on creativity and design, marketing and logistics'.[25] Similarly, the leading British garments companies have developed a strong focus on offshore production and subcontracting, notably in countries of the Mediterranean rim (such as Morocco and Tunisia), in Eastern Europe (Hungary, Poland, the Czech Republic, Romania), as well as in South and East Asia (Sri Lanka, China, India and Indonesia).

Italian firms have been the major exception to this strong shift of production to low-cost foreign locations by European producers. We noted earlier that Italy is the only major European country whose garments industry has continued to perform relatively well in the teeth of intensive global competition. In general, the Italian producers have pursued a strategy of product specialization and fashion orientation with the aim of avoiding dependence upon those types of good most strongly affected by low-cost competition. This has involved mainly small firms in a decentralized production system. 'Decentralised production has occurred in those areas of textiles where the fashion element (hence the need for greater risk and flexibility) is important. Italy's unique strength in Western Europe is to have created a kind of price-competitive mass market in fashion, where certain products often enjoy an area "trademark" (Como silk ties, Prato wool fabrics, and so on)'.[26] More recently, however, some Italian firms have established international licensing agreements for production of high-fashion and designer-label products.

The best-known Italian company to have developed an especially distinctive strategy, of course, is Benetton.[27] Benetton sees itself not as a manufacturer or retailer of garments but as a 'garments services company'. But it very much sells itself as an 'Italian' company. Whereas most European firms have shifted much of their production to Asia, around 80 per cent of Benetton's garments are still manufactured in Europe, mostly in Italy – but not by Benetton itself. The company uses around 500 subcontractors for its actual production, 90 per cent of which are located in the Veneto region of north-east Italy.[28] This system gives it considerable flexibility in responding to changing demand for its garments. Benetton itself performs only those functions – mainly design, cutting, dyeing and packing – that it considers crucial to maintain quality and cost-efficiency. So, instead of following the lead of most other European garments firms and relocating production to low-cost countries, Benetton constructed two high-technology plants at its headquarters in Italy. One plant is a state-of-the-art cutting and dyeing plant, the other is a highly computerized warehouse. These two plants employ very few workers.

The second aspect of Benetton's strategy is in the way it sells its garments. There are more than 6,000 Benetton stores in 120 countries around the world, almost all of which are franchises. Although Benetton controls what products are sold in the stores it does not carry the risk directly. Unlike the usual franchise arrangement, Benetton receives neither royalties nor takes back unsold stock. In fact, each shop is totally independent and carries the risk. Benetton, in contrast, has an assured set of high-profile outlets that it supports mainly through its often-controversial advertising campaigns. The key link between Benetton's flexible production system and its final sales through the franchised outlets is the highly sophisticated electronic point of sale (EPOS) system which transmits immediate information on the sale of every garment in every Benetton store back to the company's Italian headquarters. This allows production to be continuously and flexibly adjusted to meet expressed consumer preferences.

Until very recently, Benetton was the only major European garments firm in the fashion–basic sector to have retained its manufacturing operations in a higher-cost European location rather than relocating to low-cost Asian locations. It did so by producing a relatively limited range of garments but differentiating them primarily on the basis of colour. Most of the others, including the Swedish company H&M, have followed the Asian route. However, a much more recent, rapidly growing, garments company, Zara, has made spectacular progress by producing a very wide range of fast-changing fashion–basic garments from its domestic production base. Zara is located in La Coruña in north-west Spain, the traditional focus of the Iberian textile industry.

> At the heart of Zara's success is a vertically integrated business model spanning design, just-in-time production, marketing and sales. This gives the group more flexibility than its rivals have to respond to fickle fashion trends. Unlike other international clothing chains, such as Hennes & Mauritz (H&M) and Gap, Zara makes more than half its clothes in house, rather than relying on a network of disparate and often slow-moving suppliers. H&M, for instance, buys clothes from more than 900 firms … Starting with basic fabric dyeing, almost all Zara's clothes take shape in a design-and-manufacturing centre in La Coruña, with most of the sewing done by seamstresses from 400 local cooperatives. Designers talk daily to store managers, to discover what items are most in demand. Supported by real-time sales data, they then feed repeat orders and fresh designs into the manufacturing plant … The result is that Zara can make a new line from start to finish in three weeks, against an industry average of nine months. It produces 10,000 new designs each year; none stays in the stores for over a month.[29]

In other words, not only has Zara persisted with a manufacturing model long ago jettisoned by the major US and European garments companies (that is, producing most of its garments in-house) but also it operates within a highly volatile part of the fashion–basic sector of the industry. Zara has achieved dramatic results by combining highly efficient production and distribution logistics with a continuous monitoring of the fashion scene:

> Zara reps attend the ready-to-wear shows in Paris, New York, London and Milan, and do quick sketches as models come down the runway. Or they snap digital photos and zap them back to La Coruña, where designers are in constant contact with Zara stores around the world for additional tips on what's hot.[30]

Regionalizing production networks in the textiles and garments industries

Several broad regional shifts have occurred in the global textiles and garments industries since the 1950s:[31]

- In the 1950s and 1960s: from North America and Western Europe to Japan.
- In the 1970s and 1980s: from Japan to Hong Kong, Taiwan and South Korea.
- In the late 1980s and 1990s: from Hong Kong, Taiwan and South Korea to China, South East Asia and Sri Lanka.
- During the 1990s: towards the Americas (focused on the United States, Mexico and the Caribbean).

However, the global picture now seems to be evolving in ways that both enhance and modify the position of East Asia as a garments production complex and also consolidate regional garments complexes in other major regions, notably the Americas and Europe. The reason lies in the trade-off between labour costs on the one hand and the need for market proximity on the other:

> The growing regionalization of textile- and apparel-making may mean that skilled processes, such as cutting in apparel manufacturing and finishing operations in textiles, will remain in first-world countries like the United States, Japan, Germany, and Italy; the lower-paid occupations of assembly in apparel-making will continue to go to developing countries in Latin America, Asia, Eastern Europe, or North Africa, which can provide lower labor costs – but with a significant twist. Because time-to-market and the exigencies of short-cycle production are beginning to impact competition in retail–apparel–textile channels, three global regions are emerging: the United States plus Mexico and the Caribbean Basin; Japan plus East and South East Asia; and Western Europe plus Eastern Europe and North Africa. Each of these regions includes both advanced economies and developing areas that are close to consumer markets.[32]

In the light of these developments, let us look briefly at two of these global regions: East Asia and the Americas.

Regional restructuring of the Asian textiles–garments network[33]

The key to the internal transformation of the industry in Asia lies in the changing strategies of the northern tier East Asian NIEs. In the face of increasing competitive pressures from the newer wave of Asian producers (notably China, Malaysia, Thailand, Indonesia and, more recently, Vietnam and Cambodia), as well as restrictions on their trade with North America and Europe through the MFA, firms from Hong Kong, South Korea and Taiwan have progressively shifted their production offshore. In fact, Hong Kong firms began to shift garments production to other Asian countries as early as the mid-1960s. With the tightening grip of the MFA in the 1970s, Hong Kong firms set up plants in the Philippines, Thailand, Malaysia and Mauritius and, subsequently, in Indonesia and Sri Lanka, to get round quota restrictions. In the past decade, a huge number of investments have been made in China. East Asian firms have also begun to establish plants in Europe and North America (including the Caribbean) to serve developed country markets directly. This strategy involves more than simple geographical relocation:

> As Northeast Asian firms began moving offshore, they devised ways to coordinate and control the sourcing networks they created. Ultimately, they focused on the more profitable design and marketing segments within the apparel commodity chain to sustain their competitive edge. This transformation can be conceptualised as a process of industrial upgrading, based in large measure on building various kinds of economic and social networks between buyers and sellers.[34]

This upgrading process has consisted of three broadly sequential phases:

- simple assembly of basic garments for export trade
- subcontract manufacturing to design specified by the buyer with the product sold under the buyer's brand name (original equipment manufacturing, OEM)
- development of own-brand manufacturing (OBM) capability.

Of course, not all firms have followed this sequence. The most successful appear to have been Hong Kong firms. For example,

> The women's clothing chain, Episode, controlled by Hong Kong's Fang Brothers Group, one of the foremost OEM suppliers for Liz Claiborne in the 1970s and 1980s, has stores in 26 countries, only a third of which are in Asia. Giordano, Hong Kong's most famous clothing brand, has added to its initial base of garments factories 200 stores in Hong Kong and China, and another 300 retail outlets scattered across Southeast Asia and Korea. Hang Ten, a less-expensive line, has 200 stores in Taiwan, making it the largest foreign-clothing franchise on the island.[35]

The organization of the Asian garments production complex has a particular 'geometry': that of *triangle manufacturing*.

> The essence of triangle manufacturing … is that US (or other overseas) buyers place their orders with the NIE manufacturers they have sourced from in the past, who in turn shift some or all of the requested production to affiliated offshore factories in low-wage countries (e.g. China, Indonesia, or Guatemala). These offshore factories can be wholly owned subsidiaries of the NIE manufacturers, joint-venture partners, or simply independent overseas contractors. The triangle is completed when the finished goods are shipped directly to the overseas buyer … Triangle manufacturing thus changes the status of NIE manufacturers from established suppliers for US retailers and designers to 'middlemen' in buyer-driven commodity chains that can include as many as 50 to 60 exporting countries.
> Triangle manufacturing is socially embedded. Each of the East Asian NIEs has a different set of preferred countries where they set up their new factories … These production networks are explained in part by social and cultural factors (e.g. ethnic or familial ties, common language), as well as by unique features of a country's historical legacy.[36]

Geographical proximity is also an influential factor. For example, Hong Kong and Taiwanese firms are the major investors in China while Singaporean firms are strongly represented in Malaysia and Indonesia.

As Figure 10.8 (p. 327) shows, Asia remains the most globally connected region of garments production. A little over one-quarter of its garments trade is intra-regional compared with much higher levels in both Europe and the Americas. However, Asia has undoubtedly become a major market in its own right: in 1980 less than 5 per cent of Asian garments trade was intra-regional compared with 27 per cent in 2000.

United States-focused regional production networks in the Americas[37]

Traditionally, the US garments market was served primarily by domestic production, but in the past few decades the market has become increasingly dominated by imports from low-cost producing countries, especially in Asia. Most of these imports are organized through the buyer–retailer–supplier complex that became increasingly important in the garments industry. Table 10.3 shows a close association between the type of retailer (and the market served) and the locational sources of garments. In general, fashion-oriented companies source from Europe and first-generation NIEs, while the discount stores source from lower-cost countries.

The shifting patterns of garments production within Asia, outlined in the previous section, have seen China, in particular, eventually emerging as the dominant supplier of garments for the US market during the early 1990s. But the pattern is changing for two reasons:

- The trade-off between minimizing production costs and maximizing speed of access to consumer markets has become more critical. Proximity to markets has become a key factor in determining the geography of garments production as the dominant buyers/retailers insist on fast product turnover.
- The development of regional economic initiatives by the United States, in particular the signing of the NAFTA and the preferential arrangements with the Caribbean countries (the Caribbean Basin Initiative), has reinforced the benefits of geographical proximity driven by changing buyer–supplier relationships.

The NAFTA, in particular, is having a substantial effect on the textile and garments industries within the region. Under the NAFTA, Mexican, US and Canadian tariffs on textiles and garments are being phased out over a ten-year period. However, for garments to be eligible for tariff concessions they must be sewn with fabric that has been woven in North America. The major fear amongst US and Canadian textile and garments producers is that the huge labour cost differential between Mexico and the United States/Canada will pull more and more production across the border.

As a result, the geographical pattern of garments imports to the United States has changed dramatically since the early 1980s to one in which Mexico and the Caribbean Basin have become the dominant suppliers, with Mexico in particular

Table 10.3 Global sourcing patterns of US clothing retailers

Type of retailer	Types of order	Main sourcing countries
Fashion-oriented companies (e.g. Armani, Donna Karan, Boss, Gucci, Polo/Ralph Lauren)	Expensive 'designer' products High level of skill Orders in small quantities	Italy, France, UK, Japan South Korea, Taiwan, Hong Kong, Singapore
Department stores, specialty stores, brand-name companies (e.g. Bloomingdales, Saks Fifth Avenue, Neiman-Marcus, Macy's, The Gap, The Limited, Liz Claiborne, Calvin Klein)	Top-quality, high-priced garments sold under variety of national brands and private labels (store brands) Medium to large sized orders, often coordinated by store buying groups.	Malaysia, Indonesia, Philippines, Southern China, India, Turkey, Egypt, Brazil, Mexico, Thailand Dominican Republic, Jamaica, Haiti, Guatemala, Honduras, Costa Rica, Colombia, Chile, Poland, Hungary, Czech Republic, Bulgaria, Kenya, Zimbabwe, Mauritius, Macao, Pakistan, Sri Lanka, Bangladesh, Interior China, Tunisia, Morocco, UAE, Oman
Mass merchandisers (e.g. Sears, Montgomery Ward, JC Penney, Woolworth)	Good-quality, medium-priced goods Mostly sold under private labels Large orders	South Korea, Taiwan, Hong Kong, Singapore Malaysia, Indonesia, Philippines, Southern China, India, Turkey, Egypt, Brazil, Mexico, Thailand Dominican Republic, Jamaica, Haiti, Guatemala, Honduras, Costa Rica, Colombia, Chile, Poland, Hungary, Czech Republic, Bulgaria, Kenya, Zimbabwe, Mauritius, Macao, Pakistan, Sri Lanka, Bangladesh, Interior China, Tunisia, Morocco, UAE, Oman
Discount chains (e.g. Wal-Mart, K-mart, Target)	Low-priced goods Store-brand names Very large orders	Malaysia, Indonesia, Philippines, Southern China, India, Turkey, Egypt, Brazil, Mexico, Thailand Dominican Republic, Jamaica, Haiti, Guatemala, Honduras, Costa Rica, Colombia, Chile, Poland, Hungary, Czech Republic, Bulgaria, Kenya, Zimbabwe, Mauritius, Macao,

(Continued)

Table 10.3 (*Continued*)

Type of retailer	Types of order	Main sourcing countries
		Pakistan, Sri Lanka, Bangladesh, Interior China, Tunisia, Morocco, UAE, Oman
		Qatar, Peru, Bolivia, El Salvador, Nicaragua, Vietnam, Russia, Lesotho, Madagascar North Korea, Myanmar, Cambodia, Laos, Yap, Maldives, Fiji, Cyprus, Bahrain
Small importers	Pilot purchase and special items Sourcing done for retailers by small importers acting as 'industry scouts' in seeking out new sources of supply Relatively small orders initially but could grow rapidly	Dominican Republic, Jamaica, Haiti, Guatemala, Honduras, Costa Rica, Colombia, Chile, Poland, Hungary, Czech Republic, Bulgaria, Kenya, Zimbabwe, Mauritius, Macao, Pakistan, Sri Lanka, Bangladesh, Interior China, Tunisia, Morocco, UAE, Oman Qatar, Peru, Bolivia, El Salvador, Nicaragua, Vietnam, Russia, Lesotho, Madagascar North Korea, Myanmar, Cambodia, Laos, Yap, Maldives, Fiji, Cyprus, Bahrain

Source: Based on Gereffi, 1994: Figure 5.2 and Table 5.3

becoming exceptionally important. Table 10.4 shows this transformation very clearly. In 1981, 60 per cent of the United States' imports of garments came from three East Asian countries: Hong Kong, Taiwan and South Korea. By 1995, this share had fallen to 21 per cent and, by 2000, to 14 per cent. By 1995, almost 15 per cent of garments originated from China and a further 14 per cent from the newer South East Asian producers. However, by 2000, their combined share had declined a little.

Initially, apart from the rapid growth in Chinese imports, the other major growth area for imports into the United States was the Caribbean Basin, whose share of the total grew from less than 3 per cent in 1981 to 13 per cent in 1995. During the same period, Mexico's share of US garments imports grew from 3 per cent to 7 per cent. By 2000, however, Mexico had overtaken China to become the leading source of garments imports into the United States – a dramatic turnaround indeed.

Part of the explanation for Mexico's emergence as the United States' single most important source of garments imports lies, of course, in the signing of the NAFTA

Table 10.4 The changing origins of garments imports into the United States

	% of total imports		
Country of origin	1981	1995	2000
China	5.6	14.9	13.2
Hong Kong, Taiwan, South Korea	60.6	21.1	14.3
South East Asian producers	7.9	14.4	13.6
Caribbean Basin producers	2.8	13.0	14.5
Mexico	3.0	7.0	13.6

Source: Based on WTO (1995) *International Trade Statistics, 1995*: Table IV.57; WTO (1996) *Annual Report, 1996*: Table IV.57; WTO (2001) *International Trade Statistics, 2001*: Table IV.79

in the early 1990s. Tariffs and quotas on garments imported from Mexico to the United States have been eliminated. In addition, under the rules of origin for garments in the NAFTA, garments must be cut and sewn from fabric made from fibre originating in North America in order to qualify for duty-free access. This provides a stimulus for the development of a more integrated textiles–garments industry within North America. Thus, 'through NAFTA, apparel and textile manufacturers are acquiring the freedom and flexibility to create – duty and quota free – transborder production networks that best suit their individual needs'.[38]

Because Mexico's comparative advantage lies in garments production while the United States' comparative advantage lies in textile manufacture, synthetic fibre production and retailing, a clear division of labour is emerging between the two countries. Certainly, the combination of the NAFTA and the benefits of geographical proximity together with low production costs have stimulated a lot of garments firms in the United States to source more of their garments from Mexico. A study of garments manufacturers in Southern California, for example, found that 60 per cent of those involved in offshore sourcing in 1998 used Mexico compared with only 10 per cent in 1992.[39]

The precise form of this geographical division of labour is still evolving. Traditionally, Mexico's garments industry was dominated by *maquiladora* production: simple sewing of garments made from imported fabrics and using extremely cheap labour (see Figure 3.29). In other words, it was dominated by the very basic operations in a vertically integrated system coordinated and regulated by US manufacturers and retailers. Although this is still the dominant mode of operation, there are signs of rather more sophisticated arrangements in which Mexican firms perform some of the higher-level functions in the production chain. There is some evidence of the development of *full-package production* by US firms in which

> a local manufacturer receives detailed specifications for garments from the buyer and the supplier is responsible for acquiring the inputs and coordinating all parts of the production process: the purchase of textiles, cutting, garments assembly, laundry and finishing, packaging, and distribution.[40]

Figure 10.13 shows an example of this development in the Torreón district of Coahuila in northern Mexico.

At the same time, the geographical pattern of garments production within Mexico is changing. Initially, virtually all of the export production in the Mexican garments industry was located along the United States–Mexico border, connected with 'twin plants' across the border. However,

> there appears to be a significant shift in southern California sourcing from the western border region to central Mexico and, to a lesser extent, the northern states (excluding the border cities) and the Yucatan peninsula. The central states of Guanajuato, Puebla, Tlaxcala and the greater Mexico City area figure prominently as new production sites. Also included is the west coast state of Jalisco, located due west of Mexico City.[41]

Major reasons for the relative shift away from the border are that labour costs in the interior of Mexico tend to be lower than along the border while the quality of production tends to be higher. There is also a much lower rate of labour turnover away from the border zone.

The spectacular and rapid growth of Mexico as a source of garments imports into the United States has significant implications for the Caribbean Basin producers. As Table 10.4 shows, the take-off of the Mexican garments trade coincided with the implementation of the NAFTA, which gives Mexico a level of preferential access to the North American market denied to the Caribbean producers. In particular, the Caribbean countries must still pay import duty on the

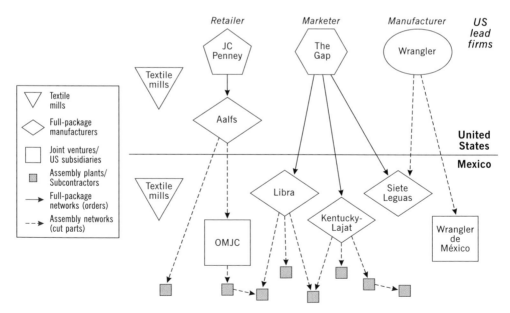

Figure 10.13   Development of 'full-package' garments production in Torreón, Mexico

Source: Based on Bair and Gereffi, 2001: Figure 2

value added in the garments assembly process, a contentious issue for these producers.

Finally, it is interesting to compare the emerging regional garments production complex in the Americas with the more established regional complex in Asia:

> In contrast to the evolution of the apparel commodity chain in Asia, which utilized East Asian NIE apparel manufacturers as the hubs of triangle manufacturing networks that knit suppliers from countries at different levels of development throughout the region, the coordinating agents in the North American apparel commodity chain are likely to be large US firms located in each of the main segments of the chain (fibers, textiles, apparel production, marketing and retailing). The main reasons for such a different outcome are various. First, Mexico and the CBI [Caribbean Basin Initiative] countries are both geographically and culturally closer to the United States than are Asian suppliers. This allows US firms to play a far more dominant role in the North American chain. Second, the role of trade policies is key here. The NAFTA pact provides Mexico at least a temporary edge over CBI suppliers, who thus far have not been granted NAFTA parity with Mexico. Even if parity is granted, Mexico has a big edge in developing a full-package supply capability because textile production in Central America and the Caribbean is virtually non-existent.[42]

Conclusion

The strategies adopted by textiles and garments producers are extremely varied and complex. The combinations of technological innovation, different types of internationalization strategy, the relationship with retailers and the constraints of the Multi-Fibre Arrangement have combined to produce a more complex global map of production and trade than a simple explanation based on labour-cost differences would suggest. Although firms in these industries do engage in foreign direct investment, this is a far less significant practice than other forms of international involvement, especially international subcontracting and licensing arrangements, often orchestrated by the large retailers and buyer groups. These are the dominant forms of international production in these industries, particularly in the manufacture of garments.

In this industry, the influence of transnational corporations tends to be more indirect than direct and the involvement of local capital and entrepreneurship greater than in many other industries. The manufacture of garments is an ideal candidate for international subcontracting. It is highly labour-intensive in the developed countries, it uses low-skill or easily trained labour, the process can be fragmented and geographically separated, with design and often cutting being performed in one location (usually a developed country) and sewing and garments assembly in another location (usually a developing country). Although international subcontracting in garments manufacture knows no geographical bounds – with designs and fabrics flowing from the United States and Europe to the far corners of Asia and finished garments flowing in the opposite direction – there are some strong and intensifying regional biases in the relationships. These industries are becoming *globally regionalized*.

Notes

1 Braudel (1984: 572).
2 UNCTAD (2000a: 82–4).
3 Abernathy et al. (1999: 9).
4 The increasingly significant role of the retailers in these industries is discussed by Abernathy et al. (1999), Gereffi (1994) and Taplin (1994).
5 Abernathy et al. (1999: 9).
6 Abernathy et al. (1999: 42, 56).
7 Taplin (1994: 211–12).
8 *The Economist* (12 February 1994).
9 *The Financial Times* (2 October 1996).
10 Klein (2000: 430).
`11 Klein (2000: 432, 433).
12 Detailed accounts of technology in the garments industry are provided by Abernathy et al. (1999).
13 Hoekman and Kostecki (1995: ch. 8) provide details of the MFA and its incorporation into the GATT Agreement in the Uruguay Round.
14 Cline (1987: 232).
15 Hoekman and Kostecki (1995: 209).
16 *The Financial Times* (10 January 1996).
17 Glasmeier, Thompson and Kays (1993: 23).
18 See Peck and Dicken (1996) for a case study of Coats.
19 *The Financial Times* (12 September 1996).
20 Glasmeier, Thompson and Kays (1993: 23–4).
21 Glasmeier, Thompson and Kays (1993: 24).
22 *The Financial Times* (23 February 1999).
23 *The Financial Times* (9 April 2002).
24 Fröbel, Heinrichs and Kreye (1980).
25 *The Financial Times* (9 January 1996).
26 Shepherd (1983: 42).
27 Jarillo (1993) and Schary and Skjøtt-Larsen (2001) provide accounts of Benetton's operations.
28 Schary and Skjøtt-Larsen (2001: 103).
29 *The Economist* (19 May 2001).
30 *Newsweek* (17 September 2001).
31 Gereffi (1999: 49) and Kessler (1999: 572–3).
32 Abernathy et al. (1999: 223).
33 Gereffi (1996a, 1999) and Khanna (1993) provide detailed analyses of the changing industry in Asia. Dicken and Hassler (2000) focus on the Indonesian garments industry from a business network perspective.
34 Gereffi (1999: 51).
35 Gereffi (1999: 56).
36 Gereffi (1996b: 97–8).
37 Abernathy et al. (1999) and Kessler (1999) provide detailed analyses of emerging garments production and trade networks focused on North America. See also Bair and Gereffi (2001); Gereffi (1999: 65–70).
38 Kessler (1999: 569).
39 Kessler (1999: 577–9).
40 Bair and Gereffi (2001: 1893).
41 Kessler (1999: 584).
42 Gereffi (1999: 70).

CHAPTER 11

'Wheels of Change': The Automobile Industry

The automobile industry was *the* key manufacturing industry for most of the middle decades of the twentieth century. The internal combustion engine was, quite literally, the major engine of growth for most of the developed market economies until the middle 1970s. Its significance lay not only in its sheer scale but also in its immense spin-off effects through its linkages with numerous other industries. Today, some 3–4 million workers are employed directly in the manufacture of automobiles throughout the world and a further 9–10 million in the manufacture of materials and components. If we add the numbers involved in selling and servicing the automobiles, we reach a total of around 20 million.

The industry came to be regarded as a vital ingredient in national economic development strategies. Consequently, the state has played an extremely important role in its evolution. In particular, trade barriers have exerted an extremely important influence in both developing and developed economies. At the same time, national governments have struggled to outbid one another in their efforts to secure the large manufacturing plants of the major automobile manufacturers (see Table 9.5). Organizationally the automobile industry is one of the most global of all manufacturing industries. It is an industry of giant corporations, many of which are increasingly organizing their activities on transnationally integrated lines. In contrast to the textiles and clothing industries discussed in the previous chapter, the world automobile industry is predominantly an industry of very large TNCs. Not surprisingly, the giant TNCs of the industry have developed consummate skills in playing governments off against one another.

The automobile production chain

The automobile industry is essentially an *assembly* industry. It brings together an immense number and variety of components, many of which are manufactured by independent firms in other industries. It is a prime example of a *producer-driven* production chain. As Figure 11.1 shows, there are three major processes prior to final assembly: the manufacture of bodies, of components and of engines and transmissions. The nature of the industry offers the possibility of organizational and geographical separation of the individual processes. The extent to which automobile manufacturers carry out the separate parts of the production sequence themselves varies considerably, although the practice of outsourcing has greatly increased in the past two decades.

We can identify a generalized sequence of development of a country's automobile industry (Figure 11.2). Although it is by no means inevitable that all countries will actually pass through the sequence, it does provide a useful indicator of possible developmental trajectories. *Stage 1*, the import of complete vehicles, tends to be limited in extent because of high transportation costs (cars are essentially large empty boxes) and by government import restrictions. *Stage 2*, local assembly of vehicles from a full 'kit' of component parts, permits transportation cost savings and provides the opportunity to make minor product modifications for the local market. *Stage 3*, assembly involving a mix of imported and locally sourced components, both depends upon, and encourages, the development of a local components industry. Not surprisingly, it is a stage strongly favoured by national governments

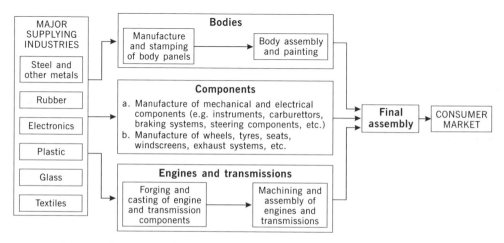

Figure 11.1 The automobile production chain

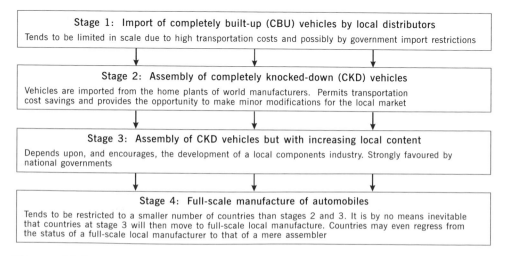

Figure 11.2 An idealized sequence of development in the automobile industry

as a potential entry to *Stage 4*, the full-scale manufacture of automobiles. This stage is restricted to a far smaller number of countries than stages 2 and 3. Indeed, it is by no means inevitable that countries that have reached stage 3 will then move to full-scale local manufacture. It is even possible that a country might regress from the status of full-scale local manufacturer to that of mere assembler.

Global shifts in the automobile industry

The geography of production

Between 1960 and 2000 world production of passenger cars increased three-fold. During those four decades, major changes occurred in the global distribution of the industry. Figure 11.3 shows that automobile production is very strongly concentrated in the global triad of Europe, East Asia and North America. These three regions account for 80 per cent of world automobile output. The United States, Japan and Germany are the three leading automobile-producing countries, accounting for almost half of the world total. Table 11.1 shows the major changes in the performance of the major producing countries since 1960.

By far the most dramatic development of the 1970s and 1980s was the spectacular growth of Japan as an automobile producer: from 1.3 per cent of the world total in 1960 to 26 per cent of the world total in 1989. By 2000, Japan's domestic automobile production had fallen back to 20.5 per cent of the world total, a reflection partly of the lengthy domestic recession in Japan but also because an increasing

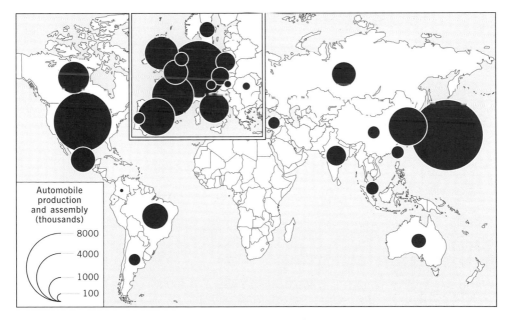

Figure 11.3 Global production and assembly of automobiles

Source: Based on material in SMMT (2001) *World Automotive Statistics*

proportion of Japanese automobile production is now carried out in overseas Japanese plants.

In 1960 the United States produced more than half of total automobile output in the world; by 2000 its share had fallen to 14 per cent. Less dramatic, though nevertheless very significant, was the decline of the United Kingdom's automobile industry. In 1960 the United Kingdom produced 10 per cent of the world total, more than eight times greater than Japan. By 2000 the United Kingdom's share was down to 4.5 per cent. The most impressive growth in automobile production in Europe occurred in Spain. In 1960 Spanish automobile output was negligible. By 2000 Spain produced 5.6 per cent of the world total.

Table 11.1 Growth of automobile production by major countries

	1960		1989		2000	
Country	Production (000 units)	World share (%)	Production (000 units)	World share (%)	Production (000 units)	World share (%)
France	1175	9.0	3409	9.6	2784	7.0
Germany	1817	14.0	4564	12.9	5309	13.4
Italy	596	4.6	1972	5.6	1410	3.6
Spain	43	0.3	1639	4.6	2208	5.6
Sweden	108	0.8	384	1.1	385	1.0
UK	1353	10.4	1299	3.7	1786	4.5
Canada	323	2.5	984	2.8	1626	4.1
USA	6675	51.4	6823	19.2	5636	14.2
Japan	165	1.3	9052	25.5	8100	20.5
Korea	NA	NA	872	2.5	2361	6.0
Malaysia	NA	NA	94	0.3	266	0.7
Taiwan	NA	NA	NA	NA	245	0.6
Argentina	30	0.2	112	0.3	225	0.6
Brazil	38	0.3	731	2.1	1103	2.8
Mexico	28	0.2	439	1.2	994	2.5
Australia	NA	NA	357	1.0	334	0.9
Czech Rep	NA	NA	184	0.5	475	1.2
Poland	NA	NA	289	0.8	573	1.5
World	12,999	100.0	35,455	100.0	39,568	100.0

NA, data not available.

Source: OECD (1983) *Long-Term Outlook for the World Automobile Industry.* Paris: OECD; SMMT (1996, 2001) *World Automotive Statistics*

Outside these leading producer countries, there are few important concentrations of automobile production. In the Americas, both Canada and Mexico are tightly enmeshed with the US automobile industry, a situation we will explore later in this chapter, while Brazil remains the major automobile production centre in Latin America. The most striking new development of recent years has been the sudden emergence of South Korea as an important producer. As recently as the early 1980s, Korea was producing only 20,000 automobiles; in 2000 Korean output was 2.4 *million* (6 per cent of the world total).

The geography of trade

This high level of geographical concentration of automobile production is also reflected in the geography of trade. The 15 leading automobile exporting countries listed in Table 11.2 accounted for 92 per cent of total automobile exports in 2000, with Germany and Japan the clear leaders, followed by the United States and Canada. However, both Germany's and Japan's share of world automobile exports declined sharply during the 1990s. Four countries considerably increased their share of exports: Canada, Mexico, Spain and South Korea. On the import side, the most striking feature is the fact that the United States absorbed nearly 30 per cent of all automobile imports in 2000, an increase of 10 percentage points over 1980. As a result, the United States had a trade deficit in automobiles of $105 billion in 2000. In contrast, Japan had a surplus of $78 billion and Germany a surplus of $50 billion.

Figures 11.4 and 11.5 illustrate the geographical pattern of world trade in automobiles at a more aggregated level. Figure 11.4 reveals the dominance of Western Europe in both exports and imports. It also shows the big trade deficit of North America (essentially the United States) and the substantial Asian surplus. The other parts of the world play a very minor role in automobile exports. Figure 11.5 analyses the connectivity of world automobile trade and also shows the varying extent to which the trade of major regions is internalized. In both North America (which includes Mexico) and Western Europe, 75 per cent of all automobile trade is *intra-regional*. North America has a large trade deficit with Asia, Western Europe and Latin America, whereas although Europe has a deficit with Asia, it has a large surplus with North America and a modest surplus with Latin America.

In summary, the following major shifts have occurred in the geography of the world automobile industry in the past four decades:

● An industry dominated in 1960 by the United States and, to a much lesser extent, Europe, was transformed during the 1970s and 1980s by the spectacular growth of Japan as a leading automobile producer. This was reflected in terms of growth of production in Japan itself, of Japanese exports to the rest of the world and of the increasing proportion of Japanese automobile production located outside Japan, especially in the United States and the United Kingdom. Japan has the most geographically extensive export network of all the major producing countries. Its tentacles spread throughout the world to a greater extent than the other major producing countries.

Table 11.2 The world's leading exporters and importers of automotive products

Exporting country	$bn	1980	1990	2000	
		(a) Exports			
Germany	92.17	21.0	21.9	16.1	
Japan	88.08	19.8	20.8	15.4	
United States	67.90	11.9	10.2	11.9	
Canada	60.66	6.9	8.9	10.6	
France	39.89	9.9	8.2	7.0	
Mexico	30.65	0.3	1.5	5.4	
Spain	28.13	1.8	3.7	4.9	
United Kingdom	25.56	5.8	4.4	4.5	
Belgium	25.01	—	—	4.4	
Italy	18.36	4.5	4.1	3.2	
South Korea	15.37	0.1	0.7	2.7	
Sweden	10.77	2.8	2.4	1.9	
Netherlands	8.66	1.1	1.5	1.5	
Austria	7.79	0.5	1.1	1.4	
Hungary	4.77	—	0.2	0.8	
Above 15	523.74	91.2	95.2	91.7	

Importing country	$bn	1980	1990	2000	Exports/Imports
		(b) Imports			
United States	172.73	20.3	24.7	29.4	−104.83
Canada	46.28	8.7	7.7	7.9	14.38
Germany	42.24	6.2	9.6	7.2	49.93
United Kingdom	36.08	5.7	7.1	6.1	−10.52
France	30.53	5.4	6.7	5.2	9.36
Spain	26.31	0.9	3.2	4.5	1.82
Italy	25.31	5.5	5.6	4.3	−6.95
Belgium	23.56	—	—	4.0	1.45
Mexico	18.82	1.8	1.6	3.2	11.83
Netherlands	12.61	2.6	2.6	2.1	3.95
Japan	9.96	0.5	2.3	1.7	78.12
Australia	8.02	1.3	1.2	1.4	
Austria	7.75	1.6	1.7	1.3	0.04
Sweden	7.57	1.5	1.4	1.3	3.2
Switzerland	6.35	1.8	1.9	1.1	
Above 15	474.09	68.8	82.8	80.6	49.65

Source: Based on WTO (2001), *International Trade Statistics, 2001*: Table IV-64

- The industry has become increasingly concentrated in the three major global regions. In the case of North America, this explains the recent fast growth of Mexico as a major automobile producer and exporter and of Canada's continued significance. In Europe, it explains the emergence of Spain as a major producer. In Asia, South Korea has emerged virtually from nowhere to become an important player, together with more limited developments, so far, in Malaysia, Thailand, Taiwan and China.

Figure 11.4

Regional shares of world trade in automotive products

Source: WTO (2001) *International Trade Statistics, 2001*: Chart IV.12

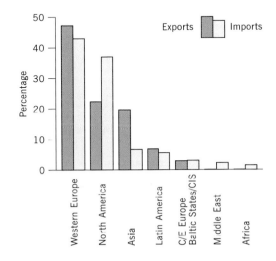

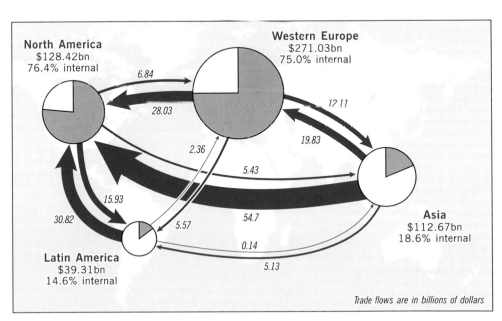

Figure 11.5 Global trade network in automobiles

Source: WTO (2001) *International Trade Statistics, 2001*: Table A10

- Outside the three major regions, the automobile industry is predominantly a simple assembly industry using mostly imported components. The major exceptions are Australia and Brazil.
- The automobile industries in the former Soviet bloc countries of the Russian Federation and Eastern Europe are currently in a phase of transition. In the latter area, the industry is becoming increasingly integrated into that of the EU.

This increasing regional concentration of the automobile industry will be analysed in more detail later in this chapter when we look specifically at the strategies of the companies themselves.

The dynamics of the market

Changes in the level, the composition and the geography of demand have all played a key role in the evolving global map of the automobile industry and in the problems facing many of the traditional producing nations and firms. Automobile production is strongly *market-oriented*. Thus, historically, production has developed within large, affluent consumer markets where high levels of demand have permitted the achievement of economies of scale. The world's three major automobile-producing regions – North America, Western Europe and Japan – are also the world's most developed consumer markets. The changing demand for automobiles has three major characteristics:

- It is highly cyclical.
- There are long-term (secular) changes in demand.
- There are signs of increasing market segmentation and fragmentation.

Demand for automobiles is highly sensitive to changes in the level of economic activity as a whole – to variations in the business cycle. Currently, there is around 30 per cent overcapacity in the world automobile industry. However, demand trends vary substantially between different parts of the world. Demand for automobiles is growing most slowly in Western Europe and North America and fastest in some of the East Asian countries and in parts of Latin America. The Asian market continues to be the major expected growth region to offset stagnation in Europe and North America: 'as disposable incomes across the region rise, the 50cc motorcycles that swarm the streets of most of the region's cities will eventually be upgraded to cars.'[1] There is also an expectation of a major surge in demand in Eastern Europe.

The major automobile manufacturers, therefore, are pinning their hopes on continuing high levels of growth in demand outside North America and Europe. In both of these mature producer regions, there is very substantial excess capacity (more than 30 per cent in Europe; around 25 per cent in the United States), amounting to several million automobiles, hence the continuing process of rationalization involving plant closures and contractions. However, the growth potential of Asia, at least in the short to medium term, may be exaggerated. The aftermath of the region's financial crisis of the late 1990s is still having an inhibiting effect on consumer demand in the region.

The slow growth in demand for automobiles in the Western European and North American markets reflects more than just cyclical forces. There are deeper *secular* or structural characteristics in these markets that limit future growth in car sales. Demand for automobiles is broadly of two kinds: new demand and replacement demand. Rapid growth in demand is associated with new demand. But as a market 'matures' and automobile ownership levels approach 'saturation', the balance between new and replacement demand alters. More and more car purchases become replacement purchases. In the mature automobile markets today some 85 per cent of total demand for automobiles is replacement demand. Replacement demand is generally slower-growing and also more variable because it can be postponed. In such mature markets manufacturers have long adopted strategies similar to those of clothing manufacturers: they regularly introduce 'new' models to entice existing owners to replace their automobiles. Often, the changes are little more than cosmetic, although promoted by massive advertising campaigns as being both significant and highly desirable. New production technologies are, however, making possible a far greater variety of automobile types.

Until relatively recently there were considerable differences between the major geographical markets, each tending to favour particular types of automobile. The greatest difference was between North America on the one hand and Western Europe and Japan on the other. Demand in the highly affluent, highly mobile, cheap-energy-based North American market was overwhelmingly for very large cars. In contrast, in Western Europe and Japan, generally lower incomes and higher energy costs plus more congested driving conditions were expressed in a demand for smaller, more fuel-efficient cars. But even within Europe, individual national markets tended to be served separately by domestically based producers. During the past 30 years, and especially since the oil crises of the 1970s, which reduced the attractiveness of large cars, these circumstances have changed dramatically.

One result of all these changes in the structure of demand has been to reduce some of the geographical differences between individual national markets. Demands have begun to *converge,* at least in the mass market sector. Regional and even global markets have become apparent. Within such markets, however, there are signs of greater market *segmentation* and *fragmentation:* demand for particular types of car for particular uses (e.g. four-wheel-drive recreation vehicles) or for a customized version of a general model. At least in the affluent consumer markets this is leading to more consumer-driven choice, which can, in turn, be satisfied because of the dramatic changes taking place in the way automobiles are made.

Production costs, technology and the changing organization of production[2]

Mass production

The basic method of producing automobiles changed very little between 1913, when Henry Ford first introduced the moving assembly line, and the 1970s, when a radically new system of production began to emerge in Japan (see Chapter 4).

The automobile industry was *the* mass production industry. It produced a limited range of standardized products for mass market customers. It produced in very large volumes in massive assembly plants, using very rigid methods, in which each assembly worker performed a highly specialized and narrow task very quickly and with endless repetition. The automobile industry appeared to have abolished craft production, apart from the small number of firms manufacturing for the luxury car market.

As developed by Ford and General Motors, and adopted by all the other Western manufacturers, this 'Fordist' method of production certainly brought the automobile as a commodity into the reach of millions of consumers. But, although very efficient in many respects, it contained one major limitation: its *rigidity*. In order to reduce costs, a particular model had to be produced in huge volumes. Assembly lines were 'dedicated' to a specific model and changing models took a great deal of time and money. Individual assembly workers were mere small cogs in the wheel of production, with no responsibilities beyond their very narrow single task.

Production of an automobile, from its initial design stage through to its appearance in the sales rooms, is an immensely complex and expensive process. A five-year development period was the norm until recently, and investment of up to $8 billion has been common for the very large-scale projects. Such enormous costs of development forced the major producers to seek very large production volumes – around 2 million vehicles per year – to achieve economies of scale in production, although the minimum efficient scale of production varied according to the particular manufacturing process involved.

For the automobile industry as a whole, labour costs may account for between one-quarter and one-third of total production costs. Their importance, like scale economies, varies between different processes, being highest for body-making and final assembly operations but significantly lower in engine and transmission production. As always, of course, it is not simply the cost of labour that is important but, rather, the overall quality and reliability of the labour force. Such considerations have played an important part in the investment-location decisions of most major transnational automobile manufacturers. Certainly they have sought out areas of surplus labour and, especially, those in which labour relations seem less likely to be problematical, both within and between individual countries. The importance of geographical differences in labour costs is, however, declining as the industry becomes more capital-intensive.

Technological change is quite obviously a major factor here. Production costs have been radically altered by developments in both product and process technology. Until recently, technological change in the industry seemed to be creating two distinct – and opposed – tendencies. One was towards even greater scale of production of *standardized* automobiles, the other was towards smaller-scale, more specialized production of automobiles aimed at particular *niches* in the market. Some automobile manufacturers have favoured one path of development while some favoured the other. However, this distinction is becoming increasingly blurred.

Product innovation in the automobile industry has generally been incremental. However, very rapid and far-reaching changes occurred after the 1970s oil crises.

The most important innovations have been those directed towards reducing fuel consumption, for example by making more efficient engines, by reducing the weight of materials used (such as substituting plastics and non-ferrous metals for steel), by reducing the size of cars (especially in North America) and by increased use of electronics to control engine performance. In addition, government safety and anti-pollution regulations in many countries have created pressures for change in car design. As yet, however, the development of cars powered by alternative fuels has been very limited.

Lean production

Many of these characteristics of the automobile industry remain very important. Since the 1970s, however, and with accelerating speed, there have been revolutionary changes in the ways automobiles are developed and manufactured. The source of such changes, which have dramatically reshaped the industry, was Japan. The term *lean production* was introduced[3] to contrast with the mass production techniques that had previously pervaded the industry. The claim was that 'lean production combines the best features of both craft production and mass production – the ability to reduce costs per unit and dramatically improve quality while, at the same time, providing an even wider range of products and even more challenging work.'[4] The major differences between craft production, mass production and lean production were discussed in Chapter 4 and are summarized in Table 4.2.

Each of the major automobile TNCs produces a wide range of models in different market segments. Historically, each model was distinctive, not only on, but also below, the surface. This posed major problems of duplication of many components and, indeed, of the basic 'platform' on which each model is built. One increasingly important strategy being used by major automobile manufacturers, therefore, is to reduce the number of different platforms (and components) to a minimal number, each of which is shared with other models within the firm's product portfolio. For example, within the German VAG group, although VW and Audi models occupy different market segments they share common platforms and many components. A similar strategy has been introduced by the major US firms, GM and Ford, and by other European producers. Such common platforms and shared components allows the achievement of economies of scale whilst, at the same time, preserving the distinctive identity of individual models through other means. An extreme version of such a strategy is the concept of *mass customization*:

> The selling of highly individual products but on a mass scale – is a logical next step in the progress of BTO ('build-to-order'), the manufacturing of goods only as and when there is an order from a customer … And that means turning a production-push industry into a demand-pull one … 'It is a the dream we are all running after,' says … the chief executive of … Fiat Auto, 'but it is difficult to bring to reality.'[5]

Such cautionary remarks provide an important check on making extravagant claims about future developments in the industry. While there is no doubt that automobile production has become much more flexible, such flexible systems do not, as is so often claimed, mean the demise of economies of scale. Rather, they facilitate more product variety and smaller scale for an individual model but within a large overall volume of production. They also make possible shorter design periods per model. What they do not do is reduce the barriers to entry for firms. The automobile industry remains a highly capital-intensive industry with very high entry barriers.

The aim, then, is to produce cars to order, rather than for inventory; to respond quickly to changing demands and for the need to switch production lines rapidly from one model to another. In this regard,

> Honda has probably gone furthest down the road to flexible global manufacturing. Not only are all its car factories capable of making several models, they are also now equipped to switch from one model to another very quickly. It takes Detroit between four and six weeks to alter models in a factory, re-jigging the robots and other tools. Honda can now do it overnight, simply by changing the software in the robots. To achieve this it has installed one single global manufacturing system.[6]

Supplier relationships

One of the most significant developments in the automobile industry in recent years has been the changing relationships between the major vehicle producers and their component suppliers. The automobile industry is an assembly industry requiring hundreds of thousands of components – ranging from entire engines and bodies down to small pieces of cosmetic trim. Historically, the big American and European manufacturers developed a particular kind of relationship with their suppliers, based on short-term, cost-minimizing contracts. The supplier–customer relationship was distant not just in functional terms but also, increasingly, in geographical terms. The major producers scoured the world for low-cost component suppliers. The close geographical proximity of customer and supplier, which had been a feature of the early years of the industry, began to break down as technological developments in transportation and communication made long-distance transactions feasible. The increased geographical distance between the assemblers and their suppliers made it necessary for the assemblers to hold huge inventories of components at their assembly plants. In this way, the possibility of the assembly line being disrupted by a temporary shortage of components (or by faulty batches) was reduced. This was the 'just-in-case' system, described in Chapter 4.

The relationship between customer and supplier in such a flexible production system must, however, be very different from that in the mass production system. (It is said that 'the traditional Detroit way of dealing with suppliers is "to beat them over the head"'[7].) The relationship has to be extremely close in functional terms, with design and production of components being carried out in very close consultation. Long-term relationships are preferable to short-term relationships. The system adopted by companies like Toyota is one of tiered suppliers, each tier having different tasks. The first-tier suppliers relate directly to the customer, second-tier suppliers to first-tier suppliers, and so on (Figure 11.6). The use of first-tier (or preferred) suppliers tends to involve a smaller number of suppliers than is common in the more remote, cost-driven, procurement systems. There is also a clear trend for the number of first-tier suppliers to be decreasing. Not only are there close functional relationships between customers and suppliers in this system but also there are closer geographical relationships. The use of just-in-time methods, discussed in Chapter 4, encourages geographical proximity although the distances involved may be very variable, ranging from suppliers located literally next door to assemblers (in 'supplier parks', for example) to those located within the broad region (such as within Europe).

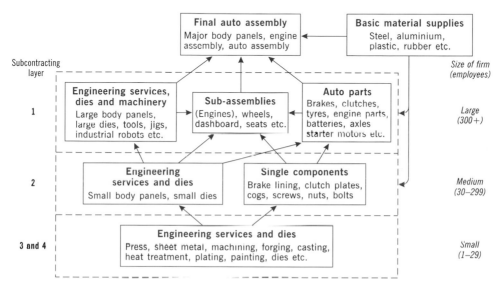

Figure 11.6   Functional tiers in the Japanese automobile supply system

Source: Based on Sheard, 1983: Figure 2

Automobile component suppliers, in fact, are now increasingly involved in different kinds of functional relationships with the automobile assemblers as the latter shift more responsibility (and risk) on to the supplier firms. The evolving relationships can take a number of different forms, as Table 11.3 shows. Each type involves rather different kinds of supplier responsibility. The most significant trend is for suppliers to manufacture entire 'sub-assemblies' rather than individual components.

Three especially ambitious projects – each in Brazil – have been established by VW, GM and Ford.

● VW's assembly plant at Resende involves component makers fitting their products directly on the assembly line, rather than simply delivering them from external factories. Suppliers assume responsibility for putting together and installing four major modules: the chassis, axles and suspension, engines and transmissions, and driver's controls. Management responsibilities are shared between VW and its main partner-suppliers.[8]

Table 11.3 Types of supplier responsibility in the automobile industry

Traditional subcontracting	Provision of component system	Parallel development	Co-development
Production of components from detailed specifications provided by automobile manufacturers	Production and assembly of complete component systems based upon technical directives of automobile manufacturers	The supplier is involved in the manufacture, assembly, and development of components	The supplier has complete responsibility for the design and production of components
The supplier has full responsibility for performance and quality	The supplier controls the quality and cost of parts procured from their suppliers	The supplier has the capability of making technical and cost adjustments during the design process	Application of simultaneous engineering of product and process
The supplier delivers the components to the assembly plant	The supplier controls the logistics chain of the component system	The automobile manufacturer retains control of development, design and prototype testing	The automobile manufacturer manages the technical interface between components

Source: Based on Laigle, 1996: Table 1

- GM's even more recent 'automotive industrial complex' (opened in 2000 at Gravataí, near Porto Alegre, in the south) consists of 17 plants, 16 of which are occupied by major suppliers. It involves much more than simply the co-location of component suppliers in a supplier park next to the GM assembly plant. Both the plant and the car were designed collaboratively. The major suppliers built their own production facilities inside the complex, each being located where its products are needed on the assembly line. Eighty-five per cent of the car produced at the plant is based on components assembled on-site compared with the usual position, where around 60 per cent of a car's components are sourced from outside.[9]
- Ford's new plant at Bahia, on the north-east coast, built to assemble a new compact vehicle, has 19 suppliers all within the factory. Each is responsible for complete modules. A further 12 component suppliers are located adjacent to the assembly plant. Together, this means that 60 per cent of the car's components are produced locally.[10]

Whether or not these 'assembly lines of the future' become accepted practice elsewhere within GM or VW, or more widely amongst other automobile manufacturers, is a matter for the future. Certainly it will be much harder to introduce such revolutionary practices in the old-established manufacturing heartlands of the automobile industry. Each of the three examples above are located well away from traditional automobile manufacturing areas. Equally uncertain is the extent to which automobile firms will utilize 'e-commerce' for purchasing supplies. In early 2000, several automobile firms (notably GM, Ford and Daimler–Chrysler), announced plans to establish an Internet-based component purchasing exchange (Covisint) with the intention of achieving huge procurement cost savings. Whether fully implemented or not the effect will to be increase the already tight squeeze on suppliers to provide ever-more sophisticated components at lower and lower costs.

In sum, new forms of assembler–supplier relationships are being forged throughout the automobile industry. They are part of the series of profound technological and organizational changes that are sweeping through the automobile industry. From being an apparently 'mature' industry in which technologies were relatively stable, the industry has 'de-matured'. These developments are having profound implications for the corporate strategies of the automobile manufacturers. Before reaching that point, however, we need to consider the influence of state policies on the industry.

The role of the state[11]

The world automobile industry is dominated by a small number of very large TNCs. Yet today, as in the past, their investment-location decisions are greatly affected by the policies of national governments towards the industry. Before looking at the strategies pursued by the major TNCs themselves, therefore, we need to look at the general ways in which national governments affect the automobile industry. Two key aspects of state policy are especially relevant:[12]

- the *degree of access* to its domestic market that the state allows to foreign firms to establish production plants there and
- the kind of *support provided by the state* to its domestic firms and the extent to which the state *discriminates* against foreign firms.

General state involvement in the automobile industry

In general, Western European governments have been more extensively involved in their domestic automobile industries than the governments of the United States and Japan. They have also gone furthest in attempting to restructure their industries. Until relatively recently, the governments of France, the United Kingdom and Italy were all involved not only in direct ownership of automobile producers but also in large-scale financial support. (Only in France does a residue of state ownership remain in the case of Renault.) Each of the state-owned enterprises had needed enormous financial subsidies. European governments have also used their regional development policies to influence the location of automobile manufacturers and to persuade both domestic and foreign firms to build plants in depressed regions of the country. Massive financial incentives have been used to establish huge assembly plants in areas of high unemployment.

Japanese government involvement in the industry was especially marked in the 1950s and 1960s. First, very tight protective barriers were placed around the domestic industry: tight import quotas and extremely high tariffs were imposed. Second, the direct involvement of foreign manufacturers in the Japanese industry was prohibited for a long period; technology was acquired through licensing agreements with foreign manufacturers. Third, MITI attempted to encourage rationalization among the major Japanese automobile producers, although with little success. Fourth, the Japanese government was heavily involved in assisting overseas marketing and exports, through financial and other assistance. These measures are no longer in force. More recently, the Japanese government's primary involvement in the automobile industry has been in negotiating 'voluntary' export restraint agreements with its trading partners.

Trade and foreign direct investment policies

Historically, the existence of national tariff barriers around sizeable consumer markets explains much of the early geographical spread of the automobile industry. Most governments in Europe and elsewhere (including Japan, Canada and Australia, for example) in the period before World War II had erected tariffs against automobile imports. These provided an obvious stimulus both to the setting up of foreign branch plants by the major automobile producers and also to the growth of domestic manufacturers. Today, few of the developed market economies levy particularly high import tariffs against automobile imports, although they remain high in some cases. The common EU tariff is 11 per cent, the US automobile tariff is only 3 per cent. Japan now has no import tariffs on automobiles, although it has been much criticized for its slowness in removing certain non-tariff barriers as the dispute with the United States exemplifies.

Although tariffs in the major developed country markets are now generally low, there has been a strong upsurge in other forms of protection, primarily against Japanese imports. In response to the growing domestic outcry from major industry pressure groups, both the United States and several Western European countries negotiated 'voluntary' export restraints with Japan during the early 1980s. In Europe, the completion of the Single European Market in 1992 replaced the bilateral agreements that had previously existed between individual member states and Japan. After lengthy and acrimonious negotiations, the European Commission negotiated an agreement with Japan which limited Japanese automobile imports to a given percentage of total EU automobile sales for a transitional period of seven years. There are now no restrictions on Japanese automobile imports into the EU.

The biggest area of controversy was over the kind of treatment to be accorded to the automobiles actually manufactured by Japanese firms within Europe itself (that is, in the UK, where, at the time, all the Japanese transplants were located). There was particularly bad feeling between the French and Italians on the one hand and the British on the other over the treatment of cars manufactured in Japanese plants in Britain. Were they sufficiently 'European' or were they really Japanese? Britain was variously described as a 'Trojan Horse' infiltrating Japanese cars into the European market or as an 'aircraft carrier' moored offshore and launching Japanese cars into Europe. Since then, as we shall see later, Toyota has built its second European plant in northern France – a smart political move (and also a smart economic move in the light of the UK's non-membership of the currency union).

State attitudes towards foreign versus domestic automobile manufactuers have varied considerably:[13]

- *France*: limited access granted to foreign firms and discriminatory intervention in favour of domestic producers
- *Britain*: unlimited access granted to foreign firms and equal treatment of both foreign and domestic producers
- *Germany*: unlimited access granted to foreign firms and discriminatory intervention in favour of domestic producers
- *United States*: unlimited access to foreign firms and equal treatment of both foreign and domestic producers.

In each of these countries, huge incentive packages have been used to attract the major automobile producers (see Table 9.5).

To this list we should add:

- *Japan*: limited access granted to foreign firms and discriminatory intervention in favour of domestic producers
- *Korea*: no access to foreign firms as producers; foreign automobile imports highly restricted by non-tariff barriers.

Local content requirements

The increased fear of Japanese imports into both North America and Western Europe also stimulated additional policy measures. One was to encourage Japanese automobile manufacturers to build production plants in their major overseas markets to displace imports. The other was to insist on specific levels of *local content* in automobile manufacture by foreign producers. Again, it was in Europe that the greatest controversy developed over this issue, both over precisely what local content means and also how it should be measured. For a specific EU member state 'local' refers to all EU countries and not just its own domestic territory.

Such measures, although relatively new in the developed country context, have long been important elements in the national policies of developing countries towards the automobile industry. Virtually all developing economies with any kind of automobile industry operate both local content requirements and various types of tariff and non-tariff import restrictions. The use of both local content and other import restrictions together with very high tariffs on automobile imports formed the basis of the strong import-substitution policies followed by a number of developing countries, particularly during the 1950s and 1960s. The aim was to build a domestic automobile industry to serve the domestic market. Each of the major industrial countries in the developing world – India, Brazil, Argentina, Mexico – as well as countries such as Spain, adopted this kind of strategy. In most cases local content requirements were set at between 50 and 90 per cent, usually on a progressively rising scale over a period of several years.

Environmental and safety legislation

A final aspect of state policy is that of environmental, fuel-efficiency and safety legislation. All governments have some legislation covering these areas, each of which has profound implications for the design, technology and materials used in cars and, therefore, in their cost. Complying with changes in legislation can be especially problematical where it involves fundamental design changes. Legislation to control noxious emissions from automobile engines has been particularly stringent for many years in states such as California and is now being implemented at similar levels in the EU. Of course, a continuing source of tension is the difference between government policy in the United States and in other countries over automobile fuel consumption and levels of taxation. For example,

> whereas the European industry has already produced cars with average fuel economy of around seven litres per 100 kilometres, North Americans continue to pump much vaster amounts of carbon dioxide into the atmosphere from vehicles consuming nearly 11 litres per 100 kilometres. For European carmakers, the disparity reflects what many describe as the 'cowardice' of the US government in not raising petroleum taxes enough to encourage fuel-efficiency and to drive 'gas guzzlers' off the streets.[14]

Vehicle emission regulations are one aspect of state involvement in the automobile industry. Another is policy towards 'end-of-life' vehicles. Here, the EU has issued a directive to come into force in 2007 under which automobile manufacturers will have to cover the cost of recycling the vehicles they have manufactured. It is estimated that the annual cost of this operation in Europe will be around 2.1 billion euros. Manufacturers will also have to ensure that recyclable components account for 85 per cent of each vehicle's weight.

Corporate strategies in the automobile industry

Increasing global concentration

In the early days of the automobile industry in North America and Western Europe there were scores of manufacturers each producing a limited range of automobiles for individual national markets. In 1920, for example, there were more than 80 automobile manufacturers operating in the United States, more than 150 in France, 40 in the United Kingdom and more than 30 in Italy. Today, each major national market is dominated by a very small number of massive corporations. In France and Sweden the entire national output of cars is produced by only two companies; in Italy one firm, Fiat, is totally dominant. But such high levels of concentration are evident not only at the national scale but also internationally: the global automobile industry is in the hands of a small number of very large firms. Table 11.4 lists the world's leading automobile manufacturers, ranked by total company sales value. A ranking based solely on car production is rather different. Table 11.4 shows that the top four car producers – produced 44 per cent of the world total while the 15 firms produced 77 per cent of world automobile output.

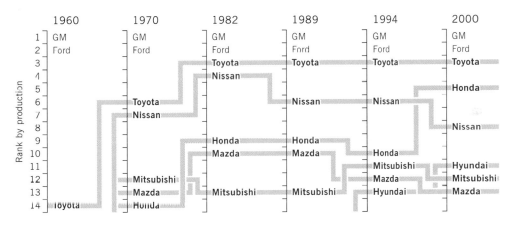

Figure 11.7 The rise of Japanese (and Korean) automobile manufacturers

For most of the past 40 years, the most dramatic shifts in relative position involved Japanese firms, particularly Toyota and Nissan (Figure 11.7). Before 1960 no Japanese manufacturer ranked among the world's top 15 automobile producers. In 1960 Toyota appeared in the top league for the first time – in fourteenth position. By 1965 Toyota had moved up to ninth position and had been joined by Nissan in eleventh place and Toyo Kogyo (Mazda) in thirteenth place. Five years later, Mitsubishi and Honda had also entered the first division while Toyota and Nissan had moved into sixth and seventh position respectively. By 1989 Toyota was challenging Ford for second place in the world league, Nissan ranked sixth, Honda ninth and Mazda tenth. By 1994, Honda had slipped down one place to tenth, Mitsubishi had climbed two places to eleventh while Mazda fell from tenth to twelfth. The most unexpected development, however, was the emergence of a Korean automobile firm, Hyundai, in the league table of leading manufacturers.

Increasing global connectedness: acquisitions, mergers and alliances

The Japanese and Korean firms grew primarily through organic growth. In contrast, the US and European producers have grown more through merger and

Table 11.4 The world league table of automobile producers, 2000

Rank	Company	Country of origin	Sales ($bn)	Car production	Share of world total (%)	Percentage produced abroad	
						2000	1989
1	GM	USA	176.6	5,472,919	13.8	61.8	41.8
2	Ford	USA	162.6	3,625,447	9.2	63.9	58.7
3	Daimler–Chrysler	Germany	151.0	1,814,006	4.6	50.0	NA
4	Toyota	Japan	119.7	3,719,895	9.4	27.5	8.3
5	VAG	Germany	70.6	4,418,018	11.2	52.5	30.5
6	Nissan	Japan	58.1	1,864,900	4.7	35.1	13.8
7	Honda	Japan	51.7	2,302,749	5.8	50.3	28.0
8	Fiat	Italy	45.2	2,118,712	5.4	33.4	7.1
9	PSA	France	37.8	1,952,385	4.9	22.6	15.4
10	Renault	France	37.6	1,854,117	4.7	27.2	17.6
11	BMW	Germany	36.7	731,289	1.9	7.0	—
12	Mitsubishi	Japan	29.1	1,062,665	2.7	29.1	—
13	Volvo	Sweden	15.1	406,244	1.0	36.7	NA
14	Hyundai	Korea	NA	1,619,541	4.1	4.5	—
15	Mazda	Japan	NA	833,426	2.1	15.4	18.3

NA, data not available

Source: Calculated from SMMT (2001) *World Automotive Statistics*; UNCTAD (2001) *World Investment Report, 2001*: Table III.1

acquisition. During the early 1990s, a number of significant acquisitions occurred: for example, Ford acquired the British firms Aston Martin and Jaguar; GM acquired Saab. Both American firms had acquired substantial equity stakes in smaller Japanese companies: Ford with Mazda, GM with Isuzu and Suzuki. The German manufacturer VAG acquired the Spanish firm Seat and the Czech company Skoda. But it was towards the end of the 1990s that some really mega-mergers occurred.

The most spectacular was the acquisition of Chrysler by the German manufacturer of Mercedes vehicles, Daimler–Benz, in 1998. In 1999 Ford acquired the Swedish company Volvo, and Renault acquired 36.8 per cent of the equity in Japan's second largest automobile firm, Nissan. In March 2000, GM announced a cross-shareholding alliance with the Italian company Fiat, a defensive response to the Daimler–Chrysler merger. Daimler–Chrysler itself acquired 34 per cent of Mitsubishi Motors of Japan. By 2002, GM had also acquired the bankrupt Korean company, Daewoo. Not all such mergers have been successful. BMW's acquisition of the British manufacturer Rover was a failure and BMW withdrew in 2000, although it retained a small-car plant and an engine plant in the UK. It sold Land Rover to Ford but acquired Rolls-Royce. Figure 11.8 is the 'organizational map' of the world's leading automobile producers: it summarizes the major equity links involving the leading producers.

However, mergers and acquisitions are not the only form of inter-firm relationship in the world automobile industry. All the world's automobile manufacturers

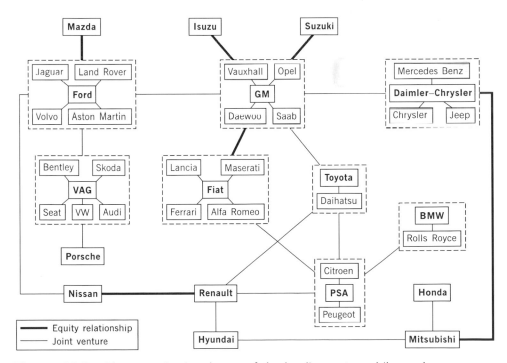

Figure 11.8 The organizational map of the leading automobile producers

are also deeply embedded in *collaborative agreements* with other manufacturers and in complex transnational sourcing arrangements for components with the major component firms. The very high level of concentration in the world automobile industry is largely the outcome of factors discussed earlier in this chapter, in particular the drive to achieve efficiency in design, production and marketing in an increasingly competitive global market. A major problem, however, is that the development of new models, which themselves have a shrinking life, requires massive investment not only in machinery and other equipment but also in research and development. Consequently, even the very largest firms are involved in collaborative ventures with other manufacturers, while the very survival of smaller firms seems to depend increasingly on inter-firm agreements to supply parts, to produce jointly under licence and to engage in joint R&D.

Consequently, a veritable transnational spider's web of strategic alliances has developed, a web that stretches across the globe:

> In recent years, there have been about 100 new alliances in the automobile industry per year. The majority of these are manufacturing joint ventures ... Around 80 per cent of the 1999 alliances (91 out of 115) were cross-border, indicating the high degree of globalization in this sector. International alliances in 1999 included 53 joint ventures, all of which (except one for marketing co-operation) were for assembly of vehicles or parts. US firms participated in 27 international alliances in 1999, followed by Germany (26), Japan (22), China (13), France (10) and Italy (8).[15]

Variations in the strategies of automobile manufacturers

Although the automobile industry is undoubtedly a global industry, there is a good deal of *variety* in the specific corporate and competitive strategies adopted by the leading companies. During the late 1970s and early 1980s two clearly defined strategic options appeared to be favoured.

- A *world car strategy* – manufacture of mass-produced cars for a world market. This was based upon transnationally integrated production between the overseas affiliates of the parent company. Component manufacture was to be located in the most favourable – that is, *least-cost* – locations at a global scale. These highly specialized plants would then supply a strategically located network of assembly plants that served specific, large-scale geographical markets. The emphasis was on standardization, with minor, often cosmetic, modifications to suit particular markets. Hence, economies of scale were a key consideration both for component manufacture and for the assembly of finished automobiles. The world car, therefore, was not merely to be sold to a world market: more significantly, it was to be *manufactured* on a world scale.

- A *niche market strategy* – production in relatively small volumes for a small market in which premium prices could be charged. Such a luxury car strategy was less likely to involve transnational production; rather the country of production (for example, Germany, Sweden) was seen as an essential element of the brand's appeal.

For some time it seemed that the world automobile industry was segmenting along these two lines. Certainly the two leading US manufacturers, GM and Ford, moved a considerable way along the world car path, while firms like BMW, Daimler–Benz, Saab and Volvo continued to operate as specialist manufacturers in luxury car niches.

In fact, by the early 1990s the picture had changed dramatically. The stark distinction between mass-produced and lower-volume cars began to disappear in the face of the Japanese development of alternative methods of 'lean' production and the intense competitive pressures that ensued. With a few exceptions, virtually all the major automobile producers shown in Figure 11.8 not only produce vehicles across a range of market segments but also do so by using many shared components (including vehicle platforms) in models with different brand identities and selling in different market segments. In addition, the automobile producers are shifting their emphasis 'downstream', recognizing that 'margins made from selling finance, insurance, parts and even mobile "infotainment" are much higher than from making and selling cars. There are also considerable savings to be achieved in sales, distribution and marketing.'[16]

Virtually all of the major automobile producers are drastically rationalizing and restructuring their operations, both domestically and overseas. In the face of the excess capacity in the industry, some plants are being closed, others are having their operations either scaled down or transformed. However, there are distinctive differences within this broad strategy according to a firm's country of origin (recall our discussion of the influence of 'place' on the nature of TNCs in Chapter 7). In analysing this situation we will first summarize the evolution of the strategies of the major producers and then examine the regional production networks that are developing in Europe, North America and East Asia.

Strategies of US automobile firms: GM and Ford

As Table 11.4 shows, Ford is the most transnational producer, with almost two-thirds of its passenger car production located outside the United States. GM's degree of transnationality, however, although slightly lower, has been increasing rapidly in recent years (in 1989, for example, only 42 per cent of GM's production was located outside the United States). Ford and GM are the world's longest-established transnational automobile producers. Ford expanded internationally mainly by opening new plants under the Ford name, GM primarily by acquiring existing foreign manufacturers and retaining their brand identity. Ford established its first foreign manufacturing plant in Canada in 1904, just across the Detroit River in Windsor, Ontario; GM acquired the Canadian automobile company McLaughlin in 1918. Ford opened its first European car assembly plant at Trafford Park, Manchester in 1911 (subsequently replaced by a massive integrated operation at Dagenham, near London, in 1931), followed by a plant at Bordeaux in 1913. Ford began to assemble cars in Berlin in 1926 and in Cologne in 1931. GM's entry into Europe began in 1925 with the acquisition of Vauxhall Motors in England, followed, in 1929, by the purchase of the Adam Opel Company in

Germany. During the 1920s both firms also established assembly operations in Latin America as well as in Australia and Japan.

These early transnational ventures were triggered primarily by the existence of protective barriers around major national markets as well as by the high cost of transporting assembled automobiles from the United States. In recent years, however, both Ford's and GM's transnational strategies have been concerned initially with expanding and integrating and subsequently rationalizing their operations on a global scale. Figure 11.9 outlines the broad geography of Ford's and GM's transnational activities. Geographically they are remarkably similar, although they have reached their current positions by rather different routes. Three-quarters of Ford's overseas production is located in Europe, with Germany, Spain, the UK and Belgium the primary nodes. Outside Europe, Ford's foreign production operations are primarily in Canada, Mexico and Brazil. GM has a global production system much like Ford's in its geographical structure, but the nature, the timing and the organization of its development have been different. Two-thirds of GM's overseas car production is located in Europe – in the same countries as Ford. Outside Europe, Canada is the most significant area of operations, followed by Brazil and Mexico.

During the past decade, both firms have engaged in substantial strategic and structural reorganization. For example, Ford's response in the mid-1990s to intensified competition in all its major markets was to embark on a hugely ambitious process of global reorganization ('Ford 2000'). The North American and European operations were merged into a single organization, with the intention of ultimately integrating its Asian, Latin American and African operations as well. Ford created a single worldwide product development organization and created five automobile centres, each having responsibility for the global development of specific products. Four of the five automobile centres are located in the United States and one – responsible for small and medium-sized, front-wheel-drive cars – in Europe (jointly in the UK and Germany). The philosophy underlying this global structure was to cut out duplication of 'automobile platforms, power-trains

Figure 11.9

The global geography of US automobile producers: GM and Ford

Source: Based on material in SMMT (2001) *World Automotive Statistics*

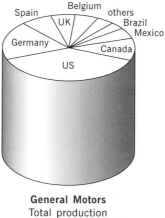

General Motors
Total production
5,472,919 cars

Ford
Total production
3,625,447 cars

(engines and transmissions) and other basic components that serve nearly identical customer needs in different markets … In the future we will have one small engine family in Europe and North America instead of two separate families that power the same car for the same kind of customer – yet are completely different and used duplicate resources in their development.'[17]

Four years later (in 1999), Ford announced yet another major reorganization of its global management structure. The aim was to try to respond more effectively to the 'global–local tension' we discussed in Chapter 7 (see Figure 7.12) by replacing the strongly global emphasis of 'Ford 2000' with a system of strategic business units organized around brands and regions: Europe, Asia–Pacific, South America and North America.[18] All of this was set within a self-conscious attempt to become what the then CEO of Ford, Jacques Nasser, called 'a relationship business' rather than a 'nuts and bolts business'. However, in late 2001, Nasser was replaced by a member of the Ford family. As a consequence, further major restructuring was implemented, involving closure of five plants and the loss of 22,000 jobs in North America, and a focus on the company's core business of producing vehicles.

Although faced with exactly the same competitive pressures, GM adopted a more cautious approach to the global reorganization of its business. In the mid-1990s, when Ford was implementing its 'Ford 2000' reorganization, GM merely established a global strategy board consisting of the heads of GM's North American and international operations and its component division but did not create a radically new structure. Although GM wanted to 'globalize' its operations it wanted to do it within its existing structure. By the end of the 1990s, in fact, GM was still more an international firm than a global firm, with a widely dispersed set of loosely connected operations.

In 1998, GM moved to merge its North American and its international operations into a single worldwide group with the aim of realizing its potentially enormous economies of scale. But the process is both massive and slow to implement. GM's purchasing operations – consisting of 27 autonomous components units – were combined into a single global operation. The engines and transmissions units were combined and then organized regionally in terms of local expertise: North America is responsible for large engines and automatic transmissions; Europe for four-cylinder engines and manual transmissions; Asia (in the form of GM's Japanese affiliate) for diesels. As in Ford, however, the process of global reorganization has been problematical, not least because of cultural differences between the centre and the geographically dispersed operations derived from what one observer called 'the ingrained Midwestern focus of GM's top executives'.[19] One major development, in 2002, was the purchase of bankrupt Korean producer Daewoo.

Strategies of Japanese automobile manufacturers

Whereas both Ford and GM have had international operations for many decades, the spectacular rise of the leading Japanese companies up the world league table was achieved almost entirely without any actual overseas production, apart from small-scale, local assembly operations, using imported kits. Beyond such operations, Toyota had no overseas production facilities for passenger cars before the early 1980s, while less than 3 per cent of Nissan's total production was located outside Japan. Paradoxically, it was one of the smaller Japanese automobile manufacturers, Honda, which was the first to build production facilities outside Asia (in Ohio in 1982). The biggest Japanese producer, Toyota, was the slowest to transnationalize its operations. For example, Toyota did not build its first European plant until 1992, six years after Nissan.

However, the transnationality of the Japanese producers changed dramatically during the 1980s (see Table 11.4). By 1989, 28 per cent of Honda's output was produced outside Japan; today the proportion is 50 per cent. Toyota tripled its overseas production share (to 28 per cent); Nissan now produces 35 per cent of its cars outside Japan compared with 14 per cent in 1989. Before the early 1980s, Japanese automobile manufacturers had shown themselves perfectly capable of serving the North American and European markets by exports from Japan. The price competitiveness of Japanese automobile exports was based upon extremely large-scale, flexibly organized and highly automated production plants in Japan. Japan's massive technological investment in automobile production during the 1950s and 1960s, aided by government measures, resulted in extremely efficient automobile production. The very high degree of vertical integration with Japanese component suppliers added to this remarkably efficient system. These production cost efficiencies more than offset any transport cost penalties arising from Japan's geographical distance from major markets.

From the early 1980s, however, the Japanese companies began to change their global strategy by locating major production plants in their major markets. The primary stimulus was the increasingly powerful political opposition, in both North America and Western Europe, towards the growth of Japanese imports. But trade protectionist measures were not the only stimulus. Given the increasing possibilities of tailoring automobiles to variations in customer demand, it was becoming more important for Japanese firms to locate close to their major markets. The primary objective of the Japanese automobile producers, therefore, has been to develop a major direct presence in each of the world's major automobile markets – the global triad – of Asia, North America and Western Europe. Figure 11.10 shows the current geographical pattern of Toyota's, Nissan's and Honda's activities. For Honda, North America is the dominant overseas production focus, with almost 75 per cent of Honda's overseas production located there. The same is true for Toyota. In the case of Nissan, however, overseas production is more evenly distributed between North America and Europe (the United Kingdom), with an important concentration also in Mexico.

After a long period of seemingly unstoppable success, the major Japanese producers began to face increasingly difficult problems during the 1990s, especially in Japan itself. The deep and prolonged recession in Japan led to steep declines in sales so that, for the first time, Japanese firms began seriously to restructure their domestic operations. The form of such restructuring has centred primarily on reducing costs: for example, by squeezing suppliers, by rationalizing procurement and, in some cases, actually closing plants in Japan (hitherto an unthinkable option).

Among the top three Japanese automobile producers, Nissan has adopted the most radical response by selling 36.8 per cent of its equity to a foreign manufacturer – Renault. Some of the smaller Japanese automobile firms already had tie-ups with foreign firms, notably Mazda with Ford, Isuzu, Subaru and Suzuki with GM, Mitsubishi with Daimler–Chrysler. But the Nissan case was the most significant. In effect, Renault became the major influence on Nissan's restructuring strategy. Five factories were to be closed between 2001 and 2002; purchasing costs were to be drastically reduced, the dealer network was to be consolidated. The two firms plan to share ten basic vehicle platforms across both their model ranges.

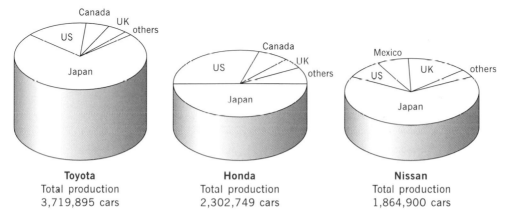

Toyota	Honda	Nissan
Total production	Total production	Total production
3,719,895 cars	2,302,749 cars	1,864,900 cars

Figure 11.10   The global geography of Japanese automobile producers: Toyota, Nissan and Honda

Source: Based on material in SMMT (2001) *World Automotive Statistics*

Strategies of European automobile manufacturers[20]

The major European automobile producers are locationally far more limited than either GM or Ford. Figure 11.11 summarizes their production geography. Of the leading European companies, VAG has pursued by far the most extensive and systematic transnational strategy. Outside Europe, VAG is a major producer in Brazil and in Mexico. Within Europe, prior to the opening up of Eastern Europe, VAG concentrated its production in two countries in a clear strategy of spatial segmentation. High-value, technologically advanced cars were produced in the former West Germany; low-cost, small cars were produced in Spain where VAG undertook a massive investment programme in Seat. During 1990, after the collapse of the Soviet-dominated system, VAG moved very rapidly to establish production of small cars in eastern Germany and to take a 70 per cent stake in the Czech firm, Skoda.

VAG embarked on a major restructuring programme which involved reducing the number of automobile platforms used by its four component companies (VW, Audi, Seat, Skoda) to ensure the maximum sharing of basic structures on which differentiated automobiles could be based. In 2001, the company announced a major reorganization of its entire operations:

> Europe's largest car-maker is attempting to transform itself from an inward-looking family of European brands into a fully-fledged global carmaker ... Seat and Skoda, acquired to give VW geographic reach in southern and eastern Europe, had outgrown their home markets and were directly competing with each other. Both were also cannibalising sales from VW ... VW and Audi, its premium car brands, were starting to compete directly against each other ... Under the reorganization, the passenger car business will be split into two ... 'the key reason they're doing this is to get more control at Wolfsburg [VW headquarters] of how the whole company is being run'.[21]

Daimler–Benz, as we have seen, has taken over the American company, Chrysler, in an arrangement that is proving extremely difficult to implement

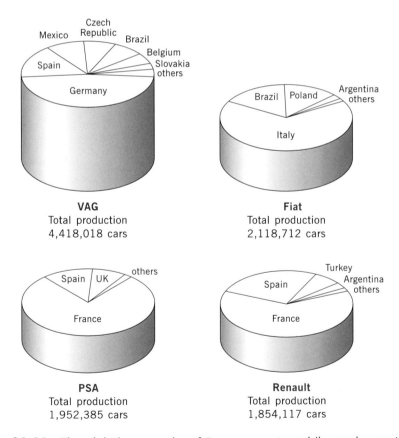

Figure 11.11 The global geography of European automobile producers: VAG, PSA, Renault and Fiat

Source: Based on material in SMMT (2001) *World Automotive Statistics*

successfully. Massive rationalization is occurring, especially in Chrysler's operations in the United States where, in 2001, 26,000 job cuts, six plant closures and cuts in supplier prices were announced. It has also strengthened its equity link with Mitsubishi. 'That means vehicles such as the next-generation Chrysler Seebring will share a platform with Mitsubishi Motors (MMC) … The company has begun work on a separate small car platform with MMC, which will be used for the replacement model for Chrysler's Neon.'[22]

BMW, Germany's third largest automobile manufacturer, remains a privately owned company producing for high-value markets. For a time in the early 1990s, BMW aspired to extend its range into the mass market segment by acquiring Rover, a deeply troubled British company. Despite huge investments both by BMW and by the British government, BMW finally withdrew from its ownership of Rover, although it retained a small-car assembly plant and an engine plant in the UK. However, BMW has extended its operations in North America and has a modest presence in Asia.

The two French automobile companies, Peugeot–Citroën (PSA) and Renault, have both traditionally been strongly home-country oriented in their production (see Figure 11.11). Peugeot–Citroën was formed by a state-induced merger of the separate Peugeot and Citroën companies in 1975. In 1978 the company purchased Chrysler's European operations. Seventy-seven per cent of Peugeot–Citroën's production is located in France with a further 12 per cent in Spain and 8 per cent in the United Kingdom. It has no production plants outside Europe and yet has managed to be very successful in serving this market. The two brands, Peugeot and Citroën have been separate in many respects but now, in line with all the other major automobile producers, they will be manufactured using common platforms and components.

Renault[23] has been, for more than 40 years, the French government's national champion, supported by massive state aid, which served to constrain its activities. State control has been reduced to 44 per cent and Renault, like Peugeot–Citroën, has been involved in major restructuring. In the early 1990s, Renault had an important strategic alliance with the Swedish company Volvo which, it was believed, would eventually result in full-scale merger of the two companies. But in late 1993, Volvo pulled out, leaving Renault in a difficult position. But, as we saw earlier, Renault pulled off a major coup in acquiring a major stake in Nissan. This alliance has become the central pillar of Renault's strategy. For Renault itself, France remains the dominant production location, with 63 per cent of its total world production. A further 26 per cent is located in Spain.

While VW was expanding its European production base to incorporate Spain in the 1980s, the Italian automobile firm Fiat initially moved in the opposite direction and reconcentrated production in its home market. Sixty-seven per cent of Fiat's production is located in Italy. A key element in Fiat's more recent strategy has been to create an extensive production network in the former Soviet Union and Eastern Europe where it has the longest-established links of any automobile producer in the world. Fiat's 'grand European design' was to build a manufacturing network extending from the Mediterranean to the Urals. At the same time, Fiat strengthened its position in Brazil. Both Brazil and Poland became the major bases for Fiat's small 'world car', the Palio, which began production in Brazil early in

1996. Fiat aimed to produce more than 50 per cent of its cars outside Italy by the year 2000 but failed to reach that target. As yet, it is too early to assess the effect of Fiat's recent strategic alliance with GM.

Even allowing for the case of VW, the most internationally oriented European automobile producer, the major European companies are far narrower in global terms than the US and, increasingly, the Japanese firms. Apart from some limited involvement in Latin America (and excluding simply local assembly plants in various countries), they are entirely European in their production networks. It is especially notable that, while the Japanese were busy building large manufacturing plants in the United States during the 1980s, the Europeans did not do this. Indeed, both VW and Renault pulled out of their earlier involvement in North America.

Strategies of Korean automobile producers

The US, Japanese and European automobile companies exert such market dominance that there have been virtually no new entrants in the industry during the past 30 years. The major exception is in South Korea, which has emerged in the space of a just a few years as a significant international force.[24] As we noted in Chapter 6, the development of a Korean automobile industry has been strongly influenced by the policies of the Korean government. 'Autos were identified as one of the priority industries in the Heavy and Chemical Industry Plan of 1973. In 1974 an industry-specific plan for automobiles was published covering the next ten years. The objectives were to achieve a 90 per cent domestic content for small passenger cars by the end of the 1970s and to turn the industry into a major exporter by the early 1980s.'[25] To meet these ambitious objectives, the Korean government took the following steps:

- It specified the three primary producers – Hyundai, Kia and what was later to become Daewoo (it was originally called GM Korea and then Saehan).
- The government had to approve the plans of each producer, having stipulated their minimum size and the maximum permitted size of engines to be manufactured.
- A promotional plan for the components industry was initiated which required the three automobile producers to cooperate in producing standard parts and to meet local content requirements.
- The three producers were later required to set export targets on a consecutive basis: South East Asia, Latin America and the Middle East, Canada 'all by way of preparation for a big push into the US market'.[26]
- The producers were encouraged to set export prices below production costs.
- The government provided them with substantial export subsidies (that is, credit).

During the early 1980s, in response to difficult economic conditions, the Korean government effectively made Hyundai the leading producer in the industry, giving it an enormous relative advantage which the firm used to grow at a very rapid rate. By 1997, there were five Korean automobile producers; today, there are just two – Hyundai and Daewoo – the others being victims of the East Asian financial crisis.

Both Hyundai and Daewoo depended for their early development on close technological and marketing relationships with US and Japanese firms. Hyundai's strategy was to compete with the Japanese in a very narrow product range and entirely on price. Its initial export success was remarkable. In 1986, 300,000 cars were exported to North America, a level more or less maintained for the following two years. On the strength of this success, Hyundai built a second new plant in Korea. More ambitiously still it built a plant in Canada, with the capacity to produce 120,000 cars and to employ 1,200 workers directly, but this plant was closed in 1991 because of quality problems.

As a result of the 1997 crisis, Hyundai was forced to undertake major restructuring and refocusing of its operations. First, it acquired Kia, one of the smaller Korean automobile firms. Second, in 2000, it separated from its parent company, the Hyundai *chaebol*. Third, it began to move away from being merely a low-price regional producer of cars primarily for the Asian market to one with much wider ambitions. In 2002, the company started construction of a 300,000-car capacity plant in the United States (in Alabama). These ambitious plans are based on Hyundai's continued rapid growth throughout the 1990s, even allowing for the effect of the 1997 crisis. The company has almost 50 per cent of the strongly protected Korean market and has greatly increased its penetration of the North American and European markets. In 1991, Hyundai's overseas sales accounted for 31 per cent of its total sales; in 2001 this had increased to 54 per cent.[27] Locationally, however, Hyundai remains overwhelmingly oriented to Korea, where 96 per cent of its production is concentrated.

In contrast, in 2002 Daewoo was taken over by GM. Daewoo, in fact, originally developed out of a joint venture with General Motors (hence its original name, GM Korea). For a time, it was 50 per cent owned by GM and was quite closely integrated into GM's global operations. In particular, Daewoo manufactured cars sold in North America by GM's Pontiac division. That relationship ended and Daewoo became an especially aggressive entrant into the European market, using innovative methods of direct distribution. Daewoo's strategy was to build

> a global network of car plants in emerging markets, where the growth rate for car sales was expected to boom. Investments poured into Poland, Romania, Uzbekistan, Ukraine, India, and Vietnam as Daewoo sought to produce 2m cars by 2000, including 1m in Korea. The goal was reached but at a heavy cost. None of the foreign plants were profitable. Car sales failed to match projections as many of these markets suffered from the financial crisis of 1997–98 in developing countries.[28]

As a consequence, almost one-quarter of Daewoo's car production was located outside Korea by the end of the 1990s (in Poland and Romania) compared with less than 5 per cent of Hyundai's production. However, Daewoo's overambitious expansion strategy was based on a very high level of external borrowing. The fragile nature of this strategy was exposed by the 1997 financial crisis and the company was declared technically bankrupt in 2000.

Regionalizing production networks in the automobile industry

Although, in one sense, the automobile industry is one of the most globalized of industries it is also an industry in which the *regionalization* of production and distribution is especially marked. Geographically, rather than attempting to organize (and reorganize) operations on a truly global scale, the tendency of most of the leading automobile producers is towards the creation of distinctive production and marketing networks within each of the three major regions of the global triad. But the precise form of such regional networks differs somewhat from region to region, according to specific political and economic circumstances.

Europe[29]

Because of its political history and its degree of political integration, Europe has the most complex regional production network. The establishment of the European Economic Community in 1957, the completion of the single market in 1992 (which removed the remaining technical and physical barriers to the flows of vehicles and components), and its subsequent enlargement to the present 15 member states have had a dramatic influence on the shape of the automobile industry within Europe. Indeed, it was in Europe that the earliest progress was made by the automobile TNCs, especially Ford and General Motors, in creating transnationally integrated production networks. At the same time, the opening up of the Eastern European region, with the collapse of the state-socialist systems in 1989, created a huge contiguous region with the potential of being both a large consumer market and also a low-cost production location for sourcing both components and finished vehicles. The political developments of the late 1980s and the early 1990s presented major strategic opportunities for automobile manufacturers operating in Europe.

As a result of both the more complete integration of the EU itself and the increasing integration with the eastern European economies the European geo-economic space has changed dramatically for the major automobile firms. Figure 11.12 maps the production of the major automobile firms within Europe. Certain countries stand out as major clusters of the production plants of several manufacturers, notably Germany, Spain, the United Kingdom and Belgium.

The actual geographical configuration of automobile production within Europe bears the very strong imprint of each firm's national origins and the history of their development. For example, Ford and GM both have a long history of manufacturing cars in Europe and had built up, over time, a multi-locational, initially nationally oriented, production network. Ford was the first automobile producer to take full advantage of the development of the European Community. In 1967 Ford reorganized its entire European operations previously separately focused on the United Kingdom, West Germany and Belgium into a single organization, Ford Europe. The separate national operations were transformed into a transnationally integrated operation, with each plant performing a specified, often specialist, role within the corporation to achieve economies of scale.

The first real product of Ford's transnationally integrated strategy in Europe was the Fiesta in the 1970s, which involved Ford building a huge assembly plant in Spain for the first time (see Chapter 9). In effect, Ford created a production system that was, in many ways, a microcosm of a globally integrated system on a European scale. But, as we saw earlier, Ford has been in the throes of massive rationalization and reorganization of its entire operations. Within Europe that has involved the closure of five plants, a 20 per cent reduction in capacity, and change in function of some plants. For example, automobile assembly has ceased at Dagenham in the UK and the plant transformed into a centre for the design and production of diesel engines both for Ford and for companies such as the French firm PSA.

In fact, the UK's position within Ford is becoming primarily that of engine and transmission production rather than car assembly (although the Jaguar is

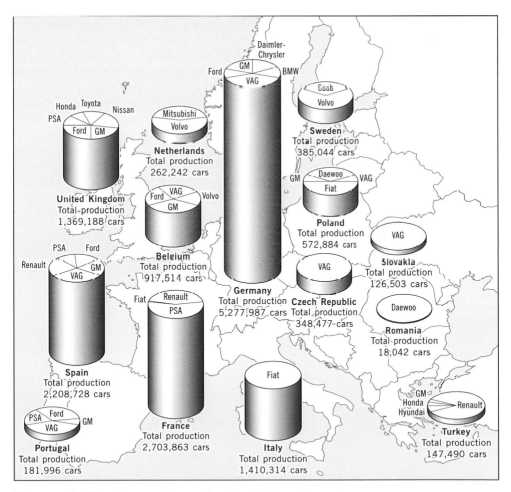

Figure 11.12   Automobile production in Europe

Source: Based on material in SMMT (2001) *World Automotive Statistics*

produced there). Ford's European vehicle assembly operations are becoming increasingly concentrated in Germany and Spain (with the likelihood of Ford's Japanese affiliate, Mazda, building cars at one or other of Ford's European plants). Volvo's Swedish operations are also becoming integrated into Ford's European operations.

GM's European operations have long been based upon two separate national subsidiaries: Vauxhall in the United Kingdom and Opel in Germany. During the 1970s Vauxhall's performance became progressively weaker as the investment emphasis shifted towards Opel. In 1979 GM drastically redrew the production responsibilities of the two European subsidiaries with the UK operations becoming less central to the company's European business. Like Ford, GM also built a major new manufacturing plant in Spain (at Zaragossa). In the 1990s, GM built a major car assembly plant near Katowice in Poland and an engine plant in Hungary.

Again, in light of the major problems facing GM, major reorganization and rationalization is occurring in Europe. In broad terms, the path being followed by GM is very similar to that of Ford, although with some individual differences reflecting the two companies' history and corporate culture. Just as Ford has closed its assembly plant at Dagenham in the UK, GM has announced the closure of its Luton plant. Production of the next generation of Vectra models will now be at a new plant being built at GM's Opel site at Russelsheim, near Frankfurt in Germany. Production is also being drastically cut back at the Antwerp, Belgium and Bochum, Germany plants. As with Ford's treatment of Volvo, GM is now integrating Saab's Swedish operations, having taken 100 per cent ownership.

Whereas the geographical configuration and reconfiguration of the US automobile manufacturers has evolved from a long-established multinational presence in Europe, the position of the Japanese car firms is very different. With no history of European car production and no inherited structure, the Japanese have been able to treat Europe as a 'clean sheet'. Beginning in the early 1980s, Japanese firms established production facilities in Europe. Interestingly, all three of the three leading Japanese firms – Toyota, Nissan, Honda – built their plants in the UK (see Table 11.5). In so doing, they avoided traditional automobile manufacturing

Table 11.5   Japanese automobile plants in Europe

Company	Location	Date established
Nissan	Sunderland, north-east England, UK	1986
Toyota	Burnaston, East Midlands, UK	1992
	Shotton, North Wales, UK	1992
	Valenciennes, France	2000
Honda	Swindon, Wiltshire, UK	1991
	Swindon, Wiltshire, UK	1992
Mitsubishi	Born, Netherlands (with Volvo)	1995

areas, opting instead for greenfield sites. Although the UK is a major market in itself, the Japanese plants in the United Kingdom were specifically oriented towards the European market. This led to political friction within the EU during the 1980s and 1990s.

Nissan was the first Japanese automobile producer to build a major assembly plant in Europe (in north-east England), which began production in 1986. Honda and Toyota followed in 1991 and 1992 respectively. Honda established its assembly and engine facilities at Swindon in southern England while Toyota built a car assembly plant near Derby in the East Midlands and an engine plant in North Wales. Significantly, given both the political friction and the fact that the UK remains outside the eurozone, Toyota's second European plant was built at Valenciennes in northern France, beginning production in 2000. Toyota has also established a joint venture with the French company PSA to develop and assemble small cars for the European market. This will be at an entirely new plant located at Kolin in the Czech Republic.

The geographical configuration of the indigenous European automobile producers is, of course, much more embedded in their national contexts. Figure 11.11 showed that, with the exception of VAG, most of the European producers are overwhelmingly oriented, in production terms, to their home territories. Only VAG has anything approaching a pan-European production network focused around the three nodes of Germany itself, Spain and its acquired Eastern European plants in the Czech Republic and in Slovakia.

Three broad geographical trends are apparent within the European automobile production network during recent years:

- First, there has been a substantial geographical rationalization of core country operations as major firms grapple with problems of over-capacity and outdated physical plant. In the case of the UK, this trend has been offset to a considerable degree by the influx of Japanese firms since the late 1980s that have, in effect, created a new automobile industry there.

- Second, starting in the 1970s and intensifying during the early 1980s, there was substantial development of automobile production in the south-western periphery of Europe, notably in Spain and, to a lesser extent, Portugal, as US, French and German manufacturers established production in lower-cost locations capable of serving the entire European market.

- Third, and most recently, there has been a significant development of automobile production (and, especially, of component production) facilities in Eastern Europe. Here, the major developments have been in the Czech Republic, Slovakia and Hungary[30]. European (notably VAG) and US firms have led this development but there have also been important entries by the Japanese firm Suzuki, in collaboration with GM, and the Korean firm Daewoo.

North America

Although political integration in general in North America is much shallower (and geographically narrower) than in the EU, in the case of the automobile industry political agreements have had profound repercussions on its geographical structure. The first major development in this respect was the *1965 Automobile Pact* between the United States and Canada. The Auto Pact reshaped the industry, producing an integrated structure in which production by the major manufacturers was rationalized and reorganized on a continental, rather than a national, basis.[31] By the early 1970s, in fact, the automobile industry in North America was fully integrated; Canadian plants performed specific functions within the larger continental production and marketing system.

The 1988 Canada–United States Free Trade Agreement (CUSFTA) also contained important provisions for the automobile industry. In particular, it redefined the level of 'North American content' necessary for a firm to be able to claim duty-free movement within the North American market. The North American Free Trade Agreement (NAFTA) had even more far-reaching implications for the automobile industry because of the lower production cost characteristics of the Mexican auto industry and the fact that it was already becoming increasingly integrated into the North American market anyway through the strategies of the American, Japanese and European producers and the changing attitude of the Mexican government.

The NAFTA regulations for the automobile industry include the following important provisions:[32]

- Tariffs on trade in automobiles between the United States and Canada on the one hand and Mexico on the other to be phased out completely by 2004.

- Trade must meet NAFTA rules of origin by the end of the transition period, that is, at least 62.5 per cent of the vehicle's value added must derive from NAFTA countries.

- Until 2004, existing limitations on passenger car imports into Mexico remain in force. This means that only the five foreign producers with existing production facilities in Mexico are allowed to import (at levels linked to their volume of local production).

Figure 11.13 shows the broad geographical structure of the North American production system. Prior to the 1980s, the North American automobile industry was totally dominated by US producers. Attempts by such European firms as Volkswagen and Renault to produce cars successfully in the United States had failed. But since the mid-1980s and through the 1990s, the position has changed dramatically. Three major waves of foreign investment can be identified.

First, each of the major Japanese firms established large-scale production facilities in the United States and Canada during the 1980s (Table 11.6). By 1991 there were more than a dozen Japanese automobile 'transplants' in place in the United States and Canada. As noted earlier, the pioneer was not one of the two largest Japanese producers, Toyota or Nissan, but Honda, which established a manufacturing plant at Marysville in Ohio in 1982. This was soon followed by the Nissan plant at Smyrna, Tennessee in 1983. The leading Japanese manufacturer, Toyota, entered North America as a producer in a very cautious manner: by establishing a 50/50 joint venture with GM to produce cars at GM's Fremont, California, plant. This plant began production in 1984; four years later Toyota began production at wholly owned plants in Georgetown, Kentucky, and Cambridge, Ontario. Honda continued to develop its American operations by opening a further plant in Ohio and one in Ontario. Mazda began production in Michigan in 1987 and Mitsubishi entered a joint venture with Chrysler – named Diamond Star – in Illinois.

Each of the major Japanese firms continued to increase its planned capacity and to make major investments in engine, transmission and components plants. As a result, by the early 1990s the Japanese transplant factories in North America had a production capacity of 2.7 million vehicles and some 25,000 employees. In other words, during the period of less than a decade an entirely new Japanese-controlled automobile industry was created in North America in fierce, direct competition with domestic manufacturers.

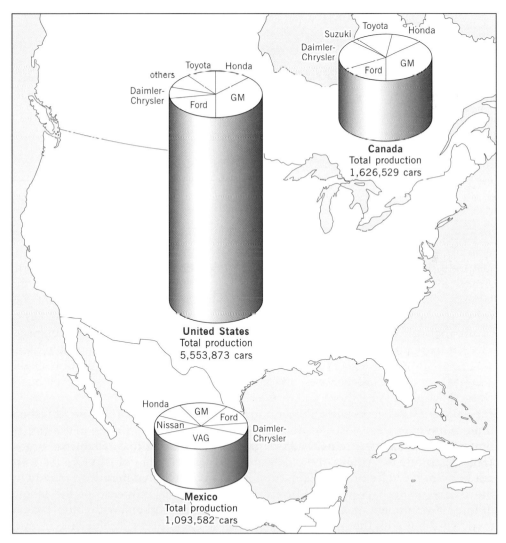

Figure 11.13 Automobile production in the Americas

Source: Based on material in SMMT (2001) *World Automotive Statistics*

Table 11.6 Japanese automobile plants in North America

Company	Location	Date established
Honda	Marysville, Ohio	1982
	East Liberty, Ohio	1989
	Alliston, Ontario	1987
Nissan	Smyrna, Tennessee	1983
	Decherd, Tennessee	1986
Toyota	Fremont, California (NUMMI with GM)	1984
	Georgetown, Kentucky	1988
	Cambridge, Ontario	1988
Mazda	Flat Rock, Michigan	1987
Mitsubishi	Normal, Illinois (with Chrysler)	1988
Subaru/Isuzu	Lafayette, Indiana	1989
Suzuki	Ingersoll, Ontario (CAMI with GM)	1989

As the Japanese plants in the United States progressively increased their North American content they were followed by a wave of Japanese component manufacturers. The locational pattern of both assemblers and suppliers in North America is quite distinctive; perhaps not surprising since, as newcomers, the Japanese have no existing plants or allegiance to specific areas. A 'transplant corridor' can be identified in the form of a 'region stretching from Southern Ontario, south through Michigan, Illinois, Indiana, Ohio, and Kentucky to Tennessee. This "transplant corridor" is organized principally along several interstate highways ... The single exception to the pattern of regional concentration is the NUMMI joint venture between Toyota and GM, which reopened a previously closed GM plant in Fremont, California'.[33]

Suppliers have generally followed the assemblers because of the use of just-in-time delivery systems. At the finer geographical scale the strong preference of Japanese automobile firms has been for greenfield sites near to small towns in rural areas. With few exceptions, the old-established automobile industry centres were not favoured. Choice of greenfield, rural locations has been determined mainly by the desire of the Japanese producers to minimize the influence of the labour unions. Recruitment has been primarily of young workers with little factory experience but with the 'right attitudes' towards the kinds of flexible labour practices employed in the plants. The advantages of being able to start from scratch, on greenfield sites, with newly designed plants and with a hand-picked workforce prepared to accept new working conditions and practices are enormous.

The second, very much smaller, wave of foreign investment in the North American automobile industry began in the mid-1990s in the form of German luxury car manufacturers – Daimler–Benz and BMW. Daimler–Benz built a new plant at Tuscaloosa, Alabama in 1993; BMW built a plant at Spartanburg, South Carolina, in 1994. Subsequently, of course, Daimler–Benz created a major shake-up of the North American automobile industry when it acquired Chrysler in the late 1990s. These incursions by the two up-market German producers were the first major European involvement in North America after the failed ventures of Volkswagen and Renault in the 1970s. As we have seen, in 2002 Hyundai established a major plant in Alabama.

The third, and potentially extremely important, development has been the large-scale investment by foreign automobile producers in Mexico.[34] Prior to the Mexican economic reforms of the 1980s and, later, the NAFTA in 1994, most of the automobile investment in Mexico was market-oriented. Subsequently,

> strategic asset-seeking and cost-reducing FDI replaced the former market-seeking FDI in the Mexican automobile industry … Registered FDI projects in the industry for 1994 alone total[led] $2.5 billion. The NAFTA impact was important here, due to the fact that the NAFTA continued for a further ten years the existing limitation of passenger car imports to the existing five auto producers in Mexico, that was contained in the 1989 Automotive Decree. This advantage and the desire to consolidate their Mexican operations into their North American production facilities is reflected in investment projects by Chrysler, General Motors and Ford worth $1.0 billion in 1994. The NAFTA rules of origin (62.5 per cent North American content) inspired investment projects by the non-US original producers (Nissan and Volkswagen) in the order of $1.2 billion in order to expand and consolidate local supplier networks. Furthermore, in spite of the advantages given to original producers, newcomers (BMW and Honda) … registered investments in the order of $246 million in 1994. These FDI figures indicate elements of Mexico's integration into global or regional production systems of many major auto TNCs.[35]

Other foreign producers continue to be attracted to Mexico as a strategy of gaining entry to the huge North American automobile market. In 2002, Toyota announced plans to build a major plant near Tijuana.

Several of these newer investments in vehicle assembly and major component plants have rather different locational characteristics from the earlier import-substituting investments. The latter, being oriented to the Mexican domestic market, tended to be concentrated in the central region of Mexico, around Mexico City. The newer investments oriented to the North American market as a whole are located primarily in states nearer the border with the United States. The major exception is VW's large integrated facility in Puebla. Rationalization of some of the former core-region plants has become evident. For example, Daimler–Chrysler recently announced the closure of two plants at its Toluca complex west of Mexico City.

By the early 2000s, then, a very different regional production network had evolved in North America compared with the one existing before the 1980s. The earlier network was dominated overwhelmingly by the US manufacturers themselves. From the mid-1960s they had begun to integrate their Canadian and

US operations. The arrival of the Japanese firms in the 1980s tended to consolidate that pattern, but with the addition of creating a new geography of production away from the old-established automobile concentration in the US Midwest. The NAFTA, together with earlier reforms within Mexico, transformed that system by incorporating into the North American regional network a production location with very low costs (and a potentially fast-growing domestic market).

East Asia

The development of distinctive regional automobile production networks in both Europe and North America reflect the combination of two forces: the size and affluence of the market and the political-economic integration of the market through the EU and the NAFTA respectively. In these circumstances, the development of a high level of intra-regional integration of supply, production and distribution becomes possible. The situation in East Asia is rather different. Although the region is regarded as potentially the fastest-growing market for cars over the next few decades, the size and composition of the East Asian market remains limited. In addition, the East Asian automobile market remains primarily a series of individual national markets, some of them very heavily protected against automobile imports. On the other hand, the undoubted potential of the East Asian market, set against the saturation of most Western markets, makes it an absolutely necessary focus for the leading automobile manufacturers. The challenge, as GM sees it, is to 'build a presence' there in order to be well-positioned to take advantage of expected market growth.[36] It is against this background that the current automobile production network in East Asia needs to be set. Figure 11.14 shows the broad geographical structure by major manufacturer.

Not surprisingly, East Asian automobile production and sales are dominated by Japanese firms. Through a network of assembly plants and joint ventures with domestic firms, Japanese cars are assembled in Thailand, Malaysia, the Philippines, Indonesia, Taiwan and China. In several of these countries, Japanese manufacturers totally dominate the automobile market. In Thailand, for example, Japanese firms have a market share of more than 90 per cent; Toyota alone controls almost 30 per cent of the Thai vehicle market. Most of these are assembled locally in individual countries to serve the local market. 'Everywhere in the region, Toyota and other Japanese car makers have, in effect, re-created a whole supply chain in order to serve the local market.[37] This is less out of choice on the part of the Japanese manufacturers than out of the necessity created by high levels of import protection in virtually all the East Asian countries, particularly those in South East Asia (notably Malaysia).

Faced with increasingly difficult circumstances in the Japanese market itself (for example, the problems created by the high value of the yen, the slowdown in demand), Japanese firms have placed increased emphasis on raising their penetration of the Asian market by beginning to develop cars specifically tailored to that market and not just versions of existing models. In the late 1990s, for example, both Toyota and Honda introduced completely new models based upon a very different approach to producing cost-efficient cars for a low-income market:

both Toyota and Honda used a new design technique that reversed the traditional process of designing a car and then reducing costs by squeezing parts suppliers. Attempting to localize production as much as possible, engineers at both Toyota and Honda focused first on what parts local companies could produce cheaply and then designed a car with those components in mind ... What Honda and Toyota came up

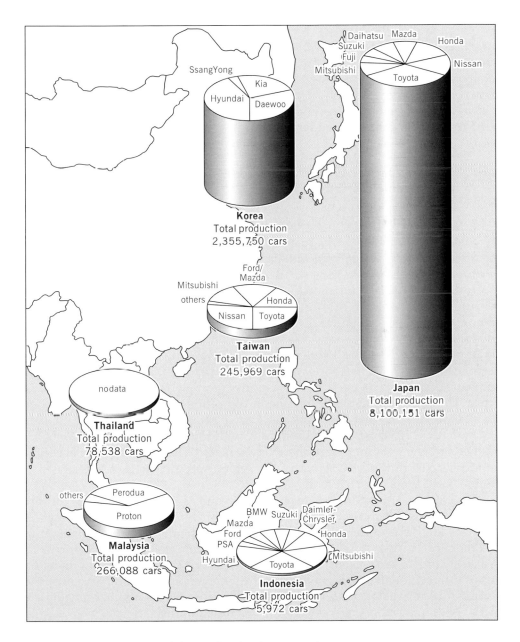

Figure 11.14   Automobile production in East Asia

Source: Based on material in SMMT (2001) *World Automotive Statistics*

with was a car priced several thousands dollars lower but only slightly smaller than the companies' previous bottom-of-the-range passenger car models ... Seventy per cent of the parts in both the Soluna and City come from within south-east Asia, against only about half for the Corolla and Civic.[38]

Outside Japan and Korea, the major automobile production foci in East and South East Asia are Malaysia, Taiwan, Thailand, China and, to a lesser extent, Indonesia and the Philippines. Malaysia has invested very heavily in a national car project: the Proton. Initiated by the government in 1985, the Proton project was based upon a close relationship with Mitsubishi. The Japanese company still retains a stake in Proton, but the Malaysian firm is now growing rapidly in its own right and is now the largest car manufacturer in the ASEAN region. It also has joint-venture assembly plants in the Philippines and Vietnam.[39] However, to protect Proton in its domestic market,

> Malaysia has decided to delay the opening of its car market to 2005, rather than the 2000 agreed by the AFTA [ASEAN Free Trade Area] regional trade grouping. Proton now benefits from preferential treatment over foreign carmakers, who must pay high import tariffs.[40]

Because of its particular 'national champion' approach to its automobile industry, Malaysia is something of a special case. In fact, it is Thailand that is currently regarded as the 'car capital' of South East Asia, with virtually all the major foreign producers having a presence there. Thailand has emerged not only as a major concentration of Japanese automobile and components production but also as the favoured point of entry of Western car manufacturers, notably GM and Ford through their Japanese partners (Isuzu/Suzuki and Mazda respectively). BMW has also recently opened a new assembly plant for its 3-Series model in Thailand. As a consequence of the large number of inward investments, Thailand has become the third largest exporter of automobile products in East Asia, after Japan and Korea; in effect it is the South East Asian export hub for both Japanese and US firms. There is a particular emphasis on the production of pick-up trucks and small, basic cars and on a high concentration of component production. 'Thailand has more than 725 components producers, with roughly 225 supplying the OEM market and the rest catering to the after-market'.[41] Thailand plays an especially important role in the production of the region-specific small cars designed by Honda and Toyota, referred to above.

Whereas Western automobile firms see Thailand as, potentially, a base for serving the whole of South East Asia, their reasons for wishing to establish operations in China are rather different. China is, in itself, potentially the mega-market for automobiles in East Asia. While all the major automobile manufacturers are extremely anxious to establish themselves in China, the Chinese government has imposed specific entry restrictions.[42] The Chinese automobile industry consists of a small number of state corporation groups together with a number of joint ventures between members of these groups and foreign firms.

> By the end of 1997, there were nearly 500 FDI-involved automotive firms in China. Among them, 80 are assembly joint ventures (including specialist vehicle assemblers), 410 are auto parts joint ventures, and 10 are wholly foreign-owned firms ... Of the total vehicle production, nearly half came from FDI-involved assemblers.[43]

VW was one of the earliest Western automobile firms to establish a joint venture in China involving a major assembly plant in Shanghai. The leading Japanese firms Toyota, Nissan and Honda all have joint ventures in China, while the Korean firm Hyundai has gained a foothold in the Chinese market through its acquisition of Kia which had a pre-existing Chinese joint venture. US firms have found entry to China rather more difficult. GM recently established a joint venture with Shanghai Automotive Industry Corporation (SAIC) to produce Buicks at a 'state-of-the-art' plant in Shanghai. Ford took considerably longer and only agreed a joint venture with the Chongqing Chang'an Automotive Group in 2001. A significant aspect of the developing Chinese automobile industry is that 'a joint venture with existing Chinese auto firms is the only available choice for FDI in assembly by MNCs in China. Thus, the location of FDI depends heavily on the locations of existing Chinese firms and the government's approval of these firms' plans for Sino-foreign joint venture projects'.[44]

Conclusion

In summary, the strategies of the major automobile producers are more diverse than is often realized, a fact not unrelated to their national origins. They are also in a condition of flux as major technological and organizational changes sweep through the industry. There is, undoubtedly, a global battle raging in the automobile industry in which there will certainly be more casualties.

The strategy of the leading Japanese companies is to establish a major integrated production system in each of the three global regions: Asia, North America and Western Europe. GM and Ford are moving in the same direction and have the advantage of already operating a sophisticated integrated network within Europe. In comparison, the European manufacturers remain far more limited geographically. Only VAG has much of an internationally integrated system. Its withdrawal from production in the United States has been replaced by its growing involvement in Mexico, where VW is the leading producer, and has preferential access to the North American market. The European picture is being changed by the new developments in Eastern Europe.

However, whilst intense competition will continue in both Europe and North America (where NAFTA is changing the regional production map) it seems likely that the site of the next car wars could well shift to Asia. Although Japanese producers currently have a dominant position, they are being threatened by the Koreans and by other low-cost regional producers, as well as by the intensified efforts being made by GM and Ford to increase their regional market penetration.

Notes

1 *The Financial Times* (9 November 2001).
2 Womack, Jones and Roos (1990) provide a detailed account of the history of technological change in the automobile industry, although their specific approach is based upon their claim that traditional mass production is being replaced by one particular form of production: 'lean' production. Williams et al. (1992) strongly contest their conclusions and predictions.
3 Womack, Jones and Roos (1990).

4 Womack, Jones and Roos (1990: 277).
5 *The Economist* (14 July 2001).
6 *The Economist* (23 February 2002).
7 *The Economist* (23 February 2002).
8 *The Financial Times* (29 March 1996).
9 *The Financial Times* (27 July 2000).
10 *The Financial Times* (10 May 2002).
11 Reich (1989) presents a detailed historical analysis of the evolution of government policy towards automobile producers in France, Germany, Britain and the United States. Sadler (1995) discusses the development of state and EC policy towards the European automobile industry.
12 Reich (1989).
13 Reich (1989).
14 *The Financial Times* (3 December 1998).
15 Kang and Sakai (2000: 24–5). Mockler (2000: Figure 1.5) provides details of the major collaborative arrangements in the automobile industry.
16 *The Financial Times* (28 February 2001).
17 Chairman of Ford, quoted in *The Financial Times* (3 April 1995).
18 *The Financial Times* (16 October 1999).
19 *The Financial Times* (4 January 1999).
20 The various chapters in Hudson and Schamp (1995) analyse the strategies of the major European automobile manufacturers.
21 *The Financial Times* (27 November 2001).
22 *The Financial Times* (19 June 2001).
23 Savary (1995) provides a detailed analysis of Renault.
24 The development of the Korean automobile industry is analysed by Amsden (1989), Kim and Lee (1994), Lee and Cason (1994) and Wade (1990a).
25 Wade (1990b: 310).
26 Wade (1990b: 310).
27 *The Financial Times* (15 November 2001).
28 *The Financial Times* (9 November 2000).
29 Hudson and Schamp (1995) present a detailed survey of automobile production in Europe.
30 See Czaban and Henderson (1998).
31 See Eden and Molot (1993); Holmes (1992)
32 UNCTAD (2000a: 131).
33 Mair, Florida and Kenney (1988: 361).
34 Mortimore (1998: 411–17); UNCTAD (2000a: 130–1).
35 Mortimore (1998: 417).
36 *The Financial Times* (17 January 2000).
37 *The Economist* (24 June 2000).
38 *The Financial Times* (11 February 1997).
39 UNCTAD (2000a: 164).
40 *The Financial Times* (11 October 2000).
41 UNCTAD (2000a: 160).
42 Sit and Liu (2000) provide a detailed account of the development of the automobile industry in China since the reforms of 1978.
43 Sit and Liu (2000: 664).
44 Sit and Liu (2000: 665).

CHAPTER 12

'Chips With Everything': The Semiconductor Industry

The microelectronics industry is, without question, today's 'industry of industries'. Its core, the semiconductor, has emerged as the dominant influence of the past four decades, extending its transformative effects into all branches of the economy and into many aspects of society at large. It is, without doubt, 'the engine of the digital age'.

> Virtually no part of modern life has been left untouched by the semiconductor revolution. Indeed, it is estimated that on an average day, a typical American interacts with more than 300 micro-controllers. Semiconductors have revolutionized the home, the office, the transport network, medical science and more ... Common everyday home appliances are becoming 'smart'. Semiconductors are invading every aspect of our lives.[1]

The first step in creating a semiconductor industry was the development of the *transistor* in the United States in 1947. The transistor replaced the thermionic valve (vacuum tube) and made possible the development of a *micro*-electronics industry. Within a few years of the initial development of the transistor, semiconductors were being manufactured commercially in the United States. The end of the 1950s saw the emergence of a second major innovation, the *integrated circuit* – a number of transistors connected together on a single piece or 'chip' of silicon. By the early 1970s it had become possible to incorporate a number of very sophisticated solid-state circuits onto a single chip the size of a fingernail to create a *microprocessor* able to perform the functions which, only two decades earlier, had taken a whole roomful of valve computers.

The progressive refinement of these basic innovations over a very short period dramatically increased the power of electronic components and also spectacularly decreased their size. It is now possible to pack millions of individual circuits onto a single chip of silicon less than one centimetre square. Increased *miniaturization* has been a fundamental development, for it permits the incorporation of electronic components into a vast range of products, from pocket electronic products to highly complex computers, industrial robots and aircraft guidance systems. In fact, it is this *increasingly pervasive application* of semiconductors which makes the industry so very significant. Its extensive ramifications, not only for other sectors of the economy but also for telecommunications and national defence, have made all governments increasingly anxious to avoid being left out of, or left behind in, what is a rapidly moving technological scene.

The semiconductor industry was the first to which the label 'global factory' could be applied because of its early use of offshore assembly. It was in the semiconductor industry that a *spatial hierarchy of production* at the global scale first became apparent, with clear geographical separation between different stages of the production process. Initially, it was the *assembly* stages that were relocated to certain developing countries. In general, the higher-level design, R&D and more complex stages of production tended either to remain in the firm's home country or to be established in other developed countries where the necessary labour skills and physical infrastructure were more readily available. However, this pattern is changing.

Semiconductors within the electronics production chain

Figure 12.1 shows where semiconductor production fits into the electronics production chain as a whole. Semiconductors themselves can be divided into two major categories:

- *memory chips* (DRAMs) which contain pre-programmed information
- *microprocessors* which are, in effect, 'computers on a chip'.

Semiconductor manufacture is a highly capital-intensive industry in which very large transnational firms tend to dominate. However, it is an industry in which, as we shall see, some parts of the production chain can be geographically separated from the other stages in the sequence.

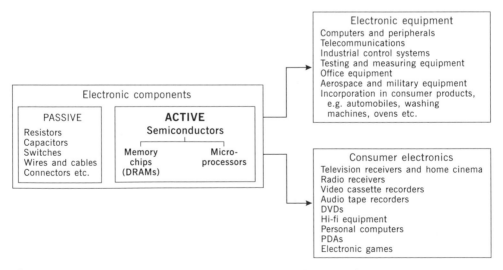

Figure 12.1 The position of semiconductors within the electronics production chain

Global shifts in the semiconductor industry

The growth of production of semiconductors has been truly phenomenal since their commercial introduction in the 1950s. Output virtually doubled every year throughout the 1970s, and, as Figure 12.2 shows, this vertiginous growth rate continued right through to the mid-1990s and then fluctuated with another major peak in 2000. Since 1980, in fact, there has been 'a twenty-fold expansion … making semiconductors by far the fastest growing industry in the world'.[2]

Figure 12.2

Growth in global semiconductor sales, 1980–2000

Source: Based on material in *The Financial Times,* 3 September 2001

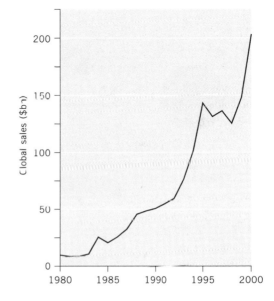

Figure 12.3 maps the geographical distribution of production of semiconductors on the global scale. It shows the extent to which the industry is dominated by the United States and East Asia, with Europe occupying a greatly inferior position (compare this pattern with that of automobiles in Figure 11.3, where the three global regions are far more equal in significance). Geographically, the commercial production of semiconductors began in the United States during the 1950s and for nearly two decades the United States dominated world production. During the 1980s, however, Japan overtook the United States to become the world's leading producer, creating, as we shall see later, a major tension between the United States and Japan.

In 1980, the United States produced just over 60 per cent of world semiconductors; by 1989 its share had fallen precipitously to 37 per cent. Conversely, by 1989, Japan was responsible for more than half the world total, having overtaken the United States in the mid-1980s. But, as Figure 12.4 shows, the United States staged a remarkable recovery. By the early 1990s it was back in the lead, while Japan's share fell very considerably. At the same time – and very much a contributory factor in the relative decline in Japan's share – some other countries had emerged as major producers in their own right. The most important of these, by far, were

South Korea and Taiwan (in Figure 12.4 these form the major part of the 'rest of the world' category). Meanwhile, Europe's position as a semiconductor producer remained more or less stagnant.

However, in interpreting these trends in the geography of semiconductor production we need to take into account changes in the product composition of the industry – especially the distinction between memory chips (DRAMs) and more sophisticated microprocessors. The basis of the United States' recovery as a

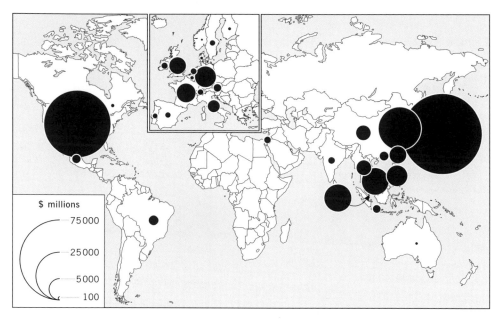

Figure 12.3 World production of active electronic components

Source: Based on data in *Yearbook of Electronic Data, 1996.* New York: Elsevier Advanced Technology

Figure 12.4

Changing shares of world semiconductor production

Source: Based on Macher, Mowery and Hodges, 1998: Figure 1

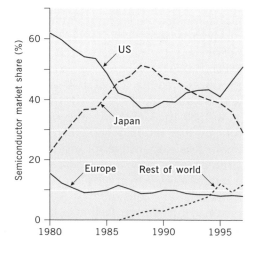

major semiconductor producer was its shift out of memory chips and into more design-intensive devices.[3] Conversely, Japanese producers made far less of a product shift than US producers. Rather, they concentrated on attempting to achieve technological leadership in 'next generation' DRAMs. The problem with this strategy is that DRAMs have essentially become 'commodities'. In such products, low production costs become especially important and this is where South Korea and Taiwan were first able to enter the semiconductor market.[4] In fact, South Korea and Taiwan emerged as significant semiconductor producers in the late 1980s/early 1990s. By 1994, they had overtaken Europe in output terms.[5]

South Korea produced less than one per cent of world semiconductor output in the mid-1980s; by the late 1990s it was the world's third largest semiconductor producer, after the United States and Japan, with more than 10 per cent of the world total.

> The Korean semiconductor industry has grown with an overwhelming focus on exports. In 1995, exports accounted for 91 per cent of production, making semiconductors the largest item in Korea's exports. Export of chips as a proportion of total output have actually been increasing, from 81 per cent of production in 1991, to 87 per cent in 1993, and 90 per cent in 1994. The 1995 proportion of 91 per cent was the peak, as world competition in DRAMs increased in the latter half of the 1990s and into the twenty-first century.[6]

Taiwan's growth as a major semiconductor producer has been similarly impressive. By the mid-1990s, Taiwan had grown to become the fourth largest semiconductor producer in the world, overtaking Germany. Yet despite the incredible growth of both South Korea and Taiwan their position needs to be kept in perspective. In the late 1990s, South Korea's semiconductor industry was producing $12 billion of output and Taiwan's was producing $8.5 billion. In comparison, the value of Japan's semiconductor output stood at $42 billion.[7] Elsewhere in East Asia, other significant centres of semiconductor production have emerged, although none is anywhere near as important as either South Korea or Taiwan. The most significant so far are Singapore, China and Malaysia, with far more modest production in the Philippines and Thailand.

The dynamics of the market

Unlike automobiles or garments, demand for semiconductors is a derived demand: it depends upon the growth of demand for products and processes in which they are incorporated. In the United States, at least initially, most of the stimulus came from the defence–aerospace sector and this explains much of the initial growth of the semiconductor industry in the United States. In Europe, and especially in Japan, defence-related demand was far less important than industrial and, particularly, consumer electronics applications. Nevertheless, governments have often been dominant customers, either directly or indirectly, throughout the semiconductor industry's short history.

As Figure 12.2 shows, global semiconductor sales in 2000 were more than $200 billion, some twenty times greater than in 1980. On average, the global market for semiconductors has been tripling every five years. Until recently, by far the biggest end-user was the computer industry and, especially, the personal computer industry. By the mid-1990s, around 60 per cent of semiconductor revenues were derived from sales to computer manufacturers. Indeed, the two industries – PCs and semiconductors – had become symbiotically locked together. On the one hand, demand by PC manufacturers has driven much of the demand for semiconductors whilst, on the other hand, the leading semiconductor manufacturers (notably Intel and AMD) created demand for more powerful PCs by introducing new generations of increasingly powerful chips. Such a cumulative process has been reinforced by the dominant position of Microsoft, the world's leading software company, whose successive new generations of software have required increasingly powerful microprocessors.

Recently, however, the market for semiconductors has begun to broaden significantly beyond the PC sector. By the turn of the millennium, the computer sector was absorbing only 40 per cent of semiconductor production whilst other markets, primarily communications equipment (including Internet infrastructure, mobile telephones) and sophisticated digital consumer electronic products (such as digital cameras, digital music players, handheld computers) had become increasingly important.

A major reason for the spectacular growth in demand for semiconductors has been the vertiginous decline in their selling price. Partly because of technological developments (discussed in the next section) and partly because of fierce price competition between producers, the price of semiconductors is a tiny fraction of that prevailing only a short time ago. Between the mid-1960s and the late 1970s, for example, 'the cost per electronic function … [fell] … by a factor of 100,000 in less than two decades'.[8] This process of continuously falling prices per unit of capacity has shown no sign of abating. Indeed, higher-capacity DRAMs now appear so quickly that each affects the price of its predecessors. In the early 1990s, for example, the price per megabyte of a DRAM was $25; by the late 1990s, the price per megabyte was down to a little over one dollar.[9]

The semiconductor industry is subject to spectacular fluctuations in demand; gluts and shortages are endemic. Supply gluts further intensify price competition; at other times, severe shortages of chips have created major problems and raised prices for end-users. For example, in the mid-1990s it was predicted that the slump in demand for semiconductors would last up to two years before the next generation of memory chips appeared to boost demand. As a result, virtually all the major semiconductor manufacturers delayed their plans to build massive new production capacity (plans which had been announced only a short time earlier) until the upturn occurred. The upturn did, indeed, occur. Headlines such as 'chip industry struggles to feed clients' appetites' and 'not enough chips for everything'[10] became commonplace, with the inevitable result that the major manufacturers began a massive investment cycle once again. In mid-2000, the semiconductor industry launched another massive wave of capital investments. Unfortunately, the demand never materialized. On the contrary, demand for semiconductors collapsed as the Y2K effect passed, as the dot.com industry haemorrhaged and as the global economy entered recession. Global PC sales fell for the first time in 15 years while the seemingly unstoppable growth in the IT industries in general slowed dramatically.

The basic problem for the semiconductor industry results from a combination of factors: the massive fixed costs of production facilities (see next section), the fact that they take at least a year to build and the fact that demand for chips can change rapidly. The situation is made more complex by the strongly *segmented* character of demand for semiconductors. There are seven major types of semiconductor, each of which has rather different patterns of future demand during the early 21st century:[11]

- *DRAMs* (standard memory devices) – predicted annual growth approximately 35 per cent
- *microprocessors* – predicted annual growth approximately 10 per cent
- *analog devices* – predicted annual growth approximately 30 per cent
- *flash memory devices* – predicted annual growth approximately 100 per cent
- *microcontrollers* – predicted annual growth approximately 25 per cent
- *logic chips* – predicted annual growth approximately 40 per cent
- *optoelectronics* – predicted annual growth approximately 50 per cent.

Geographically, the demand for semiconductors is overwhelmingly concentrated in the three major regions of the United States, Western Europe and East Asia. But, as in so many aspects of the global economy, the fastest potential growth is expected to be in China – possibly 20 per cent per year for the foreseeable future. Much of this projected growth is likely to be in the expansion of the mobile phone market, which is predicted to increase from 68 million to 250 million handsets by 2004, and in the fast-expanding local consumer electronics and related industries (for example, cars, smart toys).[12]

Production costs, technology and the changing organization of semiconductor production[13]

In the fiercely competitive environment of the global semiconductor industry the drive to push down production costs is a paramount consideration. The industry has become increasingly capital-intensive while retaining a considerable degree of labour intensity in certain parts of its production chain. The urge to minimize production costs is reflected both in the nature and the rapid rate of technological change and in the changing geography of production at various spatial scales.

Basic parameters of semiconductor production

Two considerations are especially important in the manufacture of semiconductors:

- *The reliability of the product*. A great deal of effort and expenditure is devoted to improving the 'yield' of the manufacturing process and reducing the number of rejected circuits. One important element here is the cleanliness of the environment in which wafer fabrication occurs. The very high cost of creating 'pure' environments is one of the reasons for the very high capital costs of semiconductor fabrication plants. One approach is to create 'mini-environments' which are, in effect, sealed containers to ensure that as the wafer travels through the factory from one processing step to another during its production cycle, it never comes into contact with the 'dirty' factory environment.

● *The need to pack as many circuits as possible onto a single chip.* One of the most remarkable features of the semiconductor industry is the fact that the number of components per chip has increased phenomenally. By the mid-1990s, a microprocessor could contain as many as 5.5 million individual transistors while the Pentium 4 of the late 1990s contained 42 million transistors. Compare these chip capacities with the mere 3,500 on Intel's first commercial microprocessor produced in 1971. In 2000, Intel predicted it would produce a microprocessor containing 400 million transistors within four years. In fact, the prediction by Charles Moore (co-founder of Intel) in 1965 that the number of transistors per chip would double every 18 months has been borne out (hence the description 'Moore's Law'). This 'law' is shown on the left-hand axis of Figure 12.5.

It is the combination of these two requirements – to increase both the yield and the number of circuits per chip – that accounts for the steep increase in the cost of setting up a new semiconductor plant (see the right-hand axis of Figure 12.5).

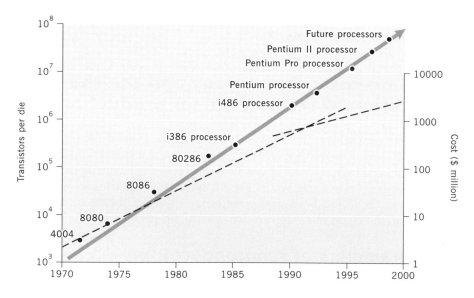

Figure 12.5 Exponential increase in the capacity of semiconductors and the cost of new manufacturing plants

Source: Based on material in *The Financial Times*, 4 February 1998

Every major increase in the number of components per chip demands far more sophisticated and expensive manufacturing equipment which, in turn, becomes obsolete more quickly. At the same time, the trend is towards larger, more capital-intensive and research-intensive production. The pace of technological change has been so rapid and far-reaching in semiconductor production that it is now necessary for production equipment to be replaced or updated every three to five years.

The cost of establishing a semiconductor plant has escalated accordingly: each new generation of semiconductors involves a doubling of plant construction costs. In the 1960s such a plant could be set up for roughly $2 million; by the early 1970s between $15 million and $20 million was needed; ten years later, in the early 1980s, an investment of between $50 million and $75 million was necessary. By the late 1980s a new chip production facility was costing $150 million. In the mid-1990s it was approaching $1 billion; today between $2 and $3 billion is needed to create a state-of-the-art chip plant. Even allowing for inflation it has obviously become a much more costly operation in which the *capital barriers to entry* have increased very substantially. Not surprisingly, therefore, the industry is one of the most capital- and research-intensive of all manufacturing industries.

The semiconductor production sequence

To understand the changing global structure of the semiconductor industry we need to look more closely at the production process itself. We also need to take into account the different types of semiconductor being produced, because the potential for mass production varies greatly between the standard device and the custom device. Production of standard devices demands large-scale mass production; manufacture of custom chips demands much smaller runs. However, the development of semi-custom and application-specific (ASIC) devices has extended the use of mass production techniques into new areas.

The general sequence of semiconductor production is shown in Figure 12.6. The differing characteristics of each stage have very important implications for the geographical organization of the industry at a global scale.

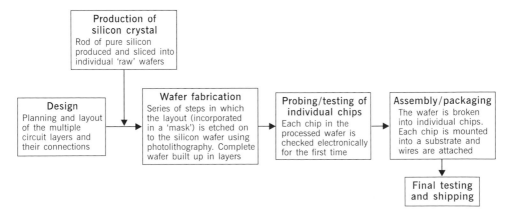

Figure 12.6 Stages in semiconductor production

- The process begins with the *design of a new circuit*. Its precise form will obviously depend upon the function it is to perform. Production of complex circuits involves the superimposition of a series of separate layers, each one being produced initially as a pattern or mask, from which the actual circuits will eventually be made.

- The production of the *silicon* from which the chip wafers will be made is a process whereby the silicon crystal is drawn out and formed into a cylindrical rod. Generally, this process is performed by specialist firms, although a few of the very large semiconductor manufacturers produce their own silicon. The silicon rods are then sliced into individual wafers 0.5 mm thick.

- The *wafer fabrication* stage consists of a number of intricate and highly precise processes in which the circuits are etched on to the wafers, layer by layer, using the masks in a photolithographic-chemical process. Each wafer contains large numbers of identical circuits.

- Each chip on the wafer is then *tested* electronically.

- The individual wafers are broken down into separate chips, *assembled* into the final integrated circuit or microprocessor using a bonding/wiring process and individually *packed*.

- They are then subjected to *final testing* and shipped to the customer for use in the final product.

A particularly important distinction exists between the design and wafer fabrication stage on the one hand and the assembly stage on the other. Each has different production characteristics and neither needs to be located in close geographical proximity to the other. Design and fabrication require high-level scientific, technical and engineering personnel while the fabrication stage itself requires an extremely pure production environment and the availability of suitable utilities (pure water supplies, waste disposal facilities for noxious chemical wastes). In contrast, the assembly of semiconductors is carried out using low-skill labour, notably female, and although there is still a need for a 'clean' production environment this is less critical than for wafer fabrication. The low-weight/high-value characteristics of semiconductors permit their transportation over virtually any geographical distance.

Hence, it is the assembly stage of the production sequence which has been most susceptible to relocation to low-labour-cost areas of the world. Overall, therefore, the manufacture of semiconductors is a highly capital- and research-intensive industry but one in which there are distinct 'breaks' in the production sequence. Such breaks are used by producers, as we shall see, to create a complex global geography of production.

Characteristics of the labour force in semiconductor production

As a consequence of the particular configuration of the production process, there is a clear *polarization* of skills in the semiconductor industry between highly trained professional and technical workers on the one hand and low-skilled production workers on the other. Increased automation has led to a much steeper relative decline in the number of production workers employed, particularly the less skilled. But there is also a clear *geographical* pattern of skills in the semi-conductor industry. Overall, professional and technical occupations are far more important in the developed countries than in the developing countries, a reflec-tion of the fact that it is mainly the more routine assembly tasks which have been located in developing countries while the more sophisticated design, R&D and wafer fabrication stages remain mostly in developed countries. Hence most, though by no means all, of the semiconductor employment in Silicon Valley and the other production clusters in the United States and Japan is in the higher-skill categories; conversely, virtually all the semiconductor employment in the EPZs of the East Asian countries and in Mexico is of production workers. However, the increasingly complex intra-regional division of labour within the semiconductor industry in East Asia is removing some of this simplicity, as we shall see later in this chapter.

In the European electronics clusters there is also a degree of spatial segmen-tation of labour skills, though in a less extreme form. For example, the Scottish electronics plants include both assembly and fabrication activities and, there-fore, have a more even mix of skills than the developing country plants. Even so, the relative absence of higher-level design and of R&D activities in the Scottish branch plants creates a relative shortage of job opportunities for particular segments of the labour force. In contrast, the electronics cluster in South East England is predominantly one of high-level administrative, design and R&D units within large firms together with highly innovative and specialized small electronics firms.

A further fundamental aspect of employment in the semiconductor industry is the predominance of female workers in the assembly stages.[14] This is apparent throughout the world and is not confined to the developing countries alone. An overwhelming majority of semiconductor assembly workers are young females on relatively low wages. In this respect, there are clear parallels with the situation in the textiles and garment industries. Female workers are the norm in the assem-bly processes wherever the plants are located. Although labour regulations are generally more stringent in developed countries, some undesirable characteristics may still apply.

The gleaming plate glass- and metal-clad plants of the modern semiconductor industry look, at first sight, to be the complete opposite of the dark, satanic mills of the 19th century and the squalid sweatshops of the garment industry. In some senses, of course, the contrast is undoubtedly what it seems. For many of the workers employed in the semiconductor industry wages are high, the work is stimulating and conditions are superb. But for others the story is rather different. At worst, conditions are just as bad, with long working hours, an unpleasant and noxious working environment and little or no job security. It is indeed paradoxi-cal that an industry which epitomizes all that is new and up-to-date at the same time harbours some of the oldest and least desirable attributes of work in manu-facturing industry.

New developments in the semiconductor industry

Developments in the technology of semiconductor production – new lithographic techniques, new methods of wafer fabrication, and the introduction of automation at all stages of the process – have created substantial changes in the organization and structure of the industry. In particular, the relative importance of labour costs has been greatly reduced, especially in the higher-value products. At the same time, there have been increased pressures to enlarge the scale of production plants. A wafer fabrication plant, in order to be profitable, needs to produce at very high volumes.

Apart from continuing attempts to introduce new materials to supplement or to replace silicon, the most important current developments in the technology of the semiconductor are related to the shift from 200 mm to 300 mm diameter wafers. This 1.5-fold increase in the diameter of a wafer will produce 2.6 times the number of identically sized chips on a single wafer.[15] If the size of each chip is also reduced, say from 0.20 microns to 0.13 microns, then the number of chips per wafer can be increased further. At the same time, such smaller chips are more efficient because the electrons have a smaller distance to travel. On the other hand, such smaller chip geometries increase the risk of 'interference' between the circuits. However, the transition to 300 mm wafers and smaller chip sizes is both difficult and expensive. It also depends upon the availability of new-generation scanners – the equipment that forms the images of the chips that are then etched onto the silicon.[16] But whatever the precise nature of these new technological developments they serve to reinforce the knowledge- and capital-intensity of the semiconductor industry and the problems of generating sufficient investment capital and scientific and technological expertise to keep up with the ever-quickening game.

The role of the state

It is easy to appreciate why governments have attempted to intervene in the development of the semiconductor industry. Semiconductor production is recognized as a key technology with enormous ramifications throughout the economy. If a country is to benefit fully from it – or if it is to avoid being left behind – it must have access to what is an expensive and rapidly changing technology.

This access may be achieved in several ways:

- by building an indigenous production capacity based upon domestically owned firms
- by attracting foreign semiconductor firms to establish production units
- by purchasing semiconductors on the open market and concentrating on developing the end-uses.

Problems are inherent in all three options. Setting up a viable domestic industry may be beyond the means of many countries. On the other hand, relying on foreign investment or the open market may lead to problems of dependency of supplies on foreign sources. Such potential vulnerability may be important not only for industrial applications but especially for defence. Semiconductor technology is at the heart of all modern defence systems, hence the sensitivity of most national governments to developments in this industry. The particular policies pursued by governments reflect their specific national circumstances, including the country's relative position in the global semiconductor industry.

US government policy towards the semiconductor industry

As we saw in Chapter 6, the United States does not have a formal industry policy. But this does not mean that the role of the federal government is unimportant. On the contrary, its role – albeit indirect in many cases – may be immense, as the case of the semiconductor industry shows. In the United States, the country in which the semiconductor industry originated, the dominant formative influences were the federal defence and aerospace sectors. In the early days of the semiconductor industry these set the direction and nature of the industry's development because they were its dominant customers. Although defence-related forces are now less important to the US semiconductor industry, their influence persists.

In the 1980s, as the United States' lead in semiconductor production began to be eroded, there was much broader criticism of the direction (even the existence) of US policy towards the semiconductor industry. The sharp deterioration in the country's global position as a semiconductor producer – seen as a threat to national security, both economic and military – led to strong political and industry lobbying for targeted policies. Two specific US policies have been significant in the semiconductor industry: *trade policy* and the facilitation and encouragement of *industry consortia*.[17]

In the 1980s, as we have seen, the major competitive threat to the dominant US semiconductor industry came from Japan. Japanese semiconductor exports were decimating not only the domestic market of US producers but also their export markets as well. The prevailing view in the United States was that Japan was engaging in 'unfair' trade practices; in particular, that Japanese producers were 'dumping' chips at excessively low prices in the United States (and, therefore, undercutting US producers) and also that the Japanese were unfairly restricting access to their domestic market. In order to combat this threat, the US government persuaded the Japanese government to sign the US–Japan Semiconductor Trade Agreement in 1986. The initial pact expired in 1991 and was renegotiated for a further five years with a particular US emphasis on access for American semiconductor manufactures to the Japanese market. On the expiry of the semiconductor pact in 1996, the US and Japan reached agreement to set up two new international industry bodies. The effects of the semiconductor pact were limited, although not insignificant for the US semiconductor industry.[18]

The second important aspect of US federal government policy towards the semiconductor industry has been its role in encouraging and facilitating interfirm collaboration through industry consortia. In 1987, SEMATECH was founded by the 14 major US semiconductor manufacturers with funding from the firms themselves and, initially, from the federal government. The focus was to be on medium-term research. The fact that much of SEMATECH's funding was defence-related, together with the political priorities underlying its establishment, meant that non-US firms were excluded from participating. 'Yet many non-US firms have been able to benefit from SEMATECH-supported R&D, either through their purchase from US equipment vendors of new products that have been improved in SEMATECH programs or through their collaborative relationships with US semiconductor manufacturing firms.'[19] More recently, the SEMATECH consortium launched a new collaborative initiative – the International 300 mm Initiative (I300I) – involving both US and non-Japanese foreign firms (European, South Korean and Taiwanese).[20]

The central role of the state in the development of the semiconductor industry in East Asia

Whereas the role of the state in the development of the semiconductor industry in the United States has been important, it has not been, in a direct sense, central to that development. In contrast, in the case of those East Asian countries that have emerged as major players in the semiconductor industry there is no doubt of the absolutely central role of the state. However, because each state has developed its semiconductor industry in rather different ways we need to look briefly at each case in turn. [21]

Japan

Recall from our discussion of the general nature of Japanese industrial policy in Chapter 6 that micro-electronics, and the related computer and information technologies, formed the central focus of the drive to develop knowledge-intensive industries. Semiconductors, in particular, were seen by the Japanese as the 'rice of industry'. Memory chips (DRAMs), being especially suitable for standardization and commodification, were chosen for targeted government policy. The Japanese government paid 'painstaking attention to the "industrial ecology" of the new sector, particularly the support and supply industries providing specialist materials and equipment'.[22] The initial aim of Japanese semiconductor policy was to avert technological dominance by the United States by discouraging direct foreign participation in the Japanese semiconductor industry and acquiring foreign technology through other means. Protection of the industry was extremely tight until the end of the 1970s, both through import controls and restrictions on inward investment.

A major Japanese government initiative was the Very Large Scale Integration (VLSI) Project, begun in 1976 on the initiative of MITI. This involved research collaboration between the five major Japanese companies – NEC, Fujitsu, Hitachi, Toshiba, Mitsubishi – to develop highly sophisticated integrated circuits for the next generation of computers. The Japanese government funded some 40 per cent of the total cost between 1976 and 1979. Indeed, it was the example of the Japanese VLSI Project that helped fuel the debate within the United States and eventually led to the creation of SEMATECH (see above).

Japanese direct financial support of the VLSI Project ceased after 1979 but the more general context of supportive policies continued, in line with overall Japanese industrial policy (see Figure 6.4). In 1996, a new initiative, Selete – encouraged, though not directly driven, by the government – was established. Like the I300I programme in the United States, Selete was a response to the problems of moving to the 300 mm wafer standard.[23] It is worth emphasizing, however, that Selete consists only of Japanese firms, unlike SEMATECH which involves both US and non-Japanese foreign firms.

South Korea

Whereas Japan's objective in stimulating its semiconductor industry was to emulate the United States, the objective of South Korea's state initiatives was to emulate both. The Korean government used a variety of devices – including its control of the telecommunications industry – to foster the entry of large Korean firms into the semiconductor industry. Indeed, the government explicitly used the major *chaebols* as the vehicles for developing a national semiconductor industry in a more or less direct imitation of the Japanese system.

Table 12.1 shows the developmental stages through which the South Korean semiconductor industry has evolved. In the first, 'preparation' stage, an important catalytic influence was the establishment of semiconductor assembly and test operations in South Korea by some of the leading US companies during the 1960s.

Table 12.1 Stages in the evolution of the South Korean semiconductor industry

Stage I Pre-1974 Preparation	Stage II 1974–81 Seeding/implantation	Stage III 1982–88 Propagation	Stage IV 1989–98 Roots of sustainability
Assembly and test operations established by foreign companies	Indigenous industry established and beginnings of IC fabrication in Korea	Incursion into mass memory chip production	Fully-fledged memory chip production and diversification/ consolidation of a semiconductor industry
KIST established	KIET leads technology leverage	VLSI take-off	Ancillary industries established
Expansion of technical education		Joint development of 4M DRAM	Collaborative R&D programmes
			Science and technology strategy as a political priority

Source: Mathews and Cho, 2000: Table 3.1

By 1974, there were nine such US-owned facilities in South Korea.[24] Domestically, major investments were made in technical education based on the Korea Institute of Science and Technology (KIST), set up in 1966. A key element in the second stage was the government's establishment of the Korea Institute of Electronics Technology (KIET) in 1976. Its responsibility was 'planning and coordinating semiconductor R&D, importing, assimilating, and disseminating foreign technologies, providing technical assistance to Korean firms, and undertaking market research'.[25] KIET played a key role in establishing an indigenous semiconductor industry in South Korea during the second half of the 1970s.

It was during the 1980s, in particular, that South Korea began to emerge as a major semiconductor producer. Again, the government's role was absolutely central. In 1982, the Long-Term Semiconductor Industry Promotion Plan was announced whereby substantial fiscal and financial incentives were made available to the four leading semiconductor *chaebols*. Within this supportive framework,

> the government … designated R&D on the 4M DRAM as a national project in October 1986. The Electronics and Telecommunications Research Institute (ETRI), a government research institute (GRI) served as a coordinator in the consortium of three *chaebol* semiconductor makers – Samsung, LG, and Hyundai – along with six universities. The objective was to develop and mass-produce the 4M DRAM by 1989 and completely close the technology gap with Japanese firms. The consortium spent $110 million for R&D over three years (1986–1989); the government shared 57 per cent of the total R&D expenditure, a disproportionately large share in comparison to other national projects.[26]

As we saw earlier, South Korea has emerged as the third largest producer of semiconductors in the world. Without doubt, the state played a critical role in making this possible.

Taiwan

Both the Japanese and South Korean semiconductor industries evolved through specific forms of state–private sector collaboration. In both cases, the result was the creation of large semiconductor operations within massive electronics groups. The Taiwanese semiconductor industry provides another variant on the manner of state involvement and private enterprise.

> Taiwan's semiconductor industry is a flourishing market-driven industry. It has no 'nationalized' firms within it: all are privately owned and managed. It is strongly export-oriented. It has never imposed any protective tariffs on its semiconductor products. There are no government 'handouts' to any of the firms involved in the industry for any of their current activities. And yet *its creation, its nurturing and its guidance have been entirely the product of government and public sector institutions.*[27]

Table 12.2 summarizes the developmental stages of the Taiwanese semiconductor industry.

As in South Korea, the initial development of semiconductor production in Taiwan derived from the decisions of US firms to locate assembly operations in low-cost East Asian locations in the 1960s. The Taiwanese government immediately capitalized on this trend by building the world's first export processing zone

(EPZ) for semiconductors in 1965. In the early 1970s, the Industrial Technology Research Institute (ITRI) was set up by the government to promote technological leverage. ITRI was a specialist semiconductor and electronics laboratory which eventually evolved into the Electronics Research Service Organization (ERSO).

> The government charged ERSO with the task of 'seeding' a semiconductor industry, through technology transfer ... ERSO was successful in signing a technology transfer agreement with ... [the US electronics firm] ... RCA. This firm agreed to transfer its obsolete 7-micron technology for a royalty charge and to train a group of up to 40 engineers in the design and fabrication of chips ... The group of young engineers ... returned to ITRI/ERSO, and with public funds provided under the Electronic Industries Development Program they were able to put together a pilot IC fabrication

Table 12.2   Stages in the evolution of the Taiwanese semiconductor industry

Stage I Pre-1976 Preparation	Stage II 1976–79 Seeding	Stage III 1980–88 Technology absorption and propagation	Stage IV 1989–98 Sustainability
Labour-intensive semiconductor back-end operations (assembly) and testing	Licensing of IC fabrication technology, and its adoption by public sector R&D institute	Technology absorption and enterprise diffusion	Entry of firms to cover all phases of semiconductor manufacturing and full product range, including DRAMs
Dominated by foreign multinationals Establishment of ITRI and ERSO	Phase I of Electronics Industry Development Program	Establishment of secure infrastructure in the form of the Hsinchu Science-based Industry Park	From VLSI to ULSI technology
		ERSO acquires skills covering all phases of semiconductor manufacturing, moving from LSI to VLSI	Submicron stage of public-sector-led R&D
		Spin-off of private companies and entry of private sector	Cooperative R&D system of innovation established

Source: Mathews and Cho, 2000: Table 4.1

plant that was soon turning out commercial-grade chips. This was 1977 and it marks the point where an IC industry in Taiwan could be said to have started – even though at this point there were no firms directly involved.[28]

Unlike Korea at a similar stage, there were no Taiwanese firms willing to enter such a capital-intensive and risky industry as semiconductors. The solution adopted by the Taiwanese government was to launch a publicly financed firm, UMC (United Microelectronics Corporation), through ITRI/ERSO. At the same time, the Hsinchu Science-based Industry Park was opened near Taipei. This has become the focus of the entire Taiwanese semiconductor industry, creating what has been called a 'Silicon Valley of the East'.[29] But even as late as 1985, when the competition in semiconductors was intensified by the entry of South Korean firms, there was only the one Taiwanese-owned fabrication plant operated by UMC.

The government's response to this serious problem was to facilitate the creation of a new joint venture company, TSMC (Taiwan Semiconductor Manufacturing Corporation), between a spin-off from ITRI and the European firm Philips. TSMC took over the existing VLSI pilot operation built by ERSO and also constructed its own advanced VLSI fabrication plant.

> The immediate effect of TSMC's launching and its success was to take Taiwan to a new level of technical sophistication and to spark the formation of dozens of small IC design houses in the Hsinchu region. Its demonstration effect also attracted serious Taiwan capital to invest in the semiconductor industry. Thus, the object of the exercise – to launch UMC and then TSMC as 'demonstration' vehicles was achieved, as skilled engineers were voting with their feet, and capitalists were voting with their finance, to enter the new industry.[30]

Today, Taiwan is the world's fourth largest semiconductor producer. It is difficult to believe that it could have achieved such a position without the focused involvement of the state. 'The semiconductor industry in Taiwan has been created and established through a process of *resource leverage* that has been *accelerated* by a judiciously constructed institutional framework.'[31]

Other East Asian semiconductor producing countries: Singapore and Malaysia[32]

Apart from the very early stages of their development, the semiconductor industries in Japan, South Korea and Taiwan have been predominantly *indigenous* industries. In Singapore and Malaysia, on the other hand, government industrialization policies have relied overwhelmingly on foreign firms to develop their semiconductor industries. Although Singapore is more advanced than Malaysia in semiconductor production, the two countries share some common features in the way they have built semiconductor industries, as Table 12.3 shows.

The focus of Singapore's strategy, implemented largely through the Economic Development Board, has been continuously to upgrade its semiconductor sector through leveraging the technologies and resources of foreign TNCs. In addition, however, the Singapore government has used one of its state-owned companies, Singapore Technologies Group (STG), to develop an indigenous presence in semiconductors. Like Taiwan, Singapore used a technology agreement with a foreign

Table 12.3 Stages in the evolution of the Singapore and Malaysian semiconductor industries

(a) Singapore

Stage I Pre-1976 Preparation	Stage II 1976–85 Seeding/implantation	Stage III 1986 to present Diffusion	Stage IV 1991 to present Roots of sustainability
Preparation by Economic Development Board	Upgrading of TNC activities and expansion of their scope	Further upgrading and expansion of TNC activities	Stock of both TNC and indigenous firms engaged in all phases of IC and wafer fabrication
Attraction of TNCs	Internal and external leverage via TNCs	Beginnings of TNC-based wafer fabrication	Framework of R&D support to upgrade activities
Assembly and test operations established by foreign companies	Beginnings of indigenous industry established as contract service providers to TNCs	Beginning of local wafer fabrication	Establishment of institutional R&D support
Expansion of technical education		Expansion of local service industries	Development of Woodlands wafer fab 'park'

(b) Malaysia

Stage I Pre-1980 Preparation	Stage II 1981 to present Seeding/implantation	Stage III 1991 to present Diffusion
Preparation by state agencies such as Penang Development Corporation	Upgrading of TNC activities and expansion of their scope	Further upgrading and expansion of TNC activities
Attraction of TNCs	Internal and external leverage via TNCs	Beginnings of TNC-based wafer fabrication
Assembly and test operations established by foreign companies	Beginnings of indigenous industry established as contract service providers to TNCs and as contract test and assembly firms	Expansion of local service firms
Expansion of technical education		Establishment of institutional R&D support Establishment of infrastructure such as Kulim High-Tech Park

Source: Mathews and Cho, 2000: Table 5.1

firm – in this case a US firm, Sierra Semiconductor – to create a state-owned IC foundry, Chartered Semiconductor Manufacturing (CSM). Subsequently, STG bought out the US partner and modelled CSM's activities on the Taiwanese foundry firm, TSMC.

> CSM expanded its production facilities rapidly in the 1990s to become one of the world's largest IC foundries. It built a second and third fabrication facility in the Woodlands semiconductor park and has entered into joint ventures with two of the world's leading semiconductor firms, HP and Lucent, to build a further two. Recently, the Singapore Technologies Group has launched a third semiconductor business, in the form of Singapore Technologies Assembly and Test Services (STATS), offering 'back-end' test and assembly contract services to complement its 'front end' IC design and IC fabrication services. By the late 1990s, STG was becoming an integrated contract IC producer.[33]

The Singapore government's direct involvement in semiconductor production through CSM is impressive but is only a small component of the country's activities. The overwhelming majority of semiconductor production in Singapore is performed by foreign TNCs. But their continued presence and development testifies to the kind of supportive environment provided by the state. A similar observation applies to Malaysia, as Table 12.3 shows.

The state and the semiconductor industry in Europe

Not surprisingly, we find a very uneven policy picture in Europe. Not only does the European semiconductor industry lag a good way behind those of the United States and Japan but also much of its production capacity is in US-, Japanese- and, more recently, South Korean-owned plants. In general, European governments have not only welcomed such investments but also, in some cases, assiduously courted the foreign semiconductor firms. European policies towards the semiconductor industry up to the mid-1980s evolved in three stages.[34]

- The period to the mid-1960s was largely one of non-intervention (apart from defence-related R&D and some bias towards national producers in government purchases).
- Between the mid-1960s and mid-1970s government focus on the computer industry gave some stimulus to semiconductor research. In neither period, however, was government involvement particularly influential.
- From the mid-1970s onwards, there was a major intensification in government involvement, this time focused on the information technologies, including micro-electronics.

Since the early 1980s European governments have been involved in supporting major collaborative ventures in information technologies in general (the ESPRIT programme) and specifically in semiconductors. The $4 billion JESSI programme – the Joint European Submicron Silicon Initiative – was concerned with developing advanced microchip technology. There has also been increasing pressure on non-European semiconductor manufacturers to locate more than merely assembly operations in Europe: the objective has been to persuade US and Japanese companies to locate more of their design and fabrication plants in Europe.

The aims of these European initiatives were three-fold: to protect European capacity in the core technologies of microelectronics; to accelerate innovation; and to encourage cross-border links between national electronics firms within Europe. A major problem, however, has been whether or not to allow non-European firms with major European operations to participate in the collaborative programmes. At the individual state level it has been the French who have been most directly involved in attempting to establish and nurture a 'national champion' semi-conductor firm.

> The chosen instrument ... was Thomson–CSF, whose semiconductor subsidiary, Sescosem, was given substantial subsidies during the 1960s and 1970s. After nationalisation in 1981, Thomson was encouraged to broaden its range of products and compete directly against the big Japanese and American companies. But despite large subventions from the state, Thomson failed to achieve the objectives the government had set for it, and the semiconductor business had become a financial burden on the rest of the group. In 1987, Thomson's semiconductor subsidiary was put into a joint venture with an Italian state-owned semiconductor firm, SCS. Over the subsequent decade ... the link with Thomson became increasingly distant, and in 1997 the French company sold its remaining shares in SGS–Thomson. What had originally been envisaged as a French national champion in semiconductors had become part of a specialised supplier focusing on market niches rather than high-volume commodity chips.[35]

Like the US federal government, the European Commission has also been active in protecting the semiconductor industry against what are seen to be unfair trading practices. In 1990 it secured a voluntary agreement with the Japanese on the minimum prices at which standard memory chips would be sold in the EC. Subsequently, the Commission initiated anti-dumping measures against South Korean semiconductor manufacturers who were alleged to be dumping chips at below acceptable prices. In effect, this was the price agreed with the Japanese, who were regarded as the lowest-cost producers. These minimum prices were re-imposed in 1997 on the biggest-selling semiconductor chips from 14 Japanese and Korean manufacturers.[36] Nevertheless, despite these national and EU-wide initiatives, Europe has not succeeded in sustaining a strong indigenous semi-conductor industry.[37]

Corporate strategies in the semiconductor industry

Increasing global concentration

Increasingly, a relatively small number of very large transnational corporations has come to dominate semiconductor production, a reflection of the technological and production characteristics of this sector (the accelerating necessity of very large capital investments brought about by the rapid and highly expensive nature of technological change). In such circumstances, small-scale operations become less and less viable, although new niches within the semiconductor production system are continuously being created and these offer possible opportunities for development.

The transformation from a fragmented industry structure, in which small and medium-sized firms were the norm, to one of large-firm dominance and to a more restricted role for the small firm, has been especially dramatic in the semiconductor industry of the United States. During the late 1950s and through much of the 1960s, entry into the semiconductor industry was relatively easy. However, the proliferation of new small firms in the US semiconductor industry as a whole slowed down as the barriers to entry increased during the 1970s and 1980s. The dominance of large firms became especially marked in the manufacture of standard semiconductor devices which depend on mass production technology. Even so, very high levels of new firm start-ups continued in the US semiconductor industry through the 1980s, especially among firms engaged in application-specific products.[38]

Figure 12.7 shows how the composition of the world's top ten leading semiconductor producers has changed since the late 1970s. The four time slices plot the fluctuating fortunes of US, Japanese and European firms, together with the rise of South Korean producers. In the late 1970s, two US companies – Texas Instruments and Motorola – were the clear industry leaders, with almost 20 per cent of global semiconductor sales. Three other US firms were also in the top ten, including – in tenth position – the subsequent industry leader, Intel. The five US firms accounted for 30 per cent of global semiconductor sales. Three of the top ten firms were Japanese (with 14 per cent of global sales) and two were European (8 per cent).

A decade later, in the late 1980s, the number of US firms in the top ten had fallen to four and their share had fallen to 18 per cent. Texas Instruments had fared especially badly, falling from first to sixth. Intel had moved up a little. On the other hand, Japanese companies had surged up the league table, clearly shattering US dominance. By the late 1980s, there were five Japanese firms in the top ten, including the three top positions (NEC, Toshiba, Hitachi). The combined Japanese share had grown to 35 per cent while the US share had fallen from 30 per cent to 18 per cent.

By the mid-1990s, the recovery of the US semiconductor industry could be discerned. Although the number of US firms in the top ten had fallen to three, their share of global semiconductor sales had improved: from 18 per cent to 20 per cent. More significant, in the light of subsequent events, was the dramatic rise of Intel from eighth position in 1989 to first in 1995. Every leading Japanese company in the top ten lost ground while their combined share of sales fell from 35 per cent to 27 per cent. The other striking feature of the mid-1990s position was the appearance of two South Korean semiconductor producers in the top ten, displacing the European firms. Indeed, in the mid-1990s, it looked as though the European semiconductor industry was in terminal decline.

However, by the late 1990s, the position had changed again. The most dramatic aspect of the current position is the dominance of Intel which, in the late 1990s,

Rank by market share	1978	Market share (%)	1989	Market share (%)	1995	Market share (%)	1999	Market share (%)
1.	**Texas Instruments**, US	*12.4*	**NEC**, Japan	*8.9*	**Intel**, US	*8.9*	**Intel**, US	*15.9*
2.	**Motorola**, US	*7.5*	**Toshiba**, Japan	*8.8*	**NEC**, Japan	*7.3*	**NEC**, Japan	*5.5*
3.	**NEC**, Japan	*5.6*	**Hitachi**, Japan	*7.0*	**Toshiba**, Japan	*6.6*	**Toshiba**, Japan	*4.5*
4.	**Philips**, Netherlands	*4.8*	**Motorola**, US	*5.9*	**Hitachi**, Japan	*6.1*	**Samsung**, S. Korea	*4.2*
5.	**Hitachi**, Japan	*4.3*	**Fujitsu**, Japan	*5.3*	**Motorola**, US	*5.9*	**Texas Instruments**, US	*4.2*
6.	**Fairchild**, US	*3.7*	**Texas Instruments**, US	*5.0*	**Samsung**, S. Korea	*5.4*	**Motorola**, US	*3.8*
7.	**Toshiba**, Japan	*3.7*	**Mitsubishi**, Japan	*4.7*	**Texas Instruments**, US	*5.2*	**Hitachi**, Japan	*3.3*
8.	**National Semiconductor**, US	*3.5*	**Intel**, US	*4.4*	**Fujitsu**, Japan	*3.6*	**Infineon**, Germany	*3.1*
9.	**Siemens**, Germany	*2.9*	**Philips**, Netherlands	*4.2*	**Mitsubishi**, Japan	*3.3*	**STM**, Italy/France	*3.0*
10.	**Intel**, US	*2.9*	**National Semiconductor**, US	*2.8*	**Hyundai**, S. Korea	*2.8*	**Philips**, Netherlands	*3.0*

Share of world total	1978 Number	1978 %	1989 Number	1989 %	1995 Number	1995 %	1999 Number	1999 %
United States	5	30	4	18	3	20	3	24
Japan	3	14	5	35	5	27	3	13
Europe	2	8	1	4	0	0	3	9
South Korea	0	0	0	0	2	8	1	4

Figure 12.7 The changing composition of the world's top ten semiconductor producers

had almost three times the sales of the second-ranking producer, NEC. As a result, the share of the three leading US companies had increased to 24 per cent while the share of the Japanese companies (three compared with five) had halved to 13 per cent. At the same time, there had been a significant European resurgence, with three companies in the top ten, whilst one of the two South Korean firms had dropped out.

Increasing global connectedness: acquisitions, mergers and alliances

In striving to compete in global markets, semiconductor firms have become increasingly interconnected through processes of acquisitions, mergers and strategic alliances. During the 1970s and 1980s there was a substantial wave of mergers and acquisitions, both domestic and international. Fast-growing smaller firms were particularly vulnerable to takeover; by 1980 only seven of 36 post-1966 start-up companies in the United States were still independent. Within Europe, the biggest merger was that between the Italian firm SGS and the French firm Thomson to form SGS–Thomson, which subsequently became STMicro-electronics. Most recently, there has been a further upsurge in acquisitions in the aftermath of the East Asian financial crisis of the late 1990s. For example, Hyundai acquired LG Semicon in 1999 and then transformed itself into Hynix. Hynix itself then became the target of the leading US memory chip producer, Micron Technology, although the merger talks failed. In Japan, a number of mergers have occurred, especially as some firms pulled out of semiconductor production. For example, Fujitsu bought the semiconductor operations of NKK.

Even more pervasive has been the use of strategic alliances, for all the reasons discussed in Chapter 8. In particular, the massive costs of R&D, the incredibly rapid pace of technological change and the escalating costs of installing new capacity all contribute towards the attractiveness of forming strategic alliances. Table 12.4 shows some of the more significant recent strategic alliances in the semiconductor industry. In addition, there has been a considerable development of multilateral alliances within consortia, often stimulated or facilitated by national governments, as we saw earlier in this chapter. Figure 12.8 shows the company membership of the two most recent consortia in the United States and Japan. In Europe an alliance of producers from different European companies formed European Silicon Structures (ES2) to manufacture custom chips.

Variations in the strategies of semiconductor producers

In such a volatile technological and competitive industry as semiconductors, firms inevitably employ a whole variety of strategies to ensure their survival and in pursuit of growth. However, corporate strategies – whether offensive or defensive in nature – increasingly are being implemented at a *global*, rather than a purely national, scale. Indeed, in the face of fierce global competition, firms have been systematically rationalizing and reorganizing both their domestic and overseas operations. A strategy common among some firms has been to specialize in specific market segments, to pursue a 'niche' strategy. Amongst other firms, however, the preferred strategy has been to increase the degree of vertical integration. Cutting across these strategies are those of increasing transnationalization of production, of automation and, overall, of rationalization and reorganization on a global scale.

It is the inexorable pursuit of such strategies – set within the context of rapidly changing market, technological and political forces – which shapes and reshapes the global map of production and trade in the semiconductor industry. Of course, many factors influence the particular mix of strategies employed. One important

Table 12.4 International strategic alliances in the semiconductor industry

Alliance partners	Purposes of the alliance
Motorola (US) – IBM (US) – Siemens (Germany) – Toshiba (Japan)	To develop next generation of memory chips, including a 1 gigabit dynamic random access (DRAM) device. Will build upon the existing alliance between IBM, Siemens and Toshiba
Siemens (Germany) – Motorola (US)	To build a new plant in the United States to make advanced memory chips
Mitsubishi (Japan) – Umax Data Systems (Taiwan) – Kanematsu (Japan)	To build a semiconductor facility in Taiwan to produce advanced memory chips
NEC (Japan) – Samsung (South Korea)	To collaborate in the production of memory chips for the European market
IBM (US) – Toshiba (Japan)	To build a semiconductor facility in the United States to produce next-generation memory chips
UMC (Taiwan) – Kawasaki Steel (Japan)	To develop ASICs
Fujitsu (Japan) – AMD (US)	To develop production of flash memory devices
Hitachi (Japan) – NEC (Japan)	To create a joint venture – Elpida – to produce DRAMs
IBM (US) – AMD (US)	To develop new technology for high-end microprocessors
NEC (Japan) – state-owned firm (China)	To produce DRAMs in China
Intel (US) – BT (UK)	To develop wireless applications
STMicroelectronics (Fr/It) – Philips (Neths) – Motorola (US)	To create R&D alliance based near Grenoble, France
AMD (US) – UMC (Taiwan)	To build $5bn fabrication plant in Singapore

variable is *size of firm*: large firms tend to operate in rather different ways from smaller firms. Another influence is, undoubtedly, a *firm's geographical origins*: the domestic context in which it has developed. There are substantial differences in behaviour between US, East Asian and European semiconductor firms.

Semiconductor firms can be classified into five broad types:

- *vertically integrated captive producers* – manufacture semiconductors entirely for their own in-house use
- *merchant producers* – manufacture semiconductors for sale to other firms
- *vertically integrated captive-merchant producers* – manufacture semiconductors partly for their own use and partly for sale to others
- *'fabless' semiconductor firms* – design semiconductors but do not manufacture them
- *foundry companies* – manufacture semiconductors to specifications of customers but do not design them.

There has been considerable shifting between the first three categories as companies have sought to reposition themselves competitively. Japanese and European firms have tended to be more vertically integrated than US firms, producing semiconductors for in-house use. This has been especially apparent in Japan, where virtually all of the major semiconductor producers were part of diversified electronics groups. Apart from such companies as IBM, the majority of US semiconductor firms developed into merchant producers. However, since the early 1990s, IBM has transformed itself into a merchant producer, completely reversing its traditional stance.[39]

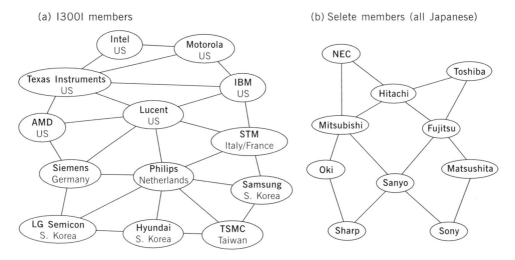

Figure 12.8 Company alliances within semiconductor consortia

Source: Based on Ham, Linden and Appleyard, 1998: Table 1

The fourth and fifth categories of semiconductor producer are more recent and reflect the significant organizational changes that are occurring in the industry. The so-called *fabless* producers emerged especially in the United States during the 1980s. Such firms are essentially design houses which

> avoided the costs of building, equipping, and operating a fab. They developed core skills in product design and development, quality assurance, marketing, sales, and customer support. Besides design, they internalized the probe and final testing steps of the manufacturing chain, to safeguard product quality, and subcontracted for raw wafer manufacturing, wafer fabrication, and chip assembly. Beyond avoiding initial capital expenses and the associated overhead burden of owning a wafer fabrication facility, the fabless strategy sidestepped equipment obsolescence risks. Fabless companies held the flexibility to take advantage of new manufacturing process technologies as they became mainstream … [they] … were confident that others were willing to shoulder equipment obsolescence risks and make foundry capacity procurable. The latter included the Korean, Taiwanese, and Singaporean governments that heavily subsidized construction of new production facilities.[40]

The fifth category of producer is that of the *semiconductor foundry*. It is, in effect, the obverse of the fabless firm. Foundries are

> third-party vendors of chipmaking services. A foundry, such as Taiwan's TSMC, does not design its own chips – instead it acts as the back end for any number of independent design houses. This breaks the traditional integrated model of a semiconductor company, which designs and manufactures its own chips … much … of the chip industry is likely to migrate to a foundry model. Chip design, the essential core of the industry's value added, will become much more widely spread among a large number of competitors. At the same time, the yield gains from experience will be collectively shared by joint use of chip foundries, helping to limit some of the disastrous pressures for price wars … the foundry model is a 'tectonic event in the development of the industry'.[41]

The essential feature of the foundry, therefore, is that it produces made-to-order chips for a whole range of external buyers. It has become an especially significant phenomenon in East Asia, notably in Taiwan, as we shall see later, and in Singapore.

Strategies of US semiconductor firms

The commercial production of semiconductors began in the United States in the 1950s. For almost three decades US firms dominated the industry before being overtaken by the Japanese. Subsequently, US firms recovered their dominance, primarily by moving out of the commodity DRAM market into more sophisticated semiconductor devices. In the view of Michael Borrus, US firms developed a new form of strategic competition known as 'Wintelism'.[42] The basis of Wintelism was the increasing use of a 'merchant producer' strategy by US firms which enabled them to specialize in developing and selling semiconductor devices to a wide variety of end-users rather than to captive parts of a vertically integrated organization.

Because their basic role was to diffuse chip technology as widely as possible, merchant semiconductor firms fostered other specialized producers throughout the electronics value-chain ... In the struggle to break loose from IBM's dominant model and to react to Japan's ascent, new product strategies emerged. The pioneering product was, of course, the PC. But the extraordinary pace of technical progress and ever-improving price/performance soon made the underlying microelectronics increasingly pervasive ... By the mid-1980s, new electronics product markets began to converge on a cost-effective, common technological foundation of networkable, microprocessor-based systems, of which the PC was only emblematic. Such systems enabled a dramatic shift in the character of electronics products: from the prior era's proprietary systems built to fully open or closed standards, to the Wintelist era's 'open-but-owned' systems built to 'restricted standards'.[43]

Thus we can see where the term 'Wintelism' comes from: a combination of the dominant software system (*Win*dows) and the dominant microprocessor producer (In*tel*).

A key element of this strategic reorientation by US semiconductor firms was its manipulation of the *geography* of production. From the beginning, the geographical pattern of the semiconductor industry was highly concentrated around two foci: the Route 128 corridor near Boston, Massachusetts and the Santa Clara Valley in California – later to be dubbed 'Silicon Valley'.[44] Subsequently, Silicon Valley became emblematic of the new semiconductor industry, although a considerable mythology has developed around the evolution of that region.[45] Through the processes of localized territorial development discussed in Chapters 2 and 4 – notably the operation of traded and untraded interdependencies and the evolution of an innovative milieu in which localized learning processes predominated – Silicon Valley became the core of the US semiconductor industry. It was not the only localized concentration, but it achieved the position of dominance.

Offshore production by US semiconductor firms first occurred in the early 1960s when a number of American firms began to seek out low-labour-cost locations for their more routine assembly operations. The initial stimulus was the intensifying competition within the United States itself, as new firms entered the industry and the need to reduce production costs accelerated. In the 1960s the differential between US labour costs and those in developing countries was especially great.

The first offshore assembly plant in the semiconductor industry was set up by Fairchild Semiconductor in Hong Kong in 1962. In 1964 General Instruments transferred some of its micro-electronics assembly to Taiwan. In 1966 Fairchild opened a plant in South Korea. Around the same time, several US manufacturers set up semiconductor assembly plants in the Mexican Border Zone. In the later 1960s US firms moved into Singapore and subsequently into Malaysia. During the following decade semiconductor assembly grew very rapidly in these Asian locations and spread to Indonesia and the Philippines. Despite some developments elsewhere (for example in Central America and the Caribbean), most of the growth of semiconductor assembly remained in East and South East Asia.

By the early 1970s, therefore, every major US semiconductor producer had established offshore assembly facilities, a tendency greatly encouraged by the offshore assembly provisions operated by the US government. Figure 12.9 shows the global distribution of US semiconductor assembly plants, indicating clusters in Mexico and the Caribbean as well as in Europe but with by far the largest concentration in East Asia. However, the structure of US semiconductor operations within East Asia has changed substantially since the early focus on low-labour-cost assembly operations. We will look at this in some detail later in the chapter.

Most of the semiconductor plants established by US semiconductor companies in Europe can be explained primarily by the import tariff imposed by the European Commission. However, the completion of the Single European Market in 1992 added a new stimulus. The increasingly stringent attitude by European governments towards full-scale local production (including the more advanced stages of design and manufacture) and the practice of using anti-dumping measures stimulated further direct investment in Europe by US semiconductor firms.

Within Europe there is a long-established US semiconductor presence with a particularly heavy concentration in Scotland and, to a much lesser extent, Ireland. US firms dominate the Scottish semiconductor sector. Scotland has a long tradition of American manufacturing investment, a good supply of both high-skilled labour from the well-developed higher education sector and of female assembly workers. In addition, the country's investment promotion agency has been a strenuous, and successful, seeker of electronics companies to an area with a substantial package of investment incentives. Ireland, too, has strong links with the United States and it, too, has adopted an aggressive strategy to attract foreign electronics firms on the basis of generous financial and tax incentives and a good

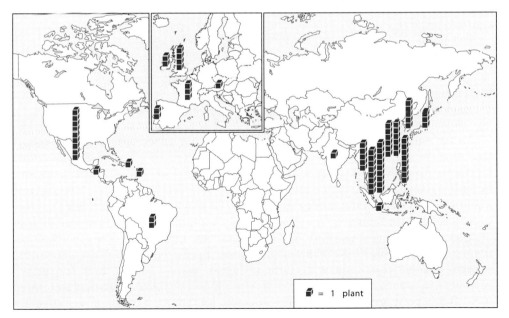

Figure 12.9   The global distribution of US-owned semiconductor assembly plants

Source: Based on Scott and Angel, 1988: Figure 5

labour supply. The need to operate behind the EC tariff wall, therefore, attracted large numbers of US semiconductor firms into Europe.

However, in the face of the deep recession in the IT industries in 2001, most of the US semiconductor firms in Scotland have been cutting back on their production and laying off workers in large numbers. Indeed, major restructuring has become endemic within the US semiconductor industry in general. In 2001, for example, IBM, Motorola, National Semiconductor, Intel and AMD all announced major job-cutting and closure programmes in their operations, both in the United States itself and overseas.

Strategies of Japanese semiconductor firms

The Japanese dominance of the global semiconductor industry in the 1980s and early 1990s rested on three basic elements:

- state involvement
- concentration on the memory chip (DRAM) market
- the fact that most Japanese semiconductor production was for in-house use, that is, within the large and diversified electronics business groups.

We discussed the central role of the state in an earlier section of this chapter. As far as concentration on the DRAM market is concerned,

> Japanese semiconductor firms' dominance of DRAM markets during the 1980s rested on low prices and high quality … Users of U.S. and Japanese devices discovered that Japanese memory products had defect rates that were one-half to one-third those of comparable U.S. memory products; in 1980, leading Japanese memory producers averaged 160 defective parts per million (PPM) while U.S. producers averaged 780 PPM for the same devices. Their skills in managing the development and introduction of new process technologies also enabled Japanese producers to 'ramp' output of new products more rapidly than their U.S. counterparts. Faster achievement of high production volumes gave Japanese firms advantages in defining product standards for leading-edge memory devices, strengthening their market position.[46]

As we saw in Chapter 8, Japanese firms are organized into complex groups (*keiretsu*). In the case of the semiconductor industry, this has meant that most Japanese production has been driven by the needs of intra-group businesses, notably consumer electronics and telecommunications. There are some obvious advantages in such vertical integration. Equally, as the success of the US merchant producers showed during the 1990s, there are dangers of insularity in such an arrangement. The revival of the US semiconductor industry on the one hand and the emergence of South Korean and Taiwanese producers on the other have put a severe squeeze on Japanese semiconductor firms. In contrast to the US producers, Japanese firms have tended to remain in the DRAM market and, therefore, have been especially heavily affected by South Korean and Taiwanese competition.

Although, Japanese semiconductor producers have concentrated most of their activities within Japan itself they have also, as Figure 12.10 shows, developed a considerable degree of offshore production. There is an especially heavy emphasis on East Asia. Seventy-one per cent of the electronic components and devices plants located outside Japan were in Asia, 17 per cent in North America (including Mexico) and 10 per cent were located in Europe. Of course, proximity to the Japanese domestic production base facilitated such developments in East and South East Asia and helped to create a complex intra-regional division of labour. Again, this is something we will look at in some detail later.

The establishment of semiconductor production facilities by Japanese firms in both Europe and North America is very much more recent than that of American firms in Europe. Until the 1970s there was virtually no direct Japanese investment in electronics in the developed countries. As Figure 12.10 shows, however, there are now substantial numbers of Japanese electronic component firms operating plants in both the United States and Europe. In the latter case more than two-thirds of the total are in the United Kingdom and Germany.

The combination of the Single European Market and the revalued yen – but especially the former – stimulated a new, and highly significant, wave of Japanese semiconductor investment in Europe. This new wave was led by Fujitsu's decision to build a $100 million chip fabrication plant in north-east England in 1989. The other leading Japanese companies followed this lead. Hitachi, for example, built an integrated semiconductor manufacturing plant in Germany. Japanese semiconductor production in the United States is much more firmly established than in Europe. In fact six out of every ten Japanese electronic component plants

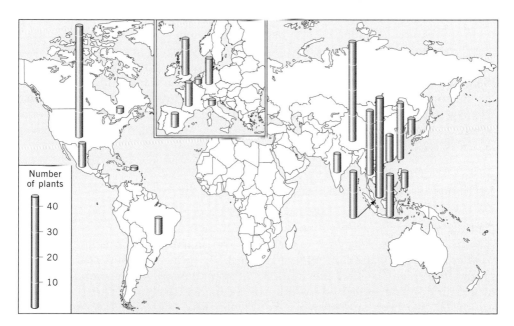

Figure 12.10   The global distribution of Japanese electronic component and devices plants

Source: Based on Electronics Industries Association of Japan (1996) *Facts and Figures on the Japanese Electronics Industry, 1996:* 91

outside Asia are located in North America (see Figure 12.10). Again, the same basic locational factors apply, with a particular emphasis on market access and proximity, and scientific and technical labour.

Faced with intensifying competition in the DRAM market, Japanese firms have recently adopted a number of strategies. One has been to diversify into such product areas as flash memories. Demand for these devices has grown very rapidly with the development of Internet-related mobile phones and related products (although, like all semiconductor products they have been affected by the recession in the IT industries in general). A second strategy pursued by several Japanese electronics firms has been to outsource more of their semiconductor production. For example, in 2000 Toshiba announced its intention to double its outsourcing of semiconductors over three years while Hitachi plans to almost quadruple its outsourcing of semiconductors to reduce costs.[47] Such moves go very much against the traditional preference of Japanese electronics firms to keep their semiconductor production in-house to control quality. As a consequence of these large-scale reductions in DRAM production a number of plants are being closed and thousands of jobs lost.

A third strategy has been to pull out of DRAM production altogether. The biggest Japanese semiconductor producer, NEC, for example, announced in 2001 that it would withdraw completely from DRAM production and close or drastically reduce the scale of several of its plants, including its biggest European facility (located in Scotland) and one of its plants in California. NEC is transferring its DRAM production to Elpida, its joint venture with Hitachi (see Table 12.4).[48] Similarly, Toshiba cut more than 18,000 jobs in 2001, sold its only US semiconductor plant to Micron Technologies, and began negotiations with Infineon of Germany about merging their memory chip operations.[49] Fujitsu announced a huge restructuring programme that will involve taking out of production three semiconductor plants. Its plan is to 'consolidate its 12 chip lines into nine and bring a flash memory factory in Oregon into a Japanese joint venture with Advanced Micro Devices'. This is part of Fujitsu's 'fundamental realignment of the Japanese operations, moving away from low-margin hardware, emphasizing instead software and services'.[50]

Strategies of European semiconductor firms

As noted already, Europe has not produced a really strong indigenous semiconductor sector. Much semiconductor production within Europe itself is either American or East Asian in ownership. The major exceptions are Infineon, STMicroelectronics and Philips (see Figure 12.7). Infineon is part of the Siemens group of Germany, STMicroelectronics is the joint Italian-French company with its origins in the state sector, while Philips is the highly diversified Dutch electronics group. Having been out-competed by US, Japanese and other East Asian firms, these three staged something of a revival in the late 1990s, primarily by vastly improving their production efficiency and by diversifying into the broader communications markets.

Siemens/Infineon, in particular, appeared to be developing into a highly successful semiconductor producer, building what was claimed in 1996 to be the world's most advanced semiconductor plant in north-east England at a cost of $1.8 billion and also establishing a $380 million memory chip plant in Portugal. Yet only one year after starting production at its north-east England plant, the company closed it down with the loss of more than 1000 jobs. The plant was subsequently taken over by the US semiconductor firm Atmel. Philips, likewise, made substantial new investments in semiconductor production in 2000. Philips purchased the IBM wafer fabrication plant in New York state, having earlier taken over VLSI, the American producer of specialist semiconductors.[51] Again, however, Philips was hit by the massive downturn in the semiconductor industry in 2001. Its large-scale rationalization plan focused especially on cutting semiconductor production.

> The big European companies have opted for different routes. Philips has been the most willing to outsource manufacturing, through its relationship with TSMC in Taiwan. It has also decided not to build any further wholly-owned fabs … Only STMicro remains big enough to build its own fabs … It has set an upper limit of outsourcing at just 15 per cent of its manufacturing to third parties.[52]

Compared with US and Japanese semiconductor producers, the European producers have not engaged in international production to anything like the same extent, although all three major European firms have assembly activities in East Asia. Philips has substantial joint venture operations in China manufacturing integrated circuits for the consumer electronics industry. It also has been increasing its sourcing of semiconductors and integrated circuits from East Asia. STMicroelectronics built a second wafer fabrication plant in Singapore to make advanced wafers, mainly for export. But none of the leading European firms operates the kinds of global production network which have been developed by the Japanese and, especially, the US producers which are so heavily dependent on East Asia.

The strategies of South Korean semiconductor firms

The speed with which the leading Korean conglomerates – Samsung, Goldstar (subsequently LG) and Hyundai – developed as extremely advanced semiconductor producers was astonishing, occurring, in fact, within a single decade. None of the Korean electronics firms was involved in semiconductors until Samsung entered the business in the mid-1970s. Goldstar followed in 1979, Hyundai and Daewoo (which were not then electronics companies) entered in 1983. Initially the Korean firms were totally dependent on technology acquired from the United States and Japan. They focused on simple types of semiconductor of low capacity – the products that were being abandoned by the US and Japanese companies. In the mid-1970s, the technology gap between Samsung and the industry standard was around 30 years; today it has long since disappeared. By 1995, as Figure 12.7 shows, not only had Samsung's aim to be in the world's top ten semiconductor producers been achieved but also Hyundai had entered the top league as well. Today, Samsung is the world's largest producer of DRAMs.

Such a dramatic, almost overnight, emergence of South Korean firms as global players was based upon 'a relentless concentration on commodity memory chips'.[53] In particular, their aims have been to drive Japanese firms out of the memory chip market (US firms, with a few exceptions, such as Micron Technology, having largely exited that market some years ago). Korean firms entered the semiconductor markets as imitators rather than innovators.[54] Samsung, for example, at various critical stages in its development sought out and purchased small firms (including US firms) with the necessary technology but which were in financial difficulties. One of its major advantages was its 'ready access to funds siphoned from cash-cow industries within the *chaebol*'.[55] In that regard, Korean firms followed the Japanese model of developing a semiconductor capability within user electronics firms.

Each of the Korean firms has established production facilities in the United States. The objective of these plants was to be close to the technological heart of the semiconductor industry – to absorb state-of-the-art technology. They then planned to move into Europe, primarily to circumvent protectionist barriers as well as to be close to their growing markets. In the mid-1990s, Samsung, Hyundai and LG all announced plans to build massive semiconductor plants in the UK. However, the financial crisis of 1997 put an end to such plans, at least for the time being. In addition, as we have seen, LG was taken over by Samsung and currently a question mark hangs over the future of Hynix (formerly Hyundai).

South Korean firms have come to dominate the memory chip industry, but they are far less prominent in other segments of the semiconductor industry. For example,

> Samsung … lags behind foreign competitors in non-memory devices, which constitute more than three-quarters of the world's semiconductor chip market … Samsung has turned its eyes to the non-memory area with an ambitious vision of ranking among the world's top ten in this market within five years. Accumulated capability from its DRAM experience appears to provide the necessary platform for venturing into non-memory technology … Samsung has been aggressive in developing strategic ties with foreign firms.[56]

The strategies of Taiwanese semiconductor firms

While South Korean firms have followed the Japanese model of development in their semiconductor businesses – focusing on DRAM production within the boundaries of huge electronics conglomerates – the approach adopted by Taiwanese firms has been very different. Again, as we saw earlier, the role of the state has been absolutely central to the industry's development but its mode of operation was different. Whereas Korean semiconductor firms developed out of the private sector *chaebol*, the Taiwanese firms TSMC and UMC were the product of government initiatives (though not as state-owned companies). Although there are a number of semiconductor firms in Taiwan, the clear leaders are TSMC and UMC. They are especially distinctive because they are *semiconductor foundries*, producing chips to order from third-party companies (including the fabless design houses). The basis of their success has been the development of massive scale economies which also accelerate technological learning. These foundries have been described as the 'workshops of the electronic world'. TSMC and UMC are the two biggest semiconductor foundry companies in the world (the third largest is the Singaporean company, Chartered Semiconductor).

So far, Taiwanese semiconductor firms have remained strongly embedded geographically within their domestic environment, although they have long had a presence in Silicon Valley in order to be at the perceived centre of the semiconductor industry. Overseas investment, otherwise, has been confined mainly to East Asia – including Japan, Singapore and, most recently, China. In 2002, the Taiwanese government ended its ban on Taiwanese firms building semiconductor plants in China but restricted such plants to 200 mm wafer technology. In order to qualify, a Taiwanese firm must already have begun 300 mm production in Taiwan; only then can it ship 200 mm equipment to China.[57]

Regionalizing production networks in the semiconductor industry: the case of East Asia

As in the other industry cases discussed in this book, the pattern of production in semiconductors has a strong regional dimension. In this industry, however, it is the position of one particular region – East Asia – that is especially important. East Asian production networks have been particularly critical in the development and redevelopment of the US semiconductor industry, both in the early stages of the industry (in the early 1960s) and also in the period of the resurgence of the US industry since the late 1980s.[58] In the 1960s, as we saw earlier, US semiconductor firms shifted the less skilled labour-intensive stages of semiconductor production to East Asia. This was a very simple geographical division of labour, albeit a critical one for the competitiveness of US firms at that time. But it was not enough to offset the immensely efficient Japanese producers of memory chips.

The US recovery in the semiconductor market was based, as we have seen, on the shift out of memory chips and into far more sophisticated devices within the framework of what Borrus called 'Wintelism'. (see above, pages 425–6).

> Wintelism could not have succeeded without the extensive inter-firm relationships with Asian-based producers that comprised the CPNs ... [cross-border production networks] ... of American-owned firms. Those cross-border ties permitted US-owned firms to exploit the growing technological sophistication and competitive strength of indigenous producers initially in Taiwan, Singapore, and Korea, and later throughout Southeast Asia, in selected cities of India, and along the coastal provinces of mainland China. The unique heterogeneity of Asia's regional economy, with different tiers of nations (Japan, Four Tigers, ASEAN, and coastal China, interior China, and India) at different stages of development provided the fertile ground for technical and production specialization that enabled the creation of CPNs; e.g. software in Bangalore, process engineering in Singapore, component assembly in Malaysia, printed circuit board (PCB) assembly in coastal China, semiconductor memory in Korea, digital design and final assembly in Taiwan.[59]

Three stages in the development of such intricate production networks can be identified:[60]

- *Stage I (1960s to late 1970s).* Low-cost production locations were sought by US firms, which established Asian affiliates as part of their transnational production network aimed at selling in advanced country markets outside Asia.

- *Stage II (1980 to 1985).* The US-owned assembly platforms were technically upgraded to encompass a wider range of production stages (for example, moving from simple assembly of chips to the more complex stage of testing). At the same time, the Asian affiliates of US companies developed extensive local relationships, particularly through increased local sourcing of components. 'The result, by the end of the 1980s, was burgeoning indigenous electronics production throughout the region, with most of it, outside Korea, under the control of overseas Chinese (OC) capital'.[61]

- *Stage III (1985 to early 1990s).* As US firms significantly shifted their focus to new product definition and design and software development they further upgraded their Asian affiliates. Local firms gained much greater manufacturing responsibilities and greater autonomy in sourcing key components within the region.

As a result of this evolutionary process of the East Asian regional production network,

> the strongest indigenous Asian producers began to control their own production networks … In sum, by the early 1990s, the division of labor between the United States and Asia, and within Asia between affiliates and local producers, deepened significantly, and US firms effectively exploited increased technical specialization in Asia.[62]

Conclusion

In a number of ways, the global pattern of semiconductor production has shifted and changed as a result of the evolving corporate strategies of the major firms and the actions of national governments. Fierce competition has forced semiconductor firms to increase their degree of functional integration, to diversify into new product lines, to relocate production in more favourable locations in terms of markets or costs, and generally to rationalize their operations on a global basis. There has been considerable geographical shift to some developing countries, mostly in East Asia. Initially, this was the assembly stage, providing a clear example of the global combination of highly capital-intensive technology with low-cost, labour-intensive production. But the simplistic global division of labour characteristic of the 1960s and early 1970s no longer applies.

Today's global map of semiconductor production, therefore, is far more complex. So, too, is the organization of the industry. The emergence of fabless design houses and semiconductor foundries has transformed the industry, creating new opportunities for producers lacking either the captive markets of the vertically integrated producers or the established customer links of the major merchant producers. One of the most striking developments has been the rise, fall and rise again of the semiconductor industry of the United States, which is, once again, the dominant player in the more advanced semiconductor product-markets.

Notes

1 *The Financial Times* (4 February 1998).
2 Mathews and Cho (2000: 32).
3 Macher, Mowery and Hodges (1998: 112).
4 Mathews and Cho (2000) provide an excellent and comprehensive analysis of the development of the semiconductor industry in East Asia.
5 Mathews and Cho (2000: 33).
6 Mathews and Cho (2000: 44).
7 Mathews and Cho (2000: Figure 1.12).
8 Siegel (1980: 3).
9 *The Financial Times* (16 July 1999).
10 *The Financial Times* (10 July 2000; 28 July 2000).
11 *The Financial Times* (14 November 2000).
12 *The Economist* (2 December 2000).
13 See Angel (1994); Mathews and Cho (2000); Ó hUallacháin (1997).
14 See Fuentes and Ehrenreich (1983); Harrison (1997); Siegel (1980).
15 Ham, Linden and Appleyard (1998: 157).
16 *The Financial Times* (6 December 2000).
17 See Angel (1994: 156–86); Ham, Linden and Appleyard (1998); Macher, Mowery and Hodges (1998).
18 Macher, Mowery and Hodges (1998: 125–6).
19 Macher, Mowery and Hodges (1998: 121–2).
20 Ham, Linden and Appleyard (1998) provide a detailed discussion of I300I.
21 Mathews and Cho (2000) provide the most recent and most comprehensive account of the role of the state in the development of the semiconductor industry in East Asia. Mathews (1997) describes the case of Taiwan and Mathews (1999) deals with Singapore.
22 Mathews and Cho (2000: 253).
23 Ham, Linden and Appleyard (1998: 151–2).
24 Mathews and Cho (2000: 112).
25 Wade (1990b: 313).
26 Kim (1997: 94–5).
27 Mathews (1997: 28; emphasis added).
28 Mathews (1997: 34).
29 Mathews (1997).
30 Mathews (1997: 36).
31 Mathews (1997: 40).
32 See Mathews (1999); Mathews and Cho (2000: ch. 5).
33 Mathews (1999: 66–7).
34 Dosi (1983).
35 Owen (1999: 274).
36 *The Financial Times* (1 April 1997).
37 Owen (1999: 275).
38 Angel (1994: 38–9).
39 Sturgeon (2002).
40 Ó hUallacháin (1997: 222).
41 *The Financial Times* (6 December 2001).
42 Borrus (2000).
43 Borrus (2000: 61–2).

44 See Saxenian (1994) for a detailed comparison of the evolution of the electronics industries along Route 128 and in Silicon Valley.
45 In addition to Saxenian (1994), see also Angel (1994), various chapters in Kenney (2000) and Siegel (1980).
46 Macher, Mowery and Hodges (1998: 113–14).
47 *The Financial Times* (21 August 2000; 28 September 2000).
48 *The Financial Times* (1 August 2001).
49 *The Financial Times* (28 August 2001).
50 *The Financial Times* (21 August 2001).
51 *The Financial Times* (22 June 2000).
52 *The Financial Times* (12 April 2002).
53 Mathews and Cho (2000: 45).
54 Kim (1997) provides a detailed account of Samsung's development as a leading semiconductor producer.
55 Kim (1997: 88).
56 Kim (1997: 96).
57 *The Financial Times* (30 March 2002).
58 See Gereffi (1996a); Borrus (2000).
59 Borrus (2000: 58–9).
60 Borrus (2000: 68–74).
61 Borrus (2000: 70).
62 Borrus (2000: 73).

CHAPTER 13

'Making the World Go Round': the Financial Services Industries

Money matters

The preceding three chapters have focused on the evolving production networks of key manufacturing sectors. In this, and the next, chapter we turn our attention to two service sectors: financial services and distribution services. Such

> commercial and financial services, which mediate and abbreviate the exchange processes within the economy, are circulation services, services ... concerned primarily with the velocity of turnover, whether of commodities, money or money capital ... service production occurs within both the sphere of production and the sphere of circulation.[1]

We start, in this chapter, with one of the most fundamental – and, at the same time, most controversial – of all economic sectors: finance.

Finance is one of the most fundamental because, as Figure 2.3 shows, it is vital to, and embedded within, all elements of production chains and production networks. All economic activities (whether they are manufactured goods or services) have to be financed at all stages of their production. Without the parallel development of systems of monetary- and credit-based exchange there could have been no development of economies beyond the most primitive organizational forms and the most geographically restricted scales.

> The geographical circuits of money and finance are the 'wiring' of the socio-economy ... along which the 'currents' of wealth creation, consumption and economic power are transmitted ... It allows – indeed provides a mechanism for – the simultaneous 'stretching' and 'compression' of social interaction across time and space, facilitating the storage, coordination and communication of the information and social power used in such interaction. Thus, money allows for the deferment of payment over time–space that is the essence of credit. Equally, money allows propinquity without the need for proximity in conducting transactions over space. These complex time–space webs of monetary flows and obligations underpin our daily social existence.[2]

Finance is also the most controversial of economic activities because of its historical relationship with state 'sovereignty'. Ever since the earliest states emerged on the scene, the creation and control of money have been regarded as being central to their legitimacy and survival.[3] Today, in the context of a globalizing world economy, this tension has become especially acute as national states see their traditional control of their monetary affairs being threatened by external market forces. The periodic financial crises that engulf some or all parts of the world economy serve to intensify these tensions and also to create legitimate fears over the accountability and responsibility of financial institutions. The increased interconnectedness and vulnerability of the world economy is tightly bound up with developments in the financial system.

In 1986, Susan Strange coined the graphic term 'casino capitalism' to describe the international financial system in which, she argued,

> every day games are played in this casino that involve sums of money so large that they cannot be imagined. At night the games go on at the other side of the world ... [the players] ... are just like the gamblers in casinos watching the clicking spin of a silver ball on a roulette wheel and putting their chips on red or black, odd numbers or even ones.[4]

Twelve years later, in 1998, she used the term 'mad money' to reflect the increased volatility of the financial system and the uncertainty it generates throughout the world economy.

International financial flows and foreign currency transactions have reached unprecedented levels. They easily dwarf the value of international trade in manufactured goods and in other services and have done so at an increasing rate over the past three decades, as Figure 13.1 shows. By 1995, the ratio of daily foreign exchange trading to world trade was a staggering 70:1, compared with only 2:1 in 1973. Of course, some of those financial transactions are directly related to international trade (and production) and are essential for that purpose. But only around 10 per cent of international financial transactions are of this kind. The remainder take the form of what are, essentially, *speculative* dealings – aimed at making short- or long-term profits as ends in themselves – through a bewildering variety of financial instruments.[5]

Figure 13.1

The growing disparity between foreign exchange trading and world trade

Source: Based on Eatwell and Taylor, 2000: 3–4

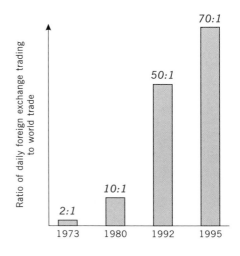

Without doubt, therefore, a distinct financial system has come into being which, although it is obviously connected to the 'real economy' of the production and trade in goods and services, has an existence of its own (see Figure 1.1). Hence, changes in the level and the geography of the flows of international portfolio investments (investments in stocks and bonds) have an enormous impact on the financial well-being of entire national economies. For example, a heavy dependence on such 'hot' money was a major factor in the Mexican peso crisis of 1994 and in the East Asian financial crisis of 1997/1998. Overall, the financial system has undergone radical and extremely rapid change in the past few decades: 'a revolution in the way finance is organized, a revolution in the structure of banks and financial institutions and a revolution in the speed and manner in which money flows around the world'.[6]

Financial services, therefore are both *circulation* services, fundamental to the operation of every aspect of the economic system, and also *commodities* or *products* in their own right, in the sense that they are produced and traded in the same way as more tangible manufactured goods are traded. They are also intrinsically *geographical* in their production and consumption. Indeed, one of the paradoxes of the global economy is that the one sector which would appear to be the most locationally 'footloose' is, in fact, highly concentrated geographically into a small number of international financial centres.

The structure of the financial services industries

> The function of the financial system as a whole is financial *intermediation*, the pooling of financial resources among those with surplus funds to be lent out to those who choose to be in deficit, that is to borrow … With financial intermediation, investors in new productive activities do not themselves have to generate a surplus to finance their projects; instead the projects can be financed by surpluses generated elsewhere within the economy.[7]

The process of intermediation has constituted the basic function of banks from the very beginning. Figure 13.2 shows the sequence of development of the banking system together with its spatial implications. The sequence shows, in effect, the increasing capacity of the banking system to create credit and the way in which this process is geographically produced. Thus, for example, the first two stages of credit provision to borrowers depend greatly upon the nature of geographically specific knowledge to legitimize lending and borrowing. The geographical scope of such knowledge and trust grows through such developments as inter-bank lending (that is, lending outside the local area) in stage 4 and, eventually, of a central bank that ultimately acts as the lender of last resort to the banking system as a whole (stage 5). Subsequent developments, especially in securitization, create a totally different scale and complexity of financial activity.

The stages of banking development	Banks and space	Credit and space
Stage 1: Pure financial intermediation Banks lend out savings Payment in commodity money No bank multiplier Saving precedes investment	Serving local communities Wealth-based, providing foundation for future financial centres	Intermediation only
Stage 2: Bank deposits used as money Convenient to use paper money as means of payment Reduced drain on bank reserves Multiplier process possible Bank credit creation with fractional reserves Investment can now precede saving	Market dependent on extent of confidence held in banker	Credit creation focused on local community because total credit constrained by redeposit ratio

Stage 3: Inter-bank lending
Credit creation still constrained by reserves
Risk of reserves loss offset by development of
  inter-bank lending
Multiplier process works more quickly
Multiplier larger because banks can hold lower
  reserves
	Banking system develops
	at national level
		Redeposit constraint relaxed
		somewhat, so can lend wider
		afield

Stage 4: Lender-of-last-resort facility
Central bank perceives need to promote
  confidence in banking system
Lender-of-last-resort facility provided if
  inter-bank lending inadequate
Reserves now respond to demand
Credit creation freed from reserves constraint
	Central bank oversees
	national system, but
	limited power to
	constrain credit
		Banks freer to respond to
		credit demand as reserves
		constraint not binding and
		they can determine volume
		and distribution of credit
		within national economy

Stage 5: Liability management
Competition from non-bank financial
  intermediaries drives struggle for market share
Banks actively supply credit and seek deposits
Credit expansion diverges from real economic
  activity
	Banks compete at
	national level with
	non-bank financial
	institutions
		Credit creation determined by
		struggle over market share
		and opportunities in
		speculative markets
		Total credit uncontrolled

Stage 6: Securitization
Capital adequacy ratios introduced to curtail
  credit
Banks have an increasing proportion of bad
  loans because of over-lending in Stage 5
Securitization of bank assets
Increase in off-balance sheet activity
Drive to liquidity
	Deregulation opens up
	international competition,
	eventually causing
	concentration in financial
	centres
		Shift to liquidity by emphasis
		being put on services rather
		than credit; credit decisions
		concentrated in financial
		centres; total credit determined
		by availability of capital,
		i.e. by central capital markets

Figure 13.2 The sequence of development of the banking system

Source: Based on Dow, 1999: Tables 1 and 2

Such increased complexity of the financial system over time has resulted in the development of a huge variety of different types of financial institution, each of which has a specific set of core functions. Figure 13.3 identifies the major types of financial service activity and their major functions within the system. In fact, the boundaries between these individual activities and institutions have become increasingly blurred. The first three financial institutions shown in Figure 13.3 are concerned with the creation and distribution of credit in various forms. The two remaining categories are concerned with different forms of risk indemnification (insurance companies) and with certifying the accuracy of financial accounts (accountancy firms).

Type	Primary functions
Commercial bank	Administers financial transactions for clients (e.g. making payments, clearing cheques). Takes in deposits and makes commercial loans, acting as intermediary between lender and borrower
Investment bank/ securities house	Buys and sells securities (i.e. stocks, bonds) on behalf of corporate or individual investors. Arranges flotation of new securities issues
Credit card company	Operates international network of credit card facilities in conjunction with banks and other financial institutions
Insurance company	Indemnifies a whole range of risks, on payment of a premium, in association with other insurers/reinsurers
Accountancy firm	Certifies the accuracy of financial accounts, particularly via the corporate audit

Figure 13.3   Major types of financial service activity

The dynamics of the market for financial services[8]

Ever since the mid- and late-1970s, financial services around the world have been characterized by a sharp intensification of competition and a rapid transformation of markets. These changes are leading to the emergence of a new, more dynamic market environment, quite unlike that which characterized the industry during the previous decades.[9]

This 'new competition' consists of four major elements:

- *Market saturation.* By the late 1970s, traditional financial services markets were reaching saturation. There were fewer and fewer new clients to add to the list; most were already being served, particularly in the commercial banking sector but also in the retail sector in the more affluent economies.

- *Disintermediation.* This is the process whereby corporate borrowers in particular make their investments or raise their needed capital without going through the 'intermediary' channels of the traditional financial institutions, particularly the bank. Instead, they have increasingly sought capital from non-bank institutions, for example through securities. A similar trend has been occurring in the retail markets, where private investors have switched funds from traditional savings deposit accounts to higher-yielding funds such as investment trusts and mutual funds.

- *Deregulation of financial markets.* Financial services markets have traditionally been extremely closely regulated by national governments. One of the most important developments of the past few years has been the increasing deregulation or liberalization of financial markets. Deregulation has been aimed primarily at three areas:
  - the opening of new geographical markets
  - the provision of new financial products
  - changes in the way in which prices of financial services are set.

- *Internationalization of financial markets.* Demand for financial services is no longer restricted to the domestic context: financial markets have become increasingly global. Three types of demand are especially significant:

  o the massive growth in international trade

  o the global spread of transnational corporations

  o the vastly increased institutionalization of savings.

Each of these forces for change in the demand for financial services is inter-related. Taken together they create a *new competitive environment*. Before looking specifically at the strategic responses of the major institutions, however, we need to consider two issues in more detail: first, technological change and, second, government regulation (and deregulation) of financial services.

Technological innovation and the financial services industries

> Travelling at the speed of light, as nothing but assemblages of zeros and ones, global money dances through the world's fiber-optic networks in astonishing volumes … National boundaries mean little in this context: it is much easier to move 41 billion from London to New York than a truckload of grapes from California to Nevada.[10]

Technological developments in communications technologies and in the broader sphere of information technology are absolutely fundamental to the financial services industries. *Information is both the process and the product of financial services.* Their raw materials are information: about markets, risks, exchange rates, returns-on-investment, creditworthiness. Their products are also information: the result of adding value to these informational inputs. In the words of one financial services executive:

> We don't have warehouses full of cash. We have *information* about cash – *that* is our product.[11]

A particularly significant piece of added value is embodied in the *speed* with which financial service firms can perform transactions and the global extent over which such transactions can be made. Not surprisingly, therefore, all the major financial services firms invest huge sums in information technologies. In the mid-1990s, for example, the top ten US banks spent more than $10 billion on information technology; the top ten European banks spent almost $12 billion in the same year.[12]

Susan Strange[13] neatly identifies the three key technologies that have transformed the financial services industries as computers, chips and satellites:

- *Computers* 'have changed the prevailing system of financial settlement out of all recognition. For hundreds of years, current payments and capital transfers were completed in negotiable cash – coins and notes – and in written promises to pay. They were recorded by handwriting in ledgers. Coins and notes gave way progressively to cheques and other promises to pay, such as letters of credit. Since the early 1970s, these too have become obsolete. *Computers have made money electronic* ... by the mid-1990s, computers had not only transformed the physical form in which money worked as a medium of exchange, they were also in the process of transforming the systems by which payments of money were exchanged and recorded.'[14]

- *Chips (microprocessors)* allow customers to pay for their purchases using plastic. The credit card revolution is one of the most pervasive of all recent social changes. Current development of 'smart cards' is extending the scope of such transactions by a huge degree as monetary value can be added to the microprocessor embedded in the card.[15]

- *Satellites* are one of the bases of global electronic communication, as we saw in Chapter 4.

Consequently, the effects of the information technologies on financial services have been immense.[16] In particular:

- They have vastly increased productivity in financial services.
- They have altered the patterns of relationships or linkages both within financial firms and also between financial firms and their clients.
- They have greatly increased the velocity, or turnover, of investment capital. For example, the ability to transfer funds electronically – and, therefore, instantaneously – has saved billions of dollars in interest payments which were formerly incurred by the delay in making transfers.
- At the international scale, they have enabled financial institutions both to increase their loan activities and also to respond immediately to fluctuations in exchange rates in international currency markets.

From a technological viewpoint, therefore, it is now possible for financial services firms to engage in global 24-hours-a-day trading – to 'follow the sun' – whether this be in securities, foreign exchange, financial and commodities futures or any other financial service.[17] As Figure 13.4 shows, the trading hours of the world's major financial centres overlap.

However, true 24-hour trading is currently limited to certain kinds of transaction partly because, although the technology is available, either the organizational structure or the national regulatory environment remain an obstacle. In addition, the nature of the product can be a limiting factor.

Global trading is encouraged by products that are homogeneous, familiar and tradable in large quantities, as, for example, trading in foreign exchange and US government bonds. A fully internationalised banking and finance system would operate on a 24-hour basis with information flowing between different countries with minimal institutional and regulatory barriers. Clearly, this situation has not been reached and there are numerous barriers still in existence.[18]

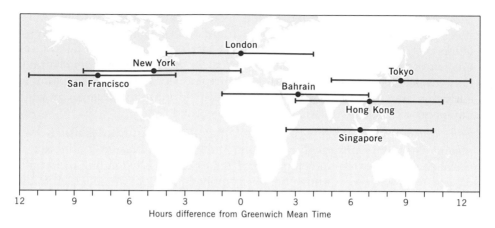

Figure 13.4 Twenty-four-hour financial trading: the trading hours of the major world financial centres

Source: Warf, 1989: Figure 5

To the extent that such electronic transactions do not require direct physical proximity between seller and buyer they are a form of 'invisible' international trade. In that sense, therefore, financial services are one form of service activity that is *tradable.* Electronic communications have also contributed greatly to the bypassing of the commercial banks and the trend towards the greater *securitization* of financial transactions. Securitization, in the broad sense, is simply the conversion of all kinds of loans and borrowings into 'paper' securities which can be bought and sold on the market. Such transactions may be performed directly by buyers and sellers without necessarily going through the intermediary channels of the commercial banks.

The *global integration of financial markets* brings many benefits to its participants: in speed and accuracy of information flows and rapidity and directness of transactions, even though the participants may be separated by many thousands of miles and by several time zones. But such global integration and instantaneous financial trading also have their costs. 'Shocks' which occur in one geographical market now spread instantaneously around the globe, creating the potential for global financial instability. Financial 'contagion' is endemic in the structure and operation of the contemporary financial system.

The influence of information technologies on the speed and geographical extent of financial transactions between sellers and buyers of financial services is one expression of the effect of technological change on the financial services industries. Another is their effect on the *internal operations* of financial services firms. For example, banks were among the earliest adopters of computer technology to automate the internal processing of financial transactions, the so-called 'back office' functions which are otherwise highly labour-intensive.

Computerization and related technologies have also been applied to the firms' 'front offices', that is, to the direct interface between firms and their customers. In retail banking, for example, counter clerks or tellers operate on-line computer terminals; automated teller machines and cash dispensers have become the norm, giving customers access to certain services outside normal banking hours; direct banking, by telephone or the Internet, has become the fastest growing sector of retailing banking business. The equivalent of such services for the large corporate customers is the development of electronic cash management systems in which the corporate customer's computers are linked directly to the bank.

These technological and organizational developments are drastically changing the nature of work at all levels. In particular, they redefine skills and increase flexibility. Figure 13.5 summarizes some of the major skills changes. In particular, there has been a

> widespread tendency for a dramatic decline in the volume of clerical processing work performed, until recently, manually by lower-tier personnel (with some assistance of mainframe computers for data crunching) ... Paralleling this transformation in data processing and data handling, increasing competition is generating new demands for both sales and assistance personnel and for specialists able to identify new markets, conceive new products, develop new systems and sell the new, often complex services (swaps, futures, etc.) ... The outcome of this profound process of skill transformation is the emergence of a new matrix of competencies that may be viewed in terms of new skills that are being substituted progressively for older ones.[19]

Old competencies	New competencies
Common emerging competencies	
• Ability to operate in well-defined and stable environment • Capacity to deal with repetitive, straightforward and concrete work process • Ability to operate in a supervised work environment • Isolated work • Ability to operate within narrow geographical and time horizons	• Ability to operate in ill-defined and ever-changing environment • Capacity to deal with non-routine and abstract work process • Ability to handle decisions and responsibilities • Group work, interactive work • System-wide understanding; ability to operate with expanding geographical and time horizons
Specific emerging competencies	
Among upper-tier workers	
• *Generalist competencies.* Broad, largely unspecialized knowledge; focus on operating managerial skills • *Administrative competencies.* Old leadership skills; routine administration; top-down, carrot-and-stick personnel management approach; ability to carry out orders from senior management	• *The new expertise.* Growing need for high-level specialized knowledge in well-defined areas needed to develop and distribute complex products • *The new entrepreneurship.* Capacity not only to manage but also set strategic goals; to share information with subordinates and to listen to them; to motivate individuals to develop new business opportunities
Among middle-tier workers	
• *Procedural competencies.* Specialized skills focused on applying established clerical procedural techniques assuming a capacity to receive and execute orders	• *Customer assistance and sales competencies.* Broader and less specialized skills focused on assisting customers and selling, capacity to define and solve problems
Among lower-tier workers	
• Specialized skills focused on data entry and data processing	• Disappearance of low-skill jobs

Figure 13.5   The changing nature of skills in banks and insurance companies

Source: Based on Bertrand and Noyelle, 1988: Table 4.1

Innovations in telecommunications and in process technologies, therefore, have helped to transform the operations of financial services firms. But there has also been a variety of *product innovations*. A whole new array of financial instruments has appeared on the scene that can be categorized into two broad types:

- those providing new methods of lending and borrowing and
- those facilitating greater spreading of risk.

Table 13.1 gives some examples of the kinds of financial innovation that have greatly increased the product diversity of financial markets. The most important product innovation since the mid-1980s has been the phenomenal growth of the so-called *derivatives* markets.[20] Derivatives are 'financial tools derived from other financial products, such as equities and currencies. The most common of these are futures, swaps, and options … The derivatives market aims to enable participants to manage their exposure to the risk of movements in interest rates, equities, and currencies'.[21]

Table 13.1 Examples of product innovations in financial markets

Type of financial instrument	Basic characteristics
Floating-rate notes (FRNs)	Medium- to long-term securities with interest rates adjusted from time to time in accordance with an agreed reference rate, e.g. the London Inter-Bank Offered Rate (LIBOR) or the New York banks' prime rate
Note issuance facilities (NIFs)	Short- to medium-term issues of paper which allow borrowers to raise loans on a revolving basis directly on the securities markets or with a group of underwriting banks
Eurocommercial paper	Non-underwritten notes sold in London for same-day settlement in US dollars in New York. More flexible than longer-term Euronotes of 1, 3 or 6 months' duration
Loan sales	The sale of a loan to a third party with or without the knowledge of the original borrower
Interest rate swaps	A contract between two borrowers to exchange interest rate liabilities on a particular loan, e.g. the exchange of fixed-rate and floating-rate interest liabilities
Currency swaps	Financial transactions in which the principal denominations are in different currencies

Without question, therefore, technological developments in telecommunications and information technology, as well as in product innovations, have transformed the financial services industries. The global integration of financial markets has become possible, collapsing space and time and creating the potential for virtually instantaneous financial transactions in loans, securities and a whole variety of financial instruments. However, completely borderless financial trading does not actually exist, for the simple reason that most financial services remain very heavily supervised and regulated by individual national governments. Let us now see how the regulatory system operates and how it is changing.

The role of the state: regulation and deregulation in financial services[22]

A tightly regulated system

> Financial systems are also *regulatory spaces*. The history of money is also a history of regulation. Money has a habit of seeking out geographical discontinuities and gaps in these regulatory spaces, escaping to places where the movement of financial assets is less constrained, where official scrutiny into financial dealing and affairs is minimal, where taxes are lower and potential profits higher.[23]

Before the 1960s there was really no such thing as a 'world' financial market. The IMF, together with the leading industrialized nations, acted to ensure a broadly efficient global mechanism for monetary management based, initially, on the post-war Bretton Woods agreement. At the national level, financial markets and institutions were very closely supervised, primarily because of concerns over the vulnerability of the financial system to periodic crisis and because of the centrality of finance to the operation of a country's economic system at both the macro- and the micro-scales.

Financial services, therefore, have been the most tightly regulated of all economic activities, certainly more so than manufacturing industries. The forms of such regulation can be divided into two major types:

- Those governing the *relationships* between different financial activities. Most countries have restricted, or even prohibited, firms from participating in a range of financial service activities. In other words, each national financial services market has been *segmented* by regulation: banks operated in specified activities, securities houses in other areas of activity; the two were not allowed to mix across the boundaries. Neither was allowed to perform the functions of the other.

- Those governing the *entry* of firms (whether domestic or foreign) into the financial sector. Restricted entry into the different financial services markets has been virtually universal, although the precise regulations differ from country to country. Most, however, have restrictions governing the entry of foreign firms into financial services. At the very least, national governments retain the discretionary power to restrict inward investments and all countries use their anti-trust/competition laws to regulate foreign acquisitions of domestic financial activities. Governments have been especially wary of a too-ready expansion of the *branches* of foreign banks and insurance companies. Unlike subsidiary companies, which have to be separately incorporated, branches are far more difficult to supervise because they form an integral part of a foreign company's activities. In almost every case, there are limits on the degree of foreign ownership permitted.

'The crumbling of the walls'

Although many of these restrictions still exist, the regulatory walls have been crumbling – even collapsing altogether in some cases. The process was relatively slow at first but accelerated rapidly after the late 1980s. Deregulatory pressures have come from several sources, most notably the increasing abilities of international firms to take advantage of 'gaps' in the regulatory system and to operate outside national regulatory boundaries.

In Susan Strange's view, the emergence of the Eurodollar (that is, offshore) markets in the 1960s was the great technological breakthrough of international finance in the mid-20th century.[24] Initially, Eurodollars were simply dollars held outside the US banking system largely by countries, like the USSR and China, that did not want their dollar holdings to be subject to US political control. 'Once it was appreciated that Eurodollars … were free of US political control, it did not take bankers long to recognise that the dollar balances were also free of US banking laws governing the holding of required reserves and controls upon the payment of interest.'[25]

The rapid growth of this new currency market outside national regulatory control was certainly one of the major stimuli towards an international financial system. It was reinforced by the revolutions in telecommunications and information technologies, discussed earlier, which made possible the internationalization of financial transactions. Pressures built up, too, from the desires of banks and other financial services institutions to operate in a less constrained and segmented manner, both domestically and internationally. The internationalization of financial services and the deregulation of national financial services markets are virtually two sides of the same coin. Forces of internationalization were one of the pressures stimulating deregulation; deregulation is a necessary process to facilitate further internationalization.

Major deregulation has occurred in all the major economies. A series of changes in the United States since the 1970s has both eased the entry of foreign banks into the domestic market and facilitated the expansion of US banks overseas. In 1981, the United States allowed the establishment of International Banking Facilities (IBFs), which created 'onshore offshore' centres able to offer specific facilities to foreign customers. Earlier, in 1975, the New York Stock Exchange had abolished fixed commissions on securities transactions. Later, the federal government introduced a major reform of the US financial system aimed at allowing banks to become involved in a whole variety of financial services and to operate nationwide branching networks. One of the paradoxes of the US system was that the 'non-bank banks', such as General Motors' Acceptance Corporation, could operate without many of the restrictions that applied to the 'proper' banks.

In the United Kingdom the so-called 'Big Bang' of October 1986 removed the barriers which previously existed between banks and securities houses and allowed the entry of foreign firms into the Stock Exchange. In France the 'Little Bang' of 1987 gradually opened up the French Stock Exchange to outsiders and to foreign and domestic banks. In Germany foreign-owned banks were allowed to lead-manage foreign DM issues, subject to reciprocity agreements. In Japan the restrictions on the entry of foreign securities houses have been relaxed (though not removed) and Japanese banks are now allowed to open international banking facilities. But the Japanese financial system remained tightly regulated. In 1996, the Japanese government announced its intention to undertake a wide-ranging deregulation of the country's financial system by 2001. That process is still ongoing in the context of Japan's continuing deep financial crisis. Even in the tightly controlled regulatory environment of Singapore, the government has progressively loosened the restrictions on the financial sector in order to maintain the country's position as a major Asian financial centre.

Increasing deregulation of financial services is an important component of both major regional economic blocs – the European Union and the NAFTA. Reforms within the EU have been directed at removing the individual national financial regulatory structures which have inhibited the creation of an EU-wide financial system so that financial services flow freely throughout the EU and financial services firms can establish a presence anywhere within the single market. The creation of a single European currency has enormous potential implications for the structure of the financial services sector in the EU. A financial services agreement was part of the Canada–United States Free Trade Agreement, while the NAFTA provides for Mexico gradually to open up its financial sector to US and Canadian firms, with all barriers to be eliminated by 2007.

In addition, negotiations within the WTO to complete an international agreement on financial services continue. The essence of those negotiations is to move towards a *multilateral*, rather than a bilateral, regulatory structure for financial services. An interim financial services agreement, involving some 30 countries, was reached by the WTO in 1995, which aimed to guarantee access to banking, securities and insurance markets. But the United States did not fully accept the agreement and reserved the right to restrict access to its own financial markets on the basis of reciprocity (although existing arrangements stayed in place). We will return to the contentious issue of a 'new financial architecture' for the world economy in Chapter 18.

The accelerating deregulation of financial services is the most important current development in the internationalization of the financial system. Even so, there remain substantial differences in the extent and nature of financial services' regulation in different countries. The international regulatory environment is highly asymmetrical. Such differences, of course, have a powerful influence on the locational strategies of international financial services firms.

Corporate strategies in financial services

The environment in which financial services companies must operate, therefore, is made up of the following interacting elements:

- shifting patterns in the demand for financial services
- technological innovations which affect how these services can be delivered
- the changing regulatory framework.

But, of course, firms are not simply passive respondents to environmental changes over which they have no influence. The processes are dynamic and interactive. Indeed, the strategies and actions of the major financial services firms – the banks, securities houses, insurance companies and the like – have themselves been highly influential in changing that environment. There is no simple cause–effect relationship but, rather, a complex interplay between actors and processes.

In this section we focus on the specialist financial services companies and exclude the in-house financial activities of TNCs in other industries. Of course, all TNCs have large-scale and sophisticated financial operations. The corporate treasury departments of major companies are significant entities in their own right and are larger than many specialist financial services companies. However, the primary functions of these in-house operations are to contribute towards the efficiency of their own corporate systems. They do this by, for example, optimizing internal flows of funds, investing surplus capital effectively on the financial markets, and hedging against the foreign currency fluctuations endemic in the operations of any TNC, whatever its line of business.

As far as the specialist financial services firms are concerned, we will explore three major strategic trends:

- *concentration and consolidation* through merger and acquisition
- *transnationalization* of operations
- *diversification* into new product markets.

Concentration and consolidation

The history of the financial services industries – particularly of banking – is one of a continuous trend towards greater concentration into a smaller number of companies. As Figure 13.2 suggests, this process is intrinsically geographical as smaller financial institutions, set up to serve geographically localized markets have become swallowed up (or have themselves swallowed up others). Within this long-term trend towards concentration, periods of frenetic merger and acquisition activity – often associated with significant regulatory changes – have greatly accelerated concentration and consolidation.

The 1990s saw a particularly heavy wave of merger activity which, together with the decline in the fortunes of Japanese banks, resulted in a major reshuffle of the top 25 banks. In 1989 (Table 13.2), no fewer than 17 of the top 25 banks were Japanese (compared with only nine in 1975), whilst all of the top seven banks in 1989 were Japanese, a reflection of Japan's spectacular economic growth during the 1980s. By 2000 (Table 13.3) the situation had changed dramatically for a number of reasons.

First, the Japanese banks suffered very badly in the collapse of the Japanese 'bubble economy' in the early 1990s. Having lent heavily (and, with hindsight,

Table 13.2 The world's top 25 banks, 1989

Rank	Bank	Home country	Assets ($million)
1	Dai-Ichi Kangyo Bank	Japan	388,900
2	Sumitomo Bank	Japan	377,200
3	Fuji Bank	Japan	365,600
4	Mitsubishi Bank	Japan	352,900
5	Sanwa Bank	Japan	350,200
6	Industrial Bank of Japan	Japan	268,500
7	Norinchukin Bank	Japan	243,600
8	Crédit Agricole	France	242,000
9	Banque National de Paris	France	231,500
10	Citicorp	United States	230,600
11	Tokai Bank	Japan	225,800
12	Mitsubishi Trust & Banking	Japan	212,000
13	Crédit Lyonnais	France	210,700
14	Mitsui Bank	Japan	206,900
15	Barclays Bank	United Kingdom	204,900
16	Deutsche Bank	Germany	202,300
17	Bank of Tokyo	Japan	198,300
18	Sumitomo Trust & Banking	Japan	197,900
19	National Westminster Bank	United Kingdom	186,500
20	Long Term Credit Bank of Japan	Japan	183,300
21	Mitsui Trust & Banking	Japan	180,000
22	Taiyo Kobe Bank	Japan	174,500
23	Yasuda Trust & Banking	Japan	171,700
24	Société Générale	France	164,800
25	Daiwa Bank	Japan	156,500

Source: Based on *Euromoney*, June 1990: 119

misguidedly) during the boom years of the 1980s, many of the major Japanese banks had to make large provisions in their balance sheets to cover losses on these big loans. Second, a number of very large bank mergers occurred in the mid-1990s. For example, by 1996 six of the world's biggest banks had been reduced in number to three through merger. In Japan itself, Mitsubishi Bank and the Bank of Tokyo (sixth and tenth in Japan respectively) merged to form the world's biggest bank. In the United States, Chemical Bank and Chase Manhattan merged to form the biggest bank in the country whilst, at the same time, California Wells Fargo took over First Interstate Bank.

In the late 1990s/early 2000s, even larger mergers occurred in the banking sector:

- Fuji Bank, Dai-Ichi Kangyo Bank and the Industrial Bank of Japan merged to form the Mizuho Financial Group – the world's largest bank in 2000.

- Bank of Tokyo, Mitsubishi Bank and Mitsubishi Trust merged to form Bank of Tokyo–Mitsubishi (ranked fifth in 2000).
- Sanwa Bank and Tokai Bank merged in 2002 to form the UFJ Bank.
- Citicorp merged with Travelers Group to form Citigroup, ranked second in 2000.
- Chase Manhattan Bank merged with JP Morgan to form JP Morgan Chase, ranked fourth in 2000.
- HSBC acquired the French bank CCF, and became the sixth largest bank in 2000.

However, not all planned mergers and acquisitions came to fruition. The most notable was the planned merger of the two major German banks, Dresdner Bank and Deutsche Bank, which was called off before completion.

Table 13.3   The world's top 25 banks, 2000

Rank	Bank	Home country	Assets ($million)
1	Mizuho Financial Group	Japan	1,259,498
2	Citigroup	United States	902,210
3	Deutsche Bank	Germany	874,706
4	JP Morgan Chase & Co	United States	715,348
5	Bank of Tokyo–Mitsubishi	Japan	675,640
6	HSBC Holdings	United Kingdom	673,614
7	Hypovereinsbank	Germany	666,706
8	UBS	Switzerland	664,560
9	BNP Paribas	France	645,793
10	Bank of America Group	United States	642,191
11	Crédit Suisse	Switzerland	603,381
12	Sumitomo Bank	Japan	540,875
13	ABN Amro Bank	Netherlands	506,687
14	Crédit Agricole Groupe	France	498,426
15	Industrial & Commercial Bank of China	China	482,983
16	Norinchukin Bank	Japan	462,593
17	Royal Bank of Scotland	United Kingdom	461,511
18	Barclays Bank	United Kingdom	458,787
19	Dresdner Bank	Germany	449,898
20	Commerzbank	Germany	427,192
21	Sanwa Bank	Japan	425,302
22	Société Générale	France	424,192
23	Sakura Bank	Japan	416,129
24	Bank of China	China	382,730
25	ING Bank	Netherlands	378,150

Source: Based on *The Banker,* July 2001: 132

These deals … are being driven by a potent mixture of fear and ambition … too many banks with too much capital are chasing too few borrowers … Banks are afraid of competition from outside the industry as well as from within it. For years now, they have been trapped in a pincer. On the one hand, mutual funds [unit trusts] … have eaten away at banks' cheap deposits … On the other hand, bankers are facing stiff competition for lending business from the financing arms of industrial firms … Worse, their traditional role as middlemen has been decimated by the growth of the capital markets, which match borrowers directly to lenders … But their motives for merging are not purely defensive. In some cases, link-ups are aimed at uniting banks' common strengths in particular areas and at producing huge cost savings.[26]

While such giant mergers have reconfigured the world banking system in spectacular ways, there is considerable doubt as to their real value. According to the Federal Reserve Bank of New York, there is no evidence that in-market mergers lead to significant improvements in efficiency.[27] One potential problem lies in the nature of financial service products themselves:

Mergers between car manufacturers or consumer goods companies allow production to be centralized because the products are essentially the same across Europe. The same is not true for many financial services … 'A bottle of beer is a real thing but financial products are intangible constructs of regulation, culture and behaviour. A current account is a different product in every country, while life insurance policies are tax-driven products and tax systems are not harmonised. Where will the synergies come from?'[28]

Nevertheless, the merger waves in the financial sector show little sign of disappearing whilst, in the process, many thousands of jobs are destroyed through post-merger rationalization actions.

Transnationalization of banking operations

Although banks have long engaged in international business – for example, through foreign exchange dealing or providing credit for trade – historically, this kind of business was carried out from their domestic locations. Any business that could not be carried out by mail or using telecommunications was handled by correspondent banks in the relevant countries. There was no need for a direct physical presence abroad.[29] A small number of banks certainly set up a few overseas operations towards the end of the 19th century. But even in the early part of the 20th century, the international banking network was very limited indeed. Almost all international banking operations were 'colonial' – part of the imperial spread of British, Dutch, French and German business activities. In 1913 the four major US banks had only six overseas branches between them. By 1920 the number of branches had grown to roughly a hundred but there was little further change until the 1960s. During that period, Citibank had by far the most extensive international network of any US bank.

As with TNCs in manufacturing industries, the most spectacular expansion of transnational banking occurred in the 1960s and 1970s. Again, the initial surge was dominated by US firms, a reflection of both the focal role of the United States in the post-war international financial and trading system and also the rapid proliferation of US TNCs. But European, and later Japanese, banks increasingly internationalized their operations. Figure 13.6 shows that the number of foreign affiliates of banks increased from 202 in 1960 to 1,928 in 1985. Growth in this international network was especially rapid in the 1960s and 1970s. At the same time the *geographical composition* of the international banking network changed. US banks became less dominant than in the 1960s and 1970s, whilst European and Japanese banks increased the size of their international branch network.

Figure 13.6

Growth in the number of overseas bank affiliates, 1960–1985

Source: Based on *OECD Observer* (1989) vol. 160, p. 36

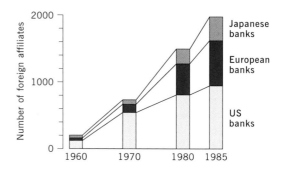

Interestingly, it is not necessarily the biggest banks in terms of total assets that are the most internationalized in their operations. Table 13.4 lists the 25 most transnational banks in 2000. A comparison with Table 13.3 shows that only six of the world's 25 largest banks have more than 50 per cent of their assets outside their home country whilst only a further four of the top banks have more than 40 per cent of their assets abroad. Most striking is the fact that no Japanese bank appears in the list of most transnational banks. In fact, even in 1989 at the peak of their dominance of world banking (see Table 13.2), Japanese banks were rather less significant in terms of their international spread than their asset values suggested.

One reason was that their roles were primarily to serve the major Japanese business corporations. Their retail business is far less extensive than that of the American and European banks. In the late 1980s, the average number of foreign operations per bank was far lower among the leading Japanese banks than in banks from the other leading source countries, although the German figure was also low. Thus, although the Japanese banks undoubtedly internationalized very rapidly, the really extensive global networks are still those operated by some of the American and European banks. Indeed, faced with financial problems from the early 1990s, many Japanese banks drastically cut back their international operations through closure of establishments. For example, in late 1998, Daiwa announced its withdrawal from all overseas business, involving the planned closure of its branches in London, Hong Kong, Shanghai, Seoul and Singapore, and the liquidation of its foreign subsidiaries.[30]

The specific reasons for, and the pace of, the transnationalization of banks and other financial services may well vary from one case to another, but there is an overall general pattern.

- In the immediate post-1945 period, the major international function of the small number of US transnational banks was the provision of finance for trade. But as the overseas operations of US TNCs in manufacturing and other activities accelerated in the early 1960s, the functions of the US transnational banks evolved to meet their particular demands. Thus, the staple business of US banks in London became that of servicing the financial needs of their major industrial clients who were rapidly extending their operations outside the United States.

- In the second half of the 1960s transnational banking functions took on a significant additional dimension with the growth of the Eurodollar market. This, as we have seen, was a market outside the regulatory control of the US government. As a consequence, US banks could raise money there for re-lending domestically as well as overseas.

Table 13.4 The world's top 25 transnational banks, 2000

Rank	Bank	Home country	Per cent of assets overseas
1	American Express Bank	United States	80.9
2	Standard Chartered	United Kingdom	79.2
3	UBS	Switzerland	76.8
4	Investec	South Africa	73.9
5	Crédit Suisse	Switzerland	72.9
6	Deutsche Bank	Germany	67.7
7	ABN Amro	Netherlands	63.0
8	HSBC	United Kingdom	62.2
9	ING Bank	Netherlands	57.8
10	Danske Bank	Denmark	54.8
11	Allied Irish Banks	Ireland	53.8
12	Bank Austria Group	Austria	52.2
13	BNP Paribas	France	47.2
14	National Australia Bank	Australia	46.0
15	KBC	Belgium	45.9
16	Bank of Montreal	Canada	45.1
17	JP Morgan	United States	45.1
18	Scotiabank	Canada	43.6
19	Dresdner Bank	Germany	43.1
20	Raiffeisen Zentralbank Österreich	Austria	42.7
21	Bank of Ireland Group	Ireland	41.5
22	Citigroup	United States	40.1
23	Banco Santander Central Hispano	Spain	40.3
24	C.I.B.C	Canada	39.7
25	Société Générale	France	38.8

Source: Based on *The Banker*, February 2001: 42

● The attractions of international banking widened progressively as both the global capital market evolved and as local capital markets overseas developed. As banks from the other industrialized economies also internationalized, the process became *self-reinforcing*. All large banks had to operate internationally; they had to have a presence in all the leading markets.

● The internationalization of financial markets received fresh impetus in the 1970s from two sources: first, the 'windfall' capital acquired by the OPEC oil-exporting countries which they sought to lend abroad, and, second, the progressive dismantling of exchange controls on capital movements by the Federal Republic of Germany, the United States, Japan and the United Kingdom.

Figure 13.7 summarizes the major phases of development of international banking. As with all such models, it should be seen as a broad general framework which captures the main features of the process, although the detail may well vary from case to case.

These broad developments, intensified by the further deregulation of financial markets in many countries and by technological innovation, have pulled more and more securities firms into international operations.

> Up to 1979/80, the US multinational investment bank had little more than a large office in London and perhaps some much smaller ones in other European countries, perhaps an Arab country and possibly (though less likely) Japan. From 1980 onward, the development of the US investment bank as a multinational changed qualitatively.[31]

The same process applied to the leading Japanese securities houses, such as Nomura, Daiwa, Nikko and Yamaichi, although with differences of detail. Finally, as both industrial and service companies spread globally a further boost was given to the increased internationalization of accountancy companies.

	Phase I National banking	Phase II International banking	Phase III International full- service banking	Phase IV World full- service banking
Internationalization of customer companies	Export – import	Active direct overseas investment	Transnational corporation	
International operations in banking	Mainly foreign exchange operations connected with foreign trade. Capital transactions are mainly short-term ones	Overseas loans and investments become important, as do medium- and longer-run capital transactions	Non-banking fringe activities such as merchant banking, leasing, consulting and others are conducted	
				Retail banking
Methods of internationalization	Correspondence contracts with foreign bank	To strengthen own overseas branches and offices	By strengthening own branches and offices, capital participation, affiliation in business, establishing non-bank fringe business firms, the most profitable ways of fund-raising and lending are sought on a global basis	
Customers of international operations	Mainly domestic customers	Mainly domestic customers	Customers are of various nationalities	

Figure 13.7 The major phases in the development of international banking

Source: Based on Fujita and Ishigaki, 1986: Table 7.6

As a result, all of the transnational financial services firms have been basing their strategy on a direct presence in each of the major geographical markets and on providing a local service based on global resources. They are, in fact, selling an *international brand image,* with the clear message that a global company can cope most easily and effectively with every possible financial problem that can possibly face any customer wherever they are located. But, again, as in the case of mergers, there is considerable doubt that such a 'one size fits all' strategy really works in a world that continues to be highly differentiated.

Diversification into new product markets

It is a short, and supposedly logical, step from this kind of global strategic orientation to the notion that global financial services should not only operate globally in their own core area of expertise but also that they should supply a *complete package of related services.* The financial conglomerate or the financial supermarket has arrived on the scene. Deregulation has been a major stimulus for such a development. It is becoming increasingly possible for banks to act as securities houses, for securities houses to act as banks, and for both to offer a bewildering array of financial services way beyond their original operations. At the same time, entirely new non-bank financial services companies have emerged. Amongst all the leading accountancy firms, for example, there has been a strong move into management consultancy, including information systems. A leading bank's current portfolio of offerings includes: clearing banking, corporate finance, insurance broking, commercial lending, life assurance, mortgages, unit trusts, travellers' cheques, treasury services, credit cards, stockbroking, fund management, development capital, personal pensions and merchant banking.

Although some of this diversification has taken place by new greenfield investment, the majority has occurred through the process of acquisition and merger. This has been particularly the case in the rush of the securities houses to gain a foothold in newly deregulated markets, like the City of London in 1986, and in the creation of totally new financial services conglomerates. The rationale for diversification, both into new products and new geographical markets, is the familiar one of economies of scale and economies of scope:

> In financial services, the arguments are that there are large fixed costs in running a network of branches or agents. These can be spread over a lot of customers in large operations ... Another supposed economy derives from the benefits of scale in participating in capital markets. Unit transaction costs tend to be lower, the larger you are. Consumer recognition is another advantage to the large company who might expect to attract customers most easily ... a large diversified company can afford to cross-subsidize price wars in one sector with the profits made in another ... As well as economies of scale, it is generally supposed that there are economies of scope in financial services. The bank doing business in all major European cities will find it easier to serve clients than a similar size bank that is locally concentrated. The international bank would be better placed to serve international clients, and would gain strategically valuable information about a range of markets from its dealings abroad.[32]

The arguments for a strategy of internationally diversified financial service operations, then, are that it enables such a company to offer an entire package of services – a 'one-stop shop' – to customers. Their supply by a large, internationally recognized, brand name firm – backed up, of course, by lavish advertising expenditure – is supposed to give a reassurance to potential customers that they will receive the highest quality of service. Whether or not economies of scale and scope really do exist to a significant extent is a matter over which there is considerable disagreement. The large financial companies themselves certainly seem to think so. Although there have been divestments as some financial services firms have shed parts of their diversified portfolio of activities, the diversification and internationalization trend continues. On the other hand, there are distinct benefits in geographical specialization:[33]

- The administrative benefits of size can be exaggerated.
- Local knowledge may be a major advantage in assessing local risks and opportunities.
- The benefits of recognition through size may be less in national markets where names have already established themselves.

In other words, 'localization might be a marketing advantage'. This, of course, has been recognized, at least in principle and in their advertising blurbs, by the major financial services companies in their emphasis on providing a locally sensitive service within a global organization. During 2002, for example, HSBC ran a series of advertisements showing examples of different national meanings given to the same behavioural symbol under the heading 'Never underestimate the importance of local knowledge'.

Geographical structures of financial services activities[34]

At first sight, technological developments in communications systems would appear to release financial services companies from the spatial constraints on the location of their activities. Financial services firms, in particular, would appear to be especially footloose. They are not tied to specific raw material locations whilst at least some of their transactions can be carried out over vast geographical distances, using telecommunications facilities. Such considerations led Richard O'Brien to entitle his 1992 book on global financial integration 'The End of Geography'. That title undoubtedly reflects the views of many people, both inside and outside the financial services industries.[35] The aim of this section is to argue that such a view is misplaced. To be fair to O'Brien, however, it should be pointed out that his position was not totally unequivocal:

> The end of geography, as a concept applied to international financial relationships, refers to a state of economic development where geographical location no longer matters in finance, or matters much less than hitherto … In theory, the end of geography should mean that location no longer matters. Yet, *'everybody has to be somewhere'*, …

Location still exists, even though electronic markets make that location more difficult to identify in traditional geographical ways. Firms, individuals, markets, even products, have to have a sense of place.[36]

There is no doubt that the revolutionary developments in the space- and time-shrinking technologies permit financial transactions to whizz around the world electronically, and that deregulation has reduced the impermeability of national boundaries to financial flows. But, far from heralding the 'end of geography', this has, in fact, made geography *more* – not less – important. Indeed, we find that, at global, national and local scales, the major financial services activities continue to be extremely *strongly concentrated geographically*. They are, in fact, more highly concentrated than virtually any other kind of economic activity except those based on highly localized raw materials. However, there are some subtle variations according to the particular function involved; there is a division of labour within financial services firms, parts of which may show a greater degree of geographical decentralization. In this section, we focus on two aspects of this issue: first, the relationship between geography and the nature of financial products, and, second, the geographical structure of financial services.

Geography and the nature of financial products

The informational content of financial products themselves is inextricably related to the spatial structure of the global finance industry.[37]

> The geography of places shapes the operation of finance markets and the global economy primarily because of the geography of information that is embedded in the provision of specific financial products. There is, in effect, a robust territoriality to the global financial industry … financial products often have a distinct spatial configuration of information embedded in their design.[38]

Table 13.5 identifies three types of financial product, each of which has distinctive, geographically derived, properties. The specific informational nature of these three types of financial product, and the fact that such information is, itself, locationally specific is reflected in a particular 'topography' of financial markets:

- *Transparent products* are produced

  where the trading volume is large enough to make economical the repetitive small markets that are associated with this form of trading. These turnover efficiencies are only met in the larger global centers, so that the design and production of transparent products provides the apex of the world's financial system.[39]

- *Translucent products* (such as balanced equity products)

  draw upon a different set of factors and create opportunities for a second (national) level of financial centers … balanced equity products [are] typically differentiated by country of origin. They all require skills in design to structure them so as to maximize returns, spread risks, and maximize product differentiation in order to give such a product an edge in the competition for other funds. Hence there are opportunities for local operators and local markets. However, the costs of assembling and maintaining the information systems necessary to identify and monitor the various investments within a fund, and the skilled personnel required

Table 13.5   Financial products and their geographical characteristics

Financial product type	Probable market scope and type of financial centre	Information intensity	Specialist expertise required	Perceived risk-adjusted return
Transparent	Global	Ubiquitous	Low (e.g. equity products like stock market indexes). Traded internationally	Low
Translucent	National	Third party market-specific	Significant. Require detailed knowledge of national/local companies. May be traded internationally	Medium
Opaque	Local	Transaction-specific	Vital. Applies to products based upon trust and long-term relationships	High

Source: Based on Clark and O'Connor, 1997: Table 4.1; 99–104

to market them, rely upon significant scale economies. These scale economies limit the effective diffusion of these products to smaller, third-level financial centers and provide the larger national centers with an important source of business ... transaction costs can be minimized by concentrating the administration and organization and flow of funds, providing a further reason for the role of the first- and second-level centers; market share can also be maximized by location in these places, as companies can use their globally distributed networks of offices ... to funnel funds to the large markets in these cities. The costs of the global networks become another source of scale economies available to the major markets.[40]

●   *Opaque products* are those

where the design and production is shrouded in some mystery to the outsider, and local knowledge is essential for confident trading. These are opaque products like the REIT or property trust, where a set of properties are packaged into a product that can be sold in units or shares ... Property trusts will often be produced by firms with access to local markets as the information about them is so specialized, thereby providing the third layer in global financial geography. But, of course, these products may be consumed at distant locations.[41]

Thus, 'different types of products produce geography' while 'different geographies imply different products … local customs, market behavior, and conventions sustain the persistence of spatial differentiation'.[42]

The global network of financial centres

Financial service activities of all kinds tend to be very strongly concentrated in key metropolitan centres. These form a complex network spanning national boundaries and connecting major cities across the world. Figure 13.8 shows one representation of this global network of international financial centres, based upon 16 statistical variables.[43] The most important are:

- the volume of international currency clearings
- the size of the Eurocurrency market
- the volume of foreign financial assets
- the number of headquarters of the large international banks.

This network of international financial centres has three major levels, with London and New York being in the top level. In this scheme, Tokyo appears in the second tier.

Although this global network has certain stable features – for example, the long-sustained hierarchical dominance of London, New York and, more recently,

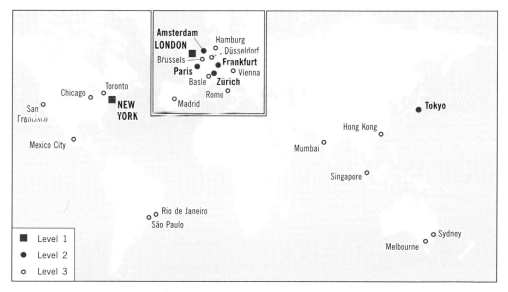

Figure 13.8   The global network of financial centres

Source: Based on Reed, 1989: Figure 1

Tokyo – it is by no means static. Financial centres may change their status in the network in response to changing conditions. Much may also depend on the measure used.[44] Figure 13.9 illustrates both of these circumstances for the top 12 banking centres. Measured in terms of aggregate income (Figure 13.9a), the effect of the Japanese financial crisis is abundantly clear. However, Tokyo's position in terms of assets (Figure 13.9b) has remained stable. Figure 13.9 also shows the emergence of several new centres, notably Beijing and Charlotte, North Carolina. Clearly, what we have is a global network of financial centres that is highly dynamic whilst, at the same time, containing elements of relative stability. Such stability reflects the operation of the kinds of path-dependent forces outlined in Chapter 2.[45] The structure of the global financial system still appears to be articulated geographically along the *tri-polar axis* of New York–London–Tokyo.

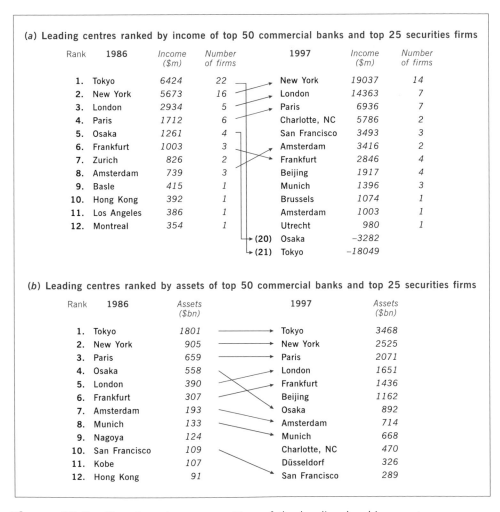

Figure 13.9 The changing composition of the leading banking centres

Source: Based on Sassen, 2001: Table 7.5

The widespread deregulation of financial services during the 1980s and 1990s had an especially significant impact on the structure of the global network of financial centres. Such developments

> made it possible for footloose financial firms to set up directly in the world's leading metropolitan centres rather than serving them from a distance ... International firms want to locate in these different metropolitan financial centres because each has a different financial specialization, and each is the hub of a different (although of course overlapping and inter-connected) continental global region. Foreign banks and related institutions have moved into these centres precisely because of geography, that is to expand their presence or gain access to specific markets, to capitalise on the economies of specialization, agglomeration and localization (skilled labour, expertise, contact, business networks, etc.) available in these centres, or to specialize their own operations and activities geographically.[46]

These basic locational attractions of international financial centres consist of four interlocking processes:[47]

- The characteristics of the *business organizations* involved in international financial centres: (a) much of the production of financial products and services occurs at the boundaries of firms and there is a strong reliance on repeat business; (b) the firms tend to be 'flattened and non-hierarchical' and based around 'small teams of relationship and product specialists'; (c) firms need to cooperate as well as to compete, as in the case of syndicated lending; (d) firms need to compare themselves with one another to judge their performance; (e) there is a need for a constant search for new business and for rapid response.

  > These shared characteristics point to two important correlates. First, in general, these firms must be sociable. Contacts are crucially important in generating and maintaining a flow of business and information about business. 'Who you know' is, in this sense, part of what you know ... 'relationship management' is a vital task for both employees and firms. Second, this hunger for contacts is easier to satisfy if contacts are concentrated, are proximate. When contacts are bunched together they are easier to gain access to, and swift access at that.[48]

- The *diversity of markets* in international financial centres: (a) their large size, which makes them both flexible in terms of entry and exit and also socially differentiated: 'more likely to consist of social "micro-networks" of buyers and sellers, whose effect on price-setting can sometimes be marked'; (b) their basis in rapid dissemination of information which may lead to major market movements; (c) their speculative and highly volatile nature. 'Again, as with the case of organizations, there are the two obvious corollaries to these characteristics: the twin needs for sociability and proximity.'[49]

- The *culture* of international financial centres: (a) such centres receive, send and interpret increasing amounts of information; (b) they are the focus of increasing amounts of expertise which arise from a complex division of labour involving workforce skills and machinery; (c) they depend on contacts and such contacts have become increasingly reflexive because of their basis in trust founded on relationships. Such cultural aspects of international financial centres are actually increasing in importance.

- The *dynamic external economies of scale* which arise from the sheer size and concentration of financial and related services firms in such centres. Such economies include: (a) the sharing of the fixed costs of operating financial markets (for example, settlement systems, document transport systems) between a large number of firms; (b) the attraction of greater information turnover and liquidity; (c) the enhanced probability of product innovations in such clusters (the 'sparking of mind against mind'); (d) the increased probability of making contacts, which rises with the number of possible contacts; (e) the attraction of linked services such as accounting, legal and computer services, which reduces the cost to the firm of acquiring such services; (f) the development of a pool of skilled labour; (g) the enhanced reputation of a centre which, in a cumulative way, increases that reputation and attracts new firms. In other words, the constellation of traded and untraded interdependencies, described in Chapter 2, tend to be especially strongly developed in international financial centres.

In effect, therefore, a small number of cities control almost all the world's financial transactions. It is a remarkable level of geographical concentration. But such cities are more than just financial centres. There is clearly a close relationship with the distribution of the corporate and regional headquarters of transnational corporations examined in Chapter 8. These global cities are, indeed, the *control points of the global economic system*. As far as financial services firms are concerned, the international financial centres – especially those at or near the top of the hierarchy – are their 'natural habitat'. Financial companies agglomerate together in these centres for the reasons outlined above.

But this does not mean that all international financial centres – even those at the top of the hierarchy – are identical. Far from it. Each has distinctive characteristics reflecting its specific historical and geographical embeddedness.[50] On those criteria measuring both the breadth and depth of global financial activity, New York and London occupy the apex of the international financial centre hierarchy. London is the more broadly based international financial centre.

> London's capacity as an international financial centre rests on three core activities: foreign exchange, international equities and exchange-traded derivatives ... London has increased its share of international foreign exchange in the last five years. It remains the largest international centre with 21.5% of business conducted in US$/DM transactions, 17% in US$/Y and 11% in US$/£ (1995 figures) ... In 1996, London's share of trade in foreign equities was 62% ... with New York accounting for 23.5% and Tokyo 0.1%.[51]

The daily turnover on the London foreign exchange market is almost as large as the turnover in New York and Tokyo put together. London's current significance as a global financial centre can be attributed to the following factors:[52]

- The historical evolution of the City as a world centre has created both a large pool of relevant skills and also an almost unparalleled concentration of linked institutions within a very small geographical area. 'Unlike the position in some other countries, insurance, commodities trading, futures markets, stock broking, bond trading and legal services in the UK are all concentrated around the City'.[53]

- Its geographical position locates it in a time zone between New York and Tokyo.
- The regulatory environment has encouraged the growth of international banking and is favourably disposed towards international bankers. Foreign banks in London can operate as 'universal' banks, a particularly important feature for US and Japanese banks. They can combine both banking and securities businesses there in ways that are prohibited so far in their domestic operations.

A major question today, however, is the extent to which London's pre-eminence as a financial centre is threatened by competition from rapidly growing European financial centres, especially Frankfurt – the location of the European Central Bank – and the UK's non-membership of the eurozone. The early signs, at least, are that London's status as a financial centre has not been adversely affected.[54]

> London's position as Europe's pre-eminent financial centre is unchallenged, in spite of the introduction of the euro and the growth of Frankfurt as a eurozone financial marketplace ... the euro had not altered the two cities' standing, even though the European Central Bank is in Frankfurt, while the UK remains outside the eurozone ... The City had retained its dominant position as Europe's leading centre for banking, capital markets, advertising, the law and management consultancy. London was 'the most connected world city' through its global service firm offices, ahead of New York, with Frankfurt in 15th place.[55]

London still has the largest concentration of foreign banks in the world, with more than 470 foreign branches, subsidiaries or representative offices located there in 2001. In comparison, Paris had 214 and Frankfurt 320. In 2001,

> London had 20 per cent of all cross-border bank lending worldwide, 52 per cent of foreign equity trading, and 31 per cent of all foreign exchange dealing: as much as its three closest rivals (New York, Tokyo, and Singapore) put together.[56]

London's strength as a financial centre, therefore, rests on the scale of its foreign exchange business and its deregulated securities markets. New York, in comparison, is by far the world's largest securities market. But it also has a huge concentration of international banks and other financial activities. London and New York still stand apart from Tokyo as truly global financial centres. The international significance of Tokyo has rested primarily on the strength of the Japanese economy itself. Tokyo's financial community has been far more domestically oriented. Japanese banks, as we saw earlier (Figure 13.6), were later entrants to international banking. The ongoing financial problems of the Japanese economy, dating from the collapse of the speculative bubble economy at the beginning of the 1990s, have not helped Tokyo's position, although it is a mistake to exaggerate the weakness of the Japanese economy.

Offshore financial centres[57]

> [s]cattered across the globe, a series of little places – islands and micro-states – have been transformed by exploiting niches in the circuits of fictitious capital. These places have set themselves up as offshore financial centres; as places where the circuits of fictitious capital meet the circuits of 'furtive money' in a murky concoction of risk and opportunity. Furtive money is 'hot' money that seeks to avoid regulatory attention and taxes.[58]

The speculative nature of financial transactions and flows and the desire to evade regulatory systems have led to the development of a number of offshore financial centres (OFCs). With few exceptions, the sole rationale for such centres is to provide these kinds of services outside the regulatory reach of national jurisdictions. They attract investors through their low tax levels and 'light' regulatory regimes.[59] For example, although more than 500 banks are 'located' in the Cayman Islands – with more than $500 billion in their accounts – only around 70 of these actually have a physical presence there. The vast majority are no more than 'a brass or plastic name plate in the lobby of another bank, as a folder in a filing cabinet or an entry in a computer system'.[60] Similarly, there are roughly 300,000 companies registered in the Virgin Islands, although 'only 9,000 of them show any signs of activity locally'.[61]

Figure 13.10 shows the geographical distribution of offshore financial centres. Each tends to fill a specific niche which it exploits in competition with other centres in the same spatial cluster and with similar niche centres elsewhere in the world. Much of the growth of such centres occurred in the 1970s, in places that were already operating as tax havens, to act as banks' 'booking centres' for their Eurocurrency transactions.

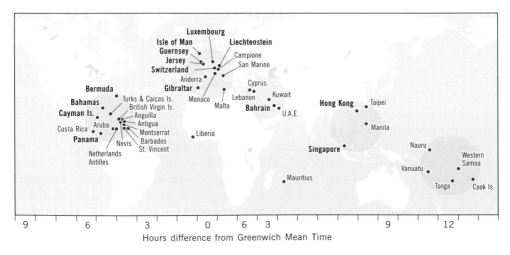

Figure 13.10 Offshore financial centres

Source: Based on Roberts, 1994: Figure 5.1

By operating offshore booking centres international banks could act free of reserve requirements and other regulations. Offshore branches could also be used as profit centres (from which profits may be repatriated at the most suitable moment for tax minimization) and as bases from which to serve the needs of multinational corporate clients.[62]

The location of these offshore centres, and especially their spatial clustering, is partly determined by time zones and the need for 24-hour financial trading.

The OECD has recently initiated a drive to make OFCs more accountable by enforcing greater disclosure and ending preferential treatment for foreigners. Not surprisingly, the 35 OFCs identified by the OECD are resisting tighter regulation. The problem is exacerbated by the fact that many OFCs are located in small, poor countries. The pressure to regulate OFCs intensified after the attacks on New York's World Trade Center on 11 September 2001 as part of US-led efforts to improve anti-money laundering controls.

Decentralization? The geographical rearrangement of 'back office' functions

All of the discussion of international financial centres would seem to suggest that the potential for other cities, outside the favoured few shown in Figure 13.8, to develop as significant centres of finance and related activities is likely to be very limited. In the United Kingdom, for example, the sheer overwhelming dominance of London makes it extremely difficult for provincial cities to develop more than a very restricted financial function. London, in that sense, is akin to the notorious upas tree, a fabulous Javanese tree so poisonous that it destroys all life for many miles around itself.

It is, of course, the 'higher order' financial and service functions which are especially heavily concentrated in the major international financial centres. So-called 'front office' functions, by definition, must be close to the customer – hence the huge branch networks of the retail banks and other financial services supplying final demand. The essence of all of the financial services, however, is the transformation of massive volumes of *information*. Much of that activity is routine data processing performed by clerical workers. Such 'back office' activity can be separated from the front office functions and performed in different locations. The early adoption of large-scale computing by banks, insurance companies and the like from the late 1950s initially led many of them to set up huge *centralized* data processing units. To escape the high costs (both land and labour) in the major financial centres such units were often relocated to less expensive centres or in the suburbs. Access to large pools of appropriate (often female) labour was a key requirement.

The introduction of microcomputers and networked computer terminals made such centralized processing units unnecessary and the tendency in recent years has been to *decentralize* back office functions more widely. In effect,

> telecommunications accelerate a spatial bifurcation within many large finance firms by enhancing the attractiveness of downtown areas for skilled managerial activities while simultaneously facilitating the exodus of low wage, back office sectors. This process

mimics the separation between headquarters and branch plant functions widely noted in manufacturing. In both cases, a vertical disintegration of production takes place, accompanied by the dispersal of standardized, capital-intensive functions and the concomitant reorganization of skilled labour-intensive functions around large, densely populated urban areas.[63]

At the same time, however, the distinction between back office and front office functions is becoming less clear as distributed computer technology has been developed further. In fact, it is not only routine back office activities that have been decentralized. It has become increasingly common for some of the higher-skilled functions to be relocated away from head office into dispersed locations, both nationally and, in some case, internationally. But centralization of back office functions by financial service companies is by no means an obsolete practice. For example, Citicorp plans to bring together all its back offices from around the world and to centralize their functions in large and more efficient centres to achieve economies of scale. Within this framework, all credit card processing for Europe is carried out in South Dakota and all statement processing for the Caribbean region is concentrated in Maryland.[64]

> Across Europe, banks have been making use of new technology to overhaul back-office operations by bringing processing out of their branches and into centralised units. For example, Lloyds TSB, the UK bank, announced an efficiency drive … which involved the loss of 3,000 jobs, stemming mainly from the centralisation of computer operations and consolidation of large-scale processing operations, including the complete removal of all back-office processing from branches … ABN-Amro bank, the Dutch bank, said it was in negotiation with at least three European banks over a possible alliance of their back offices.[65]

Even in the case of back offices, therefore, geographical centralization is far from dead.

The devastation of the heart of the US financial industry in lower Manhattan, New York City in September 2001 inevitably raised the question of whether such very high levels of geographical concentration of financial activities is prudent. After all, similar concentrations exist in other prominent financial centres, notably the City of London, Tokyo, or Frankfurt. Some degree of relocation of financial services had been happening already in New York for some years, as some firms relocated within the city or outwards to Connecticut or New Jersey. It is possible that the events of September 11 will accelerate that trend. But it seems unlikely to be on a large scale, as the following quotations from bankers after the event suggest:

> If you disperse it you lose that level of ingenuity and intellectual talent. We are committed to Manhattan. We are committed to Wall Street. But we will have facilities outside Manhattan.[66]
>
> At the end of the day, global firms need to have a presence in lower Manhattan. Prestige is still important.[67]

Conclusion

The financial services industries are central to the operation of all aspects of the economy. They are inherently embedded within all production networks and are essential to production and trade of all economic activities. At the same time, financial services firms generate 'products' in their own right on which speculative profits can be made. A striking feature of financial service activities during the past few decades is that the financial transactions essential to the operation of the 'real' economy have become increasingly dwarfed by speculative activity. Much of this activity has been facilitated by the progressive deregulation of financial markets across the world, by innovations in information technology, and by the ingenuity of firms in creating innovative financial products, such as derivatives. As a consequence of these interlocking developments, the world financial system has become increasingly volatile and unpredictable, with shocks originating in one part of the world spreading to other parts of the world at exceptional speed through the processes of 'financial contagion'.

A particularly significant feature of these industries is the fact that, although potentially highly flexible in locational terms, they continue to display an extremely high level of geographical concentration. In particular, the world financial system is articulated through a selective network of major cities. Without question, the 'end of geography' has not happened – even in such a 'footloose' activity as financial services.

Notes

1   Allen (1988: 18, 19).
2   Martin (1999: 6, 11).
3   See Braithwaite and Drahos (2000) for a discussion of the history of money and its regulation.
4   Strange (1986: 1).
5   But, as Pauly (2000: 120) points out, it is impossible to draw a clear distinction between speculative and productively necessary financial transactions.
6   Hamilton (1986: 13).
7   Dow (1999: 33).
8   Bertrand and Noyelle (1988: ch. 2) provide a concise summary of the major developments in the markets for financial services. This section draws extensively on that source.
9   Bertrand and Noyelle (1988: 16).
10   Warf and Purcell (2001: 227).
11   *The Financial Times* (16 March 1994).
12   *The Economist* (26 October 1996).
13   Strange (1998: 24–6).
14   Strange (1998: 24).
15   Cohen (2001) discusses the issue of electronic money.
16   Warf (1989).
17   See Langdale (2000).
18   Langdale (2000: 94).
19   Bertrand and Noyelle (1988: 40–1).
20   Strange (1998: 29).
21   Kelly (1995: 229).
22   There is a huge literature on the regulatory environment of the financial system. See, for example, Braithwaite and Drahos (2000); Corbridge, Martin and Thrift (1994); Eatwell and Taylor (2000); Eichengreen (1996); Martin (1999).
23   Martin (1999: 8, 9).
24   Strange (1986).
25   Lewis and Davis (1987: 225, 228).
26   *The Economist* (6 April 1996).
27   *The Financial Times* (9 February 2001).
28   *The Financial Times* (26 May 2000).

29 Grubel (1989: 61).
30 *The Financial Times* (26 October 1998).
31 Scott-Quinn (1990: 281).
32 Davis and Smales (1989: 99).
33 Davis and Smales (1989).
34 The various chapters in Corbridge, Martin and Thrift (1994) and in Martin (1999) provide varied perspectives on the geography of finance.
35 Clark and O'Connor (1997) and Martin (1994, 1999) strongly refute O'Brien's thesis.
36 O'Brien (1992: 1, 73; emphasis added).
37 Clark and O'Connor (1997).
38 Clark and O'Connor (1997: 90, 95).
39 Clark and O'Connor (1997: 101).
40 Clark and O'Connor (1997: 101, 102).
41 Clark and O'Connor (1997: 102).
42 Clarke and O'Connor (1997: 108).
43 Reed (1989).
44 See the examples given in Sassen (2001).
45 See Porteous (1999) for a discussion of this process in the development of financial centres.
46 Martin (1999: 19–20).
47 Thrift (1994).
48 Thrift (1994: 333).
49 Thrift (1994: 334).
50 See Sassen (2001). Thrift (1994) provides an illuminating account of the social and cultural structure of the City of London.
51 Budd (1999: 126, 127).
52 Lewis and Davis (1987).
53 Lewis and Davis (1987: 236).
54 Beaverstock, Hoyler, Pain and Taylor (2001).
55 *The Financial Times* (20 November 2001).
56 *The Financial Times* (8 February 2002).
57 Roberts (1994) provides a comprehensive analysis of the geography of offshore financial centres. See also Hudson (2000).
58 Roberts (1994: 92).
59 Eatwell and Taylor (2000: 189).
60 Roberts (1994: 92).
61 *The Financial Times* (7 December 2001).
62 Roberts (1994: 99).
63 Warf (1989: 267).
64 *The Financial Times* (24 October 1996).
65 *The Financial Times* (26 May 2000).
66 Quoted in *The Financial Times* (21 September 2001).
67 Quoted in *The Financial Times* (14 September 2001).

CHAPTER 14

'Making the Connections, Selling the Goods': The Distribution Industries

'Whatever happened to distribution in the globalization debate?'[1]

Good question. Whereas there is a huge literature and a continuous and often frenzied debate on the role of financial services in the processes of globalization, distribution rarely makes an appearance on the stage. It remains hidden, mainly confined to the specialist fields of supply chain management, transportation, retailing and the like. The distribution processes tend to be taken for granted. It is more or less assumed that, as transportation and communication systems have allegedly shrunk geographical distance, the problems of getting products from points of production to points of consumption have been solved. Only when something spectacular happens to upset this complacent view does the world sit up and take notice. This is what happened in the last few years of the 1990s with the sudden appearance of e-commerce, when, it was claimed, the entire system of connecting producers and customers would be totally transformed through the Internet. However, the meteoric rise of completely new 'dot.com' entrepreneurs was followed, in very short order, by 'dot.combustion' – the collapse of many of the new companies.

The focus of this chapter, then, is on those *circulation processes* that connect together the different components of the production network (including the customer). The logistics industries themselves (excluding retailing) are huge, worth around $3.4 billion in the late 1990s.[2] Such logistical processes are absolutely fundamental to the operation of all economic activities. They have become especially significant in the light of the broader forces of change discussed in earlier chapters, notably:

- changes in production methods, involving increased flexibility
- changing relationships between customers and suppliers
- changing consumer preferences
- increasing geographical complexity and extent of production networks.

In particular, *time* has come to be seen by virtually all modern businesses as the essential basis of successful competition.[3]

> Time is a fundamental business performance variable … Today, time is on the cutting edge of competitive advantage. The ways leading companies manage time – in production, in sales and distribution, in new product development and introduction – are the most powerful new sources of competitive advantage.[4]

In this context, the nature and efficiency of distribution systems become central.

> Time- and quality-based competition depends on eliminating waste in the form of time, effort, defective units, and inventory in manufacturing-distribution systems … Time- and quality-based competition requires firms to practice such logistical strategies as just-in-time management, lean logistics, vendor-managed inventory, direct delivery, and outsourcing of logistics services so that they become more flexible and fast, to better satisfy customer requirements.[5]

The structure of the distribution industries

As in the case of the financial services industries, the essential function of the distribution services is that of *intermediation* between buyers and sellers at all stages of the production chain (Figure 14.1). Without such services, a production chain simply could not function and economic needs could not be met. The distribution services involve not only the physical movement of materials and goods, using the transportation infrastructure, but also the transmission and manipulation of information relating to such movements. They both involve, above all, the organization and coordination of complex flows across increasingly extended geographical distances. In that respect, these services have been dramatically transformed by the

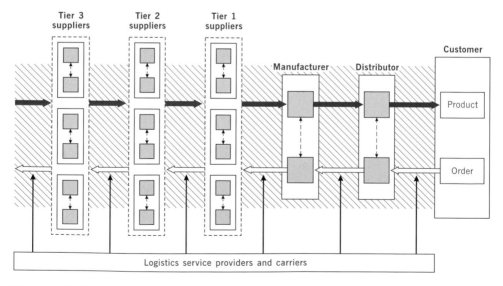

Figure 14.1 Logistics and distribution in the production chain
Source: Based, in part, on Schary and Skjøtt-Larsen, 2001: Figure 1.6

kinds of technological developments in transportation and communication (including the Internet) discussed in Chapter 4. They have also been transformed in recent years by the changing strategies of firms, in particular by the increased outsourcing of logistics and distribution services by manufacturing firms and by the emergence of new forms of logistic service providers.

In Figure 14.1 there are no political, or other, boundaries or obstacles to complicate the basic system. In reality, of course, the existence of such features greatly affects the structure and operation of logistic and distribution processes. Two kinds of 'barrier' to movement are especially significant:

- *Physical conditions* that necessitate the transfer from one transportation mode to another – for example, land / water interfaces.
- *Political boundaries* that create complications of customs clearance, tariffs, duties, administration and the like. Such barriers have become increasingly significant as economic activity has become increasingly internationalized. Figure 14.2 shows some of the elements involved in the second of these two barriers.

The logistics and distribution processes shown in Figure 14.1 can be performed in a variety of ways and by a variety of different forms of organization. At one extreme, each individual transaction may be performed by a separate firm; at the other extreme, the entire process may be carried out by a single firm or related group of firms. The major types of organization involved in logistics and distribution include:

- transportation companies
- logistics service providers
- wholesalers
- trading companies
- retailers
- e-tailers.

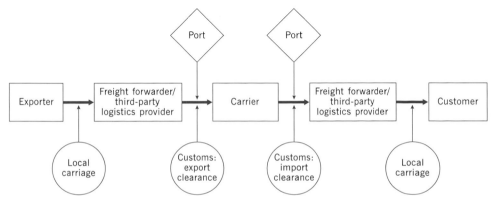

Figure 14.2 Logistics in an international context

Source: Adapted from Schary and Skjøtt-Larsen, 2001: Figure 11.7

Transportation companies (rail, road, shipping, airlines), wholesalers and retailers perform fairly clearly defined and restricted roles in the chain. On the other hand, trading companies and the more recent logistics service providers are involved in a far broader range of activities. Of course, the categories are not clear-cut. Not only are the boundaries between categories often blurred but also one form of organization may mutate into another, as in the case of some transportation and trading companies that have evolved into more comprehensive logistics service providers. At the same time, the significance of some types of intermediary has changed.

For example, traditionally, the wholesaler played a major role in collecting materials or products from a range of individual producers and then distributing them on to the next stage of the production process, or to the retailer in the case of final demand. However, the importance of the wholesaler as the key intermediary has changed substantially as the major retailers have by-passed the wholesaler to deal directly with the manufacturer, or as other forms of logistical and distribution services have developed. In a similar way, the development of e-commerce threatens to by-pass the traditional retailer as the key intermediary between producer and final consumer and to create new types of retailer.

Figure 14.3 shows one developing way in which the production/supply chain may be organized internationally. The key intermediary is the lead logistics provider, which operates

as the first tier for … [trans]national clients, offering one-stop shopping for global logistics solutions. The first tier makes subcontracts with a second tier of asset-based logistics providers, information service providers and regional carriers to perform operational logistics activities in specific regions or business areas. The second tier might further collaborate with small and medium-sized operators in local and niche markets.[6]

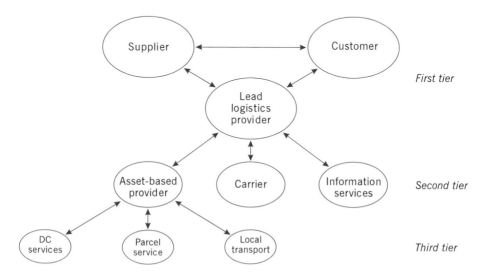

Figure 14.3   A potential way of organizing logistics services

Source: Schary and Skjøtt-Larsen, 2001: Figure 7.4

However, we should not go along with the idea that only one dominant form of logistics and distribution system will exist. As in all other forms of economic activity in the global economy, variety continues to be the norm.

The dynamics of the market

In aggregate terms, the growth of the market for distribution services is closely related to growth in the economy as a whole. Faster growth in overall economic demand will inevitably lead to an increase in demand for distribution services; slower economic growth will reduce demand for distribution services. For example, during the 1990s, while overall growth in GDP in North America, Europe and East Asia was 23 per cent, growth in logistics services (a segment of distribution services) was 18 per cent.[7] But the market for distribution services is very heterogeneous and we would not expect all parts to grow at similar rates. Demand for certain kinds of distribution services have been growing more rapidly than others, driven by the different characteristics of individual services.

Ultimately, as Figure 14.1 shows, the system is driven by the demands of the final consumer, although this influence becomes increasingly indirect the further back up the production chain we go. The primary driver of final consumer demand is, of course, the level of disposable income. As we showed in Figure 7.4, there are immense geographical variations in levels of income across the world. Overlain on the basic driver of income are many other cultural and individual factors (including age and gender) that, in combination, affect the level and nature of demand.[8] Hence, the geography of demand at both the national and international scales is actually far more complex than the distribution of income alone tends to suggest.

The intensely competitive retail markets have major repercussions on the demand for distribution and logistics services further up the supply chain. As we saw in the case of the garments industry (Chapter 10), the major retailers and buyers are exerting heavy pressure on their suppliers to deliver more rapidly, more cheaply, and in greater variety. This, in turn, creates opportunities (and challenges) for the suppliers of logistics and distribution services to provide a faster and more integrated supply system between the different tiers of the supply chain. Such demands intensify the search for new logistical systems, both in terms of their technology and their organization. In such a context, the role of electronic systems has become increasingly important. Hence, there is a link between the changing demand pressures on the suppliers of logistics and distribution services and changing technologies. Let's now look at these technologies.

Technological innovation and the distribution industries

Three criteria dominate the distribution industries: speed, flexibility and reliability. Technological innovations have impacted on each of them. At a general level, technological developments in transportation and communication have been immensely important in transforming the basic time–space infrastructure of the logistics and distribution industries. Likewise, these industries have also been transformed by the shift in process technologies from mass production to more flexible and customized production systems. The mass production systems of the late 19th and first two-thirds of the 20th century were facilitated by mass distribution systems based on the rapidly developing rail, road and ocean shipping networks.

Such mass distribution systems depended heavily on the use of large warehouses to store components and products and from which deliveries were made to customers on an infrequent basis. This was an immensely expensive system in terms of the capital tied up in large inventories. It was also, very often, a source of waste in terms of faulty products that were not discovered until they came to be put into use. As we have seen, there has been a major shift away from such systems to so-called lean systems of production. At the same time, there has been a parallel shift towards *lean systems of distribution* whose characteristics are those of minimizing the time and cost involved in moving products between suppliers and customers, including minimizing the holding of inventory at any point in the chain.

Three key elements form the core of such distribution systems:[9]

- *Electronic data interchange (EDI).* This involves the capability of the rapid transmission of large quantities of data electronically (rather than using paper documents). Such data can encompass all aspects of the logistics and distribution system throughout the supply chain, including the retailer. Information on product specifications, purchase orders, invoices, status of the transaction, location of the shipments, delivery schedules and so on can be exchanged. EDI requires a common software platform to enable data to be read by all participants in the chain. 'Without such a communications interface, information sent by a retailer might be unreadable or require extensive translation by a supplier. From the retail perspective, the reverse flow of information is even more problematic, given that many retailers use thousands of different suppliers.'[10]

- *Bar code systems.* Bar codes were first developed in the 1970s by grocery manufacturers and food chain stores to enable each item to be given an electronically readable, and unique, identity. Without doubt, bar codes were one of the major innovations of the last quarter of the twentieth century; they have now become ubiquitous throughout the supply chain and in many other areas of modern life (for example, airline baggage handling processes). 'Bar codes permit organizations to handle effectively the kind of vast product differentiation that would have been prohibitively expensive in an earlier era. They also facilitate instantaneous information at the point of sale, with significant effects on inventory management and logistics.'[11]

- *Distribution centres.* Modern distribution centres hold inventory for much shorter periods of time and turn it over very frequently. 'Four technologies have made the modern distribution centre possible: (1) bar codes and associated software systems; (2) high-speed conveyers with advanced routing and switching controls; (3) increased reliability and accuracy of laser scanning of incoming containers; and (4) increased computing capacities.'[12] In the most advanced distribution centres – such as the system used by the US retailer Wal-Mart – a method known as 'cross-docking' is used. This is a 'largely invisible logistics technique ... [in which] ... goods are continuously delivered to Wal-Mart's warehouses, where they are selected, repacked, and then dispatched to stores, often without ever sitting in inventory. Instead of spending valuable time in the warehouse, goods just cross from one loading dock to another in 48 hours or less. Cross-docking enables Wal-Mart to achieve economies that come with purchasing full truckloads of goods while avoiding the usual inventory and handling costs. Wal-Mart runs a full 85% of goods through its warehouse system – as opposed to only 50% for Kmart'.[13] Of course, few firms can afford the scale of investment required to implement such a sophisticated system.

E-commerce: a new logistics revolution

Computer-based electronic information systems are at the heart of all these technological developments in logistics and distribution systems. They have evolved over a period of 30 years, often incrementally rather than as a spectacular 'revolution'. However, the latter years of the 1990s brought into existence a new set of distribution methods based upon the Internet: *e-commerce*. In essence, e-commerce has developed out of the convergence of several technological strands: electronic data interchange, the Internet, e-mail and the medium of the worldwide web (WWW).[14] Few developments in recent years have been so heavily hyped as

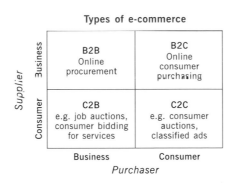

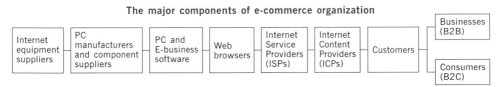

Figure 14.4   Types of e-commerce

Source: Based, in part, on Gereffi, 2001: Figure 2

the e-commerce revolution and it is unwise to extrapolate too far into the future on the basis of recent experience. But there is no doubt that the emergence of the Internet has huge implications for logistics and distribution systems. The main reason, once again, is the issue of *speed*.

Although four types of e-commerce are shown in Figure 14.4, two dominate: B2B and B2C.

- *B2B (business-to-business)*. This encompasses potentially the whole range of transactions between businesses, notably procurement of products and services and logistics. Roughly 80 per cent of all e-commerce is B2B.[15] B2B websites are electronic 'marketplaces' where firms come together to buy or sell products and services. They may be 'vertical', that is, industry-specific (e.g. the B2B procurement system, Covisint, established in 2000 by GM and Ford, together with some other automobile manufacturers, to increase the efficiency of component purchasing) or they may be 'horizontal', organized around the products and services provided rather than the industry. Connecting together large numbers of buyers and sellers through electronically automated transactions has a number of potential benefits, notably vastly increasing choice to both sellers and buyers, saving costs on transactions, and increasing the transparency of the entire supply chain. 'Three years ago… [1997] … there was no such thing as a B2B marketplace. Today … [2000] … there are more than 1,500 of them, about 85 per cent in the US.'[16] Even so, B2B transactions still constitute a minority of total business transactions, although the widespread use of EDI is likely to lead more firms to engage in B2B in the future.

- *B2C (business-to-consumer)*. B2C business is simply the selling of consumer products and services directly over the Internet by some form of Web-based firm. Although it has received far more attention, it is, so far, substantially less developed than B2B. Only around 20 per cent of total e-commerce transactions are B2C and they are a very small fraction of total retail business transactions, which still largely conform to the traditional model of store-based shopping. Probably the most often-quoted examples of B2C are Amazon, the Internet book supplier, and Dell, the PC supplier, but, of course, the dot.com revolution of the late 1990s saw the appearance of many thousands of 'e-tailers'. As in the case of B2B, the potential benefits of B2C transactions are, to the consumer, greater choice, ease of comparison of prices, instant (or very fast) delivery to the home and, to the seller, direct access to a massive potential market without the need for physical space in the form of retail outlets and the associated inventory and staffing costs.

E-commerce (both B2B and B2C) has enormous *potential* implications for the traditional intermediaries of business and retail transactions (for example, wholesalers, shippers, travel agents and the like). The early predictions were that many would disappear as their functions were displaced by direct online transactions. In fact, this has not happened to the extent predicted. Indeed, the world of e-commerce, far from abolishing the intermediary, has enhanced opportunities for such roles. Some traditional intermediaries have adapted to the new situation and found new ways of adding value as providers of logistics, information and financial services while new intermediaries have emerged.[17] Some of these are what are sometimes termed *infomediaries,* notably the Internet service and content providers shown in Figure 14.4. To some extent, these new infomediaries are increasing their relative bargaining power *vis-à-vis* producers and buyers in the conventional producer-driven and buyer-driven production chains discussed in Chapter 2.[18] More generally, Figure 14.5 shows that in the case of both physical goods and electronic goods and services, either old intermediaries transform themselves or new ones appear.

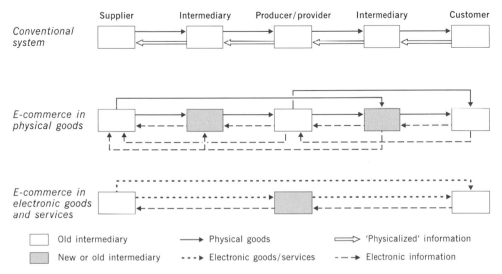

Figure 14.5 The continuing role of intermediaries in an e-commerce environment

Source: Based on Kenney and Curry, 2001: Figure 3.2

Related to the perception that e-commerce spells the end for the traditional intermediary organization is the idea that the traditional physical infrastructures will also be displaced by the new technological forms. In the case of retailing, for example, stores will be replaced by virtual transactions direct from the supplier of the product; warehouses – even the retail stores themselves – are no longer necessary. Again, this is an illusion. As Figure 14.6 shows, there are several ways of fulfilling e-commerce orders.[19]

- *'Dell' model.* This is the situation, named after the PC firm Dell Computer Corporation, where the supplier receives orders for specific products, which are then integrated directly into production. 'The customer order activates the supply chain. Customers can "design" their products from a list of options to be incorporated into a production schedule. The order then initiates a flow of component parts from suppliers to be assembled into a final product, turned over to a logistics service provider, merged with a monitor from another source and delivered to a final customer. The system avoids holding finished product inventory, providing both lower cost and more product variety.'

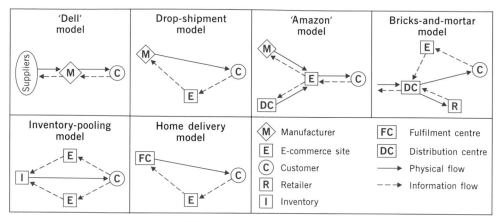

Figure 14.6   Methods of fulfilling e-commerce orders

Source: Based, in part, on Schary and Skjøtt-Larsen, 2001: Figure 4.5

- *Drop shipment model.* The e-commerce firm receives an order and passes the order on to a manufacturer for production and delivery direct to the customer.

- *'Amazon' model.* Named after the Internet bookseller Amazon.com, this is the electronic version of an old mode of direct retailing where the 'catalogue' of products is held electronically and accessed via the Internet. Customer orders may be supplied by the seller from its own distribution centre or 'drop-shipped' from the manufacturer.

- *Bricks-and-mortar model.* This is a type of system that combines both conventional retail stores, fed by distribution centres, with an Internet website that channels orders to the same distribution centres. A problem with this system is that whereas retail orders generally require large orders, individual Internet-based orders require individual units.

- *Inventory pooling model.* This enables customers in, say, a specific industry, to acquire common spare parts from an inventory pool controlled by a Web-based provider.

- *Home delivery model.* Here, customers requiring regular deliveries of, say, groceries, can place their order with a Web-based service which will deliver on a routine basis to the customer's home address.

Inherent in each of these different ways of fulfilling e-commerce business is the question of physical space: 'One of the unexpected consequences of the Internet is to make physical space *more*, not less important.'[20] In addition to warehousing space, the Internet-based system also requires another kind of dedicated physical space: the so-called *Internet 'hotels'*, which are

> centres where companies with big Internet operations, frequently telecoms carriers who look after the web traffic of many of their customers, can house their web servers … With ultra high-speed telecoms connections, massive power supplies, arctic air conditioning and squads of trained engineers, the facilities are capable of housing

thousands of computers connected to the Internet in optimal conditions ... Prime
locations are essential; the hotels should be near the customers they serve, such as City
banks, as the high-capacity fibre-optic communications pipes they require are
expensive to run far. Some of the sites are cooled by the kind of water-circulation
systems normally associated with nuclear power plants, because the computers
generate so much heat.[21]

Thus, a whole variety of technological developments has helped to transform
the nature and operations of the distribution industries. Initially, in the early 20th
century, they were the technologies that enabled mass distribution systems to
facilitate the output of the mass production systems of the day. In the last two
decades of the 20th century, the technologies were predominantly electronic. Such
technologies facilitate the operations of the more flexible production systems of
today through their ability to transmit and process vast quantities of information
on all aspects of the supply chain.

The role of the state: regulation and deregulation in the distribution industries

The primary purpose of the distribution industries is to overcome – or at least to
minimize – time–space friction by connecting together the different parts of the
supply network at maximum speed and minimum cost. However, there remains
a significant obstacle to the smooth, seamless operation of distribution systems:
the existence of *regulatory* requirements.

By definition, international distribution activities involve crossing political
boundaries. All national governments regulate, in various ways, the cross-border
movement of goods and services. We discussed one aspect of this in Chapter 5: the
variety of trade measures (both tariff and non-tariff barriers) that states use. The
existence of such regulations, for example in the form of customs requirements
and procedures, form a major discontinuity in the geographical surface over
which distribution services operate (see Figure 14.2). In this section we are concerned with a different – though related – set of regulatory structures; those affecting the basic functions of the distribution industries themselves, especially
transportation and communication systems. Such regulations are devised and
implemented at various political-geographical scales: international, regional and
national. They are also in a continuous state of flux. In the past three decades, in
particular, there have been major waves of deregulation. To illustrate these
processes we will look briefly at two areas from a regulatory perspective: transportation and communications (including the Internet), and retailing.

Regulation and deregulation of transportation and communications systems[22]

Regulation of transportation and of communications has a very long history. It has
involved a varying mix of national and international-level systems in the public

sphere, as well as of private organizations in the form of the operators themselves. Much of the regulatory system in air and sea transportation relates to issues of safety and security, whilst in communications a key issue is harmonization of standards to enable communications originating in one place to be received and understood in other places. Such international bodies as the IMO (International Maritime Organization), the ICAO (International Civil Aviation Organization) and the ITU (International Telecommunication Union) constitute the primary elements of such an international regulatory framework.

Within such an international framework, individual national states have endeavoured to exert control over their own affairs. This has especially been the case in telecommunications and air transportation. Both of these industries have had international regulatory frameworks from early in their history but they are both sectors in which states believe (or have believed until recently) that 'natural monopolies' exist. For example,

> telecommunications' regulation contains one of the earliest examples of international regulatory cooperation between states, with the creation of the International Telegraphic Union (ITgU) in 1865. But in other respects the regulation of telecommunications is a story of territorial containment. Much of the early regulatory development in the first half of the twentieth century was influenced by the economic view that telecommunications is a 'natural monopoly'. But no state thinks that there should be one world monopolist. Instead, the contours of this natural monopoly correspond with state boundaries.[23]

In most cases this involved state-ownership, although in the United States it was a private monopoly, AT&T.

Similarly in the air transportation industry,

> just as almost every nation has its telecommunications carrier (and rarely more than one), almost every nation has a flagship airline (and rarely more than one). The state controls landing rights (just as it tends to control the telecommunications infrastructure) and rations those rights, usually in ways that favour the national flag-carrier.[24]

In both cases, the operation of the regulatory framework has involved a tension between the desire on the part of most states for control over their own national spaces and the drive (primarily by business organizations) for the least possible regulation consistent with safety and efficiency. As globalizing processes intensified during the past 50 years, however, the balance shifted decisively towards greater deregulation of the nationally based systems and the privatization of state-owned companies. In the case of telecommunications, the initial impetus came with the enforced break-up of the AT&T monopoly in the early 1980s, when the company was forced to divest itself of its local Bell Telephone operating companies.

The US example stimulated a wave of European deregulation in telecommunications, led by the Thatcher government in the UK which was obsessed with the virtues of privatization.

> The Post Office monopoly over telecommunications became one of its first big targets. In 1981, British Telecom was established and in 1984 it was privatised. OFTEL (Office

of Telecommunications) was formed as the regulatory body for the new telecommunications market. Earlier in 1981, the Thatcher government had privatised Cable and Wireless, its international carrier … The shift to a more liberalized telecommunications sector occurred throughout Western Europe in the 1980s … [subsequently] … European telecommunications began to be progressively harmonized through the work of the EC.[25]

The air transportation industry, too, has experienced a similar wave of deregulation. Again, the early moves began in the United States. In the late 1970s, a new bilateral agreement was signed between the United States and the United Kingdom, which helped to undermine the cartelization of the airline industry within IATA. In 1978, the US domestic airline industry was deregulated. Subsequently,

> both the US and the UK then set about reshaping their bilateral agreements towards more liberal policies. For example, France has been the most vigorous opponent of liberalization, so the US worked at isolating France by negotiating open-skies agreements with Belgium and other countries around France … In short, the process in the 1990s is US-led liberalization that is seeing the world become gradually and chaotically more competitive. The process is chaotic because even the most liberal states, such as the US and UK … are 'liberal mercantilists' … Another chaotic element is that many European, African, and South American states support liberalization within their continents but want protection from competition outside the continent (especially from the US).[26]

The continuing tensions between states in terms of their own air spaces (and often their 'national' airlines) has important implications for the distribution industries. Two examples can be used to illustrate this. First, the continuing disagreement between the United States and the United Kingdom over mutual access over the North Atlantic route means that, on the one hand, US airlines have restricted access to London Heathrow whilst, on the other, British airlines are not allowed to fly routes onwards within the United States beyond their initial point of entry. A second example relates to the dispute between the United States and Hong Kong, which is especially important for the large express couriers (FedEx, DHL and UPS). The express couriers have developed highly sophisticated logistics operations involving strategically located hubs where cargo from other origins can be reloaded. The US company FedEx is allowed only five flights a week from Hong Kong to destinations outside the United States. This is because Hong Kong would like its airline, Cathay Pacific, to be able to be able to fly within, as well as to, the United States – which the United States refuses to allow.[27]

Some of the biggest changes in the regulatory environment have been the outcome of the emergence of regional economic blocs, such as the EU and the NAFTA. The completion of the single European market in 1992 removed virtually all major obstacles to internal movement of goods and services within the EU. For example,

> cabotage, carrying goods in domestic commerce by a foreign carrier, has gradually been deregulated since 1993. Cabotage means that a carrier from one country is allowed to perform domestic transport in another country. For example, a French

forwarder can pick up cargo between Munich and Basel on a back-haul trip from Stuttgart. Cabotage has led to increased competition between the international transport and forwarding companies, and the freight rates have been reduced on major traffic routes.[28]

Liberalization of trucking within and between the United States, Canada and Mexico was also a part of the NAFTA. Under this agreement, the United States and Mexican border states were scheduled to be opened to international trucking by the end of 1995. By January 1997, Mexican trucking companies would be allowed to operate as domestic carriers in the border states and for international cargo in the rest of the United States. By January 2000, Mexican truckers would be able to file to operate in the entire United States. In fact this has not happened. Although President George W. Bush authorized Mexican firms to transport foreign goods within the United States in June 2001, this was blocked by the US House of Representatives. The events of 11 September 2001 have made such free entry of Mexican trucks into the United States even less likely in the near future.

The difficult problem of regulating the Internet

The Internet raises some particularly tricky regulatory issues. As a medium that 'knows no boundaries' and that is allegedly (though, as we have seen, not actually) 'placeless', it involves some intractable issues as to who regulates it. The answer is far from clear, not least because of the very newness of the Internet and e-commerce and its phenomenally rapid growth.[29] The key issue is 'whose laws apply?' when e-commerce transactions transcend different national jurisdictions.

The European Union recently ruled that 'companies which "direct their activities" to consumers in other European countries can be sued in those overseas countries'.[30] In other words,

> the law of the consumer's home jurisdiction would apply in any case where a purchase was made through a web site. When a Christmas tree catches fire in a US home, can the store where the tree was bought sue the Hong Kong company if it has a presence in cyberspace? But that raises many problems for US companies, because European privacy and consumer protection laws are much tougher. Companies are forced either to set up separate web sites to comply with local laws, or one megasite which meets every conceivable national and local legal requirement … Both options are costly and defeat the supposed efficiency gains of globalized commerce. A similar collision between the territorial and the global occurs over domain names – those '.com' designations which give an e-commerce company its identity. Identical trademarks can co-exist in different countries, because trademarks are geographically limited. But domain names are global, and must be unique. Recently, an international body was established – outside the current legal system – to resolve domain name disputes.[31]

One example of the problem of regulating the Internet was the case of the French court ordering the US Internet Content Provider Yahoo! to block French customers from ordering Nazi memorabilia on its US website.

Regulating the retail sector

Retailing has been, until relatively recently, predominantly a domestic activity. With few exceptions, most retail stores were domestically owned and controlled and served individual national markets. From a regulatory perspective, this meant that the most important issues were those of planning regulations relating to the location and size of retail stores, policies towards out-of-town *vis-à-vis* city centre shopping developments, regulations regarding labelling, product safety and so on. For all of these reasons, as well as those of culture, tastes and preferences, national retail markets remained strongly differentiated. It has been quite common for countries to restrict entry via FDI of foreign retailers although this practice has declined in recent years while the rules of operation in national retail markets continue to differ substantially.

This is true even within the EU where, despite the post-1992 market integration, considerable differences still exist in the regulatory structures of individual member states. 'Each country continues to make and enforce its own rules about hours of operation, the building of large-scale retail stores, the doing of part-time and overtime work, and Sunday trading practices.'[32] In the case of retail store opening hours, for example, the UK is the most liberalizd with few restrictions, apart from limited opening hours on Sundays. But in Germany and Austria stores are closed on Sundays and this is also true of most parts of Italy. Outside the EU, the spectrum of national regulatory positions varies widely.[33] The United States has probably the most lenient regulatory system towards retailing, whereas the Japanese retail market is far more heavily regulated, although Japan has revised the Large-Scale Retail Store Law which had been a major obstacle to the growth of large retail stores. Certainly the Law had discouraged major foreign retailers from trying to enter the Japanese market. Following the financial crisis of 1997, several East Asian countries have relaxed the restrictions on foreign ownership of property.

Corporate strategies in the distribution industries

The distribution industries cover an immensely wide range of activities and consist of a mix of traditional shipping and carriage of goods through to the highly complex and sophisticated *logistics service providers* (LSPs), from trading companies to large retail chains. In this section we outline the major trends in corporate strategies as firms respond to market, technological and regulatory forces. Although there are many niche areas within the distribution sector there is a broad tendency in most activities for the size of firms to be increasing and for higher degrees of concentration in a smaller number of large firms. As in the other industries we have been examining, mergers and acquisitions have been important mechanisms in the process of consolidation. The discussion will be divided into two broad sections. In the first section we will look at trends in the global logistics sector; in the second, we will focus on transnational retailing.

Global logistics

From transportation companies to integrated logistics service providers[34]

Figure 14.7 sets out the four major types of logistics service firm, classified according to the kinds of physical and management services provided. The simplest functions are provided by the *traditional transportation and forwarding companies* (bottom left cell of Figure 14.7). The other three cells contain newer types of logistics service firms based upon the extent to which they are

- asset-based logistics providers (top left cell of Figure 14.7)
- network-based logistics providers (bottom right cell)
- skill-based logistics providers (top right cell).

Such firms are usually referred to as *third-party logistics providers*.

The *asset-based logistics providers* developed primarily from the diversification of some of the traditional transportation companies into more complex logistics service providers. For example, several of the world's leading container-shipping companies, such as Maersk–Sealand and Nedlloyd/P&O have moved in this direction.

> In 1992 … Nedlloyd/P&O (then Nedlloyd) announced the building of its new, fourteenth and largest warehouse for IBM. Since the beginning of that year all of IBM's distribution activities have been handled by Nedlloyd … The new Westport distribution centre … [in the port of Amsterdam] … receives all computer equipment

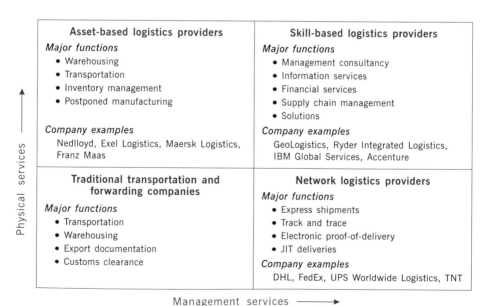

Figure 14.7 Types of logistics service providers

Source: Based, in part, on Schary and Skjøtt-Larsen, 2001: Figure 7.3

and parts from IBM factories in other European countries, North America and the Far East. Nedlloyd Districenters receive the goods, administer stocks, install software on the correct computers and test the equipment. Finally, the goods are prepared for distribution in Europe, Africa and the Middle East.

... From transport only, Nedlloyd/P&O increasingly offers complete package deals, in which the focus is on control of the flow of goods and information between manufacturers and their clients, irrespective of which part of the world producer and client are in. World-wide logistic services are the products by means of which this carrier is trying to increase its market share.[35]

Such asset-based logistics providers first emerged during the 1980s. During the early 1990s, a number of *network-based logistics providers* appeared on the international scene, notably DHL, FedEx, UPS and TNT.

These third-party logistics providers started as couriers and express parcel companies and built up global transportation and communication networks to be able to expedite express shipments fast and reliably. Supplemental information services typically include electronic proof-of-delivery and track-and-trace options from sender to receiver ... Recently, these players have moved into the time-sensitive and high-value-density third-party logistics market, such as electronics, spare parts, fashion goods and pharmaceuticals, and are competing with the traditional asset-based logistics providers in these high margin markets. As the largest transportation company in the world, UPS also offers a broad range of services from shipment tracking, shipper-accessible information services and specialized logistic services to e-commerce and financing services for overseas manufacturing.[36]

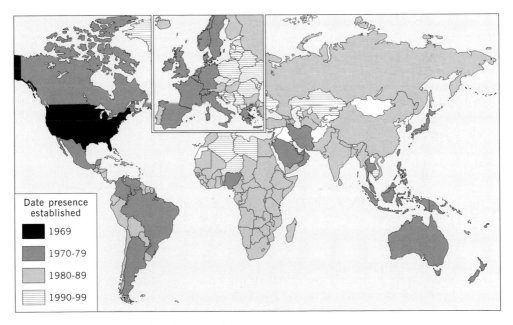

Figure 14.8   DHL's global network

Source: Based on material in DHL, *Annual Report, 2000*

The express courier service companies have built up massive networks of operations across the globe. Figure 14.8 shows the global network of DHL, which has operations in more than 630 cities in 230 countries employing 64,000 people. Its Annual Report for 2000 shows that it owns 252 aircraft and almost 19,000 other vehicles. It operates one of the largest private telecommunications networks in the world, linking the company's 3,000 stations and 35 hubs.

The fourth type of logistics service firm shown in Figure 14.7 – the *skill-based logistics providers* – became increasingly significant in the later years of the 1990s. These are firms that do not own any major physical logistics assets but provide a range of primarily information-based logistics services. Such services include consultancy services, financial services, information technology services and a range of management skills. Examples of such skill-based logistics service providers include GeoLogistics, a firm created in 1996 through the merger of three existing logistics companies (Bekins, LEP, Matrix).

Trading companies

Trading companies are one the oldest forms of organization in the distribution services. Their history goes back many hundreds of years. From the earliest days of long-distance trade they have always played an especially important role in facilitating trade in materials and products. Here we look at two important contemporary examples, both taken from East Asia.

The first example is the Japanese *sogo shosha*. These giant trading companies are undoubtedly one of the most remarkable examples of this form of organization, with a long history in the development of the Japanese economy.[37] They have also been especially influential in the development of Japanese overseas direct investment as a whole. The common translation of the term *sogo shosha* is 'general trading company'; but they are very much more than this. The six leading *sogo shosha* – Mitsubishi, Mitsui, Itochu, Marubeni, Sumitomo and Nissho-Iwai – are gargantuan commercial, financial and industrial conglomerates. They operate a massive network of subsidiaries and thousands of related companies across the globe. Figure 14.9 shows the global distribution of their offices. Their huge size and the extent of their geographical operations have given them an enormous importance both in Japan's economic affairs and in the global economy as a whole. Each of the leading *sogo shosha* handles tens of thousands of different products. In the early 1990s, they were responsible for roughly 70 per cent of total Japanese imports and for 40 per cent of Japanese exports. This is the true Japanese general trading oligopoly, each member of which has a major coordinating role within one of the Japanese *keiretsu* (see Figures 7.13 and 7.14).

Historically, the *sogo shosha* developed initially to organize exchange and distribution within the Japanese domestic market. Subsequently, they became the first Japanese companies to invest on a large scale outside Japan. These foreign investments were primarily designed to organize the flow of imports of much-needed primary materials for the resource-poor Japanese economy and to channel Japanese exports of manufactures to overseas markets. It was the particular demands of these Japanese-focused trading activities that necessitated the development of the globally extensive networks of *sogo shosha* offices around the world.

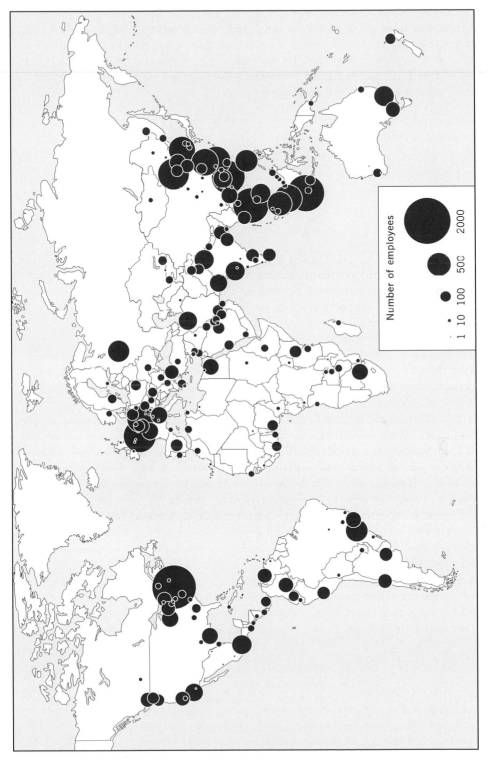

Figure 14.9 The global distribution of Japanese *sogo shosha* offices

Source: Company annual reports

In other words, they were to set up a global marketing and economic intelligence network. Once in place, this network, with all its supporting facilities, not only facilitated the growth of Japanese trade but also enabled a whole range of Japanese firms to venture overseas. Indeed, a good deal of the early overseas investment by Japanese manufacturing firms was organized by the *sogo shosha*.

They perform four specific functions:

- *trading and transactional intermediation* – matching buyers and sellers in a long-term contractual relationship
- *financial intermediation* – serving as a risk-buffer between suppliers and purchasers
- *information gathering* – collecting and collating information on market conditions throughout the world
- *organization and coordination of complex business systems* – for example, major infrastructural projects.

As the position of the Japanese economy in the global system has changed – especially with the deep and continuing recession of the 1990s – the role of the *sogo shosha* has also had to change. In addition to their traditional roles of import–export business between Japan and other parts of the world and their activities as traders in commodities markets, the *sogo shosha* have increasingly become engaged in 'third-country trade'. In this context, they act as intermediaries between firms in countries other than Japan in their trade with other parts of the world. In fact, this has been the fastest growing area of transactions of the *sogo shosha* in recent years.

The second example of a trading company, also taken from East Asia, is the specific case of the Hong Kong-based firm Li & Fung.[38] This firm is not only the biggest export trading company in Hong Kong but also – and more importantly – a highly innovative logistics company, with 35 offices spread across 20 countries (see Figure 14.10). Established in Canton in 1906, Li & Fung was originally a simple commodity broker, connecting together buyers and sellers for a fee. Today, although still a Chinese family firm, it has been transformed from the simple brokerage to an immensely sophisticated organizer of geographically dispersed manufacturing and distribution operations in a whole variety of consumer goods, but with a strong specialization in garments (see Chapter 10). As Figure 14.11 shows, Li & Fung controls and coordinates all stages of the supply chain, from design and production planning, through finding suppliers of materials and manufacturers of products, to the final stages of quality control, testing and the logistics of distribution.

> It works like this. Say a European clothes retailer wants to order a few thousand garments. The optimal division of labour might be for South Korea to make the yarn, Taiwan to weave and dye it, and a Japanese-owned factory in Guangdong Province to make the zippers. Since China's textiles quota has already been used up under some country's import rules, Thailand may be the best place to do the sewing. However, no single factory can handle such bulk, so five different suppliers must share the order. The shipping and letters of credit must be seamless and the quality assured … [organization of the supply chain] … requires knowledge. Village women with sewing-machines in Bangladesh are not on the Internet. Finding the best suppliers at

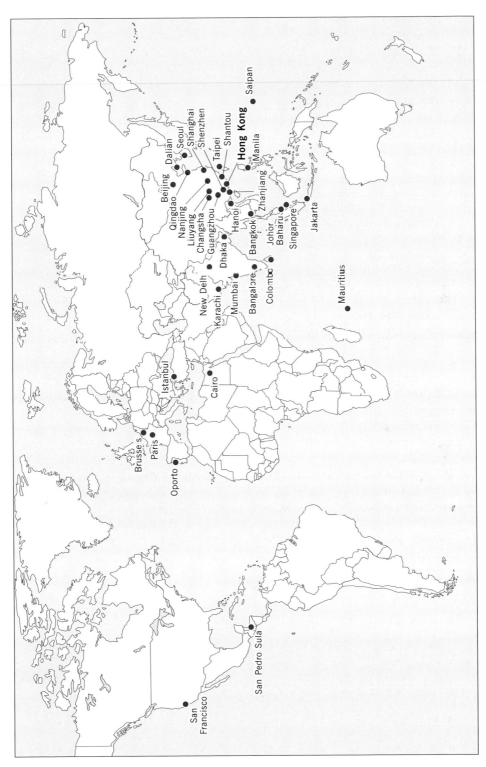

Figure 14.10 The global spread of the offices of Li & Fung

Source: Based on Magretta, 1998: 106–7

any given time, therefore, takes enormous research … [companies] … outsource the knowledge-gathering to Li & Fung, which has an army of 3,600 staff roaming 37 countries … In this sense, Li & Fung is itself a product of specialization. A company that focuses entirely on optimising supply chains for other companies is a recent phenomenon.[39]

The company is also beginning to produce private-label brands for retailers who lack the resources to do this for themselves, especially smaller companies. By organizing and managing supply chains over the Internet for such smaller retail chains Li & Fung can combine many small orders

to achieve economies of scale in production and distribution operations. A typical example is a polo shirt, made of cotton from the United States, knitted and dyed in China and sewn in Bangladesh. This shirt could be customized by adding logo and side vents, and changing collars, all managed through Li & Fung's own proprietary software.[40]

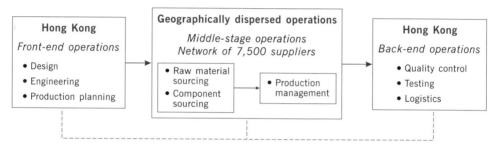

Figure 14.11   Organization of Li & Fung's logistics operations

Source: Based on Magretta, 1998: 111

Like some other traditional distribution companies, therefore, the trading companies have carved out new roles for themselves, both responding to, and creating, new demands for distribution and logistics services.

Call-centres: a new way of connecting firms and customers

A significant development of the past decade or so has been the emergence of a new form of firm–customer interface: the *call-centre*. Call-centres (often also called 'customer-response services') can be defined as:

centralized locations from which services such as sales, reservations, information provision, technical support and banking are provided to a dispersed customer base by means of telephone.[41]

The call-centre industry has grown at a phenomenal rate as more and more businesses, in almost all industries and services, have moved to centralize many of their customer-relation functions. Indeed, increasingly the telephone has become the customer's only point of contact with a company. Establishing call-centre services in low-cost locations, both domestically and overseas, has become a common business strategy. In response, economic development agencies in both developed and developing countries compete with each other to attract call-centre investments. Several British cities, especially those where traditional manufacturing has declined, have pursued strategies to attract call-centres. In North America, the province of Ontario in Canada advertises itself as 'among the leaders in Call-Centre job growth in North America'. Ireland claims to be 'the call-centre capital of Europe' – with 70 per cent of its call-centres and more than 80 per cent of call-centre employment in American-owned firms. Almost all of these are located in Dublin.[42] India is making similar claims about its call centre attractiveness since GE Capital Services opened India's first international call-centre in the mid-1990s.[43]

An especially interesting aspect of the internationalization of call-centres is the demand it places on linguistic and culturally specific skills. Most of us have had the experience of calling a customer enquiry telephone number assuming it to be located locally whereas, in fact, it may be almost anywhere in the country or even in a totally different part of the world. In this latter case, one of the problems is that of relating to customers in different countries in a manner that does not reveal that the respondent may be thousands of miles away and in a totally different time zone. To deal with this, call-centre employees in India, for example, are provided with intensive courses in the modes of speech, cultural identities and kinds of locally specific information needed to make the caller feel they are dealing with somebody they can easily relate to.

In the case of Indian call-centres connected to the United States, it is not only desirable to speak English (that is one of the major reasons why India can attract this kind of business) but to go beyond that as the following examples illustrate:

> In everything but pay … Indian call centres try to be the same as US ones. Foremost is the acquired accent of call centre workers. They are trained to speak like natives of, say, Texas or California because this is home for the millions of customers of the big US telecoms, healthcare and financial services groups that outsource their 'customer response services' to Indian companies. This makes a US accent a commercially valuable commodity in India. A British accent comes a poor second and Indians are just discovering the market potential of an Australian accent … Accent training has, therefore, emerged as an important activity at call centres. Dedicated call centre colleges have mushroomed … The second plank of tuition is US culture. This is to ensure than an Indian can appreciate, respond to and initiate conversation on US living … Employees are expected to have conversations that enhance the US client's relationship with the customer. They should know that snow is rare in Florida – and therefore not to ask a Florida caller about winter clothes – as well as be informed about bearish sentiment on the Nasdaq.[44]

The customers have to ring a number in the UK to check their mobile telephone bill, or to ask about a new product or service. They are, for the most part, spectacularly unaware that their enquiry has been routed thousands of miles away to an Indian call centre … Not, of course, that there is much to give the game away. The subterfuge is truly magnificent. Callers are greeted with a 'good afternoon' when it is already evening in India and dark. Should the caller lob in a reference to David Beckham or the Queen Mother, Indian staff are able to give a suitable off-the-cuff reply. Nothing is left to chance. This is Spectramind, one of India's newest and most sophisticated call centres ... Here recruits receive a 20-hour crash course in British culture. They watch videos of soap operas ... to accustom them to regional accents. They are told who Robbie Williams is. They learn about Yorkshire pudding. And they are taught about Britain's unfailingly miserable climate.

Each computer screen shows Greenwich Mean Time and the temperature in the UK ... 'We find showing new staff videos of Yes, Prime Minister is particularly effective,' says Spectramind's ... chief executive. 'They get a two-hour seminar on the royal family. We download the British tabloids every morning from the web to see what our customers are reading. We make sure our staff watch Premier League football games on TV. And we also explain about the weather, because British people refer to the subject so frequently.[45]

Transnational retailing

Retailing is the final link in the production chain. As such, it is extremely sensitive to the specific characteristics of the consumer markets it serves. Such markets, as we have seen, continue to have a high degree of individuality, despite the geographical spread of some types of consumer preference. Consequently, retailing has always had – and, indeed, continues to have – a predominantly domestic orientation, even though some retailing firms have become increasingly transnational in their operations. In fact, some of the world's largest retailers in terms of sales revenues continue to be entirely embedded in their domestic market. For example, although seven of the top ten retailers in the world are headquartered in the United States, only one of these firms, Wal-Mart, can be regarded as a genuinely transnational retailer. Indeed, four of the US firms in the top ten have no presence whatsoever outside North America. Nevertheless, an increasing number of retailers have expanded rapidly across national boundaries in recent years.

There are two dimensions to the transnational operations of retailing firms (Figure 14.12). The most obvious is the *selling* of products directly to the final consumer. Traditionally, this occurs within a physical structure, the retail store, but with an increasing amount occurring through the Internet (e-tailing, rather than bricks-and-mortar retailing). The other dimension is that of retailers' *sourcing* of products from different parts of the world. The big retail chains have vastly increased the geographical scale of their sourcing systems as well as exerting

Figure 14.12

The two dimensions of transnational retailing

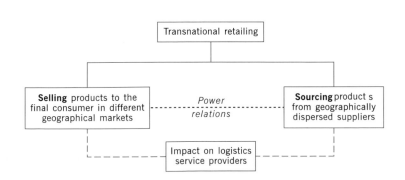

increasing *power* and influence over their suppliers (as we saw in the case of the garments industry in Chapter 10). As the scale of the leading retail chains has increased (not least through merger and acquisition), their buying power has also increased. This enhanced power affects both suppliers of products themselves and also the logistics firms responsible for getting the products to the retail stores.

> A shifting power structure in the retail trade not only changes market shares but also the structure of the distribution network. The major international retail customers ask for customized logistics solutions across borders. Apart from negotiating frame orders with significant price advantages with suppliers, the most powerful retailers also require information sharing services, such as electronic data interchange, advance shipping notices via the Internet and track and trace capabilities. They typically prefer delivery to their own distribution centers where goods are consolidated with other products for delivery to their retail stores.[46]

The world's leading transnational retailers

One indication of the rapid growth of transnational retailing is the fact that, in 1999, four retail firms were among the world's top 100 transnational corporations, whereas only a few years earlier, in 1993, no retail firm appeared in the list of the top 100. However, in terms of their transnationality index (TNI) the retailers ranked pretty low down the list, as Table 14.1 shows. Only one firm, Royal Ahold, had an index greater than 50; two firms, Metro and Carrefour, had a TNI of more than 30, while the world's biggest retailer in terms of overall sales, Wal-Mart, had a TNI of only 25. Wal-Mart actually ranked 90th in the top 100 TNCs.

From the perspective of this chapter, then, what matters is not so much the absolute size of retail firms but the degree to which they are transnational in their operations. When we look only at the retailing sector we find a very wide range of transnationality in terms of both the percentage of the firms' sales overseas and also the actual geographical extent of their retail operations. Of the top 100 retailers, more than one-third operated only in their country of origin. A further one-third can be described as 'regional' retailers, with their foreign operations confined to their immediate geographical region (for example, North American firms operating only in North America; European firms operating only in Europe,

Table 14.1 Retailers in the world's top 100 transnational corporations, 2000

Rank	TNI	Company	Country
53	52.7	Royal Ahold	The Netherlands
76	36.4	Metro	Germany
78	34.7	Carrefour	France
90	25.8	Wal-Mart	United States

Source: Calculated from UNCTAD, 2001: Table III.1

Table 14.2   The leading 20 international retailers, 2000

Rank	Company	Country	Foreign sales ($m)	%	Type of business
1	Royal Ahold	Netherlands	23,854	76.4	Supermarkets; specialty stores
2	Wal-Mart	USA	22,731	13.9	Discount stores; warehouse clubs; supercentres
3	Carrefour	France	19,834	37.7	Hypermarkets; discount stores; convenience stores; specialty stores
4	Metro	Germany	17,665	40.0	Supermarkets; shopping centres; department stores; specialty stores
5	Delhaize	Belgium	14,808	83.0	Supermarkets; drugstores; discount stores; specialty stores
6	ITM	France	13,234	36.0	Supermarkets
7	Tengelmann	Germany	12,698	47.9	Shopping centres; supermarkets; drug stores; specialty stores
8	Otto Versand	Germany	10,286	52.0	Mail order
9	Ito-Yokado	Japan	9,011	29.8	Superstores; specialty stores; supermarkets; discount stores; supermarkets
10	IGA	USA	8,800	44.9	Supermarkets
11	Aldi	Germany	8,485	32.5	Supermarkets
12	IKEA	Sweden	8,049	99.9	Specialty stores
13	Kingfisher	UK	7,184	41.0	Department stores; specialty stores; drug stores
14	Rewe	Germany	6,021	19.7	Supermarkets
15	Costco	USA	4,963	18.4	Warehouse clubs
16	Pinault Printemps	France	4,785	48.1	Mail order; department stores; supermarkets, specialty stores
17	Auchan	France	4,487	18.9	Hypermarkets
18	Sears Roebuck	USA	3,893	10.6	Department stores
19	Sainsbury	UK	3,875	15.0	Supermarkets; hypermarkets
20	Casino	France	3,258	21.0	Hypermarkets; supermarkets; convenience stores

Source: Based on www.siamfuture.com, 2002; Sternquist, 1998: 44–59

Japanese firms operating only in East Asia). Rather less than one-third of the world's top 100 retailers have what can be regarded as 'global' operations and, even here, the extent of their globality is limited – at least in terms of the geographical spread of their stores. Of course, these figures do not take into account the extent to which the major retailers source internationally. Most of them do so on a very large scale.

Table 14.2 shows some of this variation for the top 20 retailers, ranked in terms of the size of their foreign sales. It also shows the foreign percentage of the firms' total sales and the types of business they are in. Two characteristics of this group are particularly significant:[47]

- there is a relatively small 'elite' group of retailers with extensive – and fast-growing – transnational operations and
- within this group, food retailing plays an especially important role.

Only four of the leading 20 firms are US retail chains and only one, Wal-Mart, is in the top ten. But although Wal-Mart is second in terms of foreign sales, these represent only around 14 per cent of the firm's total sales. In other words, fully 86 per cent of Wal-Mart's sales are still in the United States, as are 84 per cent of its stores. Figure 14.13 shows the global distribution of Wal-Mart's stores. Within the Americas (excluding the United States) there is a substantial presence in Mexico and, to a much lesser extent in Canada, Brazil, Argentina and Puerto Rico. Wal-Mart's European presence is confined to the United Kingdom (its Asda acquisition) and Germany. In East Asia there are 11 stores in China and half that

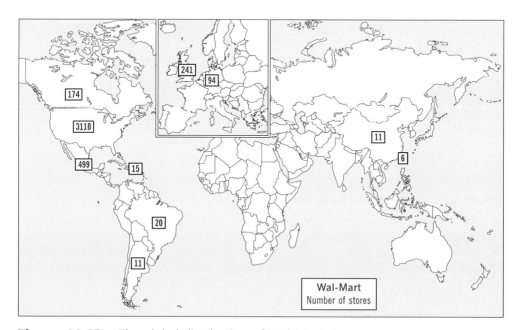

Figure 14.13 The global distribution of Wal-Mart stores
Source: Company reports

number in Taiwan. However, Wal-Mart recently acquired a stake in Seiyu, Japan's fourth largest supermarket chain, as a platform for entry into the difficult Japanese retail market.[48] Other US retailers with substantial numbers of overseas stores are in garments (The Gap, for example) or toys (Toys 'R' Us). Otherwise, of the 38 US firms among the top 100 retailers, more than half have no foreign operations whilst one-third are confined to North America. Only five of the US firms are 'global' – and even then to a very limited degree.

The extent of foreign involvement is very much higher amongst European retailers, with, in some cases, extremely high levels of transnationality. Almost 50 per cent of the European retailers in the top 100 are 'global' in the sense that they have significant operations outside Europe itself. A further one-third have operations in several European countries. Figure 14.14 shows the geographical patterns for the three leading European retailers: Royal Ahold, Carrefour and Metro. An important characteristic of all three firms is their strong presence in Eastern European countries, notably Poland and the Czech Republic. Apart from Europe, the Dutch firm Royal Ahold has a geographically extensive presence throughout the Americas, with around 1,300 stores in the United States, more than 400 in Latin America (notably in Argentina and Brazil), and some in Asia (notably Thailand).

In contrast, the French retailer Carrefour has an increasingly widespread network of stores in East Asia, including 27 in China, 26 in Taiwan, 22 in South Korea and 15 in Thailand. Carrefour also has a very large presence in Argentina (400 stores), Brazil (100) and a growing involvement in Mexico. In comparison, the geographical spread of the German retailer Metro is far more restricted. Apart from its Chinese operations, Metro's international activities are confined to Europe, although it has recently announced a new joint venture with the Japanese *sogo shosha*, Marubeni, to open cash-and-carry stores in Japan.

Japanese retailers are far less well represented in the top 100. Only ten Japanese firms appear in that list and only one of these, Ito-Yokado (owner of the 7-Eleven chain), is in the top 20. Five of the leading Japanese retailers – Ito-Yokado, Jusco, Takashimaya, Mitsukoshi and Matsuzakaya – have retail operations outside Asia; the others are essentially regional operations. There are, of course, other Japanese retailers with international operations – for example, Muji (whose name translates as 'no-brand, quality goods'). Muji currently operates 16 stores in the UK (compared with five in 1998), eight in France and one in Belgium.[49]

Motivations and market entry methods of transnational retailers

A few retailers moved into foreign markets at a relatively early stage in their development. One of the best examples is the US company F.W. Woolworth, which opened stores in Canada in 1897, in the United Kingdom in 1909 and in Germany in 1926.[50] Indeed, so familiar did Woolworths become in most big cities in the UK that few, if any, of its customers had any idea that it was a foreign firm. But this was an exceptional case on such a wide geographical scale. For the most part, retailers have been extremely hesitant entrants into foreign markets. And where they have expanded beyond their domestic borders, they have shown a very strong propensity to locate in geographically and culturally proximate countries.

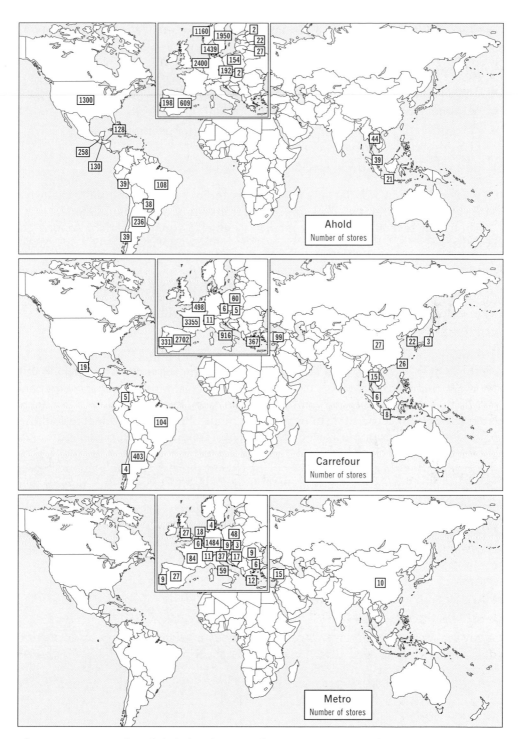

Figure 14.14 The global distribution of Royal Ahold, Carrefour and Metro stores

Source: Company reports

The motivations for the transnationalization of a retail firm's operations include the following:

- saturation of the domestic market
- intensification of competition in the domestic market
- regulatory constraints in the domestic market
- perception of profitable opportunities overseas (for example, in some fast-growing developing country markets)
- desire to exploit a firm's specific advantages in new markets.

Of course, these are not the only motivations and nor are they mutually exclusive. Most location-investment decisions derive from a mixture of motivations.

There are several ways for a retailer to enter a foreign market. Among the options are:

- to build a new store(s)
- to merge with or to acquire an existing firm in the target market
- to enter into a collaboration with a domestic firm within the target market through, for example, licensing or franchising.

The first of these is relatively unusual as a mode of initial entry, although once established in a particular country it becomes a common mode of expansion. It is far more common for retailers to enter a new territory through either merger and acquisition or a collaborative venture. In some cases, local regulations insist on the latter as a condition of entry. This is most commonly the case in developing countries (China is a prime example as well as many Latin American countries). Figure 14.15 shows the scale of cross-border merger and acquisition activity in the retail sector during the 1990s. Both the number of cases and their aggregate value increased dramatically during the second half of the 1990s.[51] One of the biggest was

Figure 14.15

Cross-border mergers and acquisitions in the retailing sector

Source: Based on *The Economist*, 19 June 1999: 83

Wal-Mart's acquisition of the British supermarket chain Asda, for almost $11 billion, and its acquisition of the German chain Wertkauf. Another was Carrefour's acquisition of the French retailer Promodès, which, amongst other things, gave it access to the firm's overseas operations. A third example is Royal Ahold's acquisition of retailers in the United States, Scandinavia, Argentina, Brazil and Chile.

The major transnational retail chains have shown a particular propensity to invest heavily, either through acquisition or joint ventures, in the emerging markets of East Asia, Latin America and Eastern Europe. Indeed,

> global retailing was characterized during the mid to late 1990s then by the efforts of an elite group of firms to leverage their increasing core-market scale and the free cash flow for expansionary investment which those markets provided, in order to secure the longer-term higher growth opportunities offered by the emerging markets.[52]

Several types of competitive advantage underpin the transnational strategies of the major retailers:[53]

- innovative retail formats
- logistics and distribution systems, particularly those that economize on inventory and distribution costs
- IT systems and supply chain management, for example the use of EDI
- access to low-cost capital for expansion
- transfer of 'best practice' knowledge
- depth of human/management capital resources giving access to a wide range of international management experience
- the ability to source supplies globally.

However, this latter advantage

> seems[s] to have been less important in the movement of the retail TNCs into emerging markets in the mid to late 1990s than has often been suggested … Yet global sourcing by retailers clearly increased significantly during the 1990s, with most retailers sourcing clothing and electronic products on a global scale … The fact remains, however, that most of the elite group of retail TNCs active in the emerging markets of the late 1990s were food and general merchandise retailers, with combination food/non-food hypermarkets being the primary vehicle for the entry of 'modern' Western-style corporate retailing into these markets. It was the non-food consumer product categories in those hypermarkets which offered retail TNCs the greatest scope to leverage global purchasing scale – the potential in food products was more limited due to differences in national tastes and preferences, perishability issues, and the more restricted overlap between countries in the food products stocked by retailers than might first appear. Nevertheless, buying centres serving the major global markets of the retail TNCs did begin to emerge … And even in the perishable fresh produce area, there were increasing examples of combined purchasing and international sourcing across several chains within a TNC.[54]

Thus, there has been very considerable growth in the transnational operations of some of the leading retail chains. But such expansion has not been without its problems. As one recent press headline put it: 'overseas expansion has been retailers' graveyard'. The report was concerned mainly with the problems of the leading UK retailer, Marks & Spencer, which has been unable to replicate its domestic-market success (although even this was also under threat in the late 1990s/early 2000s). One view was that M&S's overseas failures were 'due to the "insufferable arrogance" of management that it could replicate its UK success internationally'.[55] But Marks & Spencer is not a unique case. Carrefour failed to transfer its hypermarket model to the United States, while Benetton also had big problems in the United States, where its particular independent franchise format gave a misleading impression of related stores. A customer buying a Benetton garment in one US city but trying to exchange it in another could not do so – a real blunder in the US context. Wal-Mart has had major difficulties with its acquired German affiliates, particularly by failing to understand the fundamental differences between the German and the US retail food distribution system.

> Drawing on the US model, Wal-Mart decided it wanted to control distribution to stores rather than leave it to suppliers. The result was chaos because suppliers could not adapt to Wal-Mart's centralized demands ... Dominated by hard discounters and privately-owned businesses and suffering from overcapacity, the German food retail market has long been plagued by microscopic profit margins, oscillating between 1 and 2 per cent, compared with 4 to 6 per cent in the UK. Wal-Mart's low-price message, therefore, had none of the revolutionary ring it carried in the UK when it acquired the Asda chain ... Lack of scale has also worked against the group, by preventing it from dictating to suppliers and distributors. Although sizeable in the hypermarket segment, Wal-Mart Germany is a midget in the food retail industry as a whole, with less than 2 per cent of the market. Edeka and Rewe, Germany's two leaders, have far greater purchasing muscle.[56]

The use of local partners within a joint venture often helps to avoid the problems of misunderstanding local market conditions. But even joint ventures are not without their difficulties, especially if the foreign partner fails to learn from the knowledge embedded in the local partner.

It is also the case that, while the strength of most of the leading retailers is based on their high levels of profitability in their home market, their returns on international operations are often far lower.

> Carrefour's operating margins in France are more than 6% of sales, whereas, after operating internationally for 30 years, it still loses money in much of Asia, Latin America and even some parts of Europe. Meanwhile, Wal-Mart, which first went abroad in 1991, makes a return on capital of 5.8% on its international business, far lower than in America ... In practice ... international scale economies are hard to achieve. In the excitement of their charge into new markets, many retailers forget that the crucial ingredient of their success at home is their relative size and market share. Without enough sales and profits in a particular market, even the most long-term management will find it difficult to justify the expense of setting up a large distribution network or installing the latest technology – and without these, the international newcomer cannot compete with entrenched locals.[57]

So, the transnationalization of retailing is far from being a straightforward or unproblematical process. Competing head-to-head with local firms is particularly difficult in this sector. A major problem is that of identity. Because retailing has been, as already pointed out, very much a domestic activity, there is little knowledge of foreign retail store brands (as opposed to product brands). For many customers outside the United States, for example, Wal-Mart is an unknown quantity. The same applies to non-French residents' knowledge of Carrefour, or non-UK residents' awareness of Tesco. Yet building up a respected and trusted brand identity takes a long time. Meanwhile, local competition remains, in most cases, a very serious problem for transnational retailers. The other problem facing retailers is, of course, the development of Internet shopping.

Is e-tailing a competitive threat to retailing?

In discussing e-commerce as a technological innovation in logistics and distribution systems we concluded that its threat to established channels of distribution, including retailing, was more limited than much of the hype tends to suggest. One of the arguments was that the actual physical fulfilment of customer orders requires 'bricks' as well as 'clicks'. The other issue, especially relevant at this point, is the extent to which consumers are actually shifting their allegiance from the traditional retailer to the e-tailer. So far, the answer seems to be 'only to a very limited extent'. At the end of the 1990s, the US Department of the Census calculated the size of online retail sales (by both traditional retailers and direct Internet sales) at \$5.3 billion – a minuscule 0.64 per cent of total retail sales in the United States.[58] In the 2001 Christmas season, Internet users spent 13 per cent of their budgets online compared with 12.4 per cent a year earlier – no real change.[59] If we bear in mind that the United States is by far the most developed e-commerce market in the world, then it is clear that e-tailing is still very much in its infancy and that its quantitative significance must be kept in perspective.

The list of e-tailers that have gone out of business is enormous. Firms such as Boo.com, the seller of fashionable sports wear and the most expensive Internet start-up company in Europe in 1999, collapsed one year later. In the United States, Webvan, the Internet grocery delivery service, closed down in 2001. Etoys, the United States Internet retailer, announced the closure of its UK operations in early 2001. Of course, new Internet start-ups are happening all the time but the casualty rate remains extremely high. Even the large, well-established operations such as Amazon.com and Priceline.com have only just begun to make any profits.

> For now, the evidence suggests that online shoppers are more concerned about price than about new features or convenience – the advantage which e-tailers once thought would allow them to charge more than offline stores … Amazon.com, one of the few global names in online retailing, has made the same discovery … [its founder] has committed the group to an 'everyday low pricing model' that would not look out of place at Wal-Mart … The price pressures have been driven in part by the growth of established US discounters such as Wal-Mart and Target Online.[60]

While there is no doubt that the development of Internet retailing is a highly significant event it is a system that is very much in its infancy. It has certainly invaded the traditional retailing sector, in the form of both free-standing e-tailers and the increasingly ubiquitous Internet operations of the traditional retailers themselves. A large number of retail chains now have online facilities alongside their traditional operations. But it is likely to be a long time – if ever – before retailing becomes a virtual, rather than an actual, experience. Not least this is because, as we saw in Chapter 4, there is immense unevenness in people's access to Internet facilities. In addition, shopping has a social function for many consumers that depends upon face-to-face contact and a gregarious involvement with other shoppers. Such conditions cannot be replaced by online shopping.

Conclusion

Like the financial services industries discussed in the previous chapter, the distribution industries are circulation activities and, as such, are absolutely central to the operation of all economic activities. Their basic function is to act as intermediaries, connecting together all stages of the production chain in fast, flexible and reliable ways by overcoming the friction of geographical distance. As manufacturers and other service firms have become increasingly obsessed with time-based competition, the demands on the distribution services to 'deliver' have intensified. As in financial services, too, deregulation has been an important factor – especially deregulation of transportation and telecommunications systems. In recent years, the distribution industries have been transformed, first by such developments as electronic data interchange and bar codes and, second, by the emergence of Internet-based transactions in the form of e-commerce. The acronyms B2B and B2C have become part of the new language in these industries. They have the potential to revolutionize distribution in many respects but, as yet, it is largely an only partially realized potential.

We looked in detail in this chapter at two branches of the distribution industries: global logistics and transnational retailing. In both cases, alongside some quite dramatic changes we can see the continuation of some long-established forms of organization. In the logistics sector, for example, although the emergence of highly sophisticated third-party logistics service providers represents a major development, there continues to be a wide variety of firms engaged in providing logistics and distributions services. Similarly in retailing, although online shopping has burst on the scene with astonishing speed it has not displaced much traditional retailing activity as yet. On the other hand, there is no doubt at all that the most important force for change in the distribution industries is the further development of electronic systems in a whole variety of forms.

Notes

1 Wrigley (2000) posed this question as the starting point of his research into the globalization of retailing and rightly criticized the absence of attention to this sector in current studies of the global economy. Schary and Skjøtt-Larsen (2001) provide an excellent discussion of distribution and logistics processes from a supply chain perspective.
2 *The Financial Times* (1 December 1998).
3 See, for example, Schoenberger (1997, 2000); Stalk and Hout (1990).

4 Stalk and Hout (1990: 39).
5 Min and Keebler (2001: 265).
6 Schary and Skjøtt-Larsen (2001: 244).
7 *The Financial Times* (1 December 1998).
8 For a discussion of some of the elements involved in consumer behaviour, see Fine and Leopold (1993), O'Shaughnessy (1995) and Sklair (1995).
9 Abernathy et al. (1999: ch. 4) discuss these technologies in the context of retailing.
10 Abernathy et al. (1999: 62).
11 Abernathy et al. (1999: 61).
12 Abernathy et al. (1999: 66).
13 Stalk, Evans and Shulman (1998: 58).
14 Leinbach (2001: 15).
15 Gereffi (2001: 1628).
16 *The Financial Times* (18 October 2000).
17 US Department of Commerce (2000: 18).
18 See Gereffi (2001: 1628).
19 Schary and Skjøtt-Larsen (2001: 133–7).
20 *The Financial Times* (10 February 2000; emphasis added).
21 *The Financial Times* (1 August 2000).
22 Braithwaite and Drahos (2000) provide comprehensive treatment of the development and implementation of regulation in the telecommunications, sea and air transportation industries.
23 Braithwaite and Drahos (2000: 322).
24 Braithwaite and Drahos (2000: 454).
25 Braithwaite and Drahos (2000: 323, 324).
26 Braithwaite and Drahos (2000: 456–7).
27 *The Economist* (19 January 2002).
28 Schary and Skjøtt-Larsen (2001: 223).
29 *The Financial Times* (23 December 1999).
30 *The Financial Times* (11 December 2000)
31 *The Financial Times* (23 December 1999).
32 Sternquist (1998: 151).
33 Sternquist (1998) provides a useful account of different national regulatory practices in retailing.
34 This section is based primarily on Schary and Skjøtt (2001: 230–41). See also Beukema and Coenen (1999).
35 Beukema and Coenen (1999: 138–9).
36 Schary and Skjøtt-Larsen (2001: 231).
37 See Dicken and Miyamachi (1998).
38 See *The Economist* (2 June 2001), Magretta (1998); Schary and Skjøtt Larsen (2001: 383–5)
39 *The Economist* (2 June 2001).
40 Schary and Skjøtt-Larsen (2001: 383–4).
41 Breathnach (2000: 481).
42 Breathnach (2000).
43 *The Economist* (5 May 2001).
44 *The Financial Times* (4 April 2001).
45 *The Guardian* (9 March 2001).
46 Schary and Skjøtt-Larsen (2001: 129).
47 See Wrigley (2000: 296).
48 *The Financial Times* (15 March 2002).
49 *The Financial Times* (5 April 2001).
50 Alexander (1997: 6)
51 See Wrigley (2000: 301–4).
52 Wrigley (2000: 306).
53 See Wrigley (2000: 306–8).
54 Wrigley (2000: 307–8).

55 *The Financial Times* (30 March 2001).
56 *The Financial Times* (12 October 2000).
57 *The Economist* (19 June 1999).
58 US Department of Commerce (2000: 9).
59 *The Financial Times* (20 February 2002).
60 *The Financial Times* (20 February 2001).

PART FOUR

WINNERS AND LOSERS IN THE GLOBAL ECONOMY

CHAPTER 15

Winners and Losers: an Overview

From processes to impacts

Globalizing processes: a summary

The focus throughout the preceding chapters has been on the *patterns* and *processes* of global shift: on the *forms* being produced by the globalizing of economic activities and on the *forces* producing those forms. Since the end of World War II, as we have seen, the world economy has experienced enormous *cyclical* variation in economic activity: the unparalleled growth of the long boom lasting from the early 1950s to the mid-1970s; the deep world recession of the second half of the 1970s and the early 1980s; the impressive economic recovery of the later 1980s; the highly volatile patterns of the 1990s, with spectacular growth interspersed with unanticipated crises. The prospects for the early years of the 21st century are far from clear.

Underlying these global cyclical trends are *global structural changes* associated with the increasing globalization of economic activity. The geo-economic map has been redrawn and become much more complicated than it was only half a century ago. Although world production, trade and investment are still dominated by the developed market economies, the position of individual industrial nations has changed dramatically. Geographically, the global economy is now *multi-polar*, as new centres of production have emerged in parts of what had been, historically, the periphery of the world economy. The world is now more accurately described as a 'mosaic of unevenness in a continual state of flux'.

These transformations of the geo-economy are the outcome, primarily, of three interconnected processes:

- *Transnational corporations* are the *primary movers and shapers* of the global economy because of their potential ability to control or coordinate production networks across several countries; to take advantage of geographical differences in factor distributions; and to switch and re-switch resources globally. TNCs are both intricate organizational networks in their own right and also deeply embedded within dynamic networks of inter-firm relationships and alliances. The empirical evidence suggests increasing organizational flexibility as TNCs restructure their operations. However, it is by no means the case that TNCs are converging to a single 'global' organizational form; diversity of structures and strategies continues to exist related, in large part, to the places from which they originate – their home countries.

- *States* continue to be a major influence in the global economy through their continuing attempts to *regulate* economic transactions within and across their territorial boundaries. All states engage in such regulatory activity, although to greatly varying degrees and in very different ways according to their specific ideological stance. Some states are overtly and self-consciously 'developmental' – in that they make explicit attempts to influence the shape and direction of economic activity within and across their borders. More broadly, two sets of political forces have been especially significant in the past few years. One is the spread of 'deregulatory' forces as access to national markets has been opened up, initially to trade flows but, more recently, to foreign investment flows. The other is the proliferation of regional trade agreements that, in effect, shift the regulatory processes to a different scale.

- *Technology* is a fundamental *enabling* force in the internationalization and globalization of economic activities. It is at the dynamic heart of all economic growth and development. Technological change is, essentially, an evolutionary, learning process which occurs very unevenly through time and space. The cumulative influence of small, incremental changes tends to be overshadowed by the massive radical changes involved in the periodic creation of entirely new techno-economic paradigms which drastically shape and reshape both economy and society. Most significant have been the space-shrinking technologies and, especially, the information technologies that have come to pervade virtually all aspects of life.

As we showed in Figure 1.1, these processes form major *interconnections* within the global economy at a whole spectrum of geographical scales. The processes of globalization are not simply unidirectional, from the global to the local. Rather, all globalization processes are deeply embedded, produced and reproduced in particular contexts. Hence, the specific assemblage of characteristics of individual nations and of local communities will not only influence *how* globalizing processes are experienced but also will influence the *nature* of those processes themselves. We have repeatedly emphasized the strongly localized nature of economic activity, including that of technological change, and the continuing significance of 'place' to the nature and behaviour of transnational corporations. Both nation-states and local communities are 'containers' of distinctive cultural, social and political institutions and practices.

Of course, the real *effects* of globalizing processes are felt not at the aggregate level of the national economy but at the *local* scale: the communities within which real people live out their daily lives. It is on this scale that the physical investments in economic activities are actually put in place, restructured, or closed down. It is on this scale that most people make their living and create their own family, household and social communities. But, as Figure 1.1 shows, although the effects of globalizing processes on local communities may be direct they are, more commonly, 'refracted' through the medium of the national or regional contexts within which the particular local community is embedded.

Winners and losers: the impacts of globalizing processes

Having concentrated on processes, we now turn, in the final chapters of this book, to the question of *impact*. What do globalizing processes mean for people? Are they beneficial or are they detrimental? What can or should be done to make things better? Not surprisingly, given our discussion of the globalization debate in Chapter 2, views are strongly polarized. To its proponents,

> globalization is a savage process, but it is also *a beneficial one, in which the number of winners far outnumbers that of the losers.*[1]

To its critics, the effects are very different:

> Today, when we hear or read about the global economy, it is usually in terms of the trillions of dollars of goods, services, and investment that circle the planet, with the great increases in national wealth that accrue to states that adopt open policies. But there are other data that usually go unnoticed in these discussions. We hear less about the 100 million citizens in the industrial countries who are classified as living below the poverty line. We hear less about the 35 million in these same countries who are unemployed. We hear less about growing income inequality. And we hear still less about the 1.3 billion people in the developing world whose income level is under $1 per day. For all these people, the global economy has not yet brought either material gifts or the hope of a better life.[2]

In this brief chapter, which acts as an introduction to what follows, we examine the broad 'contours' of economic development. This will help to provide a framework within which to explore the problems facing people in developed and developing countries as they struggle to cope with the processes of global shift. The precise focus in Chapters 16 and 17 will be on the impact of globalizing processes on *jobs* and *incomes* in developed and developing countries respectively. There is a very good reason for adopting this specific focus. *Income* is the key to an individual's or a family's material well-being. However, income – or lack of it in the form of *poverty* – is not an end in its own right but, rather, a means towards what Amartya Sen[3] calls 'development as freedom'. In that sense, poverty is an 'unfreedom':

> There are good reasons for seeing poverty as a deprivation of basic capabilities, rather than merely as low income. Deprivation of elementary capabilities can be reflected in premature mortality, significant undernourishment (especially of children), persistent morbidity, widespread illiteracy and other failures.[4]

The major source of income (for all but the exceptionally wealthy) is *employment*, or *self-employment*. Hence, the question of *'where will the jobs come from?'* is a crucial one throughout the world.

In attempting to unravel this question in terms of the processes discussed throughout this book we find a very complex picture. The major employment changes that have been occurring in both developed and developing countries are the result of an intricate interaction of processes. Job losses in the developed market economies, for example, cannot be attributed simply to the relocation of production to developing countries. Although this is undoubtedly a factor, there is far more to it than this. What is clear, however, is that the industrialized economies face major problems of adjusting to the decline in manufacturing jobs. Nevertheless, the problems facing developing countries are infinitely more acute. The spectacular success of a small number of NIEs should not blind us to the fact that the majority of developing countries face enormous problems of economic survival in an increasingly globalizing economic system.

The fact that these problems, though manifested most directly at the local scale, are essentially global problems raises the question of what can, or should, be done globally to make the world a better place. In Chapter 18, therefore, we explore some major issues of global economic governance, notably those relating to finance, trade, employment, and the environment.

The contours of economic development

The contours of the development map show a landscape of great unevenness and irregularity; a landscape of staggeringly high peaks of affluence and deep troughs of deprivation interspersed with plains of greater or lesser degrees of prosperity. Geographically, the major divide is between the 'north' and the 'south' – between the developed and the developing countries, although, as we shall see, there are very significant differences in levels of development *within* each of these two categories.

The contours of world poverty

The development gap is stunningly wide. The United Nations Development Programme[5] shows that, by the end of the 1990s, the 20 per cent of the world's population living in the highest income countries had

- 86 per cent of world GDP – the bottom 20 per cent had only 1 per cent of world GDP
- 82 per cent of world export markets – the bottom 20 per cent just 1 per cent
- 68 per cent of foreign direct investment – the bottom 20 per cent had just 1 per cent
- 74 per cent of world telephone lines – the bottom 20 per cent had just 1.5 per cent.

Not only is the development gap stunningly wide – it has been getting wider. Before, the beginning of the 19th century the differences between levels of income were relatively small:

> At the dawn of the first industrial revolution, the gap in per capita income between Western Europe and India, Africa, or China was probably no more than 30 per cent. All this changed abruptly with the industrial revolution.[6]

Figure 15.1 shows how dramatically the gap between the richest and poorest countries has widened since then.

Figure 15.1

The widening income gap between countries

Source: Based on UNDP, 1999: Figure 1.6

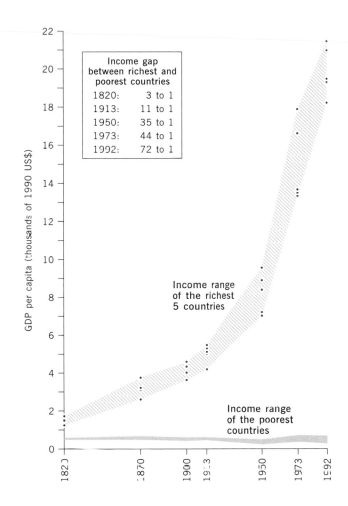

Income gap between richest and poorest countries

1820:	3 to 1
1913:	11 to 1
1950:	35 to 1
1973:	44 to 1
1992:	72 to 1

Income range of the richest 5 countries

Income range of the poorest countries

GDP per capita (thousands of 1990 US$)

Today, despite very considerable advances in some parts of the world, one in five people (around 1.2 billion) live on less than $1 per day. Nearly 70 per cent of these utterly impoverished people live in South Asia and sub-Saharan Africa (Figure 15.2). The extent to which the income gap is widening, narrowing, or staying about the same is controversial and depends on how it is measured.[7] But whichever measure is used, the fact remains that the gap between rich and poor countries is enormous and that any improvements have been relatively small compared with the sheer scale of the problem. Of course, that doesn't mean that no country has improved its position. Some – especially in East Asia – certainly have. As Figure 15.2 shows, the number of people in East Asia and the Pacific living on less than $1 per day fell by around 140 million between 1987 and 1998.

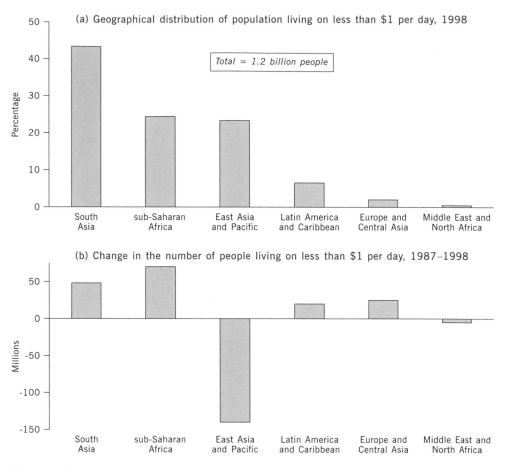

Figure 15.2   Distribution of the world's poorest population

Source: Based on World Bank, 2001: Figures 1 and 2

In general, however, the winners – and the losers – have been the usual suspects. The already affluent developed countries have sustained – even increased – their affluence, some developing countries have made very significant progress, but there is a hard core of exceptionally poor countries that remains stranded. Despite the generally rising tide associated with overall world economic growth it has not lifted all boats. Most strikingly, it is the countries of sub-Saharan Africa and parts of South Asia which have benefited least. In both cases, the number of people living on less than $1 per day increased very substantially, as Figure 15.2 shows.

Focusing the analytical lens at the country level provides a useful first cut at mapping the contours of development. But, of course, such a focus obscures the detail of the economic landscape, both at smaller geographical scales and in terms of non-geographical criteria (for example, gender, social class and so on). We will look at such criteria in more detail in Chapters 16 and 17. Here, however, we can note just two broad groups of winners and losers which cut across the broad development divide we have been discussing.

The clear winners are the elite *transnational capitalist class*[8] whose members are predominantly, although not exclusively, drawn from developed countries. The dominant group is made up of the owners and controllers of the major corporations – the globe-trotting, jet-setting TNC executives. To these we can add globalizing bureaucrats and politicians, globalizing professionals (with particular technical expertise – even including some academics), merchants and media people. Without question, these are winners in the global economy.

This transnational capitalist class (TCC) displays a number of significant characteristics:[9]

- The economic interests of its members are increasingly globally linked rather than exclusively local and national in origin.
- The TCC bases its behaviour on specific forms of global competitive and consumerist rhetoric and practice.
- Members of the TCC have outward-oriented global rather than inward-oriented local perspectives on most economic, political and cultural ideology issues.
- They tend to share similar lifestyles, especially patterns of higher education (for example, in business schools) and consumption of luxury goods and services. 'Integral to this process are exclusive clubs and restaurants, ultra-expensive resorts in all continents, private as opposed to mass forms of travel and entertainment and, ominously, increasing residential segregation of the very rich secured by armed guards and electronic surveillance'.
- They tend to project themselves as being citizens of the world as well as of their place of birth.

While transnational elites are clear winners, *women* – at least in many parts of the world – tend to be losers in the global economy. Although gender inequality has been much reduced (though certainly not eliminated) in recent years in most developed countries, the situation tends to be very different in other parts of the world. We noted earlier that one-fifth of the world's population – 1.2 billion people – live on less than $1 per day. A staggering two-thirds of these are women, living 'in abject poverty: on the margins of existence without adequate food, clean water, sanitation or health care, and without education'.[10] In Sen's terms of 'development as freedom', women are significantly more disadvantaged than men. At the same time, because of their key role in nurturing children, women hold the key to development, especially in the poorest countries of the world. The problem is that in many developing countries (as opposed to developed countries), women have a much higher mortality rate and lower survival rate than men. As a result, the female/male ratio is lower than in developed countries, implying a phenomenon of 'missing women'. Where this occurs – as in China and India, for example – the main explanation would seem to be 'the comparative neglect of female health and nutrition, especially – but not exclusively – during childhood'.[11]

The contours of world population[12]

As in the case of the distribution of income, there are dramatic divides on the population landscape. Geographical variations in population growth rates, as well as in the age composition of the population, have an extremely important influence on how globalizing processes are worked out in different places. They also relate, very clearly, to issues of poverty, to the ability of people in different places to make a living through employment, and to issues of environmental impact.

Population growth

At the beginning of the 21st century (in mid-2000), the world's population reached a total of 6.1 billion. One hundred years earlier, at the start of the 20th century, world population was less than 2 billion. Not unreasonably, then, has the 20th century been called 'the century of population' and the 'explosion of population ... [as] ... one of its defining characteristics'.[13]

> This is an absolute increase that far exceeds that which has occurred in any other period of human experience. It took until 1825 to reach one billion humans *in toto*; it took only the next 100 years to double; and the next 50 years to double again, to 4 billion in 1975. A quarter of a century later, as we were celebrating the millennium, the total jumped to 6 billion. True, the pace of increase has been slowing in the last decade or so but, like a large oil tanker decelerating at sea, that slowdown is a protracted process.[14]

The UN's medium projection is that world population in 2050 will be around 9.3 billion, although it could be as high as 10.9 billion or as low as 7.9 billion depending on what happens to fertility rates.

The most striking feature of world population growth is that it is now overwhelmingly occurring in developing countries. In 2000, 80 per cent of the world's 6 billion population was in the developing countries. Figure 15.3 shows the massive – and accelerating – divergence in population growth between developed and developing countries. 1950 was an especially significant turning point. That year marked the beginnings of the 'population explosion' brought about by the rapid fall in death rates in Africa, Asia, and Latin America coupled with continuing high fertility rates in those areas. Since then, the contrast between the very low population growth rates of the developed countries and the very high rates in many developing countries has become even more marked.

Just to replace an existing population requires a fertility rate of 2.1 children per woman. In most developed countries, fertility rates are now well below the replacement level – at 1.57 and declining – although with an expected rise in the mid-21st century to just below replacement levels. In contrast, fertility rates in the developing world as a whole are currently at 3.1. But although they are well below that in one or two cases (for example, China is now actually below replacement level) in the very poorest countries of the world fertility rates are exceptionally high.

> In 1995–2000, the 48 least developed countries had a total fertility of 5.74 children per woman, which is projected to decline to 2.51 children per woman in 2045–2050 ... In contrast, the rest of the countries in the less developed regions exhibit a total fertility of 3.06 children per woman in 1995–2000 and are projected to have 2.06 children per woman in 2045–2050.[15]

Despite high mortality rates through HIV/AIDS in many of these poorest countries, their population is expected to grow three-fold by 2050, from 658 million to 1.83 billion. For developing countries as a whole, the 2050 population is predicted by the UN to be about 8.1 billion (that compares with a *total* world population in 2000 of 6 billion).

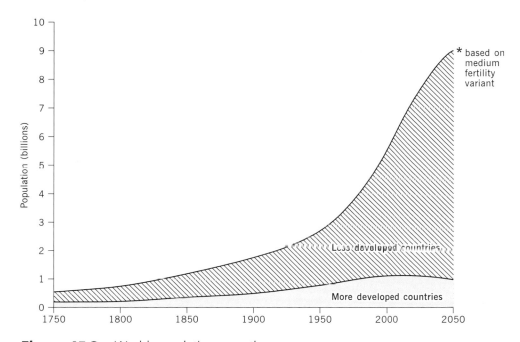

Figure 15.3   World population growth
Source: Based on data from United Nations Population Division

'Old' and 'young' populations

Such persistent unevenness in fertility rates between developed and developing countries creates significant differentials in the *age-composition* of the population. Put in a nutshell, developed countries are ageing while most developing countries continue to be youthful in population terms. Table 15.1 shows the marked geographical variations in the relative importance of different age groups. Note the huge contrast between the top part of the table and the bottom. Europe, North America and Japan all have relatively old populations, in terms of both the percentage of their populations over 65 years and the low percentage of their populations under 15 years.

Europe is the major area of the world where population ageing is most advanced. The proportion of children is projected to decline from 17 per cent in 2000 to 14 per cent in 2050, while the proportion of older persons will increase from 20 per cent in 1998 to 37 per cent in 2050. By then, there will be 2.6 older persons for every child and more than one in every three persons will be aged 60 years or older. As a result, the median age will rise from 37.5 years in 2000 to 49.5 in 2050. Japan is currently the country with the oldest population (its median age is 41 years) followed by Italy, Switzerland, Germany and Sweden, with median ages of 40 years each.[16]

Compare this with the bottom part of Table 15.1, which shows that 'young' countries (in population terms) are overwhelmingly in developing countries, particularly in Africa, which, with 43 per cent of its population below 15 years of age, is the youngest region in the world.

Such wide variations in age structure are enormously important in terms of economic and social development, consequently we will have more to say about

Table 15.1 Geographical variations in the age composition of the population

Region	Percentage of population		
	Under 15 years	15–64 years	Over 65 years
World	31	62	7
Europe	18	68	14
Northern	19	66	15
Western	17	68	15
Eastern	20	67	13
Southern	17	67	16
North America	21	66	13
Japan	15	69	16
Oceania	26	64	10
Africa	43	54	3
Northern	38	58	4
Western	45	52	3
Eastern	46	51	3
Middle	47	50	3
Southern	35	60	5
Latin America and Caribbean	33	62	5
Central America	36	60	4
Caribbean	31	62	7
South America	32	62	6
Asia	32	62	6
Western	37	59	4
South Central	37	59	4
South East	34	62	4
East	25	67	8

Source: Based on Population Reference Bureau, 1999: 2–9

them in Chapters 16 and 17. But we can note here that they impact differentially on such needs as healthcare, childcare, education, supply of and demand for employment, welfare provision, pensions and so on.

International migration

Of course, neither rates of population growth nor changes in age composition are caused solely by differences in fertility rates. That would be the case only in an entirely closed system where neither in-migration nor out-migration occurred. In fact, of course, *population migration* is a further significant process influencing the contours of the population map.[17] The number of people migrating between different parts of the world is both huge and growing, despite the fact that political barriers to migration are significantly higher than they were several decades ago.

> International migration is at an all-time high. In the mid-1990s, about 125 million people live outside their country of birth or citizenship. They account for about 2 per cent of the world's population and are expanding by 2 million to 4 million annually.[18]

However, such figures could well be underestimates. Not only is migration difficult to measure but also there are huge numbers of illegal migrants.

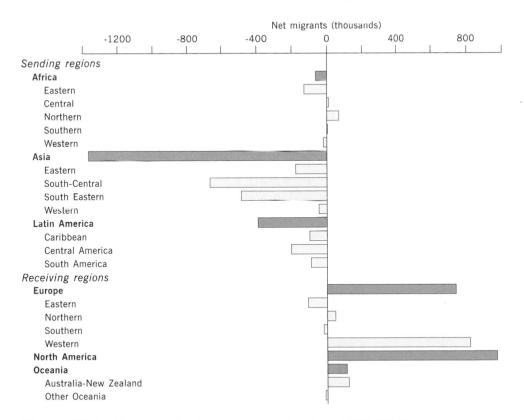

Figure 15.4   Net migration by geographical region, 1990–1995

Source: Based on Martin and Widgren, 1996: Table 1

Again, we find pronounced geographical variations in flows of international migration. Figure 15.4 shows that, in general, developing countries are the major sources of out-migration:

- Within Africa, the biggest outflows are from Eastern Africa, while there are significant inflows to Northern Africa.
- Within Asia, the biggest outflows are from South-Central and Southeastern Asia.
- Within Latin America and the Caribbean, the biggest outflows are from Central America.

Conversely, the biggest in-migrations are to developed regions, notably North America and Europe:

- North America is by far the largest net recipient of migrants.
- The pattern of migration within Europe is complex. There are huge inflows into Western Europe and large outflows from Eastern Europe.
- Oceania, notably Australia and New Zealand continue to be important destinations for international migration.

The geographical distances over which international migration occurs are enormously varied, as Figure 15.5 shows. A large proportion of migrant flows are to countries close to the place of origin – for fairly obvious reasons, including cost,

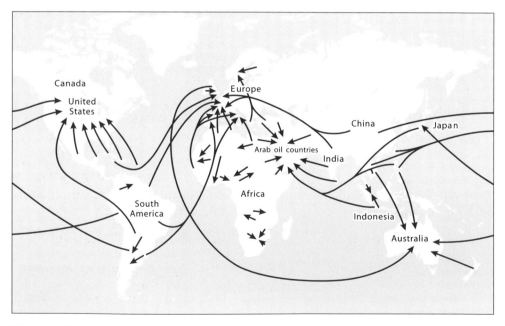

Figure 15.5 Major migration movements

Source: Based on Castles and Miller, 1993: Map 1.1

greater knowledge of closer opportunities, possibly greater cultural compatibility. But over and above such short-distance migrations are the long-distance, often inter-continental, flows. Certain migration paths are especially important. For example, there are massive movements across the Mexico–United States border and from parts of Asia to the United States. Australia has become an important focus of migration from South East Asia. In the recent past, there were huge migration flows from the Caribbean and South Asia to the United Kingdom and from countries around the Mediterranean basin to Germany.

In Chapters 16 and 17 we will explore the implications of such migrations for both sending and receiving countries. But it is worth noting at this stage that international migration has a significant effect on population growth (and its composition) in the developed countries. The UN estimates that

> without migration, the population of more developed regions as a whole would start declining in 2003 rather than in 2025, and by 2050 it would be 126 million less than the 1.18 billion projected under the assumption of continued migration.[19]

Making a living in the global economy

People strive to make a living in a whole variety of ways: for example, exchanging self-grown crops or basic handcrafted products; providing personal services in the big cities; working on the land, in factories, or in offices as paid employees; running their own businesses as self-employed entrepreneurs, and so on. For the overwhelming majority of people, *employment* (full- or part-time or as self-employment) is the most important source of income and, therefore, one of the keys to 'development as freedom.' However, not only are there not enough jobs 'to go round' – one estimate suggests that 400 million new jobs need to be created in the next ten years just to absorb newcomers to the labour market[20] – but also the volatility of employment opportunities appears to be increasing.

> Unemployment is the global problem of our times, and more than that: it is a protracted tragedy at the personal level, and destabilizing at the social level.[21]

At the end of 2000, there were approximately 160 million people unemployed in the world economy (and this figure refers only to 'open' unemployment – it does not include the millions of people suffering from 'hidden' unemployment who are not measured in the official figures). More than this,

> people living in poverty in the developing world, about 1.2 billion, are almost entirely supported by the earnings of the 500 million workers among them – the 'working poor'. If those people who work substantially less than full time, but who wish to work more, are included, then one-third of the world labour force of about 3 billion are either unemployed, underemployed or earn less than is needed to keep their families out of poverty.[22]

Even though the ILO perceived signs of improvement in employment prospects at the end of the 21st century, the situation remains volatile. And, as always, the pattern of unemployment is extremely uneven between different parts of the world (and between different parts of the same country). Serious as the unemployment position is in the industrialized nations, it pales into insignificance compared with the problems of most developing countries, particularly the least industrialized countries. At least in older industrialized countries the growth of the labour force is now easing. Only 1.1 per cent of the projected growth of the global labour force between 1995 and 2025 will be in the high-income countries. In most developing economies, on the other hand, extremely high rates of population growth mean that the number of young people seeking jobs will continue to accelerate for the foreseeable future. As Figure 15.6 shows, the low-income countries account for a growing share of the global labour force. Indeed, more than two-thirds of the projected growth in the global labour force will occur in such countries.

Figure 15.6

Distribution of the global labour force

Source: Based on World Bank, 1995: Table 1

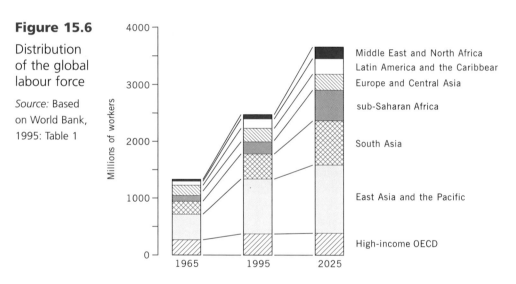

All parts of the world face major problems of adjustment to these problems of making a living. But the nature of the problem, and certainly its perception, varies according to each country's position in the global system. In this respect, the view from the older industrialized countries is very different from the view from the newly industrializing economies and different again from the least industrialized countries. But position in the global economy is only part of the picture. It is far too simplistic to 'read off' a country's or a region's problems (and solutions) solely from its place in a global division of labour. Internal circumstances – cultural, social, political as well as economic – are of enormous importance. Nevertheless, there are problems which affect older industrialized countries, newly industrializing countries and least developed countries in different ways as *groups* of countries as they grapple with the repercussions of global economic change and attempt to adjust to its employment impact. In the following two chapters we look, first, at the problems facing the developed or industrialized countries and, second, at the infinitely more intractable problems facing developing countries.

Notes

1   Micklethwait and Wooldridge (2000: ix; emphasis added).
2   Kapstein (1999: 16).
3   Sen (1999).
4   Sen (1999: 20).
5   UNDP (1999: 3).
6   Bairoch quoted in Cohen (1998: 17).
7   See Kaplinsky (2001: 48–50); World Bank (2001: ch. 1).
8   Sklair (2001).
9   Sklair (2001: 18–23).
10   Department of International Development (2000: 12).
11   Sen (1999: 106).
12   The discussion in this section is based mainly on data in United Nations Population Division (2001) and Population Reference Bureau (1999).
13   Population Reference Bureau (1999: 1).
14   Kennedy (2002: 3).
15   United Nations Population Division (2001: 2).
16   United Nations Population Division (2001: 15).
17   See Castles and Miller (1993); Martin and Widgren (1996).
18   Martin and Widgren (1996: 2).
19   United Nations Population Division (2001: vii).
20   FIET (1996: 16).
21   Luttwak (1999: 102).
22   ILO (2001: 15).

CHAPTER 16

Making a Living in Developed Countries:
Where Will the Jobs Come From?

Increasing affluence – but not everybody is a winner

As we have seen in previous chapters, on a global scale the developed countries are clearly 'winners' in the global economy. They continue to contain a disproportionate share of the world's wealth, trade, investment and access to modern technologies (especially of information technologies). But if we refocus our lens to look at what is happening *within* the developed world, we find wide variations in economic well-being, both between individual countries and – perhaps more surprisingly – within individual countries, even though most people in developed countries are significantly better-off than in the past. As Sen points out,

> it is remarkable that the extent of deprivation for particular groups in very rich countries can be comparable to that in the so-called third world. For example, in the United States, African Americans as a group have no higher – indeed have a lower – chance of reaching advanced ages than do people born in the immensely poorer economies of China or the Indian state of Kerala (or in Sri Lanka, Jamaica or Costa Rica).[1]

In this chapter we examine the position of people in developed countries in terms of their access to employment, and to the income that such employment brings, in the context of a globalizing world economy. The discussion is organized as follows:

- First, we look at trends in employment and unemployment in developed countries, focusing particularly on differences between the United States, Europe and Japan.
- Second, we look at what has been happening to employment-related incomes between different groups.
- Third, we explore the extent to which jobs and incomes vary geographically within individual countries.
- Fourth, we examine possible causes of these variations in terms of the processes we have been discussing throughout this book.
- Fifth, we look at some of the 'policy fixes' that have been proposed to alleviate these problems, particularly the question of 'where will the jobs come from?'

The jobs scene

Employment

During the past half century, two particularly important developments have occurred in the employment structure of developed economies: the displacement of jobs in manufacturing industries by jobs in services and the increasing participation of women in the labour market.

The shift from jobs in manufacturing to jobs in services

One of the most striking trends, since at least the 1960s, has been for employment in services to grow far more rapidly than employment in manufacturing. It is this trend that has led to the view that developed economies have become *de-industrialized*[2] and that they are now effectively service economies. Today, around 75 per cent of the labour force – or even more in some cases – are employed in service occupations. It was particularly during the 1970s that manufacturing employment actually declined in absolute terms in the major European economies other than the United Kingdom, where manufacturing employment began its steep fall after 1966. In recent years, virtually all the net employment growth has been in services. Between 1980 and 1993, for example,

> in Europe, just over 18 million additional jobs were gained in services … not so many less than in the US, where almost 22 million extra jobs were created, despite the much slower rate of employment growth overall (in proportionate terms, the increase was 25 per cent in Europe, 33 per cent in the US). Similarly, the expansion of jobs in services was only slightly less than in Japan (where they increased by 28 per cent).[3]

In both Europe and the United States, the number of people employed in manufacturing declined: by almost 0.4 per cent per year in Europe and by almost 0.2 per cent per year in the United States. In contrast, between 1980 and 1993, manufacturing employment growth continued in Japan (by a little more than 0.2 per cent per year). Overall, Europe's employment performance was very significantly worse than that of either the United States or Japan:

> In industry and agriculture combined … the number employed in Europe fell by just over 13 million over these 13 years, an average loss of 1 million jobs a year. In the US, the number declined by only just under 2 million and in Japan there was an increase of just over 1/2 million. In quantitative terms, therefore, the poor overall performance of the European economies in expanding employment relative to the US and Japan owes much more to the scale of job losses in the primary and secondary sectors than to the low rate of net job gains in services.[4]

A common criticism levelled at these new service jobs is that they are essentially poorly paid, low-skilled, part-time and insecure – at least compared with the kinds of jobs in manufacturing that were characteristic of the developed countries up until the 1960s. There is certainly some truth in this. Many of the new service jobs are, indeed, 'McJobs'. But that isn't the entire story. A recent OECD report suggests that most of the growth in new private services jobs in western industrialized countries is well-paid and skilled ... the expansion in service employment brought faster growth in the 1990s in high-paid than low-paid work ... 'There does not appear to be any simple trade-off between job quality and employment.' While the US has a higher proportion of its working-age population employed in low-paying jobs than in most other OECD countries, it also has a higher proportion in higher-paying jobs.[5]

Women's work

The shift in the balance of employment towards services has been closely associated with the second major structural change in the labour markets of developed economies: the increasing participation of women in the labour force. In all developed economies, the changing roles of women, away from an automatically assumed domestic role, has gone hand-in-hand with the growth of service jobs. Although women are certainly employed in manufacturing industries, their relative importance is far higher in service industries. This is especially so where there are greater opportunities for part-time work, which allows women with families a degree of flexibility to combine a paid job with their traditional gender role.

Female participation in the labour market has increased in virtually all countries. In the United States, for example, around 38 per cent of the labour force was female in 1960; today it is 60 per cent. In the United Kingdom, a similar trend is evident. In 1971, 40 per cent of the labour force was female; today it is 55 per cent. But, as Figure 16.1 shows, there is considerable variation between countries. The highest female participation rates are found in the Scandinavian countries and the Netherlands; significantly lower rates are found in Germany, France, Spain, Belgium and Italy.

Figure 16.1

Variations in female participation in the labour markets of developed countries

Source: Based on ILO, 2001: Table 2

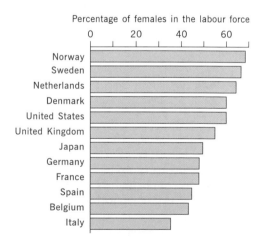

Percentage of females in the labour force

Differentials in rates of job creation

In terms of *total* jobs created, there is no doubt that the United States is in a different league from most of the other industrialized countries. During the 1990s almost 22 million new jobs were created there.[6] Between 1980 and 1993, total employment in the United States grew three times faster than in the European Union and 50 per cent faster than in Japan. However, Table 16.1 shows that there was considerable variation in the annual employment growth rate between individual developed countries during the 1990s.

Table 16.1 Differences in employment growth in selected developed economies

	Annual % change	
Country	1990–99	1995–99
Germany	3.0	1.1
Netherlands	2.1	2.9
Australia	1.6	2.1
United States	1.3	1.7
Canada	1.3	2.2
Norway	1.2	2.1
Belgium	1.1	1.3
United Kingdom	0.3	1.3
France	0.3	0.6
Denmark	0.2	1.0
Switzerland	0.1	0.8
Japan	0.1	−0.5
Italy	−0.2	0.8
Sweden	−1.1	0.4

Source: Based on ILO, 2001: Table 1.10

Unemployment

Unemployment rates

The obverse of employment growth is, of course, *unemployment*. Compared with the 1960s and early 1970s, unemployment rates in the industrialized countries have increased dramatically. Figure 16.2 shows the trends for the 1960–1996 period. The graph shows as a horizontal line the average unemployment rate for the 1960–1973 period – the so-called 'golden age of growth' (see Chapter 3). During that period, unemployment rates were highest in the US and lowest in Japan, with the EC12 and the UK falling in between. But the post-1973 experience was very different (see Figure 16.2). Whereas both Japan and the United States experienced an increase in the general level of their unemployment – more so in the case of the United States – the European experience was appalling, with average unemployment rates in the EC12 rising to above 12 per cent in the early 1990s compared with less than 4 per cent in Japan and a little over 5 per cent in the United States. Within these aggregate figures, there is a particular problem for those people who have been out of work for long periods of time.

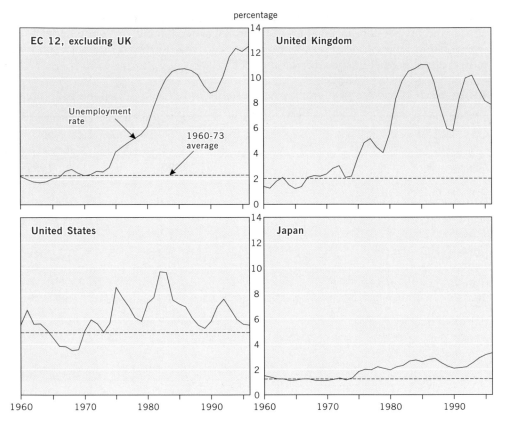

Figure 16.2 Unemployment rates 1960–1996 compared with the average rate for 1960–1973

Source: Based on ILO, 1997: Figure 3.1

However, the late 1990s saw a considerable improvement in the unemployment situation in most developed countries, as Table 16.2 indicates. In every case – with the exception of Japan and France – the unemployment rate at the end of the 1990s was below that of 1997. But such rates are very volatile. The sharp economic slow-down in the United States, coupled with the economic shock of the events of 11 September 2001, was rapidly reflected in waves of corporate job reductions that will inevitably be reflected in unemployment rates. The positions of Germany and Japan in Table 16.2 are interesting. The level of unemployment in Germany in the 1990s was heavily influenced by the effects of political reunification which brought major problems of economic adjustment reflected in increased levels of unemployment, especially in the former East Germany. The unemployment situation in Japan at the end of the 1990s represents a particularly massive change. Historically, unemployment rates in Japan have been extremely low. A combination of a rapidly growing economy and a very strong orientation towards job security in the large-company sector of the economy sustained lower rates of unemployment than in any other industrialized country. But the burst of the bubble economy at the end of the 1980s and persistent domestic recession throughout the 1990s has changed this. The lifetime job system is crumbling – the days of the 'salaryman' appear to be numbered. As a result – and much to the dismay of Japanese society – Japanese unemployment rates are now higher than those in the United States.

Table 16.2 Rates of unemployment (%) in developed countries, 1995 and 1999

Country	1995	1999
Netherlands	7.1	3.6
United States	5.6	4.2
Japan	3.2	4.9
United Kingdom	8.6	6.0
Australia	8.1	7.0
Canada	9.4	7.6
Germany	8.1	8.7
Italy	11.5	11.3
France	11.6	11.8

Source: Based on ILO, 2001: Table 1.10

The selective impact of unemployment

Although the general level of unemployment, including long-term unemployment, remains very high in most industrialized countries, its actual incidence is extremely uneven. Job loss is a *socially selective* process. For example, males aged between 25 and 54 years, with a good education and training, are far less likely to be unemployed, on average, than women, younger people, older workers and minorities. Most of these latter categories tend to be unskilled or semi-skilled workers. The vulnerability of women and young people to unemployment reflects two major features of the labour markets of the older industrialized countries. First, as we have seen, the increased participation of women in the labour force – particularly married women – has increased dramatically. A large proportion of these are employed as part-time workers in both manufacturing and services, especially the latter. Second, *youth unemployment* during the 1980s partly arose from the entry on to the labour market of vast numbers of 1960s 'baby boom' teenagers. In most industrialized countries, therefore, unemployment rates among the young (under 25 years) have been roughly twice as high as that for the over-25s. In some cases youth unemployment is three times higher than adult unemployment.

On the other hand, the demographic trends in the developed countries discussed in Chapter 15 mean that, over the medium and long term, the unemployment problem should ease. The ageing of the population and the slowdown in the number of new entrants into the labour market should ease labour market pressures in general, although there will continue to be substantial variations between and within individual countries. In fact, there is likely to be an increasing need for immigrant workers to fill the gaps created by the 'greying' of the populations in all the developed economies. This, of course, raises big social and political problems.

Unemployment tends to be especially high among *minority groups* within a population. In the United States, for example, unemployment among black youths can be 150 per cent higher than among white youths. Similarly, unemployment rates among Hispanic youths are at least 50 per cent higher than among white youths. In Western Europe the problem of minority group unemployment reflects the large-scale immigration of labour in the boom years of the 1960s. Relatively easily absorbed – indeed welcomed as 'guestworkers' – in the good times, the migrant workers – and their children born in Europe – now face enormous problems both in times of economic recession and also because of longer-term decline in the demand for certain kinds of worker. In continental Europe most of the migrant labour came from the Mediterranean rim. In the United Kingdom, where the nature of the immigration was different because of Commonwealth obligations, most migrants came from South Asia (India, Pakistan, Bangladesh) and the Caribbean. In the United States the major sources of new migrants were Mexico, parts of the Caribbean and parts of East Asia.

As we shall see in Chapter 17, such out-migration has been enormously important for the countries of origin. For the European host countries, too, the migrants have performed an extremely significant role. In the 1960s there were severe labour shortages as the European economies grew very rapidly. One response was to recruit migrant labour on temporary contracts. The migrant workers were overwhelmingly young, male, unaccompanied – and unskilled. Their numbers grew spectacularly. However, during the 1980s, the number of jobs available to low-skilled workers fell dramatically and this had a major adverse effect on immigrant workers.[7]

Incomes

The third significant trend in the jobs picture in the older industrialized countries concerns changes in *incomes* received by individuals in the labour force. There are two aspects to this. One is the overall growth in income levels; the other is the distribution of incomes between different segments of the population. Throughout the long economic boom of the 1960s and early 1970s, there was both a progressive increase in incomes in developed countries and also a general reduction in the income gap between the top and bottom segments of the population. For the first 25 years or so after World War II, the general trend was for the earnings gap between the top and the bottom segments of the labour force to narrow whilst, at the same time, the overall level of per capita income increased substantially. In other words, most people became better off. This is no longer the case, especially in the United States but also in the United Kingdom where

> the pre-1973 decline in inequality has been more than reversed, with the USA experiencing a dramatic surge in the upward redistribution of income, surpassed only by the United Kingdom. The beneficiaries of the shift have been the richest 5 per cent, but particularly the richest 1 per cent whose pre-tax incomes grew by 93 per cent from 1977 to 1995. This growth in inequality is a result of a shift of income from wages to capital income (profits and interest) and a growing inequality among wage earners.[8]

Figure 16.3 shows the trends in the dispersion of earnings for the leading industrialized countries between 1979 and 1995. In 1995, the ratio of the earnings of the highest 10 per cent of the labour force to that of the lowest 10 per cent rose in the United States from 3.2 to 4.4 and in the United Kingdom from 2.4 to 3.4. The average income of the top 5 per cent of US households was roughly seven times that of the bottom 40 per cent of households in the early 1970s. In the mid-1990s, the top 5 per cent earned on average ten times more than the bottom 40 per cent.

As in the case of employment, it is the less skilled workers who have been most adversely affected. Conversely,

> as compensation has fallen for the unskilled worker, it has increased mightily for highly educated workers … in 1979 male workers with a college degree earned on average about 50 per cent more than unskilled workers; by 1993 that difference was nearly 90 per cent. To put the inequality problem in its starkest terms, between 1979 and 1994 the upper 5 per cent of American families captured *99 per cent* of the nation's per capita gains in gross domestic product! That is, with a mean family gain over this period of $4,419, $4,365 went to the upper 5 per cent.[9]

The pattern is more mixed across other industrialized countries. It is apparent from Figure 16.3, for example, that the same degree of increasing income dispersion within the labour force did not occur in many of the continental European countries. In some cases, indeed, the gap narrowed rather than widened. On the other hand, these countries have experienced much higher levels of unemployment than the United States in particular and even the United Kingdom. This suggests that labour market adjustments are occurring in different ways in different countries. In the United States and the United Kingdom adjustment has been primarily in the form of a relative lowering of wages at the bottom of the scale and a consequent *increase in income inequality*. In Western Europe, on the other hand, such wage levels may have been maintained at the expense of jobs with an *increase in unemployment*.

> Indeed, it is said – probably too simplistically – that the United States has accepted greater inequality, whereas Europe has accepted more unemployment.[10]

Figure 16.3

Trends in the dispersion of earnings for male workers, 1979–1995

Source: Based on ILO, 1997: Figure 3.5

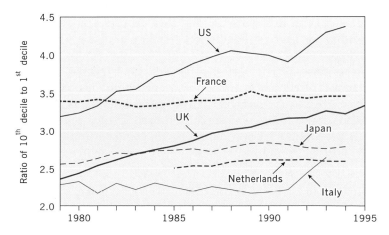

An uneven geography

As well as being socially differentiated, employment, unemployment and incomes are also *geographically uneven* within individual countries. For example, within the older industrialized countries, three broad trends are apparent:

- *Broad interregional shifts* in employment opportunities, as exemplified by the relative shift of investment from 'Snowbelt' to 'Sunbelt' in the United States, from north to south within the United Kingdom.

- *Relative decline of the large urban-metropolitan areas* as centres of manufacturing activity and the growth of new manufacturing investment in non-metropolitan and rural areas. Such a trend is apparent in both North America and many parts of Western Europe as part of a quite powerful decentralization tendency.

- *Hollowing-out of the inner cities of the older industrialized countries.* In virtually every case, the inner urban cores have experienced massive employment loss as the focus of economic activity shifted first from central city to suburb and subsequently to less urbanized areas.

Thus, de-industrialization has been most dramatically experienced in the older industrial cities of all the developed market economies as well as in those broad regions in which the decline of specific industries (including agriculture) has been especially heavy. The physical expression of these processes is the mile upon mile of industrial wasteland; the human expression is the despair of whole communities, families and individuals whose means of livelihood have disappeared. One outcome of these cataclysmic changes has been the growth of an *informal* or *hidden economy*, a world of interpersonal cash transactions or payments in kind for services rendered, a world much of which borders on the illegal and some of which is transparently criminal.

Figures 16.4 to 16.6 illustrate the enormous geographical unevenness characteristic of the US, European and Japanese economies.

The geography of economic distress in the United States

Figure 16.4 maps, at the county level, a composite index of 'economic distress' based upon per capita market income, transfer payments as a share of total income, labour force participation rates and unemployment levels.[11] It shows the pattern for two time periods, 1960 and 1997. In 1960, the most seriously distressed counties were concentrated in a relatively few regions. Appalachia and the Mississippi Delta alone contained almost 50 per cent of the distressed counties. The remainder were more dispersed: the upper peninsula of Michigan, some counties along the US–Mexico border, some natural resource communities in both east and west, and Indian reservations. Altogether in 1960, 518 counties fell into the lowest category. By 1997, the number of counties in this category had increased by almost 50 per cent to 774.

> Of the 774 distressed counties, 37 per cent were concentrated in Appalachia and the Mississippi Delta, with the rest fanning out across the nation. These communities are

found in timber, agricultural, and mineral and energy resource areas in the heartland of the nation, in strong manufacturing areas of the past century, as well as on Indian reservations, historic New Mexican and Native American communities, and along our borders with Mexico. More recently, a number of new entrants have begun to experience economic distress with the collapse of their post-war low wage manufacturing economies. In both cases, these are the communities whose natural wealth and human

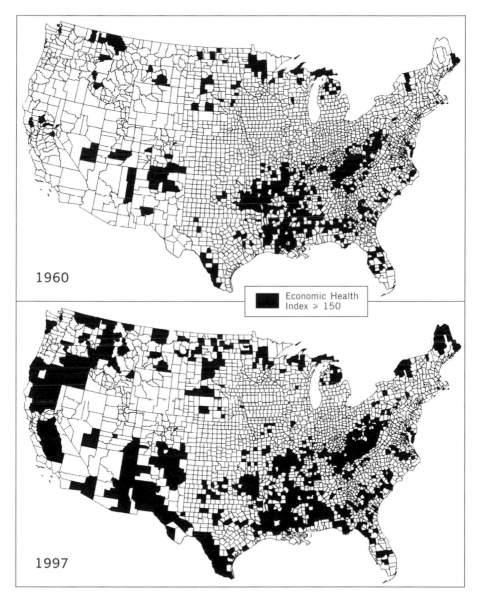

Figure 16.4 Counties within the United States with high levels of economic distress, 1960 and 1997

Source: Based on Glasmeier, 2002: Figure 1

capital helped build the American economy of the 20th century. After having been a reservoir of resources tapped to make America an industrial powerhouse, many of these communities are now at risk of being left behind.[12]

Regional inequalities in Europe

Figure 16.5 shows the distribution of regional GDP per capita at purchasing power standards (PPS) relative to the EU average for sub-national statistical units (known, in EU-speak, as NUTS III). The degree of dispersion in per capita incomes is striking, both between and within individual countries. Virtually all EU member states have relatively poor areas within them.

> At the opposite extreme, within the EU a significant number of German regions and two French regions were located above the 200 per cent of the average EU per capita income … More generally, a large share of the regions located at more than 125 per cent above the average were West German, and most were clustered around an axis (the so-called 'blue banana') extending from Greater London through Belgium and the Netherlands along the Rhine and into Lombardy and Emilia Romagna in the north of Italy.[13]

A glance back to Figure 3.27 will help to locate this geographically uneven pattern of economic development within Europe.

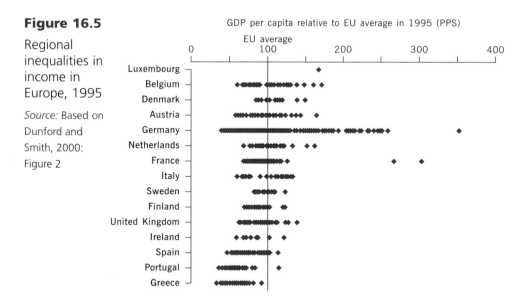

Figure 16.5

Regional inequalities in income in Europe, 1995

Source: Based on Dunford and Smith, 2000: Figure 2

Regional unemployment in Japan

We noted earlier (Table 16.2) the recent rise in unemployment in Japan from levels that were historically far below those in the West (even allowing for problems of measurement). The average unemployment rate in Japan stood at around 5.2 per cent in 2001 but, as Figure 16.6 shows, there were marked regional variations within Japan. The range, in fact, was from 3.2 per cent in Nagano to 8.4 per cent in Okinawa. Levels well above the national average were apparent in Osaka (7.2 per cent), Kyoto (6.3 per cent), Fukuoka and Hyougo (both 6.2 per cent). The lowest regional unemployment levels were in Nagano (3.2 per cent), in Yamanashi and Ishikawa (both 3.3 per cent), and in Shizuoka and Shiga (both 3.8 per cent). The range of such unemployment disparities indicates that Japan's continuing economic recession is experienced increasingly unevenly as the ability of the national economy to spread the benefits of economic well-being is lessened.

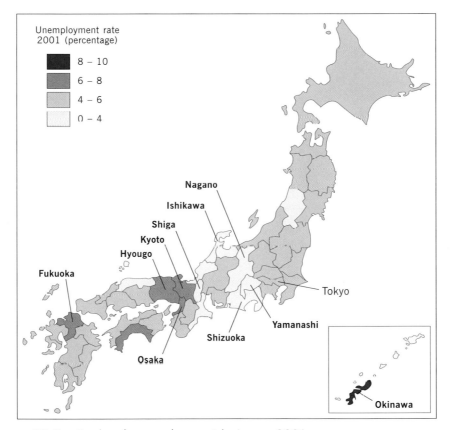

Figure 16.6   Regional unemployment in Japan, 2001

Source: Japanese Ministry of Labour

Inequalities within the global cities[14]

Within the older industrialized economies, the sharpest contrasts in levels of income, nature of employment, unemployment and minority participation in the labour force occur in the major cities, especially the so-called 'global cities', of which New York and London are the prime exemplars. In Chapter 8 we examined the tendency for high-level corporate functions (such as headquarters and high-level service functions) to concentrate in the major cities shown in Figure 8.1. In

Chapter 13, we explored the significance of such cities as global financial centres (Figure 13.8). New York and London are the prime examples of cities whose economies and associated social structures extend way beyond their national bases. As such, they constitute microcosms of many of the processes and tensions inherent in processes of globalization. The same may be said of some other big cities but not to the same degree.

Because of their particular functions in the global economy as the 'control points' of global financial markets and of transnational corporate activity, cities like New York and London contain both highly sophisticated economic activities, with their highly paid, cosmopolitan workforces, and also large supporting work-forces in low- and medium-level services. The result is a high degree of social and spatial *polarization* within these cities. Figure 16.7 shows the general ways in which such processes may operate.

> Spatial polarization arises from class polarization. And in world cities, class polarization has three principal facets: huge income gaps between transnational elites and low-skilled workers, large-scale immigration from rural areas or from abroad and structural trends in the evolution of jobs ... The whole comprises an *ecology* of jobs ... the restructuring process ... involves the *destruction* of jobs in the high-wage, unionised sectors ... and job *creation* in ... the production of global control capability.

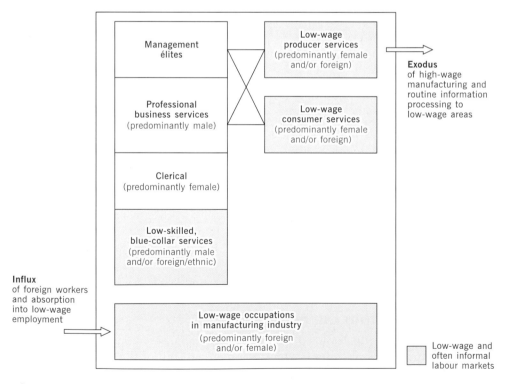

Figure 16.7   Processes of restructuring in global cities

Source: Based on Friedmann, 1986: Figure 2

Linked to these dynamic sectors are certain personal services (employing primarily female and/or foreign workers), while the slack in manufacturing is taken up by sweatshops and small industries employing non-union labour at near the minimum wage. It is this structural shift which accounts for the rapid decline of the middle-income sectors during the 1970s ... The rapid influx of poor workers into world cities – be it from abroad or from within the country – generates massive needs for social reproduction, among them housing, education, health, transportation and welfare ... In this competitive struggle, the poor and especially the new immigrant populations tend to lose out.[15]

Although this description of the processes involved is rather over-generalized, and the precise form they take depends on the history and circumstances of each individual city, it provides a useful device for understanding a very complex situation. Empirical evidence certainly bears out its major features.[16]

- *Employment trends.* In both London and New York, the proportion of the labour force employed in manufacturing declined from over 20 per cent in 1977 to well under 10 per cent in the mid-1990s.

- *Income inequalities.* During the 1990s, income inequality in New York City grew much more sharply than in the United States as a whole. 'New York has the worst income inequality in the US. The average annual income, adjusted for inflation, of the richest one-fifth of NY families rose to a level 20 times above the income of the bottom fifth of families in the 1996–1998 period, more than double this gap in the late 1970s. The city's middle class also lost ground to the richer fifth.'[17] Similarly, earnings differentials within London increased substantially between the mid-1980s and the late 1990s. At the same time, the differential between overall earnings in London and the South East and the rest of the United Kingdom widened.

- *Casual and informal labour markets.* 'There has been a pronounced increase in casual employment and in the informalization of work in both New York and London ... In addition, jobs that were once full-time ones are now being made into part-time or temporary jobs ... A majority are low-wage jobs ... The growth of service jobs is crucial to the expansion of part-time jobs ... Part-time jobs can recruit women more easily, create greater flexibility in filling various shifts, and reduce labor costs by avoiding various benefits and overtime payments required by full-time workers ... There is also evidence pointing to an expansion of the underground economy.'[18] The precise forms of casualization vary between individual cities: in New York the emphasis has been on informal work whereas in London most growth has been in part-time work.[19]

- *Race and nationality in the labour market.* In New York, 'Blacks and Hispanics increased their share of jobs, while whites lost share. Half of all resident workers in New York are now minority ... Blacks and Hispanics are far less likely than whites to hold the new high-income jobs and far more likely than whites to hold the new low-wage jobs. Thus, one can infer a trend toward replacement of whites in existing lower-income jobs by Blacks and Hispanics as well as these minorities' entry into new lower-income jobs.'[20] London is the major focus of Asian and Afro-Caribbean populations in the United Kingdom. 'In inner and outer London, 29 per cent and 22 per cent respectively of residents are from ethnic minorities in 2000.'[21]

Why is it happening?

These highly uneven trends in employment, unemployment and incomes in the industrialized economies cannot be explained simplistically in terms of a single set of causes. Insofar as globalization contributes to these trends it does so precisely because globalizing processes are themselves highly complex and intrinsically uneven. As we observed in Chapter 2, globalization is a 'syndrome' of processes and activities rather than a single unified phenomenon. Most importantly, globalizing processes are not universal processes, stamping an identical imprint across the economic landscape. Rather, they are *contingent* processes whose outcomes depend fundamentally on their interaction with locally specific circumstances. The fact that there are very substantial differences both between individual countries and also between parts of the same country suggests the operation of both general and specific forces.

For example, the most general explanation of an overall high level of unemployment in the older industrialized countries between the early 1970s and mid-1980s and, again, in the 1990s is the effect of world recession. Recession, whatever its causes, drastically reduces levels of demand for goods and services. By this explanation, the bulk of unemployment in the older industrialized countries as a whole is *demand-deficient* unemployment. But the general force of recession does not explain the *geographical variation* in unemployment between and within countries. In fact, a whole set of interconnected processes operates simultaneously to produce the changing map of employment, its reverse image, unemployment, and the increasingly uneven map of income.

In this section we examine the structural effects of 'global shifts' – the processes of globalization – on the uneven economic well-being within industrialized countries. We cannot provide a precise quantitative assessment of this influence; we can merely suggest the kinds of influence involved. Let us now recapitulate the ways in which the processes we have been concerned with throughout this book may help to explain the current situation in the older industrialized countries, given the specific circumstances (social, cultural, institutional) which exist in individual countries.

Technological change

Technological developments in products and processes are widely regarded as being a major factor in changing both the number and the type of jobs available. In general, *product innovations* tend to increase employment opportunities overall as they create new demands. On the other hand, *process innovations* are generally introduced to reduce production costs and increase productive efficiency. They tend to be labour-saving rather than job-creating. Such process innovations are characteristic of the mature phase of product cycles and became a dominant phenomenon from the late 1960s onwards.

The general effect of process innovations, therefore, is to increase labour productivity and to permit the same, or an increased, volume of output from the same, or even a smaller, number of workers. Each of the case studies in Part III demonstrated this tendency. But, again, the impact of such technological change on jobs tends to be uneven. In most cases, it has been the semi- and unskilled workers who have been displaced in the largest numbers. It is manual workers rather than professional, technical and supervisory workers whose numbers have been reduced most of all, although the spread of the new information technologies is changing this situation.

There can be little doubt that changes in process technology have adversely affected the employment opportunities of less skilled members of the population. However, there is much disagreement about the overall contribution of technological change to unemployment. Some argue that the 'end of work' is nigh and that much of this is due to the job-displacing effects of technological change. On the basis of his calculation that three out of four US jobs (including both white- and blue-collar jobs) could be automated, and that thousands of jobs have already disappeared from major US corporations, Rifkin predicted that hundreds of millions of workers will be without jobs by the middle of the 21st century.[22] The explosive spread of new information and communication technologies (ICT) would seem to confirm such apocalyptic views.

But do they? Not in the opinion of the ILO. In a recent report on the employment implications of ICT they draw the following conclusions:[23]

- Initial employment trends suggest that ICT does not destroy jobs.
- In the 'core' ICT sector, the jobs being lost in manufacturing are more than compensated for by rapid growth in the services segment of these industries (software, computer and data processing services).
- There is huge potential for high-cost economies to move up the value chain and to create higher-skill jobs.
- For most workers, employment stability remains the norm.
- However, these changes in employment tend to reinforce gender inequalities.

But although technological change may continue to create, in net terms, more jobs than it destroys, the problem is the actual *distribution* of such new jobs in relation to those destroyed. Although

> human labour in general has never yet become obsolete new technologies often displace particular workers and work hardships upon them, and upon whole industries and regions.[24]

Global restructuring and the transnationalization of production

Fundamentally, the highly uneven pattern of employment and unemployment change reflects the outcome of the complex industrial restructuring processes which have been occurring at the global scale since the mid-1970s. However, the problems created for particular social groups and particular geographical areas are greatly exacerbated in periods of economic recession and by demographic trends in the size and composition of the labour force. Restructuring of operations by business enterprises, especially the large TNCs, is a continuous process, as we argued in Chapter 8. But its effects are more keenly felt in times of recession than in times of growth. At the same time, recession intensifies the efforts of enterprises to rationalize and restructure their activities to sustain their profitability.

As we have seen, a key feature of the post-war period has been the development and intensification of global competition in virtually every industry, both old and new. In such circumstances, business enterprises have adopted a variety of strategies to ensure their survival and to grow. Geographical and product diversification have greatly increased, as have efforts to cut back production costs through processes of intensification, investment in new, labour-saving technology and rationalization. New, more flexible work practices are being adopted on an extensive scale. Within many major corporations in the West, especially in the United States and the United Kingdom, the late 1980s and early 1990s witnessed the twin phenomena of 'downsizing' and 'corporate re-engineering'. In both cases, the aim was drastically to reduce the numbers of workers employed within companies.

Most significant, however, is the fact that many of these restructuring processes have themselves become increasingly global in their extent. Virtually all large firms, and many medium-sized ones, have become *transnational* enterprises. They engage in transnational, rather than merely domestic, production, whether directly through the establishment of overseas branch plants or indirectly through alliances and subcontracting. As such, their strategies impinge upon both their home country and also the host countries in which they operate. But the process itself is extremely complex.

Both the reasons for engaging in transnational production and also the effects of such operations vary greatly between different types of firm in different industries and according to the characteristics of home and host countries. But there is no doubt that a good deal of the changing level and distribution of employment opportunities in the older industrialized countries is associated with the actions of TNCs. Some of these involve the location and relocation of production units in developing countries, but the majority are still within the older industrialized countries themselves. In one sense, therefore, TNCs can indeed be blamed for some of the loss of manufacturing jobs in their home countries. But as we argued in Chapter 9, the question is very complex. What would have been the realistic alternative to such overseas investment? Could the specific investment have been made at home? In an increasingly global economy is it possible for firms to opt out of international production? Would the absence of overseas investment have increased employment at home or would it have resulted in even greater employment loss? Sweeping generalizations – pro or con – cannot be made. Nevertheless, where the volume of outward investment is very high in relation to domestic production, there are bound to be adverse effects on the domestic economy and, especially, on employment.

Import penetration from developing countries

One of the most contentious explanations for the employment and income problems facing workers in the older industrialized countries is the contribution of imports of manufactured goods from developing countries, notably the NIEs. The rapid development of manufacturing production in a small number of NIEs, and their accelerating involvement in world trade, has been a major theme of this book. It is one of the most striking manifestations of global shifts in the world economy. In some cases, a major driving force has been the direct involvement of foreign TNCs; in others, although the *direct* participation of TNCs may have been relatively small, their *indirect* effects are invariably large – especially through the power of the major buyers and outsourcers.

The common element in virtually all the NIEs, however, is a very strong government involvement in guiding or directing the economy. The adoption and pursuit of vigorous export-oriented industrial strategies were discussed in detail in Chapters 5 and 6 and in the case studies of Part III. By enhancing their initial comparative advantage of a large and cheap labour force through the provision of basic physical infrastructure (including EPZs), by substantial financial and tax incentives and, most of all, by their 'guarantee' of a malleable labour force, the NIE governments created a powerful new force on the world economic scene.

The basic question is: how far has the industrialization of these fast-growing economies – as expressed through *trade* – contributed towards the de-industrialization of the older industrialized countries, to the increased levels of unemployment, and to the pauperization of workers at the bottom end of the labour market? As we showed in Chapter 3, imports from developing countries represent only about 20 per cent of total imports into developed countries, although they tend to be concentrated into specific sectors (for example, textiles, clothing, footwear, toys, electronics).

There is a wide range of views on the relationship between developing country trade and employment and income changes in industrialized countries.[25] Wood, for example argues that trade with developing countries has had a considerable impact, especially in widening the gap between skilled and unskilled workers:

> Countries in the South have increased their production of labour-intensive goods (both for export and domestic use) and their imports of skill-intensive goods, raising the demand for unskilled but literate labour, relative to more skilled workers. In the North, the skill composition of labour demand has been twisted the other way. Production of skill-intensive goods for export has increased, while production of labour-intensive goods has been replaced by imports, reducing the demand for unskilled relative to skilled workers … up to 1990 the changes in trade with the South had reduced the demand for unskilled relative to skilled labour in the North as a whole by something like 20 per cent … Thus expansion of trade with the South was an important cause of the de-industrialization of employment in the North over the past few decades. However, it does not appear to have been the sole cause … [26]

The general conclusion of the ILO is that the results of the many economic studies of the relationship between trade and wage and income inequality in the older industrialized countries are 'inconclusive':

> although international trade has contributed to income inequality trends to some extent, it has not played a major role in pushing down the relative wage of less-skilled workers … [in the case of the United States] employment patterns in industries least affected by trade moved in the same direction as those in trade-affected manufacturing

industry, increasing the share of high-wage employment. This pattern of change in the employment structure is not well explained by the argument relying on the trade effect.[27]

The basic problem in all of the individual factor explanations – technology, transnational corporate restructuring, or import penetration – is the fact that they are treated *independently* of one another. It is as though changes in one of the variables are unrelated to the others. But this is clearly not the case. For example, although the *direct* effects of trade may be relatively small, the *indirect* effects may be larger because of the ways in which firms respond to the threat of increased international competition. They may, for instance, invest in labour-saving technologies to raise labour productivity and to reduce costs. This would appear as a 'technology effect' whereas the underlying reason for such technological change may be quite different: a response to low-cost external competition. In fact, the decline in manufacturing employment in the older industrialized countries is primarily the result of increased productivity. But this has affected the labour force differentially with the greatest relative losses of jobs and of income falling on the least skilled, least educated workers.

Even if we accept that the impact of import penetration may be rather less than sometimes claimed – or, at least, that the evidence is inconclusive – such a position is based upon past experience. But are we, as Kaplinsky argues, entering a new phase exemplified, in particular, by the entry of China (and potentially India) into the world market?

> In each of these major markets, China is no longer a marginal supplier, but plays a major role in determining overall product availability, and hence the prices realized for these products; its presence therefore also affects the terms of trade of all economies trading in these product categories. China's low production costs arise from and are coupled with growing industrial competence … In 1995, China alone accounted for just over one-fifth of global population, and almost one-quarter of the global labour force. Together with India these proportions rise respectively to over two-thirds of the global total. Therefore, developments in these labour forces, when these economies are integrated into the global labour force, have the capacity to significantly affect global wage levels.
>
> It is not just the wages of unskilled labour in the global economy which are being and will increasingly be undermined by the size of the labour reservoir in China (and India). One of the most striking features of the Chinese labour market is its growing level of education and skilling.[28]

A balance sheet of positive and negative effects

The processes of globalization, which are, themselves, extremely complex, produce complex employment and income effects. Table 16.3 summarizes the major positive and negative interpretations of the effects of globalization on the older industrialized countries. It reminds us that we need to beware of making universal generalizations about impact and that we need to examine cases with care. However, that does not mean sitting on the fence where the evidence of negative (or positive) effects is clear. The problem is that, in many cases, such clarity does not exist.

Table 16.3   Positive and negative effects of globalization processes on employment in older industrialized countries

Positive effects	Negative effects
Cheaper imports of relatively labour-intensive manufactures promote greater economic efficiency through the demand side while releasing labour for higher productivity sectors	Particularly in relatively labour-intensive industries, the rising imports from developing countries, together with competition-driven changes in technology and other factors, lead to inevitable losses in employment and/or quality of jobs, including real wages. This increases inequality between skilled and unskilled workers, and causes extreme redeployment difficulties
Growth in developing countries through industry relocation and export-generated income leads to (a) increased demand for industrialized country exports and (b) shifts in production in industrialized countries from lower- to higher-valued consumer goods, to more capital- and/or skill-intensive manufacturing and services	Employment gains from rising industrialized country exports are unlikely to compensate fully for the job losses, especially if (a) industrialized country wages remain well above those of the NIEs and other emerging developing countries and (b) the rates of world economic growth are relatively low, and/or excessively concentrated in East and South East Asia
Employment growth and job quality improvement for skilled workers are likely to be significant in the short and medium term, even though in the long run the effects are unclear	The employment growth and job quality improvement for skilled workers will dwindle in the long run, as a result of relatively cheaper and more productive skilled labour in the NIEs
Relocation of production and/or imports causes negative short-term effects on workers but promotes labour market flexibility and efficiency through greater mobility of workers within countries (and, to a lesser extent, within regional economic spaces) to economic activities and areas with relative scarcities of labour	Increased trade will further reduce demand for unskilled labour. This exacerbates unemployment because, in a world of mobile capital, the industrialized countries no longer retain a capital-based comparative advantage

Source: Based on ILO, 1996b: Table Int.1

Policy responses

Removing 'obstacles' to adjustment

The period of low unemployment during the 1960s and early 1970s in the industrialized countries was associated with growing demand for manufactured products and also, in most cases, with a particular system of labour regulation.

> Based on the principle of full (male) employment, this system sought to maintain a balance between the normalization of aggregate demand, containment of class conflict, expansion of social welfare, and regulation of social reproduction.[29]

However, the intensification of globalizing processes, the relative shifts of production away from the older industrialized economies, and intensified global competition have dramatically changed the labour policy environment. The overriding concern is now with the problems of *adjustment* to intensified global competition and global shift.

State policies vary from one country to another, as we saw in Chapter 6. While there are elements of convergence in the policies of industrialized countries – notably a trend towards privatization and deregulation – considerable differences also remain, particularly between the United States and the continental European countries (with the United Kingdom occupying a position between the two, but closer to the United States). Even so, a new conventional wisdom has emerged which is very different from the old. Its essence is that of removing what are seen to be *rigidities* in the labour markets of the older industrialized countries. Its aim is to make labour markets more *flexible,* in tune with what are seen to be the dominant characteristics of a globalizing world economy. The 'flexibilization' of labour markets through deregulation involves greatly increased pressures and restrictions on labour organizations, the drastic cutting back of welfare provisions, and the move away from welfare towards *workfare*.

> The essence of workfarism ... involves the imposition of a range of compulsory programs and mandatory requirements for welfare recipients with a view to *enforcing work while residualizing welfare*. ... workfare systems privilege dynamic transitions from welfare to *work*, typically through the combined use of 'carrots' in the form of work and job-search programs and 'sticks' in the form of benefit cuts for the noncompliant.[30]

The process has gone furthest in the United States. Its apparent success in continuing to create large numbers of jobs (albeit with the widening of income gaps) has 'been the most persuasive argument for neo-liberal policies'.[31] It certainly stimulated the United Kingdom government to move along the same path. As yet, the countries of continental Europe have not moved as far, or as fast, down the flexibilization path. Most European governments are concerned that the social costs of reducing unemployment using the US model may be politically unacceptable in a system in which the social dimension of the labour market is very strongly entrenched. But there are clear signs of change as governments become increasingly concerned about the financial costs of sustaining existing practices and the continuing loss of competitive edge.

As a result, a variety of labour market measures, employed in various combinations in different European countries, has emerged. These include:

- the use of more temporary and fixed-term contracts
- the introduction of different forms of flexible working time
- moves to encourage greater wage flexibility by getting the long-term unemployed and the young to take low-paid jobs
- increased vocational training to provide more transferable skills
- reforms in state employment services
- incentives to employers to take on workers
- measures to encourage workers to leave the labour market
- reductions in the non-wage labour cost burdens on employers
- specific schemes to target the long-term unemployed.

In 2000, the EU summit in Lisbon announced no fewer than 15 goals for modernizing welfare systems, strengthening investment in education and combating social exclusion.

> The leaders expected their measures to bring the EU closer to the US in terms of employment and competitiveness. 'The overall aim of these measures should be … to raise the employment rate from an average of 61 per cent today to as close as possible to 70 per cent by 2001 and to increase the number of women in employment from an average of 51 per cent today to more than 60 per cent by 2010' … Mr Blair said that the summit marked 'a sea change in EU economic thinking'. The EU was taking a 'new direction, away from the social regulation of the 1980s towards enterprise, innovation, competitiveness and employment'.[32]

One of the biggest problems facing displaced workers is that there is frequently a geographical mismatch between the destruction of old jobs and the creation of new ones. In such circumstances,

> Education and training and employment subsidies will be for naught if workers are unable or unwilling to take new jobs where they exist. Studies to date show that high-skilled workers are much more willing than the unskilled to move to a new work location. This would indicate that removal of the obstacles that block unskilled workers from moving has an important role in labor policy.[33]

But there are also pronounced geographical differences in the propensity for labour to move. On the whole, labour is geographically less mobile in Europe than in North America, for reasons which lie deep in social and economic history and in cultural attitudes. There are also many real obstacles to the geographical mobility of labour. Not only must deep community and family ties be broken but also the nature of the housing market may make relocation extremely difficult. For example, selling a house in an area of industrial decline may be virtually impossible; getting accommodation at a realistic price in an area of growth may be equally difficult. Rigidities in the public housing sector may also inhibit geographical mobility. The whole question of the geographical mobility of labour in the older industrialized countries is more complex than is often supposed, especially by politicians. The issue of labour mobility is particularly paradoxical in the context of the European Union where all citizens of member states have the automatic right to work in any other member state. In fact, the actual level of labour mobility between states is minuscule.

Quite apart from efforts by governments to remove what they see as obstacles to economic expansion without increasing inflation, four types of policy have received a great deal of attention in the older industrialized countries:

- developing new technologies
- attracting foreign investment
- promoting entrepreneurship and small firms
- protecting domestic industries against imports.

Developing new technologies

As we have seen at several points in this book, governments have become universally involved in attempting to stimulate technological developments within their own national territories, either directly or indirectly. But controversy has always raged over the employment implications of introducing new technology. Does new technology create or destroy jobs? If it creates new jobs are they different from the old jobs either in terms of skills required or in their geographical location? As noted already, historical experience seems to show that, *in aggregate,* new technologies create more jobs than they destroy, at least over the longer term. This seems to occur because they create new demands for goods and services, many of which could not have been foreseen. However, we should beware of too ready an adoption of the notion that investing heavily in new technology will automatically increase employment opportunities.

For example, the new wave of information technologies may well not create the number of jobs some of its advocates suggest. Much of the growth associated with the new technologies may well be 'jobless growth'. The new information technologies are also having a major effect on many service industries and may well reduce the capacity of the services sector to absorb employees displaced from manufacturing as they have done in the older industrialized countries since the 1960s. What is certain is that new technologies redefine the nature of the jobs performed, the skills required and the training and qualifications needed. They alter the balance of the labour force between different types of worker. To some writers the outcome is the de-skilling of the labour force, but this is by no means a universal outcome. Re-skilling and multi-skilling are also significant outcomes of technological change.

A further complication is that the new technologies, and the industries based upon them, will not necessarily emerge in the same geographical locations as the old industries. The terms 'sunrise' and 'sunset' industries themselves (probably unconsciously) imply a geographical distinction (the sun does not rise and set in the same place!). This geographical dimension of technological change is apparent at both the inter- and the intra-national scales. At the international scale, for example,

Europe lags behind both the United States and Japan in a number of key sectors, not least in the new information technologies. Hence, one of the objectives of EU policy, following the 2000 Lisbon summit, is for the EU to 'become the most competitive and dynamic knowledge-based economy in the world'.[34] Within Europe, too, there are considerable technological gaps between individual nations, although these tend to vary by industry. In all cases, however, it is clear that much of the new economic activity based on new technologies has a different geographical profile from that of the obsolescent industries which are being displaced (and relocated overseas). The 'anatomy of job creation' is rather different from the 'anatomy of job loss'.

Attracting foreign investment

A second kind of adjustment policy pursued by the older industrialized countries in an attempt to cope with employment decline has been one of attracting foreign investment. As we saw in Chapter 9, rivalry for the investment favours of TNCs has become intense and often pursued at the highest governmental and state level. Prime ministers and presidents exert their influence either overtly or covertly to persuade the major TNCs to locate new, job-creating investment in their particular countries. Whether such a policy is beneficial in the long term is a matter of debate. From a political point of view, of course, the 'foreign investment fix' has the advantages of having a high profile and being relatively quick. Undoubtedly new jobs are created by inward investment but whether there is a gain in *net* terms depends on its impact on existing firms and on the effect of foreign investment on the country's technological development (see Chapter 9). In any case, there is a limited amount of internationally mobile investment to go round and since much of it will be in higher-technology industries anyway, the number of jobs created may well be far less than in the past.

Promoting entrepreneurship and small firms

There has been a remarkable swing of the pendulum in attitudes towards firms of different sizes. In the 1960s the key to economic (and employment) growth was seen to be the very large firm and the very large plant. Only in such organizations, it was argued, could economies of scale be achieved to enhance competitiveness. Many governments encouraged mergers between enterprises towards this end. By the 1970s, in complete contrast, disillusionment with the large enterprise had set in and the employment panacea was seen to be the small firm with its supposed dynamism and lack of rigidity. This view was strongly reinforced by claims that the vast majority of all net new jobs created in the United States during the 1980s had been in firms employing fewer than 20 workers.

However, reservations have been expressed about the job-creating significance of small firms.[35] It is certainly true that a large number of jobs are located in small firms in most industrialized countries. However, it would take many thousands of small firms to replace the jobs lost through the rationalization processes of even a few large firms, let alone to create additional jobs. There is no doubt that small firms are an important source of growth in an economy. But it should not be forgotten that the majority of small firms are far from dynamic, that most depend upon large firms for their markets, and that the failure rate of small firms is very high. Thus, the 'small firm fix' is likely to be limited in its impact on unemployment:

> small, per se, is neither unusually bountiful nor especially beautiful, at least when it comes to job creation in the age of flexibility.[36]

Protecting domestic industries against imports

Running in parallel with attempts to attract foreign investment have been increasing measures to restrict imports of manufactured goods. This 'protectionist fix' is especially contentious not only in terms of its likely effects on the older industrialized countries themselves, but also globally because of its implications for the economic well-being of developing countries. The kinds of trade protection measures which governments can pursue were discussed in some detail in Chapter 5. Trade frictions have undoubtedly increased across the entire spectrum of countries. 'Managed' trade or 'strategic' trade policies have become more widespread in most of the older industrialized countries. Quite apart from the continuing frictions between the United States and the EU in certain industries (at the time of writing the issue was steel imports to the United States), and even between individual members of the EU, there have been two major targets for this new wave of protectionism: Japan and the NIEs.

The crux of the issue is this: how far should those industries in which the older industrialized countries no longer have a comparative advantage be allowed to run down and be allowed to develop elsewhere? The neo-liberal view is that the answer is self-evident. The older industrialized countries should move out of what are, for them, obsolete activities involving products that can be manufactured more cheaply in developing countries and move to higher-technology products and into the more sophisticated service industries. But it is not as simple as this. The basic argument for protection against imports is that it is necessary to give domestic industry time to adjust: it provides a breathing space. There can be no doubt of the justification of such temporary measures in those sectors where international competition has intensified very rapidly. But if the breathing space is used, as it should be, to restore an industry's competitiveness or to shift into new activities, it will not necessarily preserve employment. As we have seen, new investment is likely to be labour-saving; new activities may be located in different places. It may also be the case that protection against imports in the sectors most sharply affected by NIE growth will not prevent an inevitable decline in employment in the older industrialized countries. There is also the broader question of the effects on customers of protecting domestic industries. This is not just a problem for individual purchasers of consumer goods. Keeping domestic prices high through protection means that products used by other industries will also cost more – and this may itself lead to job reductions in those industries.

The thorny problem of migration

It may seem paradoxical to think of migration as one of the ways of helping to solve the adjustment problems of the older industrialized countries. After all – especially in Europe – we have been talking about not enough jobs to meet the demands of the existing populations. To add further to what appears to be an over-supplied labour market seems perverse to say the least. It is such considerations, together with fears of social unrest between indigenous and immigrant populations, that have made current immigration policies in most developed countries so rigid. But, as ever, things are not as simple as aggregate figures suggest. Quite apart from any humanitarian concern for refugees, asylum seekers, or people simply trying to improve their lives, there are two reasons why developed countries need to create a sensible policy toward in-migration. One is immediate, the other is longer-term.

The immediate reason for asserting the need for more enlightened immigration policies is the fact that in many parts of the economy there is a *severe shortage of labour*. This applies as much in high-skill sectors such as IT and healthcare as in some low-skill service sectors.

Unemployment may remain serious in Germany, where nearly 10 per cent of the workforce does not have a job, but the German government is now offering term-limited employment opportunities for as many as 30,000 people recruited from outside the European Union. The hope is to attract workers with high technology skills from areas such as Bangalore, southern India's 'Silicon Valley'.
The US has made itself more open to foreign talent by increasing the number of people who can take advantage of a work permit. But a growing number of observers believe the cumbersome US immigration system needs to relax more and offer permanent rather than temporary work visas for those with intellectual and business abilities. This may not mean a return to the pre-1914 world of open frontiers but it would go a long way towards ensuring that the new economy is not held back by a shortage of people able to exploit its opportunities.[37]

The fear of such a policy among domestic workers is, of course, that they would be squeezed out of employment by lower-paid migrant workers. And, of course, this does happen. But not invariably so. A recent research report in the UK argued as follows:

Native wages have *not* been depressed in the UK. Immigrants have tended to perform three types of job. They have worked in public services, especially health, where pay is determined by the government. Wages are well below market levels and the effect of newcomers is to reduce shortages … At the other extreme, 'in relatively low paid and insecure sectors [such as] catering and domestic services, unskilled natives are simply unwilling … to take on the large number of available jobs … if migrants do not fill these jobs they simply go unfilled or uncreated.' An estimated 70 per cent of catering jobs are filled by migrants.
There are also the highly skilled information technology workers. According to the Home Office study, the inflow of these technicians has enabled the IT sector to grow faster rather than to depress pay in it.[38]

The other, longer-term reason why the older industrialized countries need seriously to reconsider their policies towards immigration is the very fact that the populations of such countries are *getting older* (see Table 15.1). Their active populations are shrinking. There will not be enough people of working age to support future dependent populations. For both short- and longer-term reasons, then, there is a pressing need to rethink immigration policies. But, of course, there are major political obstacles to doing so. There is enormous resistance among some sections of the population to any relaxation of immigration controls. Fears (sometimes justified, often not) of being squeezed out of jobs by incomers or of local cultures and practices being diluted by 'foreign ways' generate powerful forces of opposition. Such fears are easily exploited by political groups of the extreme right, as can be seen today in many European countries, as well as in the United States. But the need for an influx of new workers will not go away. On the contrary, given the demographic trends in all the developed countries, the need will increase.

Conclusion

Although on a global scale the developed economies are clearly 'winners' in the global economy, that isn't true for everybody. There are undoubtedly 'losers' alongside the winners in terms of both jobs and incomes. While some – perhaps the majority – of developed country populations are better off than ever in material terms, others are relatively disadvantaged. There is both a social and a geographical bias to this. Certain groups, certain kinds of geographical area, have fared substantially worse than others. The effects of globalizing processes within the developed countries are extremely uneven. In these respects, the older industrialized countries continue to face considerable difficulties in adjusting to the intensified competitive environment of a global economy. Preserving, let alone creating, jobs for their active populations has become infinitely more complex in today's highly interconnected world. The imperatives that drive nation-states to strive to enhance their international competitiveness will not always be job-creating. However, the problems facing the affluent industrialized economies pale into insignificance when set beside the problems facing the world's developing countries. It is to these countries that we turn in the next chapter.

Notes

1 Sen (1999: 21).
2 See Rowthorn and Ramaswamy (1997).
3 European Commission (1996: 101–2).
4 European Commission (1996: 103).
5 *The Financial Times* (22 June 2001).
6 *The Economist* (15 January 2000).
7 Castles and Miller (1993: 191).
8 Faux and Mishel (2000: 102).
9 Kapstein (1999: 101).
10 Kapstein (2000: 379).
11 Glasmeier (2002).

12 Glasmeier (2002: 163).
13 Dunford and Smith (2000: 178, 180).
14 See Friedmann (1986); Sassen (2001).
15 Friedmann (1986: 76, 77, 79).
16 Sassen (2001: chs 8 and 9) provides much detailed empirical data on London, New York and Tokyo.
17 Sassen (2001: 270).
18 Sassen (2001: 289, 290).
19 Sassen (2001: 294).
20 Sassen (2001: 306).
21 Sassen (2001: 309).
22 Rifkin (1995).
23 ILO (2001: 140–1).
24 Tobin (1984: 83).
25 See, for example, Cline (1997); ILO (1997); Kaplinsky (2001); Kapstein (1999, 2000); Wood (1994).
26 Wood (1994: 8, 11, 13).
27 ILO (1997: 71, 73).
28 Kaplinsky (2001: 56–7).
29 Peck (1996: 240).
30 Peck (2001: 10).
31 Faux and Mishel (2000: 101).
32 *The Financial Times* (25 March 2000).
33 Kapstein (1999: 156–7).
34 *The Financial Times* (25 March 2000).
35 For a stimulating polemic against the small firm phenomenon, see Harrison (1997).
36 Harrison (1997: 52).
37 *The Financial Times* (27 April 2000).
38 *The Financial Times* (25 October 2001).

CHAPTER 17

Making a Living in Developing Countries: Sustaining Growth, Enhancing Equity, Ensuring Survival

Some winners – but mostly losers

In large part, though by no means entirely, the economic progress and material well-being of developing countries are linked to what happens in the developed economies. A continuation of buoyant economic conditions in the industrialized economies, with a general expansion of demand for both primary and manufactured products, would undoubtedly help developing countries. But the notion that 'a rising tide will lift all boats', while containing some truth, ignores the enormous variations that exist between countries. The shape of the 'economic coastline' is highly irregular; some economies are beached and stranded way above the present water level. For such countries there is no automatic guarantee that a rising tide of economic activity would, on its own, do very much to re-float them. The internal conditions of individual developing countries – their history, cultures, political institutions, forms of civil society, resource base (both natural and human) – obviously influence their developmental prospects. However, despite the claims of the 'neo-environmental determinists', low levels of development cannot be explained simplistically in terms of the natural environment (for example, climatic conditions).[1] As always, it is a specific combination of external and internal conditions which determines the developmental trajectory of individual countries.

For the developing world as a whole, as we saw in Chapter 15, the basic problems are those of extreme poverty, continuing rapid population growth and a lack of adequate employment opportunities. Apart from the yawning gap between developed and developing countries as a whole, however, there are enormous disparities within the developing world itself. It is these internal variations that we address in this chapter. The discussion is organized in the following way:

- First, we look at the heterogeneity of the developing world focusing, in particular, on issues of income, employment and population migration.
- Second, we look at the 'winners' among developing countries: the so-called NIEs. Two related questions are addressed: how can economic growth be sustained and how can this be achieved while sustaining some degree of equity?
- Finally, we focus on the clearest of the 'losers' – the poorest developing countries whose populations exist on the very margins of survival.

Heterogeneity of the developing world

Income differentials

Using broad terms like 'developing countries' or the 'third world' implies a degree of homogeneity which simply does not exist within this group of countries. In this respect, the World Bank makes a useful distinction between three groups of developing countries based on per capita income level. In 2000, the income thresholds for each category, together with that for the high-income group, were as follows:

- *low-income countries*: $755 or less
- *lower middle-income countries*: $756 – $2,995
- *upper middle-income countries*: $2,996 – $9,255
- *high-income countries*: over $9,255

Figure 17.1 shows the geographical distribution of each of these groups of developing countries. There are some striking features, most notably the very heavy concentration of the lowest-income countries in sub-Saharan Africa, which contains no less than 60 per cent of the world's low-income countries. The whole of the Indian sub-continent (with the exception of Sri Lanka) also falls into the low-income category, as do a number of the transitional economies of the former Soviet Union. In contrast, most East Asian countries are in the middle-income categories. Two – South Korea and Malaysia – are in the upper middle-income group, while Singapore, Taiwan and Hong Kong have sufficiently high per capita incomes to fall into the high-income category. Apart from Haiti and Nicaragua (which are in the low-income category), all countries in Latin America and the Caribbean are in the middle-income group, with the major Latin American economies being in the upper middle-income group.

The averages for three development indicators – per capita income, fertility rates and infant mortality rates – give some impression of the heterogeneity within the developing world as well as a stark indication of the gap between these countries and the high-income countries of the world (Table 17.1). The income disparities are especially marked. The average per capita income of the 64 poorest countries is a mere $410 compared with $1,200 in the lower middle-income group of developing countries, $4,900 in the upper middle-income countries, and $25,730 for the high-income countries. The average for the upper middle-income group is roughly 12 times greater than that of the low-income group. However, the average income gap between the upper middle-income group and the high-income group is substantially less than this, at roughly five times. Note also the very wide range of incomes within each of the broad categories, a reminder that putting countries into boxes, while helpful in certain respects, has its limitations.

This income gradient is, not surprisingly, reflected in the data for fertility rates and infant mortality rates. There is a broadly systematic negative relationship between income levels and these two indicators. As average incomes rise, both infant mortality and fertility rates fall. But it is also noteworthy that these rates have fallen in all income groups over the past two decades. But, of course, the

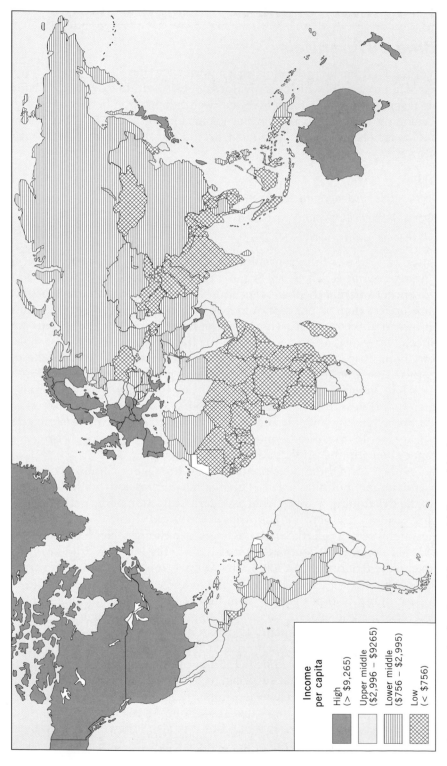

Figure 17.1 Distribution of low-, middle- and high-income countries

Source: Based on World Bank, 2001: 334–5

range of values of these two indicators is enormous. For example, among the low-income group of countries, with an average fertility rate in the late 1990s of 3.1, there were rates as high as 7.3.

Table 17.1 Variations in income and other indicators within the developing world

Country group	Per capita income ($) (mean and range)	Fertility rate (births per woman)	Infant mortality rate (per 1000 live births)	Life expectancy at birth (years) M	F	Adult illiteracy rate (% popn aged 15 and above) M	F
Low-income (64 countries)	410 100–720	3.1	68	59	61	30	49
Lower middle-income (55 countries)	1,200 760–2,900	2.5	35	67	72	10	23
Upper middle-income (38 countries)	4,900 3,070–8,490	2.4	26	67	74	9	11
High-income (25 countries)	25,730 9,890–38,350	1.7	6	75	81	NA	NA

NA, no data available.

Source: Based on material in World Bank, 2001: Tables 1, 2, and 7

Employment, unemployment and underemployment

The basic problem: labour force growth outstrips the growth of jobs

Although the employment structure of developing countries has undergone marked change (Table 17.2), the fact remains that most developing countries are predominantly agricultural economies. As Table 17.2 shows, an average of 69 per cent of the labour force in the lowest-income countries was employed in agriculture in 1990 (in some countries the figure was above 90 per cent) compared with only 5 per cent in the industrial market economies. Even in the upper middle-income group (in which most industrial development has occurred), agriculture employs around 20 per cent of the labour force. In each category the relative importance of agriculture has declined even though in absolute terms the numbers employed in agriculture continued to grow. The balance of employment has shifted towards the other sectors in the economy: industry and services.

These broad sectoral changes in employment in developing countries have to be seen within the broader context of growth in the size of the labour force. The contrast with the experience of the industrialized countries in the 19th century is especially sharp. During that earlier period the European labour force increased by less than 1 per cent per year on average; in today's developing countries the labour force is growing at more than 2 per cent every year. Thus, the labour force in the developing world has doubled roughly every 30 years compared with the 90 years taken in the 19th century for the European labour force to double. Hence, it is very much more difficult for developing countries to absorb the exceptionally rapid growth of the labour force into the economy. The problem is not likely to ease in the near future because labour force growth is determined mainly by past population growth with a lag of about 15 years. As we saw in Chapter 15, virtually all of the world's population growth since around 1950 – more than 90 per cent of it – has occurred in the developing countries.

There is, therefore, an enormous difference in labour force growth between the older industrialized countries on the one hand and the developing countries on the other. But the scale of the problem also differs markedly between different parts of the developing world itself. By far the greatest problem exists in low-income Asian countries, where the projected increase of 250 million in the labour force between 1975 and 2000 was twice that of the region's labour force growth rate between 1950 and 1975. More generally,

> every year another 80 million or more people reach working age and join the 2.1 billion-strong workforce in the developing world. Mexico, Turkey, and the Philippines must create 500,000 to 1 million new jobs annually to accommodate the youths entering their workforce. Developing countries such as El Salvador and China already have 20 to 40 per cent of their work forces unemployed or underemployed.[2]

Table 17.2 The changing structure of the labour force in developing countries

| | Percentage of the labour force (mean and range) in | | | | | |
| | Agriculture | | Industry | | Services | |
Country group	1960	1990	1960	1990	1960	1990
Low-income	77	69	9	15	14	16
	54–95	18–94	1–18	2–31	3–37	NA
Lower middle-income	71	36	11	27	18	37
	39–93	13–79	2–26	7–48	5–39	NA
Upper middle-income	49	21	20	27	31	52
	8–67	6–51	9–52	16–46	19–69	NA
High-income	18	5	38	31	44	64
	4–42	1–18	25–50	25–38	27–57	NA

NA, no data available.
'Industry' includes mining, manufacturing, construction, electricity, water, gas.

Source: Based on World Bank, 1995: Tables 1 and 3

Of course, pressure on the labour market is lessened where lower population growth rates occur. Although this is undoubtedly occurring in many parts of the developing world (see Chapter 15), fertility rates are still sufficiently high to ensure that developing countries' labour forces will continue to increase for the next few decades at least. The basic dilemma facing most developing countries, therefore, is that the growth of the labour force vastly exceeds the growth in the number of employment opportunities available.

Formal and informal sectors in developing country labour markets

It is extremely difficult to quantify the actual size of the unemployment problem in developing countries. There are three main reasons for this. One is the simple lack of accurate statistics. A second is the nature of the unemployment itself, which tends to be somewhat different from that in the developed economies. A third reason is the structure of most developing country economies, particularly their division into two distinctive, though closely linked, sectors: *formal* and *informal*. Published figures in developing countries tend to show a very low level of unemployment, in some cases lower than those recorded in the industrialized countries. But the two sets of figures are not comparable. Unemployment in developing countries is not the same as unemployment in industrial economies. To understand this we need to appreciate the strongly segmented nature of the labour market in developing countries.

- *The formal sector* is the sector in which employment is in the form of wage labour, where jobs are (relatively) secure and hours and conditions of work clearly established. It is the kind of employment which characterizes the majority of the workforce in the developed market economies. But in most developing countries the formal sector is not the dominant employer, even though it is the sector in which the modern forms of economic activity are found.

- *The informal sector* encompasses both legal and illegal activities, but it is not totally separate from the formal sector: the two are interrelated in a variety of complex ways. The informal sector is especially important in urban areas; some estimates suggest that between 40 and 70 per cent of the urban labour force may work in this sector. But measuring its size accurately is virtually impossible. By its very nature, the informal sector is a floating, kaleidoscopic phenomenon, continually changing in response to shifting circumstances and opportunities.

In a situation where only a minority of the population of working age are 'employed' in the sense of working for wages or salaries, defining unemployment is thus a very different issue from that in the developed economies. In fact, the major problem in developing countries is *underemployment*, whereby people may be able to find work of varying kinds on a transitory basis, for example, in seasonal agriculture, as casual labour in workshops or in services.

Urban–rural contrasts

Not only are the variations in material well-being between developing countries much greater than those between industrialized countries but also variations between *different parts of the same developing country* tend to be much greater. In particular, the differential between urban and rural areas is especially great. Although the low- and lower middle-income developing countries are still predominantly rural, the trend towards increasing urbanization is clear, as Table 17.3 shows. It is especially marked within the upper middle-income group of countries where, by the end of the 1990s, three-quarters of the population lived in urban areas.

In complete contrast to the older industrialized countries, therefore, where a growing *counter-urbanization* trend was evident for some years, urban growth in most developing countries has continued to accelerate.[3] The highest rates of urban growth are now in developing countries where the number of very large cities has increased enormously.

> Latin America and the Caribbean already have 75 per cent city dwellers, while in contrast, only one-third of the population of Africa and Asia live in urban areas … By 2015, 153 of the world's 358 cities with more than one million inhabitants will be in Asia. Of the 27 'megacities' with more than 10 million inhabitants, 15 will be in Asia …
>
> Currently, three-quarters of global population growth occurs in the *urban areas* of developing countries, causing hypergrowth in the cities least capable of catering for such growth … urbanization processes in the South do not merely recapitulate the past experience of the developed nations. Contemporary urban growth and rural–urban shifts in the South are occurring in a context of far higher absolute population growth, at much lower income levels, with much less institutional and financial capacity, and with considerably fewer opportunities to expand into new frontiers, foreign or domestic.[4]

The sprawling shanty towns endemic throughout the developing world are the physical expression of this explosive growth. In the older industrialized countries, most industrial growth now occurs away from the major urban centres. In the developing countries, the reverse is the case: virtually all industrial growth is in the big cities. Like labour force growth in general, there is a stark contrast with the experience of the growing cities of the 19th century industrial revolution.

Table 17.3   Urbanization trends

Country group	% population urban	
	1980	1999
Low-income	24	31
Lower middle-income	31	43
Upper middle-income	64	76
High-income	75	77

Source: Based on World Bank, 2001: Table 2

United Nations data show that in Africa as a whole, in the mid-1990s, 29 per cent of the urban population lived in 'absolute' poverty compared with 58 per cent of the rural population. In Asia and Latin America the urban–rural differential was less but still substantial. In Asia, 34 per cent of the urban population and 47 per cent of the rural population lived in absolute poverty; in Latin America the figures were 32 per cent and 45 per cent respectively.[5] However, the explosive growth of cities throughout the developing world ensures that, increasingly, very high levels of poverty tend to be concentrated in urban areas. Whereas rural dwellers may be able to feed themselves and their families from the land, such an option is not available in the cities. In addition, there is a whole syndrome of urban pathologies to contend with:

> About 220 million urban dwellers, 13 per cent of the world's urban population, do not have access to safe drinking water, and about twice this number lack even the simplest of latrines. Women suffer the most from these deficiencies ... poverty also includes exposure to contaminated environments and being at risk of criminal victimization ... Poverty is closely linked to the wide spread of preventable diseases and health risks in urban areas.[6]

Of course, there are major problems in the urban areas of the developed countries. However,

> urban poverty in the Third World is on a scale quite different to that in the developed countries ... In the Third World city the relative poverty of the black Baltimore slum dweller is accentuated by absolute material deprivation. Some poor people in the United States suffer from malnutrition. Most of the poor in Indian cities fall into this category. Overcrowded tenement slums and too few jobs are abhorrent, but the lack of fresh water, medical services, drainage, and unemployment compensation adds to this problem in most Third World cities.[7]

Underemployment and a general lack of employment opportunities are widespread in both rural and urban areas in developing countries. There is a massive underemployment and poverty crisis in rural areas arising from the inability of the agricultural sector to provide an adequate livelihood for the rapidly growing population and from the very limited development of the formal sector in rural areas. Some industrial development has occurred in rural areas, notably in those countries with a well-developed transport network. Mostly this is subcontracting work to small workshops and households in industries such as garment manufacture. But most modern industries are overwhelmingly concentrated in the major cities or in the export processing zones.

It is in the big cities that the locational needs of manufacturing firms are most easily satisfied. Yet despite the considerable growth of manufacturing and service industries in the cities the supply of jobs in no way keeps pace with the growth of the urban labour force. Not only is natural population increase very high in the cities of many developing countries, but also migration from rural areas has reached gigantic dimensions. The pull of the city for rural dwellers is directly related to the fact that urban employment opportunities, scarce as they are, are much greater than those in rural areas. In developing countries (and in general), cities are centres of power and privilege ... Certainly, many urban dwellers live in desperate conditions ... [but] even those in the poorest trades reported that they were better off than they had been in the rural areas ... The urban areas, and especially the major cities, invariably offer more and better facilities than their rural hinterlands.[8]

Labour migration as a 'solution' to the jobs problem

Despite its considerable growth in at least some developing countries, manufacturing industry has made barely a dent in the unemployment and underemployment problem of most developing countries. Only in the very small NIEs, such as Hong Kong and Singapore – essentially city-states with a minuscule agricultural population – has manufacturing growth absorbed large numbers of people. Indeed, Singapore has a labour shortage and has had to resort to controlled in-migration while Hong Kong firms have had to relocate most of their manufacturing production into southern China. In all other cases, however, the problem is not so much that large numbers of people have not been absorbed into employment – they have – but that the *rate of absorption* cannot keep pace with the growth of the labour force.

Not surprisingly, in the face of lack of employment opportunities at home, large numbers of potential workers have migrated to other countries. As we saw in Chapter 15, the number of people living outside their country of birth or citizenship runs into the tens of millions. Of course, some of these are political or religious refugees, fleeing persecution at home. But a very large number – more than 40 million – are migrant workers. And remember, these are pretty conservative estimates – much migration is illegal and, therefore, undocumented. At one level, the decision to migrate abroad in search of work is an individual decision, made in the context of social and family circumstances. When successful, that is when the migrant succeeds in obtaining work and building a life in a new environment, the benefits to the individual and his/her family are clear (although there may be problems of dislocation and emotional stress). There is invariably, as well, discrimination against migrant workers in host countries. In many cases, migrants are employed in very low-grade occupations; they may have few, if any, rights and their employment security is often non-existent. They may also be subject to abuse and maltreatment.

But what about the effects of out-migration on the exporting country? They are both positive and negative.

- *Positive effects of out-migration*. Out-migration helps to reduce pressures in local labour markets. In particular, the money sent back home by migrant workers makes a massive contribution not only to the individual recipients and their local communities but also to the home country's balance of payments position and to its foreign exchange situation. By the late 1990s, total remittances sent home by migrants totalled almost $80 billion. Indeed, in many cases the value of foreign remittances is equivalent to a very large share of the country's export earnings, as Figure 17.2 shows.

Figure 17.2

Migrant workers' remittances as a percentage of home country exports

Source: Based on World Bank, 1995: Tables 13, 17

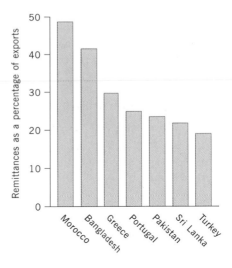

In El Salvador, the *remesas* (remittances), the money sent home by emigrants to their families, has become the principal source of income, far ahead of coffee exports. The one million Sri Lankan emigrants, two-thirds of whom are in the Middle East, are thought to have sent some 830 million dollars back to the country in 1997. In Bangladesh, too, emigration is one of the most important sources of income (nearly 1.5 billion dollars in 1997) ... in some cases ... emigration may help defuse a tense social situation created by unemployment and the lack of future prospects. Proof of this is the eagerness of some countries to encourage the emigration of their unemployed workers. Vietnam for example: 12,661 Vietnamese worked abroad in 1996, and the government wants to increase this number. It has taken measures to liberalise the recruitment agency sector and to train future emigrants. The same applies to Sri Lanka, which hopes to increase emigration by 20 per cent. Thailand, which imports labour from Burma, Cambodia, Laos, etc., sends workers abroad ... and is trying to increase their number ... Elsewhere, in Guatemala, for example, economists stress that the young should be taught English as 'they will constitute our principal export product'.[9]

However, most developed countries are now far less open to in-migration than they have been in the past. This is despite the fact that, as we saw in Chapter 16, there is an increasing need for migrants to offset the decline in population. The reason for such relative closure, of course, is essentially domestic-political; the fear that far-right nationalist groups will foment unrest (as they do). But from a developing country point of view, this is a disastrous policy. The value of remittances is now equivalent to more than 1.5 times the value of direct development aid.

- *Negative effects of out-migration.* The other side of the coin is less attractive for the labour-exporting countries. The migrants are often the young and most active members of the population. Further,

returning migrants are rarely bearers of initiative and generators of employment. Only a small number acquire appropriate vocational training – most are trapped in dead-end jobs – and their prime interest on return is to enhance their social status. This they attempt to achieve by disdaining manual employment, by early retirement, by the construction of a new house, by the purchase of land, a car and other consumer durables, or by taking over a small service establishment like a bar or taxi business; there is also a tendency for formerly rural dwellers to settle in urban centres. There is thus a reinforcement of the very conditions that promoted emigration in the first place. It is ironic that those migrants who are potentially most valuable for stimulating development in their home area – the minority who have acquired valuable skills abroad – are the very ones who, because of successful adaptation abroad, are least likely to return. There are also problems of demographic imbalance stemming from the selective nature of emigration. Many villages in Southern Europe have been denuded of young men, with consequences not only for family formation and maintenance but also for agricultural production.[10]

A balance sheet of positive and negative effects

There is no question that the magnitude of the employment and unemployment problem in developing countries is infinitely greater than that facing the older industrialized countries, serious as their problem undoubtedly is. The biggest problem in developing countries is *underemployment* and its associated *poverty*. The high rate of labour force growth in many developing countries continues to exert enormous pressures on the labour markets of both rural and urban areas. Such pressures are unlikely to be alleviated very much by the development of manufacturing industry alone. With one or two exceptions among the NIEs, industrial growth has done little to reduce the severe problems of unemployment and underemployment – with their resulting poverty – in developing countries. Globalizing processes, whilst offering some considerable employment benefits to some developing countries, are, again, a double-edged sword, as Table 17.4 shows.

Sustaining growth and ensuring equity in newly industrializing economies

The spectacular economic growth of a relatively small number of developing countries – especially the NIEs of East Asia – has been, as we have seen, one of the most significant developments in the world economy. In particular, the four East Asian 'dragons' or 'tigers' have 'succeeded in industrializing and joining the ranks of middle-income, emerging market countries in the span of 30 years. This is a performance unmatched in the 19th and 20th centuries by Western countries and, for that matter, by Latin American economies'.[11] The spread of the growth processes to encompass some other East Asian countries during the 1980s and 1990s has meant that

> already, in little more than a generation, hundreds of millions of people have been lifted out of abject poverty, and many of these are now well on their way to enjoying the sort of prosperity that has been known in North America and Western Europe for some time … While some groups have been obliged to bear the negative consequences of development far more than others … it seems clear that the development project in

Table 17.4 The positive and negative effects of globalization processes on employment in developing countries

Positive effects	*Negative effects*
Higher export-generated income promotes investment in productive capacity with a potentially positive local development impact, depending on intersectoral and inter-firm linkages, the ability to maintain competitiveness, etc	The increases in employment and/or earnings are (in contradiction to the supposed positive effects) unlikely to be sufficiently large and widespread to reduce inequality. On the contrary, in most countries, inequality is likely to grow because unequal controls over profits and earnings will cause profits to grow faster
Employment growth in relatively labour-intensive manufacturing of tradable goods causes (a) an increase in overall employment and/or (b) a reduction of employment in lower-wage sectors. Either of these outcomes tends to drive up wages, to a point which depends on the relative international mobility of each particular industry, labour supply–demand pressure and national wage–setting/ bargaining practices	Relocations of relatively mobile, labour-intensive manufacturing from industrialized to developing countries, in some conditions, can have disruptive social effects if – in the absence of effective planning and negotiations between international companies and the government and/or companies of the host country – the relocated activity promotes urban-bound migration and its length of stay is short. Especially in cases of export assembly operations with very limited participation and development of local industry and limited improvement of skills, the short-term benefits of employment creation may not offset those negative social effects
These increases in employment and/or wages – if substantial and widespread – have the potential effect of reducing social inequality if the social structure, political institutions and social policies play a favourable role	Pressures to create local employment, and international competition in bidding for it, often put international firms in a powerful position to impose or negotiate labour standards and labour management practices that are inferior to those of industrialized countries and, as in the case of some EPZs, even inferior to the prevailing ones in the host country
Exposure to new technology and, in some industries, a considerable absorption of technological capacity leads to improvements in skills and labour productivity, which facilitate the upgrading of industry into more value-added output, while either enabling further wage growth or relaxing the downward pressure	

Source: Based on ILO, 1996b: Table Int.1

the Asia–Pacific region holds out the promise of a scale of generalized prosperity unknown in human history.[12]

Certainly, there can be no doubting the remarkable economic progress of the NIEs, especially those in East Asia, although these 'economic miracles' are not without their serious internal difficulties. But what of the future? What are the major problems facing the NIEs? Two are especially significant:

- sustaining economic growth and
- ensuring that such growth is achieved with equity for the countries' own people.

Sustaining economic growth

Measured in terms of increased per capita income, larger shares of world production and trade the East Asian NIEs, in particular, have been phenomenally successful. But can such spectacular growth rates be maintained in the future? Although each of the four leading NIEs has managed to sustain very high rates of growth for a very long period there were signs in the mid-1990s that perhaps the 'miracle' was coming to an end. Press headlines such as 'Asia's precarious miracle', 'is it over?', 'roaring tiger is running out of breath' had begun to appear. A few academic observers had also begun to question the sustainability of East Asian growth.[13] But none had predicted the suddenness of the economic crisis that hit many parts of East Asia in 1997, a crisis signalled by the devaluation of the Thai *baht* in early July of that year and which spread rapidly over the ensuing months. Virtually overnight, it seemed, the East Asian miracle economies had become 'basket cases'.

Millions of words have been written on the causes of the 1997 East Asian crisis.[14] To many writers, especially in the West, the causes lay in structures and practices inside the affected countries, notably corruption (so-called 'crony capitalism') and the failure of banking systems and financial controls. While there is no doubt that such problems did exist, they were not entirely unique to East Asia, as the ignominious collapse of two massive US companies, LTCM and Enron, amidst charges of the same faults of cronyism and corruption, demonstrates. To most East Asian observers, on the other hand, much of the fault was seen to lie in the practices of those financial investors who had surged into East Asia on a wave of speculative anticipation and who then left just as quickly when things seemed to go wrong. The real explanation of the crisis, of course, lies in specific combinations of external and internal processes. In fact, despite the gloom-laden predictions of many writers at the time of the crisis, most of the East Asian economies have recovered much (though by no means all) of their growth momentum. Whether that recovery is sustainable is the key issue.

Maintaining export growth

Economic growth in the East Asian NIEs has been based primarily upon an aggressive export-oriented strategy. It is no coincidence that the take-off of the first wave of East Asian NIEs occurred during the so-called 'golden age of growth' in the 1960s and early 1970s, or that it was made possible by the relative openness of industrialized country markets. The growth and openness of such markets is, therefore, vital for the continued economic growth and development of the NIEs. During the 1960s the conditions were indeed favourable; future prospects look far less propitious as the older industrialized countries have reduced their demands for NIE exports, partly through the deliberate operation of protectionist trade measures. From the NIEs' viewpoint, therefore, the macroeconomic expansion of the industrialized economies is vital. But this will be effective only if trade barriers – especially non-tariff barriers – are also removed or at least lowered. The present political climate in the older industrialized countries makes both possibilities somewhat remote.

Trade tensions, particularly between the United States, their biggest export market and the leading Asian NIEs are palpable and show little sign of disappearing. Countries like South Korea and Taiwan – as well as more recent fast-growing economies like China – are regarded by the United States as being less open to industrialized country imports than they might be. Of course, this is changing, as South Korea has joined the OECD and China has joined the WTO. In both cases, liberalization of domestic markets is inevitably occurring. However, in the light of less-favourable external conditions for trade, the NIEs will need to develop further their own domestic markets. There is a problem for the smaller NIEs, like Singapore, which have very limited domestic markets. However, it is significant that the regional market has become increasingly important.[15] The East Asian NIEs' share of global GDP virtually doubled between 1976 and 1995 (to 25 per cent of the world total) whilst its share of world imports grew from 10 per cent in 1984 to 17 per cent in the mid-1990s. Hence, not only are the first-tier NIEs big producers they are also, together with their regional neighbours, increasingly major global markets. Of course, one effect of the 1997 crisis was to reduce consumer demand in the region, although this is beginning to recover.

Facing intensifying competition

A second problem facing the longer-established NIEs in sustaining economic growth arises from the *intensifying competition from other developing countries*. There are further 'tiers' of potential NIEs which have also been growing substantially as manufacturing centres. These include such countries as China, Malaysia, the Philippines, Thailand, Pakistan, Indonesia, Colombia, Chile, Peru, Turkey and also the transitional economies of Eastern Europe. Competition from these lower-wage countries has intensified as labour costs in the first tier of NIEs have risen. The competition is obviously most severe in the lower-skill, labour-intensive activities on which NIE industrialization was originally based. Indeed, one of the major problems is that

> most East Asian economies have been following growth trajectories which involve ever-intensifying competition in external product markets … as the pace of globalisation speeds up … firms and economies constantly need to run faster just to stand still, fighting to raise productivity and product innovation faster than the decline in margins as competitive pressures intensify. What is scarce today – for

example, the ability to fabricate semiconductor chips, or to assemble automobiles – will become common tomorrow. Thus the ability to sustain income growth depends upon the capability to respond flexibly to changing competitive circumstances … [however] … the ability to physically transform inputs into outputs is increasingly widespread and in many sectors involves few barriers to entry … there are two possible responses to growing competitive pressures. The 'high road' is to upgrade production in such a way as to capture rents, to create barriers to entry, and hence to escape the competitive pressures. The 'low road' is to lower prices as a way of maintaining market share.[16]

It is in this context that the entry of China into the equation has to be seen.[17] China has become, or is becoming, the leading exporter of many of the manufactured products in which developing countries compete. Thus, the development of competition from other developing countries, together with trends in the automation of some labour-intensive processes in the industrialized countries, has added to the pressures on the leading NIEs to shift to more skill-intensive and capital-intensive products and processes. As we have seen at various points in this book, they have been very successful in making this transition so far. But the competitive pressure continues to intensify as other newly industrializing economies not only take on the less skilled functions but also, themselves, strive to upgrade their economies even further.

However, although China has undoubtedly become the biggest competitive threat to other East Asian countries – indeed to all countries competing in similar economic sectors – it faces massive internal problems of its own. Its spectacular economic growth since its opening up in the early 1980s has created vast inequalities between different parts of the country (see Chapter 6). Most problematical is the

vast pool of the unemployed and underemployed created by the restructuring of China's inefficient state industries and the continuing exodus from agriculture. The World Bank estimates 36 million workers have lost their jobs over the last three years. There are at least 100 million workers already displaced into insecure jobs in the informal sector and 70 million more expected to leave farms over the next ten years.

Just to keep pace with its growing workforce, the Chinese economy needs to create about 8 to 9 million jobs a year … The World Bank thinks the real unemployment rate is at least 10 per cent and in some of the rust-belt provinces of the north-east … it could be as high as 25 per cent.[18]

Prospects for the other NIEs outside Asia depend very much on their specific internal and external circumstances. One of the internal features which differentiates the leading East Asian NIEs from many others is the big difference in educational levels. Largely following the example of Japan, the East Asian countries have invested heavily, and from a very early stage, in education – 'and they have reaped as they have sown'.[19] In terms of external factors, regional context matters a great deal. In the case of Mexico, for example, the key issue is its ability to prosper within the NAFTA. Whilst its access to the US and Canadian markets is now guaranteed, its own economy is now feeling the full force of external competition within its own borders. Similarly, the southern European NIEs (Spain, Portugal and Greece), which are full members of the EU, have unfettered access to the entire EU market. However, not only are their own economies open to the full force of competition from the industrialized economies of the EU but they also now face increasing competition from the opening up of Eastern Europe (as, of course, do the NIEs in general).

Ensuring economic growth with equity

Sustaining economic growth is only one of the difficulties facing the NIEs (both existing and potential) in today's less favourable global environment. Sustaining growth *with equity* for the populations of the NIEs themselves is also a major problem. Two aspects of this issue are especially important: income distribution and the socio-political climate within individual countries.

Income distribution

A widely voiced criticism of industrialization in developing countries has been that its material benefits have not been widely diffused to the majority of the population. There is indeed evidence of highly uneven income distribution within many developing countries, as Table 17.5 reveals. In countries such as Brazil, Chile, Mexico and Malaysia, for example, the share of total household income received by the top 20 per cent of households was very much higher than that in the industrial market economies. However, this pattern does not apply in all cases. For example, South Korea has a household income distribution very similar to that of the industrial market economies. Of course, the question of income distribution is very much more complex than these simple figures suggest and is the subject of much disagreement among analysts. The fact remains, however, that in

Table 17.5 Distribution of income within selected developing and developed countries

Country (year)	Lowest 20%	Highest 20%
Brazil (1996)	2.5	63.8
Chile (1994)	3.5	61.0
Mexico (1995)	3.6	58.2
Malaysia (1995)	4.5	53.8
Philippines (1997)	5.4	52.3
South Korea (1993)	7.5	39.3
India (1994)	8.1	46.1
France (1995)	7.2	40.2
Germany (1994)	8.2	38.5
Sweden (1992)	9.6	34.5
United Kingdom (1991)	6.6	43.0
United States (1997)	5.2	46.4

Source: Based on World Bank, 2001: Table 5

general the Asian NIEs have a more equitable income distribution than the Latin American countries.

Without doubt, the more equitable income distribution in countries like South Korea, compared with Latin America, reflects the specific historical experiences of these countries. In particular, it reflects the different patterns of land owner-ship and reform. In South Korea and in Taiwan, for example, post-war reform of land ownership had a massive effect, increasing individual incomes through greater agricultural productivity, expanding domestic demand, and contributing to political stability.[20]

> In 1953, while Taiwan was still recovering from World War II, the island had a level of income inequality that was about the level found in *present-day* Latin America. Ten years later it had dropped to the level *now found* in France. At the same time, growth rates in this period were of the order of 9 per cent per annum. How did these developments coincide?
>
> According to a definitive study, this outcome was due primarily to improved income distribution in Taiwan's agriculture sector. This improvement, in turn, rested on a specific set of governmental policies, that focused in the first instance on agricultural reforms – especially land reform, infrastructure investment, and price reform – coupled with a rapid proliferation of educational opportunities for Taiwanese students at all levels.
>
> In terms of distributive measures, land reform was of greatest significance.[21]

Democratic institutions

Income distribution is one aspect of the 'growth with equity' question. Another is the broader social and political issue of democratic institutions, civil rights and labour freedom. Although the degree of repression and centralized control in NIEs may sometimes be exaggerated, the fact is that such conditions have existed in a number of cases.[22] The very strong state involvement in economic manage-ment in most NIEs has brought with it often draconian measures to control the labour force. Labour laws tend to be extremely stringent and restrictive; in many instances strikes are banned.

> With regard to the maintenance of low-wage labour reserves, the proletarianized segment of labour in all Asian NIEs has suffered from state intervention to depress workers' wages below market rates in order to make exports competitive on the international market ... But internal conditions and state strategies varied ... In both Korea and Singapore, the government has been actively involved in the creation of a 'hyperproletarian' segment of the labour market that has been largely filled by women and is characterised by a high turnover of labour, institutionalised job insecurity, and low wages ... The absence of either state or community restraint on exploitation of workers in Korea is manifested by extremely poor working conditions for this segment. Labour laws have been arbitrarily enforced and favour employers over workers, particularly in the case of heavy industry and automobile production, the sectors in which labour has become most militant.[23]

Although we have treated the questions of sustaining growth and sustaining growth with equity as separate, they are, in fact, closely related. It is an open question as to how far the various forms of the developmental state, which are manifested in different NIEs, can continue to provide the basis for future economic development.

In this respect, it is significant that, from the late 1980s, both South Korea and Taiwan began to democratize. In general, the signs are positive, although there is still some way to go. It remains to be seen, for example, whether China can continue to develop as a market economy while still retaining an authoritarian communist political system.

The conventional view in the West is that economic development and democracy must go together; that the first depends on the second. Of course, there is a view – expressed, for example, by the founder of the Singaporean state, Lee Kuan Yew, and by the Malaysian Prime Minister, Mahathir Mohamad – that 'Asian' democracy is different from the Western model. In Lee's words, 'I do not believe that democracy necessarily leads to development. I believe what a country needs to develop is discipline more than democracy.[24] His view is that Asian value systems are different from those of the West in their emphasis on collective responsibility rather than individualism and on the roles and responsibilities of the state, which is seen as essentially paternalistic.

Figure 17.3 sets out the major components of this concept of 'Asian values' which, in effect, 'recast "Asia" as a moral opposite of the West. Thus ... the Asian penchant for hard work, frugality and love of the family are unproblematically figured as things the West lacks or has lost'.[25] Whether such opinions reflect the situation across the whole of East Asia (let alone of Asia as a whole) is unclear to say the least. But insofar as they reflect at least some of the social and political characteristics of some East Asian economies, they form a considerable contrast with the situation in other parts of the world and help to explain some of the differences between East Asia's and, say, Latin America's, economic development trajectories.

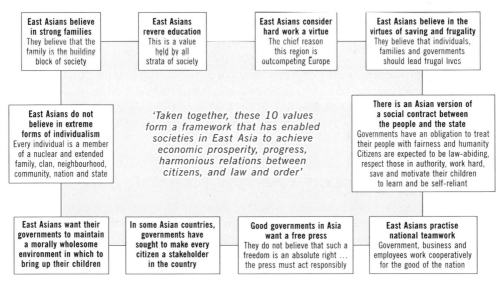

Figure 17.3 'Asian values'

Source: Based on material in Koh, 1993

Environmental problems

Environmental degradation is also a major social problem facing many of the NIEs. In the cases of South Korea and Taiwan, for example, Bello and Rosenfeld write of the 'toxic trade-off' and 'the making of an environmental nightmare' respectively. Although extensive environmental damage is certainly not confined to the NIEs,

> their single-minded pursuit of rapid economic growth has caused particularly severe environmental degradation. Much of the countryside in both South Korea and Taiwan is severely and perhaps irreparably damaged. South Korean rural areas suffer from extensive deforestation, related problems of soil erosion and flooding, and widespread chemical contamination of ground water … Rising environmental costs in both urban and rural areas are materializing in poor health, physical damage, loss of amenities, and other problems that demand extensive remedial spending … In order to stimulate rapid growth, the NIEs have used up significant environmental capital that can only be restored, if at all, at considerable cost to future generations.[26]

Similar comments can be made about the situation in the Mexico–United States border zone, where decades of unregulated pollution by the *maquiladora* factories has led to disastrous levels of toxicity in the water supplies and the atmosphere. It was for such reasons – and the United States' fear of 'environmental dumping' – that a specific environmental side-agreement was incorporated into the NAFTA.

Ensuring survival and reducing poverty in the least developed countries

Although the NIEs certainly face problems, they are not of the same magnitude or seriousness as those facing the least developed, lowest-income countries. As Table 17.1 demonstrated, the poorest sixty or so developing countries are poor not just in terms of income but also in virtually every other aspect of material well-being. They are the countries of the deepest poverty, several of which face mass starvation. For large numbers of people in the low-income countries (and in some of the higher-income countries, too) life is of the lowest material quality. In fact, their position is worse than the average figures in Table 17.1 suggest. Figure 17.4 maps the 33 poorest countries in the world. These are the countries with per capita incomes less than the average for the low-income group of $410. No fewer than 24 of the 33 most impoverished countries are in Africa.

Overdependence on a narrow economic base

There is no single explanation for the deep poverty of low-income countries (and of some of the lower middle-income countries too). But in the context of the global economy, one factor is extremely important: an overdependence on a very narrow economic base together with the nature of the conditions of trade. We saw earlier (Table 17.2) that the overwhelming majority of the labour force in low-income countries is employed in agriculture. This, together with the extraction of other primary products, forms the basis of these countries' involvement in the world economy. Approximately 80 per cent of the exports of developing countries are of primary products compared with less than 25 per cent for the developed economies. Apart from the more successful of the East Asian NIEs,

> the exports of developing countries are still concentrated on the exploitation of natural resources or unskilled labour; these products generally lack dynamism in world markets.[27]

In the classical theories of international trade, based upon the comparative advantage of different factor endowments, it is totally logical for countries to specialize in the production of those goods for which they are well endowed by nature. Thus, it is argued, countries with an abundance of particular primary materials should concentrate on producing and exporting these and import those goods in which they have a comparative disadvantage. This was the rationale underlying the 'old' international division of labour in which the core countries produced and exported manufactured goods and the countries of the global periphery supplied the basic materials. According to traditional trade theory, all countries benefit from such an arrangement. But such a neat sharing of the benefits of trade presupposes some degree of equality between trading partners, some stability in the relative prices of traded goods and an efficient mechanism – the market – which ensures that, over time, the benefits are indeed shared equitably.

In the real world – and especially in the trading relationships between the industrialized countries and the low-income, primary-producing countries – these conditions do not hold. In the first place, there is a long-term tendency for the composition of demand to change as incomes rise. Thus, growth in demand for manufactured goods is greater than the growth in demand for primary products. This immediately builds a bias into trade relationships between the two groups of

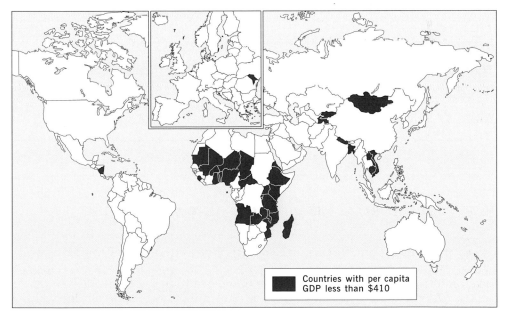

Figure 17.4   Distribution of the world's poorest countries

Source: Based on World Bank, 2001: Table 1

countries, favouring the industrialized countries at the expense of the primary producers.

Over time, these inequalities tend to be reinforced through the operation of the *cumulative* processes of economic growth. The prices of manufactured goods tend to increase more rapidly than those of primary products and, therefore, the *terms of trade* for manufactured and primary products tend to diverge. (The terms of trade are simply the ratio of export prices to import prices for any particular country or group of countries.) As the price of manufactured goods increases relative to the price of primary products, the terms of trade move against the primary producers and in favour of the industrial producers. For the primary producers it becomes necessary to export a larger quantity of goods in order to buy the same, or even a smaller, quantity of manufactured goods. Although the terms of trade do indeed fluctuate over time, there is no doubt that they have generally and systematically deteriorated for the non-oil primary-producing countries over many years.

Table 17.6 illustrates the seriousness of this problem for some of the African countries. Each of them is heavily dependent on a single commodity for export earnings. In some cases – for example Uganda and Zambia – a single commodity accounts for more than half of the country's total exports. As the table shows, there was a dramatic deterioration of the terms of trade for these countries at the end of the 1990s (although, as suggested above, this was nothing especially new). The figures show how the effects of the deepening slowdown in Western economies at the end of the 1990s/early 2000s were transmitted to these commodity producers.

> Shocks to global demand are often transmitted swiftly through the commodity markets, compounding the effects of increases in supply in recent years. A small recovery in global non-energy commodity prices in 2000 … has been snuffed out. Coffee prices have fallen by half in two years and cotton by two-thirds since 1995.[28]

Low levels of investment; high levels of debt

The world's poor countries do not attract high levels of foreign direct investment. In some cases, they attract *hardly any* at all. As we saw in Chapter 3, the vast bulk

Table 17.6 Deteriorating terms of trade for primary-producing countries

Country	1998	1999	2000	Main commodity	Exports (%)
Uganda	−5	−17	−34	Coffee	56
Zambia	−20	−26	−25	Copper	56
Mali	−11	−23	−28	Cotton	46
Rwanda	6	−11	−25	Coffee	45
Chad	−6	−15	−20	Cotton	42
Burkina Faso	−4	−16	−25	Cotton	39
Guyana	0	7	−14	Gold	16
Tanzania	1	−7	−13	Coffee	11

Source: Based on material in *The Financial Times*, 30 January 2002

Figure 17.5 The map of developing country debt

Source: Based on World Bank, 2001: Table 21

of the world's FDI goes to developed economies while the share going to developing countries is concentrated in a relatively small number of countries, notably the NIEs (see Table 3.9). Neither do these poorer countries attract much portfolio investment because there are few attractive investment opportunities in the domestic economy. For the very poorest countries, the major external flow of income is through aid programmes. But these are not only far below the necessary levels, they are also declining. For the group of low-income countries as a whole, aid amounted to just $7 per capita in 1998 (down from $9 in 1990). It represents a mere 1.3 per cent of GNP (compared with 2.6 per cent in 1990).[29] In fact,

> rich nations have slashed development assistance to poor countries over the past decade by a total of $3.5bn, according to Oxfam. Fewer than half a dozen are giving 0.7 per cent of gross domestic product in aid, the UN target (0.2 per cent for LDCs).[30]

Although the world's poor countries receive relatively little financial investment, many of them still face a crippling external *debt problem*. In fact, the problem of debt affects more than just the poorest countries. Much of the economic growth of the NIEs has been financed by overseas borrowing (as was that of the industrializing countries in the 19th century). A number of even the more successful NIEs have experienced periodic debt problems and financial crisis (for example, several of the East Asian NIEs after 1997, Mexico in 1995, Argentina in 2002). Hence, the highest absolute levels of debt are found in some of the NIEs, such as South Korea, China, Brazil and Argentina, as well as in the Russian Federation.

However, absolute debt levels matter less than *relative* levels. What matters is the scale of a country's debt relative to the size of its overall economy. Figure 17.5 shows the global distribution of debt as a proportion of developing countries' GNP. Measured in this way, we can see clearly that most of the highly indebted countries are also among the poorest. For example, the three countries with external debt of more than 200 per cent of their GNP are Angola (279), the Republic of Congo (280) and Nicaragua (262). In comparison, the huge absolute debt of countries such as Brazil, China, South Korea or Mexico looks more reasonable. Brazil's $232 billion debt is equal to 29 per cent of its GNP, China's is 15 per cent, South Korea's is 43 per cent, Mexico's is 39 per cent.

To be 'globalized', or not to be 'globalized'?

> [I]s the problem actually globalization or not-globalization? Is the difficulty being part of the system or not being part of it? How can globalization be the source of problems for those excluded from it?[31]

It is abundantly clear that the relationship of many of the world's poorest countries is highly marginal in terms of the global economy. The usual prescription for those developing countries poorly integrated into the global system is that they should open their economies more, for example, by positively encouraging exports and by liberalizing their regulatory structures.

> For policymakers around the world, the appeal of opening up to global markets is based on a simple but powerful promise: international economic integration will

improve economic performance. As countries reduce their tariff and non-tariff barriers to trade and open up to international capital flows, the expectation is that economic growth will increase. This, in turn, will reduce poverty and improve the quality of life for the vast majority of the residents of developing countries.

The trouble is … that there is no convincing evidence that openness, in the sense of low barriers to trade and capital flows, systematically produces these consequences. In practice, the links between openness and economic growth tend to be weak and contingent on the presence of complementary policies and institutions.[32]

'Openness', then, is the name of game. But this will work only if the playing field is relatively level – which it clearly is not. And it also has to work both ways – which clearly it does not. Tariffs imposed by the developed countries on imports of many developing country products remain very high. It is common for tariffs to increase with the degree of processing (so-called tariff escalation), so that higher-value products from developing countries are discriminated against. At the same time, agricultural subsidies make imports from developing countries uncompetitive. In other words, the odds are stacked against them.[33]

The human costs of unfair trade are immense. If Africa, South Asia, and Latin America were each to increase their share of world exports by one per cent, the resulting gains in income could lift 128 million people out of poverty …

When developing countries export to rich-country markets, they face tariff barriers that are four times higher than those encountered by rich countries. Those barriers cost them $100bn a year – twice as much as they receive in aid.[34]

Simply opening up a developing economy on its own will almost certainly lead to further disaster. There is the danger of local businesses being wiped out by more efficient foreign competition before they can get a toehold in the wider world. Hence a prerequisite for positive and beneficial engagement with the global economy is the development of robust internal structures.

The severity of the situation facing the developing countries in general, and the low-income, least industrialized countries in particular, has led to successive demands, over the past three decades, for a radical change in the workings of the world economic system. In fact, very little progress has been made on most of these demands. The very poor developing countries, those at the bottom of the well-being league table, have benefited least from the globalization of economic activities. Both their present and their future are dire. Ways have to be found to solve the problems of poverty and deprivation.

In such a highly interconnected world as we now inhabit, and as our children will certainly inhabit, it is difficult to believe that anything less than global solutions can deal with such global problems. But how is such collaboration to be achieved? Can a new international economic order be created or is the likely future one of international economic disorder? There can be no doubt that the present immense global inequalities are a moral outrage. The problem is one of reconciling what many perceive to be conflicting interests. For example, one of the biggest problems facing the older industrialized countries is unemployment. Many in the West believe that imports from low-cost developing countries are a major cause of such unemployment. In such circumstances, the pressure on Western governments to adopt restrictive trade policies becomes considerable.

The alternative is to adopt policies that ease the adjustment for those groups and areas adversely affected and to stimulate new sectors. But this requires a more positive attitude than most Western governments have been prepared to adopt. For the poorer developing countries it is unlikely, however, that industrialization will provide the solution to their massive problems. Despite its rapid growth in some developing countries, manufacturing industry has made barely a small dent in the unemployment and underemployment problems of developing countries as a whole. For most, the answer must lie in other sectors, particularly agriculture, but the seriousness of these countries' difficulties necessitates concerted international action. This, in turn, is part of a much larger debate about the overall governance of the world economy. This is the focus of the next, and final, chapter.

Notes

1 One of the surprising developments of recent years has been the return to the old – and totally discredited – 'environmental' explanations of differences in economic development by the historian Landes (1998) and the economist Sachs (1997).
2 Martin and Widgren (1996: 10).
3 See United Nations (1996) and United Nations Centre for Human Settlements (2001).
4 United Nations for Human Settlements (2001: 3).
5 United Nations (1996: 113).
6 United Nations for Human Settlements (2001: 14).
7 Gilbert and Gugler (1982: 23, 25).
8 Gilbert and Gugler (1982: 50, 52).
9 ICFTU (1998: 13).
10 Jones (1990: 250).
11 Ito (2001: 77).
12 Henderson (1998: 356–7).
13 Bello and Rosenfeld (1990) were strongly of the view that the first tier of NIEs was in serious crisis.
14 See, for example, Haggard and MacIntyre (1998); Jomo (1998); Kaplinsky (1999); Stiglitz and Yusuf (2001).
15 Dicken and Yeung (1999: 109–11).
16 Kaplinsky (1999: 4, 5).
17 See Kaplinsky (1999, 2001).
18 *The Guardian* (8 April 2002).
19 Sen (1999: 41).
20 Kapstein (1999: 118).
21 Kapstein (1999: 119).
22 See Bello and Rosenfeld (1990); Brohman (1996); Douglass (1994).
23 Douglass (1994: 554).
24 Lee (2000: 342).
25 Yao (1997: 238).
26 Brohman (1996: 126, 127).
27 UNCTAD (2002: 53).
28 *The Financial Times* (30 January 2002).
29 World Bank (2001: Table 21).
30 *The Financial Times* (21 May 2001).
31 Mittelman (2000: 241).
32 Rodrik (1999: 136–7; emphasis added).
33 See UNCTAD (2002).
34 Oxfam (2002: 1).

CHAPTER 18
Making the World a Better Place

'The best of all possible worlds'?

Voltaire, the 18th century French writer, wrote a wonderful satirical novel, *Candide*, in which the eponymous hero lives in a world of immense suffering and hardship yet whose tutor, Dr Pangloss, insists that Candide's world is 'the best of all possible worlds, where everything is connected and arranged for the best'.[1] This Panglossian view of the world is not far removed from those to whom an unfettered capitalist market system – based on the unhindered flow of commodities, goods, services, and investment capital – constitutes the 'best of all possible worlds'.

The evidence discussed in the preceding chapters of Part IV suggests a very different reality for a substantial proportion of the world's population, particularly in the poorest countries and regions and among certain sectors of the population in many countries (notably women, minority groups and the unskilled) who do not benefit from the overall rise in material well-being. While there has, indeed, been immense growth in the production and consumption of goods and services and, through the mechanism of international trade, a huge increase in the variety of goods available, there remains a vast inequality between the haves and the have-nots. If anything, that gap has been widening, despite the operation of precisely those globalizing processes that are supposed to create benefits for everybody.

In Chapter 2 we identified three key questions in the debates over globalization. Throughout this book our focus has been on the first two of those questions: What precisely is happening? What does it mean? The aim of this final chapter is to address the third question: What can or should be done about it? In other words, how can the world be made a better place for all? There is, of course, no simple answer to such a disarmingly simple question. Much depends on one's political and ideological point of view. Ultimately, we all have to decide for ourselves and act accordingly in our daily lives.

The chapter is organized as follows:

- First, we outline some of the different perspectives on the problems created by, and the solutions advocated to deal with, the effects of globalization.
- Second, we examine existing global governance structures in the areas of finance and trade together with the so far unsuccessful attempts to regulate transnational corporations.

- Third, we focus on two highly contentious problem areas: labour standards and environmental regulation.
- Finally, we ask what the future might be and what it should be.

Globalization and its 'discontents'

In early December 1999, most people in Western countries were probably thinking about their Christmas shopping. Certainly, very few would have had the issues of 'globalization' at the front of their minds – or in their minds at all. Then, overnight, the TV screens were full of pictures of street protesters in Seattle objecting to the attempts being made to initiate a new round of international trade negotiations within the WTO. The protests became the story. Suddenly, people in their sitting-rooms at home became aware of an anti-globalization movement – of a broad coalition of 'discontents'. For the next two years – until the autumn of 2001 – similar protests (both peaceful and violent) occurred at every international meeting of government leaders and of bodies such as the IMF, the WTO, the World Bank, the European Union, or the World Economic Forum. Only the events of 11 September 2001 in New York City gave rise to a pause. Subsequently, the protest movements have become more muted in some respects although, for some groups, the campaign has merged into a broader anti-war campaign with the outbreak of conflicts in the Middle East.

What is striking about the anti-globalization movements is their diversity of both form and substance, despite their shared focus on the phenomenon of 'globalization'. They include the long-established NGO pressure groups such as Oxfam, Greenpeace and Friends of the Earth; more recent ones like Jubilee 2000; organized labour unions like the AFL–CIO or the TUC; labour support organizations like Women Working Worldwide or the Maquila Solidarity Network; organizations focused primarily on TNCs and big corporations (like Corporate Watch or Global Exchange); right-wing nationalist/populist groups (exemplified by such figures as Pat Buchanan in the United States or Jean-Marie Le Pen's extreme right party in France); anti-capitalist groups (like ATTAC, or the Socialist Workers Party); and various anarchist groups.[2]

Within these groups' general focus on the costs of globalization there is huge variation both in their agendas and in how these agendas are pursued. For the anti-capitalist groups, nothing less than the replacement of the capitalist system is advocated, although precisely what the alternative should be varies between groups. For some, it would be a democratically elected world government; to others, a structure in which the means of production were controlled by a nationally elected government. For others, it would be a system of locally self-sufficient communities in which long-distance trade would be minimized. This is the position, for example, of the 'deep green' environmental groups. For some, the focus is on 'fair', rather than 'free', trade – although who decides what is 'fair' varies a good deal. For the more nationalist–populist groups, and for some of the labour unions, the agenda is one of protecting domestic industries and jobs from external competition (especially from developing countries) and restricting immigration. For some, the objective is removing the burden of debt from the world's poorest countries or improving labour standards in the developing world (especially of child labour).

The problem is that, very often, these agendas are contradictory. There are some very unholy alliances involved. Nevertheless, the anti-globalization movements force people – including politicians – to recognize and to engage with the uncomfortable reality that both the benefits and the costs of globalization are very unevenly distributed and that there are severe and pressing problems that need resolution. Some response from the global institutions, such as the World Bank and the WTO, has been achieved. There is now more of a dialogue between the non-violent protest groups and the institutions. One potentially significant recent development is the emergence of the World Social Forum, consisting of many of the interest groups set up in response to the World Economic Forum (an annual meeting of the world's leading companies which, essentially, promotes a neo-liberal economic agenda). The World Social Forum defines itself as

> a global solidarity movement, united in our determination to fight against the concentration of wealth, the proliferation of poverty and inequalities, and the destruction of our Earth.

At its 2002 meeting in Porto Alegre, a number of proposals were agreed, including:

- forgiveness of debt for the poorest nations
- tariff exemption for products from the poorest nations
- increased development aid to 0.7 per cent of wealthy countries' GDP
- 'flexibilization' of the WTO's laws on intellectual property rights
- access to basic drugs at the lowest possible price
- larger reduction in carbon emissions by industrialized nations
- implementation of ILO principles on labour, particularly the right to organize and the prohibition of child and forced labour
- creation of taxes on global financial transactions and abolition of tax havens.[3]

Such proposals imply an essentially reformist, rather than revolutionary, agenda. A major focus is the need to reform (or even replace) existing global regulatory institutions – notably the IMF, the World Bank and the WTO – which are seen to be excessively dominated by the leading industrialized countries (especially the United States).

Global governance structures[4]

> While the world has become much more highly integrated economically, the mechanisms for managing the system in a stable, sustainable way have lagged behind.[5]

More than at any time in the past 50 years, virtually the entire world economy is now a *market economy*. The collapse of the state socialist systems at the end of the

1980s and their headlong rush to embrace the market, together with the more controlled opening-up of the Chinese economy since 1979, has created a very different global system from that which emerged after World War II. Virtually all parts of the world are now, to a greater or lesser extent, connected into an increasingly integrated system in which the parameters of the market dominate.

The acceleration and intensification of technological change, and the emergence of transnational corporations, with their intricate internal and external networks, together ensure that what happens in one part of the world is very rapidly transmitted to other parts of the world. The massive international flows of goods, services and, especially, of finance in its increasingly bewildering variety, have created a real world whose rules of governance have not kept pace with such changes. As Susan Strange[6] argues, power has shifted

- upwards, from weak states to stronger states with global or regional reach beyond their frontiers
- sideways from states to markets and, hence, to non-state authorities which derive their power from their market shares and
- some power has 'evaporated' in so far as no one exercises it.

Thus, although, as we argued in Chapters 5 and 6, the nation-state remains a highly significant actor in the world economy, it is abundantly clear that its role has been changing and that it faces increasingly intractable problems in regulating its domestic economy in a flow-intensive, market-dominated, globalizing world.

There is no doubt that the market can be a highly effective mechanism for facilitating economic growth and development. All markets have to operate within socially defined rules. Totally unregulated markets are neither sustainable nor socially equitable; the unfettered market cannot be relied upon to create outcomes that maximize benefits for the many rather than just the few. To do so demands a regulatory or governance system legitimated by individual nation-states and by the communities and interest groups which constitute them.

A confusion of governance institutions

The current international economic governance system is, in fact, made up of several levels operating at different, but interconnected, geographical scales:[7]

- *international regulatory bodies* established by agreement by nation-states to perform specific roles; examples include the IMF and the WTO (formerly the GATT)
- *international coordinating groups* with a broader, but less formal, remit; examples include the groups of leading industrialized countries (G3, G5, G7, etc.)
- *regional blocs* such as the EU or NAFTA
- *national regulatory bodies* operating within individual nation-states
- *local agencies* operating at the level of the individual community.

These five levels are interdependent to a considerable degree. Effective governance of economic activities requires that mechanisms be in place at all five levels, even though the types and methods of regulation are very different at each level ... [However] the different levels and functions of governance need to be tied together in a division of control that sustains the division of labour ... The governing powers (international, national and regional) need to be 'sutured' together into a relatively well integrated system ... The issue at stake is *whether* such a coherent system will develop ...[8]

Such a 'confusion' of governance structures and institutions is shown in Figure 18.1.

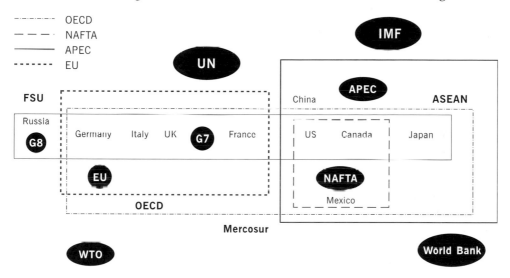

Figure 18.1   A 'confusion' of governance structures and institutions

Source: Cable, 1999: Figure 3.1

In this discussion, we focus specifically on the *international* scales of governance. Within a volatile global economy, there are many issues which pose very serious problems for all states and communities throughout the world and which need to be addressed at the international or global scale. In the following sections, we outline some of the problems associated with governance in three areas, each of which is central to our concerns throughout this book:

- international finance
- international trade
- transnational corporations.

Regulating the international financial system[9]

The 'architecture' of the international financial system

The regulatory 'architecture' of the modern international financial system came into being formally at an international conference at Bretton Woods, New

Hampshire, in 1944. It resulted in the creation of two international financial institutions: the *International Monetary Fund* (IMF) and the *International Bank for Reconstruction and Development* (later renamed the *World Bank*). The International Monetary Fund's primary purpose was to encourage international monetary cooperation among nations through a set of rules for world payments and currencies. Each member nation contributes to the fund (a quota) and voting rights are proportional to the size of a nation's quota. A major function of the IMF has been to aid member states in temporary balance of payments difficulties. A country can obtain foreign exchange from the IMF in return for its own currency which is deposited with the IMF. A condition of such aid is IMF supervision or advice on the necessary corrective policies – the *conditionality* requirements. The World Bank's role is to facilitate development through capital investment. Its initial focus was Europe in the immediate post-war period. Subsequently, its attention shifted to the developing economies.

The primary objective of the Bretton Woods system was to stabilize and regulate international financial transactions between nations on the basis of fixed currency exchange rates in which the US dollar played the central role. In this way it was hoped to provide the necessary financial 'lubricant' for a reconstructed world economy. However, through a whole series of developments (discussed in Chapter 13) the relatively stable basis of the Bretton Woods system was progressively undermined, particularly after the early 1970s. In effect, we have moved from a 'government-led international monetary system (G-IMS) of the Bretton Woods era to the market-led international monetary system (M-IMS) of today'.[10]

As we saw in Chapter 13, Susan Strange coined the graphic term 'casino capitalism' to describe the international financial system:

> Every day games are played in this casino that involve sums of money so large that they cannot be imagined. At night the games go on at the other side of the world ... [the players] are just like the gamblers in casinos watching the clicking spin of a silver ball on a roulette wheel and putting their chips on red or black, odd numbers or even ones.[11]

What we do not have, therefore, is a comprehensive and integrated global system of governance of the financial system. Instead, there are various areas of regulation performed by different bodies which are strongly *nationally* based:[12]

- The G3 (the United States, Japan, EU) takes an overall view of the monetary, fiscal and exchange rate relationships between the G3 countries although, in practice, this has been confined primarily to attempts to determine the global money supply and to manipulate exchange rates. This has not resolved the basic problem of 'an institutional gap between the increasingly international nature of the financial system and the still predominantly "national" remits of the major central banks and the wider nationally located regulatory mechanisms for financial markets and institutions'. The problem is that the G3, as well as the broader G5 and G7, has no real institutional base. It is a largely informal arrangement structured around periodic summits of national leaders.

- The international payments system is operated through the national central banks rather than through an international central bank. 'While this central banking function remains unfulfilled at the international level the risks of default increase and disturbances threaten to become magnified across the whole system … [there is] a growing network of cooperative and coordinative institutionalized mechanisms for monitoring, codifying and regulating such transactions (centred on the Bank for International Settlements and headed by the Group of Experts on Payment Systems).'

- The supervision of financial institutions themselves is carried out through the Bank for International Settlements (BIS), established in 1975 and subsequently based upon the 1988 Basel Committee's Capital Accord (which is currently being revised). 'Much like the monetary summits of the G3 to G7 the Basel Committee was initially designed as a forum for the exchange of ideas in an informal atmosphere with no set rules or procedures or decision-making powers. But, although it maintains this original informal atmosphere, its evolution has been toward much more involvement in hard-headed rule-making and implementation monitoring.'

Thus, there are several regulatory bodies in operation at the international level. Yet the fear remains that, in the absence of a more coordinated and institutionalized system, the international financial system could easily spiral out of control. Indeed, this is what appeared to be happening following the East Asian financial crisis of 1997 and its subsequent spillover effects on countries like Russia and Brazil. Suddenly, there was a chorus of calls for a new, or reformed, *financial architecture*. There was particular concern over the volatile nature of international capital flows:

> In financial markets … bubbles and crashes are endemic. This instability has many causes. Markets may lack information. Banks are prone to mismatch their assets and their liabilities. And, most important, the herd-like behaviour of investors can worsen volatility. Asset prices at times depend less on economic fundamentals than they do on investors' expectations of how other investors will act.[13]

Consequently, there have been renewed calls for the introduction of a tax on international financial transactions, such as the one proposed in the early 1970s by James Tobin. The purpose of the 'Tobin tax' was to 'throw some sand in the wheels' of cross-border financial transactions by levying a small tax on each one. The idea was to discourage excessive flows of 'hot' money – short-term capital flows which can so easily de-stabilize financial systems, especially of weaker countries. So far, no such tax has been implemented, although in 2001, several European political leaders expressed renewed interest in the concept and the IMF acknowledged the need to look again at such proposals.

Developing countries in the international financial system

For obvious reasons, developing countries tend to be particularly vulnerable to the volatilities of capital flows. One of the weaknesses of the various reforms to the international financial architecture which have occurred since the breakdown of the Bretton Woods system is that they have tended to maintain a separation between the problems facing developed countries and those facing developing countries, instead of seeing them as inextricably linked together:

despite the initial emphasis of some policy makers in the leading industrial economies on the need for systemic reform, moves in that direction have subsequently stalled. Instead of establishing institutions and mechanisms at the international level to reduce the likelihood of such crises and better manage them when they do occur, there has been a very one-sided emphasis on reforming domestic institutions and policies in developing countries.

Efforts in the past few years have focused on measures designed to discipline debtors and provide costly self-defence mechanisms. Countries have been urged to better manage risk by adopting strict financial standards, improving transparency, adopting appropriate rate regimes, carrying large amounts of reserves, and making voluntary arrangements with private creditors to involve them in crisis resolution. While some of these reforms undoubtedly have their merits, they presume that the cause of crises rests primarily with policy and institutional weaknesses in the debtor countries and accordingly place the onus of responsibility for reform firmly on their shoulders. By contrast, little attention is given to the role played by institutions and policies in creditor countries in triggering international financial crises.[14]

There is a widespread view that the IMF/World Bank's conditionality 'medicine' may make the patient worse rather than better. By imposing massive financial stringency on countries in difficulty – including raising domestic interest rates, insisting on increased openness of the domestic economy, reducing social spending, and the like – it becomes extremely difficult for countries to help themselves out of difficulty. For example, during the East Asian crisis of the late 1990s, Indonesia was especially seriously affected. But although the effect of a heavily devalued *rupiah* made Indonesian exports potentially more competitive, the astronomical interest rates meant that many businesses could not raise liquid capital to buy materials to take advantage of such increased international competitiveness.[15] No less a figure than a former Chief Economist of the World Bank, Joseph Stiglitz, has strongly criticized the policy of conditionality. He poses the question:

> was imposing conditionality an effective way of changing policies? There is increasing evidence that it was not … there is a concern that the way the changes were effected undermined democratic processes.[16]

Conditionality requirements are also frequently employed in the distribution of *aid* to developing countries and in attitudes towards *debt*. As we saw in Chapter 17, aid levels are very low whereas levels of debt are extremely high. At the 2002 UN Conference on Financing International Development held in Monterrey, Mexico, the United States belatedly offered to increase its aid to developing countries – but it still ranks among the lowest aid donors in the world. The UN target of aid equivalent to 0.7 per cent of developed countries' GDP is far from being fulfilled. Insistence on certain conditions has become more common as it has in the case of debt relief. In the view of one NGO representative at Monterrey:

> the Monterrey consensus (the statement of intent adopted by the conference) 'is just the Washington consensus wearing a sombrero'.[17]

On the other hand, there has been some considerable progress in relieving the poorest countries of the burden of debt. Not least because of the persistent lobbying of such pressure groups as Jubilee 2000, Oxfam and other aid agencies, there is now a programme to alleviate the debt burden of the highly indebted poor countries (HIPCs). A growing number of countries, led by the United Kingdom, have agreed to cancel the debt of the poorest countries in the world. But there is still a very long way to go. Oxfam estimates that 'at least 13 LDCs will emerge from the highly indebted poor countries (HIPC) debt relief programme with an unsustainable debt burden'.[18]

Regulating international trade

Compared with the international financial system, the governance of international trade in commodities and manufactured products is much clearer. In 1947, the General Agreement on Tariffs and Trade (GATT) was established as one of the three international institutions formed in the aftermath of World War II (the others being the IMF and the World Bank discussed above). Establishment of the GATT reflected the view that the 'beggar-my-neighbour' protectionist policies of the 1930s should not be allowed to recur. The objective was to be free trade based upon the *principle of comparative advantage*, first introduced by David Ricardo in 1817. This states that a country (or any geographical area) should specialize in producing and exporting those products in which it has a comparative, or relative cost, advantage compared with other countries and should import those goods in which it has a comparative disadvantage. Out of such specialization, it is argued, will accrue greater benefit for all.[19] The purpose of the GATT was to create a set of rules which would facilitate such free trade, through the reduction of tariff barriers and other types of trade discrimination.

The GATT was in fact introduced as a temporary framework in 1947 in anticipation of the establishment of a fully fledged International Trade Organization (ITO). However, because of disagreements between leading economic powers, this never happened. Initially there were just twenty-three signatories to the GATT. In 2001, there were 139 members of the World Trade Organization (WTO), the successor to the GATT established in January 1995, whilst another 30 states were seeking to join. More than 90 per cent of all world trade is now covered by the WTO framework.[20]

Between 1947 and 1994, when the Uruguay Round was finally concluded, there were eight rounds of multilateral trade negotiations. Table 18.1 summarizes the major characteristics and outcomes of each of the rounds while Figure 18.2 shows the downward trend in average tariff levels achieved in successive GATT rounds. Prior to the mid-1960s, GATT was most concerned with trade between the developed nations. As a result, widespread dissatisfaction emerged among the developing countries with the state of world trade in respect of their own commodities and products. This led to the establishment, in 1964, of UNCTAD (the United Nations Conference on Trade and Development), whose major role was to promote the trading interests of the developing nations.

A particularly sensitive issue was the access of developing country exports to developed country markets. Pressure led, in 1965, to the adoption within GATT of a generalized system of preferences (GSP) under which exports of manufactured and semi-manufactured goods from developing countries would be granted preferential access to developed country markets. In fact, there were a number of exclusions from the GSP, of which one of the most important was textiles and garments. As we saw in Chapter 10, such trade is strongly affected by special trading agreements. Nevertheless, the GSP did mark a major shift in international trade policy, particularly in manufacturing industry.

Table 18.1 The evolution of the GATT/WTO international trade framework

Round	No. of countries	Major outcomes
Geneva Round (1947)	23	Concessions on 43 tariff lines
Annecy Round (1949)	29	Modest tariff reductions
Torquay Round (1950–51)	32	8,700 tariff concessions
Geneva Round (1955–56)	33	Modest tariff reductions
Dillon Round (1960–61)	39	Tariff reductions following formation of EEC 4,400 tariff concessions exchanged
Kennedy Round (1963–67)	74	Average tariff reduction of 35% by developed countries Some 30,000 tariff lines bound Agreement on anti-dumping and customs valuation Moves to incorporate preferential treatment for developing countries
Tokyo Round (1973–9)	99	Average tariff reduction of one-third by developed countries Codes of conduct established for interested GATT members on specific non-tariff measures
Uruguay Round (1986–94)	103 (1986) 117 by end 1993 124 by early 1995	Average tariff reduction of one-third by developed countries Agriculture, textiles and clothing brought into the GATT Creation of the WTO Agreements on services (GATS), intellectual property (TRIPs), trade-related investment (TRIMs) Most Tokyo Round codes enhanced and made part of GATT-1994, i.e. apply to all members of the WTO

Source: Based on Hoekman and Kostecki, 1995: Table 1.2

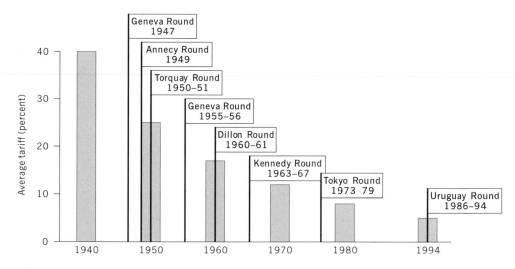

Figure 18.2 Reduction in tariffs and the sequence of GATT negotiating rounds

The Tokyo Round of the 1970s involved almost 100 countries and achieved a reduction in tariffs comparable to the previous Kennedy Round. It also created a series of 'codes' to deal with specific issues. However, it was the Uruguay Round, started in 1986 and eventually concluded in 1994 after a series of delays and near breakdowns, which constituted the most ambitious and wide-ranging of all the GATT rounds. For the first time, the Uruguay Round addressed a number of additional trade issues. It brought agriculture, textiles and garments into the GATT, while special agreements were concluded in services (GATS – the General Agreement on Trade in Services), intellectual property (TRIPs – Trade-Related Aspects of Intellectual Property Rights) and trade-related investment (TRIMs – Trade-Related Investment Measures). There was a further large reduction in overall tariff levels. The major organizational change was the creation of a new World Trade Organization (WTO) which came into being in January 1995, almost 50 years after the original proposal to create an International Trade Organization foundered. The WTO incorporates the GATT itself, together with the new areas of responsibility.

The WTO represents a *rule-oriented* approach to multilateral trade cooperation rather than one which is based upon results (e.g. market share, volume of trade, etc.).

> Rule-oriented approaches focus not on outcomes, but on the rules of the game, and involve agreements on the level of trade barriers that are permitted as well as attempts to establish the general conditions of competition facing foreign producers in export markets.[21]

The fundamental basis of the WTO (and the GATT) is that of *non-discrimination*. This has two components:

- The *most-favoured nation principle* (MFN) states that a trade concession negotiated between two countries must also apply to all other countries; that is, all must be treated in the same way. The MFN is 'one of the pillars of the GATT' and 'applies unconditionally, the only major exception being if a subset of Members form a free-trade area or a customs union or grant preferential access to developing countries'.[22]

- The *national treatment rule* requires that imported foreign goods are treated in the same way as domestic goods.

In November 1999, a WTO meeting was held in Seattle to try to initiate a new round of trade negotiations. The meeting did not succeed, for reasons outlined by the United Nations' Secretary-General, Kofi Annan:

> According to popular myth, the trade negotiations in Seattle in November 1999 were blocked by the peoples of the world joining together in the streets to defend their right to be different, against a group of faceless international bureaucrats ... The truth is more prosaic. Those of us who had hoped to see the launch of a 'development round' that would at last deliver to the developing countries the benefits they have so often been promised from free trade, instead saw governments – particularly those of the world's leading economic powers – unable to agree on their priorities. As a result, no round was launched at all, development or otherwise. The developing countries played a more active and united role than in previous conferences, but the industrialized countries remained locked in arguments among themselves. The governments all favour free trade in principle, but too often they lack the political strength to confront those within their own countries who have come to rely on protectionist arrangements. They have not yet succeeded in putting across to their peoples the wider interest that we all share in having a global market from which everyone, not just the lucky few, can benefit.[23]

It was not until the end of 2001 that a new global trade round was announced at Doha. However, the WTO continues to be subject to widespread criticism from many directions and from interest groups in both developed and developing countries. For example, unilateralist groups within the United States regard the WTO as a basic infringement of the country's national sovereignty. Among developing countries, there is resentment over what is regarded as the bullying and unfair behaviour of the powerful industrialized countries. To the anti-globalization protesters, the WTO is an undemocratic, self-interested body acting primarily in the interests of global corporations.

There are certainly problems inherent in the workings of the WTO, but we should not forget that the WTO itself

> is an inter-governmental organization comprising almost 140 sovereign governments acting on behalf of their constituents in accordance with multilaterally agreed rules that have been adopted by consensus. If the WTO is accused of acting against public interests, it is this collectivity of governments that is doing so ... Nevertheless, there is certainly a problem relating to the systematic absence of many developing countries (particularly those that are small) from both informal and formal meetings in the WTO. Most small delegations from developing countries do not have the appropriate

resources either in Geneva or at home to service the increasingly frequent, complex, and resource-intensive negotiation process at the WTO …

However, knowledge and resources are not enough for all countries to be effective in WTO negotiations. *An important reality is that the WTO rules do not entirely remove the inequality in the power of nations. It remains the case that countries with big markets have a greater ability than countries with small markets to secure market access and to deter actions against their exporters.*[24]

Whether or not the Doha Round will succeed in meeting the needs of countries in a fair manner is far from certain for several reasons.[25]

- A big jump has to be made from the vague statements of intent which formed the basis of the Doha agreement to start a new trade round and the real negotiating positions themselves.

- It is far from clear how far the developed countries will really be prepared to go in meeting the demands of developing countries when it comes down to justifying such actions to domestic pressure groups.

- The problem of agriculture will be a major stumbling block, particularly because of the EU's prevarication in reforming the Common Agricultural Policy and its attitude towards export subsidies.

- The entry of China into the WTO will significantly alter the balance of political power while China faces immense problems in meeting its new obligations under the WTO.

- The trade tensions between the world's two dominant economic powers – the United States and the European Union – will continue to spill over into broader trade negotiations.

(Not) regulating transnational corporations[26]

In the case of foreign direct investment and transnational corporations there is no international body comparable to the WTO. However, the Uruguay Round negotiations of the GATT included efforts to agree on a set of *trade-related investment measures* (TRIMs). Within the TRIMs framework, some of the industrialized countries, led by the United States, wished to prohibit or restrict a number of the measures listed in Figure 5.6, notably local content rules, export performance requirements and the like. The advocates of TRIMs argued that such measures restrict or distort trade. The opponents to TRIMs, including many developing countries, saw such measures as essential elements of their economic development strategies. They, in turn, wished to see a tightening of the regulations against the restrictive business practices of transnational corporations. Similarly, organized labour groups were generally opposed to measures that they felt would increase the ability of TNCs to affect workers' interests or to switch their operations from country to country. As a result, the TRIMs agreement was 'not surprisingly, a compromise. It explicitly affirms that GATT disciplines … apply to investment policies in so far as this directly affects trade flows … Thus, TRIMs that violate the GATT's national treatment rule or its prohibition on the use of QRs [quantitative restrictions] are banned.'[27]

In fact, there is quite a long history of attempts to introduce an international framework relating to FDI and TNCs (apart from those agreed bilaterally or within the context of regional trade blocs). Examples include:

- the OECD Guidelines for Multinational Enterprises (first introduced in 1976)
- the International Labour Organization Tripartite Declaration of Principles Concerning Multinational Enterprises and Social Policy (1977)
- the United Nations Code of Conduct for Transnational Corporations (initiated in 1982).

More recently, in the mid-1990s, an attempt was made, within the OECD, to introduce a *Multilateral Agreement on Investment* (MAI). The main provisions of the MAI included the following:

- Countries were to open up all sectors of the economy to foreign investment or ownership (the ability to exempt key sectors in the national interest was to be removed).
- Foreign firms were to be treated in exactly the same manner as domestic firms.
- Performance requirements (for example, local content requirements) were to be removed.
- Capital movements (including profits) were to be unrestricted.
- A dispute resolution process would enable foreign firms to be able to sue governments for damages if they felt that local rules violated MAI rules.
- All states were to comply with the MAI.

Not surprisingly, the MAI generated a huge amount of opposition and was eventually blocked. Opposition was drawn from a very broad spectrum indeed, across both developed and developing countries.

> The choice of the OECD as the venue for the negotiations was a serious mistake because the OECD is a rich-country club and many LDCs were excluded from the discussions; why would LDCs accept an agreement that they had no part in formulating and that protected the interests of [T]NCs? Even many OECD countries objected to rules that would harm their own interests; France and Canada, for instance, wanted cultural affairs (broadcasting, films, etc.) excluded and the United States wanted restrictions on farm lands. The EU did not want interference with some of its policies. Labor and environmentalists objected that MAI would give [T]NCs license to disregard workers' interests and pollute the environment. Many critics charged that no protection was provided against the evils committed by [T]NCs. Even official American enthusiasm cooled when people realised that the MAI dispute mechanism could be used against the United States and its [T]NCs.[28]

The major dilemma in any attempt to establish a global regulatory framework for FDI and TNCs is the sharp conflict of interest inherent in the process. Should the focus be on regulating the *conduct* of TNCs (the viewpoint of most developing countries, some developed countries, labour and environmental groups) or should it be concerned with the *protection* of TNCs' interests? Despite the defeat of the MAI, it seems that 'the main emphasis in international negotiations is today placed on formulating standards concerning the treatment (or protection) of TNCs rather than standards regarding their conduct'.[29] Certainly the 'trade-related investment measures' (TRIMs) of the Uruguay Round of the GATT were primarily concerned with ensuring fair treatment for TNCs.

Nevertheless, the pressures on TNCs to recognize their social responsibilities and to conform to acceptable ethical standards continue to intensify. Anti-globalization protesters routinely target high-profile TNCs like Nike, McDonald's, Shell, Starbucks, or The Gap. In industries such as garments and footwear, there have been sustained campaigns aimed at forcing TNCs to comply with decent labour standards (an issue we look at in more detail in the next section). As a result, a number of *voluntary codes of conduct* have appeared. Most leading TNCs are now falling over themselves to include corporate social responsibility statements in their annual reports and in their public relations efforts.

> In April 2000, Starbucks Corporation announced it would buy coffee beans from importers who pay above market prices to small farmers (so-called fair trade beans) and sell them in more than 2,000 of its shops across the United States. In August of the same year, the McDonald's Corporation sent a letter to the producers of the nearly 2 billion eggs it buys annually, ordering them to comply with strict guidelines for the humane treatment of hens or risk losing the company's business. And in 1998, De Beers Consolidated Mines, the company that control two thirds of the world trade in uncut diamonds, began investing heavily in Canada to distance itself from the controversy surrounding 'blood diamonds' – gems sold to finance warring rebel factions in Africa.
>
> Are these episodes sudden attacks of conscience on the part of the world's top CEOs? Not quite. Under increasing pressure from environmental and labor activists, multilateral organizations, and regulatory agencies in their home countries, multinational firms are implementing 'certification' arrangements – codes of conduct, production guidelines, and monitoring standards that govern and attest to not only the corporations' behavior but also to that of their suppliers.[30]

An example of a broader voluntary coalition created recently is the Global Alliance for Workers and Communities, which involves Nike and The Gap together with the World Bank and the International Youth Foundation. Its mission is 'to improve the lives and future prospects of workers involved in global production and service supply chains'.[31] A rather higher-profile voluntary code of good corporate conduct was introduced by the UN Secretary-General in 2000. This *Global Compact* involves some fifty companies and twelve NGOs and is based on nine principles which business leaders are asked to adopt (see Figure 18.3).

Inevitably, there is a good deal of scepticism about such voluntary codes, whether at the individual firm or collective level. In one sense, of course, anything which contributes to better conditions for people and communities should be welcomed. But without some degree of compulsion – and the monitoring of compliance – there is always the danger that such codes will amount to little more than a gesture or that companies will be able to influence how the process works. Nevertheless, they do represent a response to pressure from NGOs as well as some governments and consumer groups. The two areas where there has been greatest activity in this regard are labour standards and the environment.

Human rights	
Principle 1	Support and respect the protection of international human rights within their sphere of influence
Principle 2	Make sure their own corporations are not complicit in human rights abuses

Labour standards	
Principle 3	The freedom of association and the effective recognition of the right to collective bargaining
Principle 4	The elimination of all forms of forced and compulsory labour
Principle 5	The effective abolition of child labour
Principle 6	The elimination of discrimination in respect of employment and occupation

Environment	
Principle 7	Support a precautionary approach to environmental challenges
Principle 8	Undertake initiatives to promote greater environmental responsibility
Principle 9	Encourage the development and diffusion of environmentally friendly technologies

Figure 18.3 Principles of the UN Global Compact

Two key concerns: labour standards and environmental regulation

A fundamental question is the extent to which international differences in labour standards and regulations (such as the use of child labour, poor health and safety conditions, repression of labour unions and workers' rights) and in environmental standards and regulations (such as industrial pollution, the unsafe use of toxic materials in production processes) distort the trading system and create unfair advantages. In both cases, the basic argument is that firms – as well as individual countries – may be able to undercut their competitors by capitalizing on cheap and exploited labour and lax environmental standards. Much of the focus of this concern is on the export processing zones (EPZs) which, as we saw in Chapter 6, have proliferated throughout the developing world.

These two issues of labour and the environment were explicitly addressed in the negotiations for the NAFTA, in which the United States insisted on the signing of two side agreements to protect its domestic firms from low labour and environmental standards in Mexico. More recently, a group of countries led primarily by the United States, but also including some European countries, made a concerted attempt formally to incorporate the issue of *labour standards* into the WTO at its ministerial meeting in Singapore in December 1996. The attempt failed, partly because not all industrialized countries supported it but also because the developing countries were vehemently opposed. The argument of those opposed to its inclusion within the WTO's remit is that labour standards are the responsibility of the International Labour Organization (ILO). Indeed, all members of the ILO have agreed to the set of core principles shown in Figure 18.4. The counter-argument is that the ILO lacks any powers of enforcement. It is also notable that the United States, despite its current position on including labour standards in trade agreements, has 'signed only one of the five core labour standards conventions issued by the International Labour Organization – and ratified only 12 of the total 176 ILO conventions. It says that though it respects their spirit, the UN body's conventions do not mesh with its own laws.'[32]

The controversy over labour standards

There is no doubt that stark differences do exist in labour standards in different parts of the world. As we have seen at various points in this book, basic workers' rights are denied in many countries. Working conditions, especially in the EPZs – but not only in these zones – are often appalling.[33] As far as child labour is concerned, the ILO calculates that around 73 million children aged between 10 and 14 years are employed throughout the world, approximately 13 per cent of that age group. In Africa, one-quarter of children aged 10–14 are working. If children under 10 are included, as well as young girls working full-time at home, the ILO estimates that there are probably 'hundreds of millions' of child workers in the world. However, the ILO also points out that 90 per cent of children work in agriculture or linked activities in rural areas and that most are employed within the family rather than for outside employers. Even so, there is substantial evidence that, in many cases, young children are employed by manufacturers (whether in factories or as outworkers) in such industries as garments, footwear, toys, sports goods, artificial flowers, plastic products and the like. Their wages are a pittance and their working conditions often abysmal.

From the viewpoint of many developing countries, however, there is a strong feeling that the labour standards stance of many developed countries is merely another form of protectionism against their exports and, as such, an obstacle to their much-needed economic development. There is a suspicion, for example, that at least some of the developed country lobbies are pressing for international

---

Core principles on labour standards agreed by all members of the ILO

- All workers to be free to join a labour union
- Labour unions to be free from interference from public authorities
- No discrimination against workers on grounds of: race, colour, gender, religion, political opinion, national extraction or social origins

- Forced or compulsory labour to be illegal
- Employment of children under the age of completion of compulsory education to be illegal
- Equal pay regardless of gender for jobs of equal value
- Slaving and child prostitution to be illegal

Figure 18.4   The core principles of the International Labour Organization

agreements on a minimum wage in order to lessen the low labour cost advantages of developing countries. By incorporating such labour standards criteria into the WTO framework, it is believed, developed countries could use trade regulations to enforce such indirect protectionism. But even where labour standard campaigns are based upon principles of human rights, the casualties may be the very people the campaigners are attempting to help. For example,

> In 1993 American television broadcast pictures of Bangladeshi children making clothes for Wal-Mart. After a public outcry the store cancelled its contracts in Bangladesh and a senator proposed a bill to block imports of goods produced by child labour. Faced with losing one of its biggest markets, the Bangladesh garment manufacturers association announced it would eliminate the use of child labour within four months. Thousands of children were sacked, many of whom then found work in more dangerous industries. Few returned to school because their families simply could not afford it.
>
> In the end the campaign against child labour in Bangladesh did have some positive results. Unicef, the United Nations' children's agency, negotiated a voluntary code of standards with the factories and provided money to send some of the children back to school.
>
> Yet the incident illustrates another weakness of using boycotts of exports to raise labour standards in the third world: workers in the export sector usually enjoy better conditions than the vast majority of other third world workers who earn their livelihoods in the informal sector where there are no employment rights or health and safety provisions …
>
> Keith Watkins of Oxfam says that a majority in the developing world are not working in the formal sector and they are not producing goods for export. 'There is no case you can point at where sanctions have improved labour standards. It's about a labour aristocracy in the north protecting its jobs and conditions.'[34]

There is clearly a basic dilemma for the international community. On the one hand, ethical considerations *must* be a basic component of international trade agreements. There is a clear moral responsibility to ensure that workers are not excessively exploited or ill-treated. On the other hand, there is a real danger of some threatened developed country interest groups using labour standards issues as a device to protect their own commercial interests. And even where this may not be the objective, there is always the possibility of unintended consequences.

Globalization, trade and the environment

A similar dilemma is central to the other globalization and trade-related question: that of the *environment*. To what extent should variations in environmental standards be incorporated into international trade regulations? How far should developing countries be expected to forgo possibilities of development through industrialization because of the imposition of environmental regulations which the industrialized countries did not have to face at a similar stage in their development? If they should, then who should foot the bill? Again, there is no doubting the existence of huge differences in the nature, scope and enforcement of environmental regulations across the world. There is no doubt, either, that the highest incidence of low environmental standards is in developing countries. The existence of such an *environmental gradient* certainly constitutes a stimulus for some firms at least to take advantage of low standards. These may be domestic firms or they may be foreign firms.

Although it does not necessarily follow that such firms have relocated to environmentally lax areas simply to avoid more stringent regulations and higher costs, some undoubtedly do so. But the fact that they do operate there is a major problem. Again, one of the most notorious, and best-documented, cases is that of the *maquiladoras* of the United States–Mexico border zone:

> Matamoros, with more than 100 *maquiladoras*, has only three government health and safety inspectors, who normally give advance warning of factory visits. Air quality and industrial effluents are tested only once a year. Senator Richard Gephardt … [noted] '21st century technology combined with 19th century living and working conditions. We drove by industrial parks where companies continue to dump their toxic wastes at night into rivers. We saw furniture plants using highly toxic solvents and finishes that once operated in California and throughout the US, and which had moved to Mexico because of lax environmental enforcement'.[35]

But there are also broader environmental issues associated with global production and trade. As we saw in Chapter 2, the production process is a system of materials flows and balances in which there are, inevitably, negative consequences for the natural environment (see Figure 2.12). Many of these negative environmental effects are not confined within national boundaries but spill over to affect other communities. The 'acid rain' produced by certain types of energy-creation is carried by the wind way beyond its points of origin to create environmental damage. The damage to the ozone layer in the earth's stratosphere is caused by the use of certain chemicals that retain their stability over long periods of time and move upwards into the stratosphere, expelling chlorine that destroys ozone molecules. Again, such chemicals have an effect way beyond their points of use. Most notably, the inexorable increases in greenhouse gases affect the atmospheric balance (and especially the temperatures) of all parts of the world.

Growing realization of the seriousness and pervasiveness of environmental problems associated with economic activity has led to attempts to devise global regulatory mechanisms.[36] The most important agreements concluded in the 1990s included:

- the Montreal Protocol on Substances that Deplete the Ozone Layer (1989–1992)
- the Convention on Biological Diversity (1992)
- the Framework Convention on Climate Change (FCCC) (1992)
- the Kyoto Protocol on Climate Change (1997).

Particular attention has focused on the climate change negotiations because of the concern over global warming. A key objective of the FCCC was to achieve

> the stabilization of greenhouse gas concentrations in the atmosphere at a level that would prevent dangerous anthropogenic interference with the climate system.[37]

The FCCC was based upon *voluntary* reduction of carbon dioxide levels. The lack of success led to the signing of the Kyoto protocol in 1997, in which the signatories to the FCCC agreed on a set of *binding* emissions targets for developed countries based on 1990 levels. The specific limits varied between countries. For the EU, the agreed target was 8 per cent below the 1990 level, for the United States it was 7 per cent, and for Japan, 6 per cent. However, the agreement was extremely complex and contained many detailed ambiguities. In addition, the new Bush administration in the United States announced in 2001 that it would not, after all, sign the Kyoto agreement.

Nevertheless, 186 countries (excluding the United States) eventually agreed in Bonn in 2001 to implement the Kyoto agreement (subject to national ratification). In addition to agreeing on specific target emissions by 2010, the deal also included the following:[38]

- Countries must submit plans for reducing greenhouse gas emissions and update progress in meeting their targets.
- An international system of trading in carbon emissions will be initiated.
- Industrialized countries will be able to claim credits for removing carbon dioxide from the atmosphere through such actions as planting and managing forests and changing farming practices.
- Countries failing to meet their first set of targets by 2012 will have to add the shortfall to the next commitment period plus a 30 per cent penalty. They will also be excluded from carbon trading and will have to take corrective measures at home.
- The industrialized countries would provide some financial assistance to developing countries to help them to adapt to climate change and to provide clean new technologies.

Despite this agreement, immense problems remain, not least in the extent to which developing countries can or should be expected to comply with emissions targets. It seems likely that greenhouse gas emissions from developing countries will reach the level of those of developed countries by between 2015 and 2020. Yet the costs to developing countries of introducing new clean technologies will be huge, even with assistance from developed countries. There is also the political issue – the feeling among developing countries that the imposition of carbon emission targets is another form of protectionism.

In this respect, a closely related issue is the link between environmental standards and their regulation and international trade. At one level, the problem is exactly the same as that of labour standards. If a country allows lax environmental standards, it is argued, then it should not be able to use what is, in effect, a subsidy on firms located there to be able to sell its products more cheaply on the international market. The question then becomes one of whether the solution lies in using international trade regulations or in some other forms of sanction. The proponents and opponents of the 'trade solution' are the same as those discussed above in the case of labour standards.

However, there is an even more extreme position adopted by some environmentalists which is that the pursuit of ever-increasing international trade – which is clearly encouraged by a free trade regime like the WTO – should be totally abandoned, not merely regulated. The argument here is basically that *sustainable development* is incompatible with the pursuit of further economic growth and, especially, with an economic system which is based upon very high levels of geographical specialization, since such specialization inevitably depends upon, and generates, ever-increasing trade in materials and products.

One of the criticisms voiced by environmentalists is that the energy costs of transporting materials and goods across the world are not taken into account in setting the prices of traded goods and that, in effect, trade is being massively subsidized at a huge short-term and long-term environmental cost. Free trade, it is argued, injects new inefficiencies into the system:

> more than half of all international trade involves the simultaneous import and export of essentially the same goods. For example, Americans import Danish sugar cookies, and Danes import American sugar cookies. Exchanging recipes would surely be more efficient.[39]

For example, it has often been demonstrated that the typical meal of the Western family often involves ingredients that have travelled many thousands of miles through the complex logistics channels of the supermarkets (see Chapter 14) when many could have been obtained locally. Not surprisingly, one recent newspaper article was headlined the 'barminess of the long-distance dinner'.[40]

But by no means all environmentalists agree with this kind of viewpoint, as the following quotation demonstrates.

> Unquestionably, there are environmental problems inherent in the existing trading system. But there is also extensive confusion in the environmentalist critique of free trade ... Given the potentially large gains to be obtained from free trade, adopting restrictions on trade for environmental purposes is a policy that needs to be approached with caution. Most importantly, all other approaches to reducing environmental damage should be exhausted before trade policy measures are contemplated ... the policy implication of a negative association between freer trade and environmental degradation is not that freer trade should be halted. What matters is the adoption of the most cost-effective policies to optimize the externality. Restricting trade is unlikely to be the most efficient way of controlling the problem ... The losses can best be minimized by firm *domestic* environmental policy design to uncouple the environmental impacts from economic activity ... The 'first' best approach to correcting externalities is to tackle them directly through implementation of the polluter pays principle (PPP), not through restrictions on the level of trade. Where the PPP is not feasible (e.g. if the exporter is a poor developing country), it is likely to be preferable to engage in cooperative policies, e.g. making clean technology transfers, assisting with clean-up policies etc., rather than adopting import restrictions.[41]

As in the case of labour standards, there is wide divergence of opinion over both the impact of globalization processes and trade on the environment and also on what should (or can) be done.[42] There are, in other words, very different 'shades of green', ranging from the position that human ingenuity and new technologies will find the solutions without necessitating a change in lifestyles (the Panglossian view) through to the 'deep green' arguments that only a return to a totally different, small-scale, highly localized mode of existence will suffice. But such a path, rather than improving the position for the poor in the world economy, 'would condemn the vast majority of people to a miserable future, at best on the margins of the bare minimum of physical existence'.[43] It is not a socially acceptable policy:

> The alternative to an economic system that involves trade is not bucolic simplicity and hardy self-sufficiency, but extreme poverty. South Korea has plenty of problems, but not nearly so many as its neighbour to the north.[44]

This is not to argue for the status quo but, rather, for a policy towards the environment that recognizes the needs and aspirations of people within an equitable context. In other words,

> development that meets the needs of the present without compromising the ability of future generations to meet their own needs.[45]

What might the future be?
What *should* the future be?

It is always tempting to look at recent trends and to extrapolate them into the future. There is, of course, some logic in this. After all, as we pointed out in Chapter 2, there is a strong element of *path-dependency* in human affairs. But it isn't as simple as that. Path-dependency does not imply *determinacy*. All paths have branching points – some go off in unexpected directions, others into a dead-end. This means that we need both a good map and a clear sense of the direction we wish to travel. The first requires a better understanding of how the world actually works (that's what this book has been about). The second requires a sense of moral vision. It is about values. It is about where we want to be.

One thing is clear: we're not very good at making predictions. It is very difficult to identify which contemporary events and circumstances are likely to have long-lasting effects. For example, when the East Asian financial crisis broke with such suddenness in 1997 (to the surprise of most of the world, as we have seen), the popular – and much of the academic – literature was full of predictions of doom: the end of the East Asian 'miracle' had arrived. The future of the region was dire. Few would make those same predictions today, even though problems remain. Similarly, looking a little further back in time, who, from the standpoint of 1960, would have predicted that Japan would soon challenge the United States as an economic power and, in some respects, overtake it? Who would have predicted that South Korea would become one of the world's most dynamic economies within the space of twenty years or so? After all, in 1960, South Korea was one of the poorest countries in the world, with a per capita income comparable with that of Ghana. Which observer in the early 1970s would have predicted that China would open up its economy to the world? Or that the command economies of the Soviet Union and Eastern Europe would, by the end of the 1980s, begin to be transformed into capitalist market economies?

Such anecdotal examples about the problems of prediction should make us wary. But they raise a much bigger question. Will the tendency towards an increasingly highly interconnected and interdependent global economy intensify? Is 'globalization' an inexorable and unstoppable force? Not inevitably, as the experience of the period between the two World Wars shows. During that time, the unprecedented openness of the world economy that had come into being in the period between 1870 and 1913 was largely reversed through the actions of states responding to recession through increased protectionism. It took several decades to return to a similar degree of openness, by which time the world was a very different place. However, the interconnections within the global economy are now much deeper than in the past because of the ways in which the whole process of production and distribution have been transformed. On the one hand, the highly interconnected global economy has generated unprecedented wealth and material opportunity; on the other it has perpetuated immense disparities between people in different parts of the world.

So, the key question is not so much what the world might be like in the future but what it *should* be like. After all, 'globalization is not a force of nature; it is a social process'.[46] The major global challenge is to meet the material needs of the world community as a whole in ways that reduce, rather than increase, inequality and which do so without destroying the environment. That, of course, is far easier said than done. It requires the involvement of all major actors – business firms, states and international institutions – in establishing mechanisms to capture the gains of globalization for the majority and not just for the powerful minority.

Such a system has to be built on a world trading system that is *equitable*. This must involve reform of such institutions as the WTO, the World Bank and the IMF. As we saw earlier, the exercise of developed country *power* through the various kinds of conditionality and trade-opening requirements imposed on poorer countries has seriously negative results. Without doubt, trade is one of the most effective ways of enhancing material well-being, but it has to be based upon a genuinely fairer basis than at present. The poorer countries must be allowed to open up their markets in a manner appropriate to their needs and conditions. After all, that is precisely what the United States did during its phase of industrialization, as did Japan and the East Asian NIEs during theirs. At the same time, developed countries must operate a fairer system of access to their own markets for poor countries. Even the journal *Business Week* acknowledges this when it states that

> the flaws of trying to force every country into the same template have become clear … it is time to forge a more enlightened consensus.[47]

Of course, this will cause problems for some people and communities in the developed countries and these should not be underestimated. They are reflected most clearly in the various lobbying groups that attempt to influence government trade policy. Often they succeed in doing so, as the Bush administration's implementation of tariffs on steel imports in 2002 showed, and as the EU's continuing distortion of agricultural trade through its common agricultural policy demonstrates. It is highly paradoxical that the two major trading groups at the forefront of demands for a more open trading system are also all too ready to protect their own markets.

As we saw in Chapter 16, there are, indeed, many losers in the otherwise affluent economies. Their governments must design and implement appropriate adjustment policies for such groups if such trade policies are to be acceptable politically. Equally, governments of developing countries must engage in their own internal reforms: to strengthen domestic institutions, enhance civil society, increase political participation, raise the quality of education and reduce internal social polarization. Although difficult, such policies are not impossible if the social and political will is there. The imperatives are both practical and moral. In practical terms, the continued existence of vast numbers of impoverished people across the world – but who can see the manifestations of immense wealth elsewhere through the electronic media – poses a serious threat to social and political stability. But the moral argument is, I believe, more powerful. It is utterly repellent that so many people live in such abject poverty and deprivation whilst, at the same time, others live in immense luxury. This is not an argument for levelling down but for raising up. The means for doing this are there. What matters is the *will* to do it. We all have a responsibility to ensure that the contours of the global economic map in the 21st century are not as steep and uneven as those of the 20th century.

Notes

1 Voltaire (1947: 8).
2 See Bircham and Charlton (2001); Gilpin (2000: ch. 10); Rupert (2000).
3 *The Financial Times* (5 February 2002).
4 General discussions of issues of global governance are provided by Cable (1999), Commission on Global Governance (1995), Hirst and Thompson (1999), Ould-Mey (1999), Prakeesh and Hart (1999) and Strange (1996).
5 Commission on Global Governance (1995: 135–6).
6 Strange (1996: 189).
7 Hirst and Thompson (1999:191–2).
8 Hirst and Thompson (1999: 192, 269–70).
9 Problems of governance of the international financial system are dealt with by Braithwaite and Drahos (2000: ch. 8), Eatwell and Taylor (2000), Eichengreen (1996), Hirst and Thompson (1999: 202–9) and Martin (1999).
10 Hirst and Thompson (1999: 202).
11 Strange (1986: 1).
12 Hirst and Thompson (1999: 202–4).
13 *The Economist* (23 May 1998).
14 UNCTAD (2001: vi–vii).
15 Dicken and Hassler (2000: 278–9).
16 Stiglitz (1999: F591).
17 *The Financial Times* (25 March 2002).
18 *The Financial Times* (21 May 2001).
19 All the standard texts on international economics contain explanations of the principle of comparative advantage as do most texts on international business. See for example, Krugman and Obstfeld (1994: ch. 2).
20 Hoekman and Kostecki (1995) provide an extremely comprehensive account of the evolution of the international regulatory framework for trade, from the GATT through to the WTO. Sampson (2001) examines the role of the WTO in global governance.
21 Hoekman and Kostecki (1995: 24).
22 Hoekman and Kostecki (1995: 26).
23 Kofi Annan in Sampson (2001: 19).
24 Sampson (2001: 7–8; emphasis added).

25 *The Financial Times* (1 February 2002). See UNCTAD (2002, ch. II) for a developing country perspective on the next WTO trade round.

26 Attempts to regulate FDI and TNCs are discussed in Braithwaite and Drahos (2000: ch. 10), Gilpin (2000: ch. 6) and UNCTAD (1996b: Part 3).

27 Hoekman and Kostecki (1995: 121).

28 Gilpin (2000: 184–5).

29 UNCTC (1988: 360).

30 Gereffi, Garcia-Johnson and Sasser (2001: 56).

31 *The Financial Times* (22 February 2001).

32 *The Financial Times* (20 June 1996).

33 See, ICFTU (1998) and ILO (1996a) for examples of labour conditions in developing countries.

34 *The Guardian* (16 February 2001).

35 *The Financial Times* (6 June 1997).

36 See Braithwaite and Drahos (2000: ch. 12); Pearce (1995).

37 Quoted in Pearce (1995: 149).

38 *The Guardian* (24 July 2001).

39 Daly (1993: 25).

40 *London Evening Standard* (12 December 2001).

41 Pearce (1995: 74, 77, 78).

42 See Hudson (2001: 315–21); Turner, Pearce and Bateman (1994: ch. 2).

43 Hudson (2001: 315).

44 Elliott (2002). See also Elliott (2000).

45 United Nations World Commission on the Environment and Development (1987: 43).

46 Massey (2000: 24).

47 *Business Week* (6 November 2000: 45).

BIBLIOGRAPHY

Abernathy, F., Dunlop, J., Hammond, J.H. and Weil, D. (1999) *A Stitch in Time: Lean Retailing and the Transformation of Manufacturing: Lessons from the Apparel and Textile Industry*. New York: Oxford University Press.

Abo, T. (ed.) (1994) *Hybrid Factory: The Japanese Production System in the United States*. New York: Oxford University Press.

Abo, T. (1996) The Japanese production system: the process of adaptation to national settings, in R. Boyer and D. Drache (eds), *States Against Markets: The Limits of Globalization*. London: Routledge. Chapter 5.

Abo, T. (2000) Spontaneous integration in Japan and East Asia: development, crisis, and beyond, in G.L. Clark, M.P. Feldman and M.S. Gertler (eds), *The Oxford Handbook of Economic Geography*. Oxford: Oxford University Press. Chapter 31.

Agnew, J. and Corbridge, S. (1995) *Mastering Space*. London: Routledge.

Alexander, N. (1997) *International Retailing*. Oxford: Blackwell.

Allen, J. (1988) Service industries: uneven development and uneven knowledge, *Area*, 20: 15–22.

Amin, A. (ed.) (1994) *Post-Fordism: A Reader*. Oxford: Blackwell.

Amin, A. (2000) The European Union as more than a triad market for national economic spaces, in G.L. Clark, M.P. Feldman and M.S. Gertler (eds), *The Oxford Handbook of Economic Geography*. Oxford: Oxford University Press. Chapter 33.

Amin, A. and Robins, K. (1990) The re-emergence of regional economies? The mythical geography of flexible accumulation, *Environment and Planning, D: Society and Space*, 8: 7–34.

Amin, A. and Thrift, N.J. (1992) Neo-Marshallian nodes in global networks, *International Journal of Urban and Regional Research*, 16: 571–587.

Amin, A. and Thrift, N.J. (1994) Living in the global, in A. Amin and N.J. Thrift (eds), *Globalization, Institutions and Regional Development in Europe*. Oxford: Oxford University Press. Chapter 1.

Amsden, A. (1989) *Asia's Next Giant: South Korea and Late Industrialization*. Oxford: Oxford University Press.

Anderson, M. (1995) The role of collaborative integration in industrial organization: observations from the Canadian aerospace industry, *Economic Geography*, 71: 55–78.

Anderson, S. and Cavanagh, J. (2000) *Top 200: The Rise of Corporate Global Power*. Washington, DC: Institute for Policy Studies.

Angel, D. (1994) *Restructuring for Innovation: The Remaking of the US Semiconductor Industry*. New York: Guilford.

Aoki, M. (1984) Aspects of the Japanese firm, in M. Aoki (ed.), *The Economic Analysis of the Japanese Firm*. Dordrecht: North Holland. pp. 3–46.

Aoki, A. and Tachiki, D. (1992) Overseas Japanese business operations: the emerging role of regional headquarters, *RIM Pacific Business and Industries*, 1: 28–39.

Archibugi, D., Howells, J. and Michie, J. (eds) (1999) *Innovation Policy in a Global Economy*. Cambridge: Cambridge University Press.

Archibugi, D. and Michie, J. (eds) (1997) *Technology, Globalization, and Economic Performance*. Cambridge: Cambridge University Press.

Badaracco, J.L. Jr (1991) The boundaries of the firm, in A. Etzioni and P.R. Lawrence (eds), *Socio-Economics: Towards a New Synthesis*. Armonk: M.E. Sharpe. pp. 293–327.

Bailey, P., Parisotto, A. and Renshaw, G. (eds) (1993) *Multinationals and Employment: The Global Economy of the 1990s*. Geneva: ILO.

Bair, J. and Gereffi, G. (2001) Local clusters in global chains: the causes and consequences of export dynamism in Torreón's blue jeans industry, *World Development*, 29: 1885–1903.

Barnet, R.J. and Cavanagh, J. (1994) *Global Dreams: Imperial Corporations and the New World Order*. New York: Simon & Schuster.

Barnet, R.J. and Muller, R.E. (1975) *Global Reach: The Power of the Multinational Corporation*. London: Cape.

Bartlett, C.A. and Ghoshal, S. (1998) *Managing Across Borders: The Transnational Solution*. New York: Random House.

Batty, M. and Barr, R. (1994) The electronic frontier: exploring and mapping cyberspace, *Futures*, 26: 699–712.

Baylin, F. (1996) *World Satellite Yearbook*, 4th edn. Boulder, CO: Baylin Publications.

Beaverstock, J.V., Hoyler, M., Pain, K. and Taylor, P.J. (2001) *Comparing London and Frankfurt as World Cities: A Relational Study of Contemporary Urban Change*. London: Anglo-German Foundation for the Study of Industrial Society.

Behrman, J.N. and Fischer, W.A. (1980) *Overseas R&D Activities of Transnational Companies*. Cambridge, MA: Oelgeschlager, Gunn & Hain.

Beechler, S.L. and Bird, A. (eds) (1999) *Japanese Multinationals Abroad: Individual and Organizational Learning*. New York: Oxford University Press.

Bell, D. (1974) *The Coming of Post-Industrial Society*. London: Heinemann.

Bello, W. and Rosenfeld, S. (1990) *Dragons in Distress: Asian Miracle Economies in Crisis*. San Francisco: Institute for Food and Development Policy.

Benewick, R. and Wingrove, P. (eds) (1995) *China in the 1990s*. London: Macmillan.

Berger, S. and Dore, R. (eds) (1996) *National Diversity and Global Capitalism*. Ithaca, NY: Cornell University Press.

Bernard, M. and Ravenhill, J. (1995) Beyond product cycles and flying geese, *World Politics*, 47: 171–209.

Bertrand, O. and Noyelle, T. (1988) *Human Resources and Corporate Strategy: Technological Change in Banks and Insurance Companies*. Paris: OECD.

Beukema, L. and Coenen, H. (1999) Global logistic chains: the increasing importance of local labour relations, in P. Leisenk (ed.), *Globalization and Labour Relations*. Cheltenham: Edward Elgar. Chapter 7.

Bircham, E. and Charlton, J. (eds) (2001) *Anti-Capitalism: A Guide to the Movement*. London: Bookmarks Publications.

Birkinshaw, J. and Morrison, A.J. (1995) Configurations of strategy and structure in subsidiaries of multinational corporations, *Journal of International Business Studies*, 26: 729–755.

Blanc, H. and Sierra, C. (1999) The internationalisation of R&D by multinationals: a trade-off between external and internal proximity, *Cambridge Journal of Economics*, 23: 187–206.

Borrus, M. (2000) The resurgence of US electronics: Asian production networks and the rise of Wintelism, in M. Borrus, D. Ernst and S. Haggard (eds), *International Production Networks in Asia: Rivalry or Riches?* London: Routledge. Chapter 3.

Borrus, M., Ernst, D. and Haggard, S. (eds) (2000) *International Production Networks in Asia: Rivalry or Riches?* London: Routledge.

Braithwaite, J. and Drahos, P. (2000) *Global Business Regulation*. Cambridge: Cambridge University Press.

Braudel, F. (1984) *Civilization and Capitalism, 15th–18th Centuries*, 3 vols. London: Collins.

Breathnach, P. (2000) Globalization, information technology and the emergence of niche transnational cities: the growth of the call centre sector in Dublin, *Geoforum*, 31: 477–485.

Brenner, N. (1998) Between fixity and motion: accumulation, territorial organization and the historical geography of spatial scales, *Environment and Planning D: Society and Space*, 16: 459–481.

Brenner, N. (2001) The limits to scale? Methodological reflections on scalar structuration, *Progress in Human Geography*, 25: 591–614.

Brohman, J. (1996) Postwar development in the Asian NICs: Does the neoliberal model fit reality? *Economic Geography*, 72: 107–131.

Brown, L.R. (ed.) (2002) *State of the World, 2001*. London: Earthscan.

Brunn, S.D. and Dodge, M. (2001) Mapping the 'worlds' of the World Wide Web, *American Behavioral Scientist*, 44: 1717–1739.

Brunn, S.D. and Leinbach, T.R. (eds) (1991) *Collapsing Space and Time: Geographic Aspects of Communication and Information*. New York: HarperCollins.

Buckley, P.J. and Casson, M. (1976) *The Future of the Multinational Enterprise*. London: Macmillan.

Budd, L. (1999) Globalization and the crisis of territorial embeddedness of international financial markets, in R. Martin (ed.), *Money and the Space-Economy*. Chichester: Wiley. Chapter 6.

Bunnell, T.G. and Coe, N.M. (2001) Spaces and scales of innovation, *Progress in Human Geography*, 25: 569–589.

Cable, V. (1999) *Globalization and Global Governance*. London: Royal Institute of International Affairs.

Cable, V. and Henderson, D. (eds) (1994) *Trade Blocs? The Future of Regional Integration*. London: Royal Institute of International Affairs.

Cairncross, F. (1992) *Costing the Earth*. Boston, MA: Harvard Business School Press.

Cairncross, F. (1997) *The Death of Distance: How the Communications Revolution Will Change Our Lives*. London: Orion Business Books.

Cantwell, J. (1997) The globalization of technology: what remains of the product cycle model?, in D. Archibugi and J. Michie (eds), *Technology, Gobalization, and Economic Performance*. Cambridge: Cambridge University Press. Chapter 8.

Cantwell, J. and Iammarino, S. (2000) Multinational corporations and the location of technological innovation in the UK regions, *Regional Studies*, 34: 317–332.

Casson, M. (ed.) (1983) *The Growth of International Business*. London: Allen & Unwin.

Castells, M. (1989) *The Informational City*. Oxford: Blackwell.

Castells, M. (1996) *The Information Age: Economy, Society and Culture*, 3 vols. Oxford: Blackwell.

Castells, M. and Hall, P. (1994) *Technopoles of the World: The Making of 21st Century Industrial Complexes*. London: Routledge.

Castles, S. and Miller, M.J. (1993) *The Age of Migration: International Population Movements in the Modern World*. New York: Guilford Press.

Cecchini, P. (1988) *The European Challenge 1992: The Benefits of a Single Market*. Aldershot: Wildwood House.

Cerny, P.G. (1991) The limits of deregulation: transnational interpenetrations and policy change, *European Journal of Political Research*, 19: 173–196.

Chang, H-J. (1998a) South Korea: the misunderstood crisis, in K.S. Jomo (ed.), *Tigers in Trouble: Financial Governance, Liberalization and Crises in East Asia*. London: Zed Books. Chapter 10.

Chang, H-J. (1998b) Transnational corporations and strategic industrial policy, in R. Kozul-Wright and R. Rowthorn (eds), *Transnational Corporations and the Global Economy*. London: Macmillan. Chapter 7.

Chant, S. and McIlwaine, C. (1995) Gender and export manufacturing in the Philippines: continuity or change in female employment? The case of the Mactan Export Processing Zone, *Gender, Place and Culture*, 2: 147–176.

Chesnais, F. (1986) Science, technology and competitiveness, *Science Technology Industry Review*, 1: 85–129.

Clark, G.L. (1994) Strategy and structure: corporate restructuring and the scope and characteristics of sunk costs, *Environment and Planning, A*, 26: 9–32.

Clark, G.L. and O'Connor, K. (1997) The informational content of financial products and the spatial structure of the global finance industry, in K. Cox (ed.), *Spaces of Globalization: Reasserting the Power of the Local*. New York: Guilford. Chapter 4.

Clark, G.L. and Wrigley, N. (1995) Sunk costs: a framework for economic geography, *Transactions of the Institute of British Geographers*, 20: 204–223.

Cline, W.R. (1987) *The Future of World Trade in Textiles and Apparel*. Washington, DC: Institute for International Economics.

Cline, W.R. (1997) *Trade and Income Distribution*. Washington, DC: Institute for International Economics.

Coase, R. (1937) The nature of the firm, *Economica*, 4: 386–405.

Cohen, B.J. (2001) Electronic money: new day or false dawn? *Review of International Political Economy*, 8: 197–225.

Cohen, D. (1998) *The Wealth of the World and the Poverty of Nations*. Cambridge, MA: MIT Press.

Cohen, R.B. (1981) The new international division of labour, multinational corporations and the urban hierarchy, in M. Dear and A.J. Scott (eds), *Urbanization and Urban Planning in Capitalist Society*. London: Methuen. Chapter 12.

Cohen, S.S. and Zysman, J. (1987) *Manufacturing Matters: The Myth of the Post-Industrial Economy*. New York: Basic Books.

Commission on Global Governance (1995) *Our Global Neighbourhood*. New York: Oxford University Press.

Corbridge, S., Martin, R. and Thrift, N.J. (eds), (1994) *Money, Power and Space*. Oxford: Blackwell.

Crane, G.T. (1990) *The Political Economy of China's Special Economic Zones*. Armonk: M.E. Sharpe.

Crook, C. (2001) Globalization and its critics: a survey of globalization, *The Economist*, 29 September.

Czaban, L. and Henderson, J. (1998) Globalization, institutional legacies and industrial transformation in Eastern Europe, *Economy and Society*, 27: 585–613.

Daly, H.E. (1993) The perils of free trade, *Scientific American*, November: 24–29.

D'Aveni, R.A. (1994) *Hypercompetition: Managing the Dynamics of Strategic Manoeuvring*. New York: The Free Press.

Davis, E. and Smailes, C. (1989) The integration of European financial services, in E. Davis et al. (eds), *1992: Myths and Realities*. London: London Business School. Chapter 5.

Department of International Development (2000) *Eliminating World Poverty: Making Globalization Work for the Poor*. London: DoID.

Deyo, F.C. (1992) The political economy of social policy formation: East Asia's newly industrialized countries, in R.P. Appelbaum and J. Henderson (eds), *States and Development in the Asian Pacific Rim*. London: Sage. Chapter 11.

Dicken, P. (2000) Places and flows: situating international investment, in G.L. Clark, M.P. Feldman and M.S. Gertler (eds), *The Oxford Handbook of Economic Geography*. Oxford: Oxford University Press. Chapter 14.

Dicken, P., Forsgren, M. and Malmberg, A. (1994) The local embeddedness of transnational corporations, in A. Amin and N.J. Thrift (eds), *Globalization, Institutions, and Regional Development in Europe*. Oxford: Oxford University Press. Chapter 2.

Dicken, P. and Hassler, M. (2000) Organizing the Indonesian clothing industry in the global economy: the role of business networks, *Environment and Planning A*, 32: 263–280.

Dicken P., Kelly P.F., Olds, K. and Yeung, H.W-c. (2001) Chains and networks, territories and scales: towards a relational framework for analysing the global economy, *Global Networks*, 1: 99–123.

Dicken, P. and Lloyd, P.E. (1990) *Location in Space: Theoretical Perspectives in Economic Geography*. 3rd edn. New York: Harper & Row.

Dicken, P. and Malmberg, A. (2001) Firms in territories: a relational perspective, *Economic Geography*, 77: 345–363.

Dicken, P. and Miyamachi, Y. (1998) 'From noodles to satellites': the changing geography of the Japanese *sogo shosha*, *Transactions of the Institute of British Geographers*, 23: 55–78.

Dicken, P. and Thrift, N.J. (1992) The organization of production and the production of organization: why business enterprises matter in the study of geographical industrialization, *Transactions of the Institute of British Geographers*, 17: 279–291.

Dicken, P., Peck, J.A. and Tickell, A. (1997) Unpacking the global, in R. Lee and J. Wills (eds), *Geographies of Economies*. London: Edward Arnold. Chapter 12.

Dicken, P. and Yeung, H.W-c. (1999) Investing in the future: East and Southeast Asian firms in the global economy, in K. Olds, P. Dicken, P.F. Kelly, L. Kong and H.W-c. Yeung (eds), *Globalization and the Asia Pacific: Contested Territories*. London: Routledge. Chapter 7.

Dicken, P., Tickell, A. and Yeung, H.Y-c. (1997) Putting Japanese investment in Europe in its place, *Area*, 29: 200–212.

Donaghu, M.T. and Barff, R. (1990) Nike just did it: international subcontracting and flexibility in athletic footwear production, *Regional Studies*, 24: 537–552.

Dore, R. (1986) *Flexible Rigidities: Industrial Policy and Structural Adjustment in the Japanese Economy, 1970–1980*. Stanford, CA: Stanford University Press.

Doremus, P.N., Keller, W.W., Pauly, L.W. and Reich, S. (1998) *The Myth of the Global Corporation*. Princeton, NJ: Princeton University Press.

Dosi, G. (1983) Semiconductors: Europe's precarious survival in high technology, in G. Shepherd, F. Duchene and C. Saunders (eds), *Europe's Industries: Public and Private Strategies for Change*. London: Pinter. Chapter 9.

Dosi, G. (1999) Some notes on national systems of innovation and production, and their implications for economic analysis, in D. Archibugi, J. Howells and J. Michie (eds), *Innovation Policy in a Global Economy*. Cambridge: Cambridge University Press. Chapter 3.

Dosi, G., Freeman, C., Nelson, R., Silverberg, G. and Soete, L. (eds) (1988) *Technical Change and Economic Theory*. London: Pinter.

Douglass, M. (1994) The 'developmental state' and the newly industrialized economies of Asia, *Environment and Planning, A*, 26: 543–566.

Dow, S.C. (1999) The stages of banking development and the spatial evolution of financial systems, in R. Martin (ed.), *Money and the Space-Economy*. Chichester: Wiley. Chapter 2.

Doz, Y. (1986a) Government polices and global industries, in M.E. Porter (ed.), *Competition in Global Industries*. Boston, MA: Harvard Business School. Chapter 7.

Doz, Y. (1986b) *Strategic Management in Multinational Companies*. Oxford: Pergamon.

Dunford, M. and Kafkalas, G. (1992) The global–local interplay, corporate geographies and spatial development strategies in Europe, in M. Dunford and G. Kafkalas (eds), *Cities and Regions in the New Europe: The Global–Local Interplay and Spatial Development Strategies*. London: Belhaven Press. Chapter 1.

Dunford, M. and Smith, A. (2000) Catching up or falling behind? Economic performance and regional trajectories in the 'new Europe', *Economic Geography*, 76: 169–195.

Dunning, J.H. (1979) Explaining changing patterns of international production: in defence of the eclectic theory, *Oxford Bulletin of Economics and Statistics*, 41: 269–296.

Dunning, J.H. (1980) Towards an eclectic theory of international production: some empirical tests, *Journal of International Business Studies*, 11: 9–31.

Dunning, J. H. (1992) The competitive advantages of countries and the activities of transnational corporations, *Transnational Corporations*, 1: 135–168.

Dunning, J.H. (1993) *Multinational Enterprises and the Global Economy*. Reading, MA: Addison–Wesley.

Dunning, J.H. (ed.) (2000) *Regions, Globalization and the Knowledge-Based Economy*. Oxford: Oxford University Press.

Eatwell, J. and Taylor, L. (2000) *Global Finance at Risk: The Case for International Regulation*. Cambridge: Polity Press.

Eden, L. and Molot, M.A. (1993) Insiders and outsiders: defining 'who is us' in the North American automobile industry, *Transnational Corporations*, 2, 3: 31–64.

Eden, L. and Monteils, A. (2000) Regional integration: NAFTA and the reconfiguration of North American industry, in J.H. Dunning (ed.), *Regions, Globalization and the Knowledge-Based Economy*. Oxford: Oxford University Press. Chapter 7.

Eichengreen, B. (1996) *Globalizing Capital: A History of the International Monetary System*. Princeton, NJ: Princeton University Press.

Elliott, L. (2000) Free trade, no choice, in B. Gunnell and D. Timms (eds), *After Seattle: Globalization and its Discontents*. London: Catalyst. Chapter 2.

Elliott, L. (2002) Morals of the brothel, *The Guardian*, 15 April 2002.

Ernst, D. and Kim, L. (2001) Global production networks, knowledge diffusion, and local capability formation, *Paper presented to DRUID Conference, Aalborg, Denmark*.

European Commission (1996) *Employment in Europe*. Brussels: European Commission.

Faux, J. and Mishel, L. (2000) Inequality and the global economy, in W. Hutton and A. Giddens (eds), *On the Edge: Living with Global Capitalism*. London: Jonathan Cape. pp. 93–111.

FIET (1996) *A Social Dimension to Globalization*. Geneva: FIET (International Federation of Commercial, Clerical, Professional and Technical Employees).

Fine, B. and Leopold, E. (1993) *The World of Consumption*. London: Routledge.

Freeman, C. (1982) *The Economics of Industrial Innovation*. London: Pinter.

Freeman, C. (1987) The challenge of new technologies, in OECD, *Interdependence and Cooperation in Tomorrow's World*. Paris: OECD. pp. 123–156.

Freeman, C. (1988) Introduction, in G. Dosi, C. Freeman, R. Nelson, G. Silverberg and L. Soete (eds), *Technical Change and Economic Theory*. London: Pinter. Chapter 1.

Freeman, C. (1997) The 'national system of innovation' in historical perspective, in D. Archibugi and J. Michie (eds), *Technology, Globalization and Economic Performance*. Cambridge: Cambridge University Press. Chapter 2.

Freeman, C., Clark, J. and Soete, L. (1982) *Unemployment and Technical Change*. London: Pinter.

Freeman, C. and Perez, C. (1988) Structural crises of adjustment, business cycles and investment behaviour, in G. Dosi, C. Freeman, R. Nelson, G. Silverberg and L. Soete (eds), *Technical Change and Economic Theory*. London: Pinter. Chapter 3.

Friedman, T. (1999) *The Lexus and the Olive Tree*. New York: HarperCollins.

Friedmann, J. (1986) The world city hypothesis, *Development and Change*, 17: 69–83.

Fröbel, F., Heinrichs, J. and Kreye, O. (1980) *The New International Division of Labour*. Cambridge: Cambridge University Press.

Fruin, W.M. (1992) *The Japanese Enterprise System*. Oxford: Clarendon Press.

Fuentes, A. and Ehrenreich, B. (1983) *Women in the Global Factory*. Boston, MA: South End Press.

Fuentes, N.A., Alegria, T., Brannon, J.T., James, D.D. and Lucker, G.W. (1993) Local sourcing and indirect employment: multinational enterprises in northern Mexico, in P. Bailey, A. Parisotto and G. Renshaw (eds), *Multinationals and Employment: The Global Economy of the 1990s*. Geneva: ILO. Chapter 6.

Fujita, M. and Ishigaki, K. (1986) The internationalisation of commercial banking, in M.J. Taylor and N.J. Thrift (eds), *Multinationals and the Restructuring of the World Economy*. London: Croom Helm. Chapter 7.

Gabriel, P. (1966) The investment in the LDC: asset with a fixed maturity, *Columbia Journal of World Business*, 1: 113–120.

Gamble, A. and Payne, A. (eds) (1996) *Regionalism and World Order*. London: Macmillan.

Garrett, G. (1998) Global markets and national politics: collision course or virtuous circle? *International Organization*, 52: 787–824.

Gereffi, G. (1990) Paths of industrialization: an overview, in G. Gereffi and D.L. Wyman (eds), *Manufacturing Miracles: Paths of Industrialization in Latin America and East Asia*. Princeton, NJ: Princeton University Press. Chapter 1.

Gereffi, G. (1994) The organization of buyer-driven global commodity chains: how US retailers shape overseas production networks, in G. Gereffi and M. Korzeniewicz (eds), *Commodity Chains and Global Capitalism*. Westport: Praeger. Chapter 5.

Gereffi, G. (1996a) Commodity chains and regional divisions of labor in East Asia, *Journal of Asian Business*, 12: 75–112.

Gereffi, G. (1996b) Global commodity chains: new forms of coordination and control among nations and firms in international industries, *Competition & Change*, 1: 427–439.

Gereffi, G. (1999) International trade and industrial upgrading in the apparel commodity chain, *Journal of International Economics*, 48: 37–70.

Gereffi, G. (2001) Shifting governance structures in global commodity chains, with special reference to the Internet, *American Behavioral Scientist*, 44: 1616–1637.

Gereffi, G., Garcia-Johnson, R. and Sasser, E. (2001) The NGO-industrial complex, *Foreign Policy*, July–August: 56–65.

Gereffi, G. and Wyman, D.L. (eds) (1990) *Manufacturing Miracles: Paths of Industrialization in Latin America and East Asia*. Princeton, NJ: Princeton University Press.

Gerlach, M. (1992) *Alliance Capitalism: The Social Organization of Japanese Business*. Berkeley, CA: University of California Press.

Gertler, M.S. (1988) The limits to flexibility: comments on the post-Fordist vision of production and its geography, *Transactions of the Institute of British Geographers*, 13: 419–432.

Gertler, M.S. (2001) Best practice? Geography, learning and the industrial limits to strong convergence, *Journal of Economic Geography*, 1: 5–26.

Gibb, R. and Michalak, W. (eds) (1994) *Continental Trading Blocs: The Growth of Regionalism in the World Economy*. Chichester: Wiley.

Giddens, A. (1999) *Runaway World: How Globalization is Shaping our Lives*. London: Profile Books.

Gilbert, A. and Gugler, J. (1982) *Cities, Poverty and Development: Urbanization in the Third World*. Oxford: Oxford University Press.

Gilpin, R. (2000) *The Challenge of Global Capitalism: The World Economy in the 21st Century*. Princeton, NJ: Princeton University Press.

Gilpin, R. (2001) *Global Political Economy: Understanding the International Economic Order*. Princeton, NJ: Princeton University Press.

Glasmeier, A. (2002) One nation pulling apart: the basis of persistent poverty in the United States, *Progress in Human Geography*, 26: 155–174.

Glasmeier, A., Thompson, J.W. and Kays, A.J. (1993) The geography of trade policy: trade regimes and location decisions in the textile and apparel complex, *Transactions of the Institute of British Geographers*, 18: 19–35.

Glassner, M.I. (1993) *Political Geography*. New York: Wiley.

Glyn, A. and Sutcliffe, B. (1992) Global but leaderless? The new capitalist order, in R. Miliband and L. Panitch (eds), *New World Order: The Socialist Register*. London: Merlin Press. pp. 76–95.

Gomes-Casseres, B. (1996) *The Alliance Revolution: The New Shape of Business Rivalry*. Cambridge, MA: Harvard University Press.

Gordon, D.M. (1988) The global economy: new edifice or crumbling foundations? *New Left Review*, 168: 24–64.

Govindarajan, V. and Gupta, A. (2000) Analysis of the emerging global arena, *European Management Journal*, 18: 274–284.

Grabher, G. (1995) The elegance of incoherence: economic transformation in East Germany and Hungary, in E. Dietrich et al. (eds), *Industrial Transformation in Europe*. London: Sage.

Graham, S. and Marvin, S. (1996) *Telecommunications and the City: Electronic Spaces, Urban Places*. London: Routledge.

Granovetter, M. (1985) Economic action and social structure: the problem of embeddedness, *American Journal of Sociology*, 91: 481–510.

Granovetter, M. and Swedberg, R. (eds) (1992) *The Sociology of Economic Life*. Boulder, CO: Westview Press.

Greider, W. (1997) *One World, Ready or Not: The Manic Logic of Global Capitalism*. London: Penguin.

Gretschmann, K. (1994) Germany in the global economy of the 1990s: from player to pawn? in R. Stubbs and G.R.D. Underhill (eds), *Political Economy and the Changing Global Order*. London: Macmillan. Chapter 29.

Grubel, H.G. (1989) Multinational banking, in P. Enderwick (ed.), *Multinational Service Firms*. London: Routledge. Chapter 3.

Grunsven, L. van, Egeraat, C. van and Meijsen, S. (1995) New manufacturing establishments and regional economy in Johor: production linkages, employment and labour fields, *Department of Geography of Developing Countries, University of Utrecht Report Series*.

Guibernau, M. (1999) *Nations Without States: Political Communities in a Global Age*. Cambridge: Polity Press.

Guisinger, S. (1985) *Investment Incentives and Performance Requirements*. New York: Praeger.

Hachten, W.A. (1974) Mass media in Africa, in A. Wells (ed.), *Mass Communications: A World View*. Palo Alto, CA: National Press Books.

Hagedoorn, J. and Schakenraad, J. (1990) Inter-firm partnerships and cooperative strategies in core technologies, in B. Dankbaar, J. Groenewegen and H. Schenk (eds), *Perspectives in Industrial Economics*. Dordrecht: Kluwer. pp. 47–65.

Haggard, S. (1995) *Developing Nations and the Politics of Global Integration*. Washington, DC: The Brookings Institution.

Haggard, S. and MacIntyre, A. (1998) The political economy of the Asian economic crisis, *Review of International Political Economy*, 5: 381–392.

Haig, R.M. (1926) Toward an understanding of the metropolis, *Quarterly Journal of Economics*, 40: 421–433.

Hall, P. and Preston, P. (1988) *The Carrier Wave: New Information Technology and the Geography of Innovation, 1846–2003*. London: Unwin Hyman.

Hall, P.A. and Soskice, D. (2001) *Varieties of Capitalism: The Institutional Foundations of Comparative Advantage*. Oxford: Oxford University Press.

Ham, R.M., Linden, G. and Appleyard, M.M. (1998) The evolving role of semiconductor consortia in the United States and Japan, *California Management Review*, 41, 1: 137–163.

Hamill, J. (1993) Employment effects of the changing strategies of multinational enterprises, in P. Bailey, A. Parisotto and G. Renshaw (eds), *Multinationals and Employment: The Global Economy of the 1990s*. Geneva: ILO. Chapter 3.

Hamilton, A. (1986) *The Financial Revolution*. Harmondsworth: Penguin.

Hamilton, F.E.I. (1979) *The Planned Economies*. London: Macmillan.

Hamilton, G.G. and Feenstra, R.C. (1998) Varieties of hierarchies and markets: an introduction, in G. Dosi, D.J. Teece and J. Chytry (eds), *Technology, Organization and Competitiveness*. Oxford: Oxford University Press. pp. 105–146.

Harrison, B. (1997) *Lean and Mean: The Changing Landscape of Corporate Power in the Age of Flexibility*. New York: Guilford.

Harvey, D. (1982) *The Limits to Capital*. Oxford: Blackwell.

Harvey, N. (ed.) (1993) *Mexico: Dilemmas of Transition*. London: Institute of Latin American Studies.

Harzing, A-W. (2000) An empirical analysis and extension of the Bartlett and Ghoshal typology of multinational companies, *Journal of International Business Studies*, 31: 101–120.

Hawkins, R.G. (1972) Job displacement and the multinational firm: a methodological review, *Center for Multinational Studies, Occasional Paper 3*. Washington, DC.

Hedlund, G. (1986) The hypermodern MNC – a heterarchy, *Human Resource Management*, 25: 9–35.

Heenan, D.A. and Perlmutter, H. (1979) *Multinational Organizational Development: A Social Architecture Perspective*. Reading, MA: Addison–Wesley.

Held, D., McGrew, A., Goldblatt, D. and Perraton, J. (1999) *Global Transformations: Politics, Economics and Culture*. Cambridge: Polity Press.

Helou, A. (1991) The nature and competitiveness of Japan's *keiretsu*, *Journal of World Trade*, 25: 99–131.

Henderson, J. (1998) Danger and opportunity in the Asia–Pacific, in G. Thompson (ed.), *Economic Dynamism in the Asia–Pacific*. London: Routledge. Chapter 14.

Henderson, J. and Castells, M. (eds) (1987) *Global Restructuring and Territorial Development*. London: Sage.

Henderson, J., Dicken, P., Hess, M., Coe, N. and Yeung, H.W-c. (2002) Global production networks and the analysis of economic development, *Review of International Political Economy*.

Herod, A. (1997) From a geography of labor to a labor geography: rethinking conceptions of labor in economic geography, *Antipode*, 29: 1–31.

Herod, A. (2001) *Labor Geographies: Workers and the Landscape of Capitalism*. New York: Guilford.

Higgott, R. (1993) Competing theoretical approaches to international cooperation: implications for the Asia Pacific, in R. Higgott, R. Leaver and J. Ravenhill (eds), *Pacific Economic Relations in the 1990s*. London: Allen & Unwin. Chapter 14.

Higgott, R. (1999) The political economy of globalization in East Asia: the salience of 'region building', in K. Olds, P. Dicken, P.F. Kelly, L. Kong and H.W-c. Yeung (eds), *Globalization and the Asia–Pacific: Contested Territories*. London: Routledge. Chapter 6.

Hill, H. (1993) Employment and multinational enterprises in Indonesia, in P. Bailey, A. Parisotto and G. Renshaw (eds), *Multinationals and Employment: The Global Economy of the 1990s*. Geneva: ILO. Chapter 7.

Hirsch, S. (1967) *Location of Industry and International Competitiveness*. Oxford: Clarendon Press.

Hirsch, S. (1972) The United States electronics industry in international trade, in L.T. Wells Jr (ed.), *The Product Life Cycle and International Trade*. Boston, MA: Harvard Business School. pp. 39–54.

Hirst, P. and Thompson, G. (1992) The problem of 'globalization': international economic relations, national economic management and the formation of trading blocs, *Economy and Society*, 21: 357–396.

Hirst, P. and Thompson, G. (1999) *Globalization in Question: The International Economy and the Possibilities of Governance*, 2nd edn. Cambridge: Polity Press.

Hoekman, B. and Kostecki, M. (1995) *The Political Economy of the World Trading System: From GATT to WTO*. Oxford: Oxford University Press.

Hofstede, G. (1980) *Culture's Consequences*. London: Sage

Hofstede, G. (1983) The cultural relativity of organizational practices and theories, *Journal of International Business Studies*, Fall: 75–89.

Hollingsworth, J.R. (1997) Continuities and changes in social systems of production: the cases of Japan, Germany, and the United States, in J.R. Hollingsworth and R. Boyer (eds), *Contemporary Capitalism: The Embeddedness of Institutions*. Cambridge: Cambridge University Press. Chapter 9.

Hollingsworth, J.R. and Boyer, R. (eds) (1997) *Contemporary Capitalism: The Embeddedness of Institutions*. Cambridge: Cambridge University Press.

Hollingsworth, J.R., and Boyer, R. (1997) Coordination of economic activities and social systems of production, in J.R. Hollingsworth and R. Boyer (eds), *Contemporary Capitalism: The Embeddedness of Institutions*. Cambridge: Cambridge University Press. Chapter 1.

Holmes, J. (1992) The continental integration of the North American automobile industry; from the Auto Pact to the FTA, *Environment and Planning A*, 24: 95–120.

Holmes, J. (2000) Regional economic integration in North America, in G.L. Clark, M.P. Feldman and M.S. Gertler (eds), *The Oxford Handbook of Economic Geography*. Oxford: Oxford University Press. Chapter 32.

Hood, N. and Young, S. (1982) US multinational R&D: corporate strategies and policy implications for the UK, *Multinational Business*, 2: 10–23.

Hood, N. and Young, S. (2000) Globalization, corporate strategies, and business services, in N. Hood and S. Young (eds), *The Globalization of Multinational Enterprise Activity and Economic Development.* London: Macmillan. Chapter 4.

Hotz-Hart, B. (2000) Innovation networks, regions, and globalization, in G.L. Clark, M.P. Feldman and M.S. Gertler (eds), *The Oxford Handbook of Economic Geography.* Oxford: Oxford University Press. Chapter 22.

Howells, J. (2000) Knowledge, innovation and location, in J.R. Bryson, P.W. Daniels, N. Henry and J. Pollard (eds), *Knowledge, Space, Economy.* London: Routledge. Chapter 4.

Hu, Y-S. (1992) Global firms are national firms with international operations, *California Management Review,* 34: 107–126.

Hu, Y-S. (1995) The international transferability of the firm's advantages, *California Management Review,* 37: 73–88.

Hudson, A. (2000) Offshoreness, globalization and sovereignty: a postmodern geo-political economy? *Transactions of the Institute of British Geographers,* 25: 269–283.

Hudson, R. (2001) *Producing Places.* New York: Guilford.

Hudson, R. and Schamp, E. (eds) (1995) *Towards a New Map of Automobile Manufacturing in Europe?* Berlin: Springer.

Humbert, M. (1994) Strategic industrial policies in a global industrial system, *Review of International Political Economy,* 1: 445–464.

Humphrey, J. and Schmitz, H. (2001) Governance in global value chains, *IDS Bulletin,* 32, 3: 19–29.

Hymer, S.H. (1972) The multinational corporation and the law of uneven development, in J.N. Bhagwati (ed.), *Economics and World Order.* London: Macmillan. pp. 113–140.

Hymer, S.H. (1976) *The International Operations of National Firms: A Study of Direct Foreign Investment.* Cambridge, MA: MIT Press.

ICFTU (1998) *Migration and Globalization: The New Slaves.* Brussels: ICFTU (International Confederation of Free Trade Unions).

ILO (1981a) *Employment Effects of Multinational Enterprises in Developing Countries.* Geneva: ILO.

ILO (1981b) *Employment Effects of Multinational Enterprises in Industrialized Countries.* Geneva: ILO.

ILO (1984) *Technology Choice and Employment Generation by Multinational Enterprises* in *Developing Countries.* Geneva: ILO.

ILO (1988) *Economic and Social Effects of Multinational Enterprises in Export Processing Zones.* Geneva: ILO.

ILO (1996a) *Child Labour: What Is to Be Done?* Geneva: ILO.

ILO (1996b) *Globalization of the Footwear, Textiles, and Clothing Industries.* Geneva: ILO.

ILO (1997) *World Employment, 1996/97: National Policies in a Global Context.* Geneva: ILO.

ILO (1998) *Labour and Social Issues Relating to Export Processing Zones.* Geneva: ILO.

ILO (2001) *World Employment Report 2001: Life at Work in the Information Economy.* Geneva: ILO.

Ito, T. (2001) Growth, crisis, and the future of economic recovery in East Asia, in J.E. Stiglitz and S. Yusuf (eds), *Rethinking the East Asia Miracle.* New York: Oxford University Press. Chapter 2.

Jarillo, J.C. (1993) *Strategic Networks: Creating the Borderless Organization.* London: Butterworth–Heinemann.

Jessop, B. (1994) Post-Fordism and the State, in A. Amin (ed.), *Post-Fordism: A Reader.* Oxford: Blackwell. Chapter 8.

Johnson, C. (1982) *MITI and the Japanese Economic Miracle: The Growth of Industrial Policy, 1925–1975.* Stanford, CA: Stanford University Press.

Johnson, C. (1985) The institutional foundations of Japanese industrial policy, *California Management Review,* XXVII: 59–69.

Jomo, K.S. (1998) *Tigers in Trouble: Financial Governance, Liberalisation and Crises in East Asia.* London: Zed Books.

Jones, H.R. (1990) *A Population Geography,* 2nd edn. London: Paul Chapman.

Kang, N-H. and Johansson, S. (2000) Cross-border mergers and acquisitions: their role in industrial globalization, *OECD Working Paper 2000/1.*

Kang, N-H. and Sakai, K. (2000) International strategic alliances: their role in industrial globalization, *OECD STI Working Paper 2000/5.*

Kao, J. (1993) The worldwide web of Chinese business, *Harvard Business Review,* March–April: 24–36.

Kaplinsky, R. (1999) 'If you want to get somewhere else, you must run at least twice as fast as that!': the roots of the East Asian crisis, *Competition and Change*, 4: 1–30.

Kaplinsky, R. (2001) Is globalization all it is cracked up to be? *Review of International Political Economy*, 8: 45–65.

Kapstein, E. (1999) *Sharing the Wealth: Workers and the World Economy*. New York: W.W. Norton.

Kapstein, E. (2000) Winners and losers in the global economy, *International Organization*, 54: 359–384.

Kelly, P.F. (1999) The geographies and politics of globalization, *Progress in Human Geography*, 23: 379–400.

Kelly, R. (1995) Derivatives: a growing threat to the international financial system, in J. Michie and J. Grieve-Smith (eds), *Managing the Global Economy*. Oxford: Oxford University Press. Chapter 9.

Kennedy, P. (2002) Global challenges at the beginning of the twenty-first century, in P. Kennedy, D. Messner and F. Nuscheler (eds), *Global Trends and Global Governance*. London: Pluto Press.

Kenney, M. (ed.) (2000) *Understanding Silicon Valley: The Anatomy of an Entrepreneurial Region*. Stanford, CA: Stanford University Press.

Kenney, M. and Curry, J. (2001) Beyond transaction costs: e-commerce and the power of the Internet dataspace, in T.R. Leinbach and S.D. Brunn (eds), *Worlds of E-Commerce: Economic, Geographical, and Social Dimensions*. Chichester: Wiley. Chapter 3.

Kenney, M. and Florida, R. (1989) Japan's role in a post-Fordist age, *Futures*, 21: 136–151.

Kessler, J.A. (1999) The North American Free Trade Agreement, emerging apparel production networks and industrial upgrading: the southern California/Mexico connection, *Review of International Political Economy*, 6: 565–608.

Khanna, S.R. (1993) Structural changes in Asian textiles and clothing industries: the second migration of production, *Textile Outlook International*, September: 11–32.

Kim, H.Y. and Lee, S-H (1994) Commodity chains and the Korean automobile industry, in G. Gereffi and M. Korzeniewicz (eds), *Commodity Chains and Global Capitalism*. Westport, CT: Praeger. Chapter 14.

Kim, L. (1997) The dynamics of Samsung's technological learning in semiconductors, *California Management Review*, 39, 3: 86–100.

Kindleberger, C.P. (1969) *American Business Abroad*. New Haven, CT: Yale University Press.

Klein, N. (2000) *No Logo*. London: Flamingo.

Kobrin, S.J. (1987) Testing the bargaining hypothesis in the manufacturing sector in developing countries, *International Organization*, 41: 609–638.

Kogut, B. (1999) What makes a company global? *Harvard Business Review*, January–February: 165–170.

Koh, T. (1993) 10 values that help East Asia's economic progress and prosperity, *Singapore Straits Times*, 14 December.

Koo, H. and Kim, E.M. (1992) The developmental state and capital accumulation in South Korea, in R.P. Appelbaum and J. Henderson (eds), *States and Development in the Asian Pacific Rim*. London: Sage. Chapter 5.

Korten, D.C. (1995) *When Corporations Rule the World*. West Hartford, CT: Kumarian Press.

Korzeniewicz, M. (1994) Commodity chains and marketing strategies: Nike and the global athletic footwear industry, in G. Gereffi and M. Korzeniewicz (eds), *Commodity Chains and Global Capitalism*. Westport, CT: Praeger. Chapter 12.

Kozul-Wright, R. (1995) Transnational corporations and the nation-state, in J. Michie and J. Grieve-Smith (eds), *Managing the Global Economy*. Oxford: Oxford University Press. Chapter 6.

Kozul-Wright, R. and Rowthorn, R. (1998) Spoilt for choice? Multinational corporations and the geography of international production, *Oxford Review of Economic Policy*, 14: 74–92.

Krugman, P. (ed.) (1986) *Strategic Trade Policy and the New International Economics*. Cambridge, MA: MIT Press.

Krugman, P. (1990) *Rethinking International Trade*. Cambridge, MA: MIT Press.

Krugman, P. (1994) Competitiveness: a dangerous obsession, *Foreign Affairs*, March–April: 28–44.

Krugman, P. (1995) *Development, Geography and Economic Theory*. Cambridge, MA: MIT Press.

Krugman, P. (1998) What's new about the new economic geography? *Oxford Review of Economic Policy*, 14: 7–17.

Krugman, P. and Obstfeld, M. (1994) *International Economics: Theory and Policy*. New York: Harper Collins.

Laigle, L. (1996) New relationships between suppliers and car makers: towards development cooperation, *EUNIT Discussion Paper*, 2.

Lall, S. (1994) Industrial policy: the role of government in promoting industrial and technological development, *UNCTAD Review*, 1994: 65–90.

Lall, S. and Streeten, P. (1977) *Foreign Investment, Transnationals and Developing Countries*. London: Macmillan.

Landes, D.S. (1998) *The Wealth and Poverty of Nations*. New York: W.W. Norton.

Langdale, J.V. (2000) Telecommunications and 24-hour trading in the international securities industry, in M.I. Wilson and K.E. Corey (eds), *Information Tectonics: Space, Place, and Technology in an Electronic Age*. Chichester: Wiley. Chapter 6.

Lasserre, P. (1996) Regional headquarters: the spearhead for Asia Pacific Markets, *Long Range Planning*, 29: 30–37.

Law, J. (1986) On the methods of long-distance control: vessels, navigation and the Portuguese route to India, *Sociological Review Monograph*, 32: 234–263.

Lawrence, R.Z. (1996) *Regionalism, Multilateralism, and Deeper Integration*. Washington, DC: The Brookings Institution.

League of Nations (1945) *Industrialization and Foreign Trade*. New York: League of Nations.

Lee, K-Y. (2000) *From Third World to First. The Singapore Story: 1965–2000*. Singapore: Times Publishing Group.

Lee, N. and Cason, J. (1994) Automobile commodity chains in the NICs: a comparison of South Korea, Mexico, and Brazil, in G. Gereffi and M. Korzeniewicz (eds), *Commodity Chains and Global Capitalism*. Westport, CT: Praeger. Chapter 11.

Leinbach, T.R. (2001) Emergence of the digital economy and e-commerce, in T.R. Leinbach and S.D. Brunn (eds), *Worlds of E-Commerce: Economic, Geographical and Social Dimensions*. Chichester: Wiley. Chapter 1.

Leonard, H.J. (1988) *Pollution and the Struggle for World Product: Multinational Corporations, Environment and International Comparative Advantage*. Cambridge: Cambridge University Press.

Lewis, M.K. and Davis, J.T. (1987) *Domestic and International Banking*. Oxford: Philip Allan.

Leyshon, A. (1992) The transformation of regulatory order: regulating the global economy and environment, *Geoforum*, 23: 249–267.

Liao, S. (1997) ASEAN model in international economic cooperation, in H. Soesastro (ed.), *One South East Asia in a New Regional and International Setting*. Jakarta: Centre for Strategic and International Studies.

Lim, C.H. (ed.) (1988) *Policy Options for the Singapore Economy*. Singapore: McGraw–Hill.

Lim, L.Y.C. and Pang, E.F. (1986) *Trade, Employment and Industrialization in Singapore*. Geneva: ILO.

Linden, G. (2000) Japan and the United States in the Malaysian electronics sector, in M. Borrus, D. Ernst and S. Haggard (eds), *International Production Networks in Asia: Rivalry or Riches?* London: Routledge. Chapter 8.

Lundvall, B-Å. (ed.) (1992) *National Systems of Innovation*. London: Pinter.

Lundvall, B-Å. and Maskell, P. (2000) Nation-states and economic development: from national systems of production to national systems of knowledge creation and learning, in G.L. Clark, M.P. Feldman and M.S. Gertler (eds), *The Oxford Handbook of Economic Geography*. Oxford: Oxford University Press. Chapter 18.

Luttwak, E. (1999) *Turbo-Capitalism: Winners and Losers in the Global Economy*. London: Orion Business Books.

Lyons, D. and Salmon, S. (1995) World cities, multinational corporations, and urban hierarchy: the case of the United States, in P.L. Knox and P.J. Taylor (eds), *World Cities in a World-System*. Cambridge: Cambridge University Press. Chapter 6.

Macher, J.T., Mowery, D.C. and Hodges, D.A. (1998) Reversal of fortune? The recovery of the U.S. semiconductor industry, *California Management Review*, 41, 1: 107–136.

Maddison, A. (2001) *The World Economy: A Millennial Perspective*. Paris: OECD.

Magaziner, I.C. and Hout, T.M. (1980) *Japanese Industrial Policy*. London: Policy Studies Association.

Magretta, J. (1998) Fast, global, and entrepreneurial: supply chain management, Hong Kong style. An interview with Victor Fung, *Harvard Business Review*, September–October: 103–114.

Mair, A., Florida, R. and Kenney, M. (1988) The new geography of automobile production: Japanese transplants in North America, *Economic Geography*, 64: 352–373.

Malmberg, A. (1996) Industrial geography: agglomeration and local milieu, *Progress in Human Geography*, 20: 386–397.

Malmberg, A. (1999) The elusive concept of agglomeration economies: theoretical principles and empirical paradoxes, *SCASSS Seminar Paper*.

Malmberg, A. and Maskell, P. (1997) Towards an explanation of regional specialization and industry agglomeration, *European Planning Studies*, 5: 25–41.

Malmberg, A., Sölvell, Ö. and Zander, I. (1996) Spatial clustering, local accumulation of knowledge and firm competitiveness, *Geografiska Annaler*, 76B: 85–97.

Malnight, T.W. (1996) The transition from decentralized to network-based MNC structures: an evolutionary perspective, *Journal of International Business Studies*, 27: 43–63.

Mansfield, E. and Milner, H.V. (1999) The new wave of regionalism, *International Organization*, 53: 589–627.

Markusen, A. (1996) Sticky places in slippery space: a typology of industrial districts, *Economic Geography*, 72: 293–313.

Martin, P. and Widgren, J. (1996) International migration: a global challenge, *Population Bulletin*, 51.

Martin, R. (1994) Stateless monies, global financial integration and national economic autonomy: the end of geography?, in S. Corbridge, R. Martin, and N.J. Thrift (eds), *Money, Power and Space*. Oxford: Blackwell. Chapter 11.

Martin, R. (1999) The new economic geography of money, in R. Martin (ed.) *Money and the Space-Economy*. Chichester: Wiley. Chapter 1.

Mason, M. (1994) Historical perspectives on Japanese direct investment in Europe, in M. Mason and D. Encarnation (eds), *Does Ownership Matter? Japanese Multinationals in Europe*. Oxford: Clarendon Press. Chapter 1.

Massey, D. (2000) The geography of power, in B. Gunnell and D. Timms (eds), *After Seattle: Globalization and its Discontents*. London: Catalyst.

Mathews, J.A. (1997) A Silicon Valley of the East: creating Taiwan's semiconductor industry, *California Management Review*, 39, 4: 26–54.

Mathews, J.A. (1999) A Silicon Island of the East: creating a semiconductor industry in Singapore, *California Management Review*, 41, 2: 55–78.

Mathews, J.A and Cho, D S. (2000) *Tiger Technology. The Creation of a Semiconductor Industry in East Asia*. Cambridge: Cambridge University Press.

McConnell, J. and Macpherson, A. (1994) The North American Free Trade Agreement: an overview of issues and prospects, in R. Gibb and W. Michalak (eds), *Continental Trading Blocs: The Growth of Regionalism in the World Economy*. Chichester: Wiley. Chapter 6.

McGrew, A.G. (1992) Conceptualizing global politics, in A.G. McGrew and P.G. Lewis (eds), *Global Politics: Globalization and the Nation-State*. Cambridge: Polity Press. Chapter 1.

McHale, J. (1969) *The Future of the Future*. New York: George Braziller.

McNeill, J. (2000) *Something New Under the Sun: An Environmental History of the Twentieth Century*. London: Allen Lane/The Penguin Press.

Mentzer, J.T. (ed.) (2001) *Supply Chain Management*. Thousand Oaks, CA: Sage.

Metcalfe, J.S. and Dilisio, N. (1996) Innovation, capabilities and knowledge: the epistemic connection, in J. de la Mothe and G. Paquet (eds), *Evolutionary Economics and the New International Political Economy*. London: Pinter. Chapter 3.

Michalet, C-A. (1980) International subcontracting: a state of the art, in D. Germidis (ed.), *International Subcontracting: A New Form of Investment*. Paris: OECD.

Micklethwait, J. and Wooldridge, A. (2000) *A Future Perfect: The Challenge and Hidden Promise of Globalization*. New York: Crown Business.

Miles, R. and Snow, C.C. (1986) Organizations: new concepts for new forms, *California Management Review*, XXVIII: 62–73.

Miles, R., Snow, C.C., Mathews, J.A. and Miles, G. (1999) Cellular-network organizations, in W.E. Halal and K.B. Taylor (eds), *Twenty-First Century Economics: Perspectives of Socioeconomics for a Changing World*. New York: St Martin's Press. Chapter 7.

Min, S. and Keebler, J.S. (2001) The role of logistics in the supply chain, in J.T. Mentzer (ed.), *Supply Chain Management*. Thousand Oaks, CA: Sage. Chapter 10.

Mirza, H. (2000) The globalization of business and East Asian developing-country multinationals, in N. Hood and S. Young (eds), *The Globalization of Multinational Enterprise Activity and Economic Development*. London: Macmillan. Chapter 9.

Mittelman, J.H. (2000) *The Globalization Syndrome: Transformation and Resistance*. Princeton, NJ: Princeton University Press.

Mockler, R.J. (2000) *Multinational Strategic Alliances*. Chichester: Wiley.

Morris, D. and Hergert, M. (1987) Trends in international collaborative agreements, *Columbia Journal of World Business*, XXII: 15–21.

Morrison, A.J. and Roth, K. (1992) The regional solution: an alternative to globalization, *Transnational Corporations*, 1: 37–55.

Mortimore, M. (1998) Mexico's TNC-centric industrialization process, in R. Kozul-Wright and R. Rowthorn (eds), *Transnational Corporations and the Global Economy*. London: Macmillan. Chapter 13.

Myant, M., Fleischer, F., Hornschild, K., Vintrová, R., Zeman, K. and Souček, Z. (eds), (1996) *Successful Transformations? The Creation of Market Economies in Eastern Germany and the Czech Republic*. Cheltenham: Edward Elgar.

Myrdal, G. (1958) *Rich Lands and Poor*. New York: Harper & Row.

Mytelka, L.K. (2000) Locational tournaments for FDI: inward investment into Europe in a global world, in N. Hood and S. Young (eds), *The Globalization of Multinational Enterprise Activity and Economic Development*. London: Macmillan. Chapter 12.

Nelson, R.R. (ed.) (1993) *National Innovation Systems: A Comparative Study*. New York: Oxford University Press.

Nixson, F. (1988) The political economy of bargaining with transnational corporations: some preliminary observations, *Manchester Papers in Development*, IV: 377–390.

Nohria, N. and Ghoshal, S. (1997) *The Differentiated Network: Organizing Multinational Corporations for Value Creation*. San Francisco: Jossey–Bass.

Nolan, P. (2001) *China and the Global Business Revolution*. London: Palgrave.

O'Brien, R. (1992) *Global Financial Integration: The End of Geography*. London: Royal Institute of International Affairs.

OECD (1979) *The Impact of the Newly Industrializing Countries on Production and Trade in Manufactures*. Paris: OECD.

Offe, C. (1996) *Varieties of Transition: The East European and East German Experience*. Cambridge: Polity Press.

Office of Technology Assessment (1994) *Multinationals and the US Technology Base*. Washington, DC: Office of Technology Assessment.

Ogden, M.R. (1994) Politics in a parallel universe: is there a future for cyberdemocracy? *Futures*, 26: 713–729.

Ohmae, K. (1985) *Triad Power: The Coming Shape of Global Competition*. New York: Free Press.

Ohmae, K. (ed.) (1995) *The Evolving Global Economy: Making Sense of the New World Order*. Boston: Harvard Business Review Press.

Oinas, P. (1997) On the socio-spatial embeddedness of business firms, *Erdkunde*, 51: 23–32.

Oinas, P. (1999) Voices and silences: the problem of access to embeddedness, *Geoforum*, 30: 351–361.

Ó hUallacháin, B. (1997) Restructuring the American semiconductor industry, *Annals of the Association of American Geographers*, 87: 217–237.

Okimoto, D.I. (1989) *Between MITI and the Market: Japanese Industrial Policy for High Technology*. Stanford, CA: Stanford University Press.

Orrù, M., Biggart, N.W. and Hamilton, G.G. (1997) *The Economic Organization of East Asian Capitalism*. Thousand Oaks, CA: Sage.

O'Shaughnessy, J. (1995) *Competitive Marketing: A Strategic Approach*, 3rd edn. London: Routledge.

Ostry, S. (1990) *Governments and Corporations in a Shrinking World*. New York: Council on Foreign Relations Press.

Ould-Mey, M. (1999) The new global command economy, *Environment and Planning D: Society and Space*, 17: 155–180.

Owen, G. (1999) *From Empire to Europe: The Decline and Revival of British Industry Since the Second World War*. London: HarperCollins.

Oxfam (2002) *Rigged Rules and Double Standards: Trade, Globalization, and the Fight Against Poverty*. London: Oxfam.

Palloix, C. (1975) The internationalization of capital and the circuit of social capital, in G. Radice (ed.), *International Firms and Modern Imperialism*. Harmondsworth: Penguin. Chapter 3.

Palloix, C. (1977) The self-expansion of capital on a world scale, *Review of Radical Political Economics*, 9: 1–28.

Patel, P. (1995) Localized production of technology for global markets, *Cambridge Journal of Economics*, 19: 141–153.

Patel, P. and Pavitt, K. (1998) Uneven (and divergent) technological accumulation among advanced countries: evidence and a framework of explanation, in G. Dosi, D.J. Teece and J. Chytry (eds), *Technology, Organization and Competitiveness*. Oxford: Oxford University Press. pp. 289–317.

Pauly, L.W. (2000) Capital mobility and the new global order, in R. Stubbs and G.R.D. Underhill (eds), *Political Economy and the Changing Global Order*. Oxford: Oxford University Press. pp. 119–129.

Pauly, L.W. and Reich, S. (1997) National structures and multinational corporate behavior: enduring differences in the age of globalization, *International Organization*, 51: 1–30.

Pearce, D.E. (1995) *Capturing Environmental Value*. London: Earthscan.

Peck, J.A. (1996) *Work-Place: The Social Regulation of Labour Markets*. New York: Guilford.

Peck, J.A. (2001) *Workfare States*. New York: Guilford.

Peck, J.A. and Dicken, P. (1996) Tootal: internationalization, corporate restructuring and 'hollowing out', in J-E. Nilsson, P. Dicken and J.A. Peck (eds), *The Internationalization Process: European Firms in Global Competition*. London: Paul Chapman. Chapter 7.

Peck, J.A. and Miyamachi, Y. (1995) Regulating Japan? Regulation theory versus the Japanese experience, *Environment and Planning D: Society and Space*, 12: 639–674.

Perez, C. (1985) Microelectronics, long waves and world structural change, *World Development*, 13: 441–463.

Phillips, D.R. and Yeh, A.G.O. (1990) Foreign investment and trade: impact on spatial structure of the economy, in T. Cannon and A. Jenkins (eds), *The Geography of Contemporary China: The Impact of Deng Xiaoping's Decade*. London: Routledge. Chapter 9.

Picciotto, S. (1991) The internationalisation of the state, *Capital & Class*, 43: 43–63.

Piore, M.J. and Sabel, C.F. (1984) *The Second Industrial Divide: Possibilities for Prosperity*. New York: Basic Books.

Pitelis, C. (1991) Beyond the nation-state? The transnational firm and the nation-state, *Capital and Class*, 43: 131–152.

Pitelis, C. and Sugden, R. (eds) (1991) *The Nature of the Transnational Firm*. London: Routledge.

Polanyi, M. (1962) *Personal Knowledge: Towards a Post-Critical Philosophy*. Chicago: University of Chicago Press.

Poon, J.P. (1997) The cosmopolitization of trade regions: global trends and implications, 1965–1990, *Economic Geography*, 73: 390–404.

Poon, J.P.H., Thompson, E.R. and Kelly, P.F. (2000) Myth of the triad? The geography of trade and investment 'blocs', *Transactions of the Institute of British Geographers*, 25: 427–444.

Population Reference Bureau (1999) *World Population Data Sheet*. Washington, DC: Population Reference Bureau.

Porteous, D. (1999) The development of financial centres: locations, information externalities and path dependence, in R. Martin (ed.) *Money and the Space-Economy*. Chichester: Wiley. Chapter 5.

Porter, M.E. (1990) *The Competitive Advantage of Nations*. London: Macmillan.

Porter, M.E. (1998) Clusters and the new economics of competition, *Harvard Business Review*, November–December: 77–90.

Porter, M.E. (2000) Locations, clusters, and company strategy, in G.L. Clark, M.P. Feldman and M.S. Gertler (eds), *The Oxford Handbook of Economic Geography*. Oxford: Oxford University Press. Chapter 13.

Porter, M.E., Takeuchi, H. and Sakakibara, M. (2000) *Can Japan Compete?* London: Macmillan.

Prahalad, C.K. and Doz, Y. (1987) *The Multinational Mission*. New York: Free Press.

Prakeesh, A. and Hart, J.A. (eds) (1999) *Globalization and Governance*. London: Routledge.

Rabach, E. and Kim, E.M. (1994) Where is the chain in commodity chains? The service sector nexus, in G. Gereffi and M. Korzeniewicz (eds), *Commodity Chains and Global Capitalism*. Westport, CA: Praeger. Chapter 6.

Raikes, P., Jensen, M.F. and Ponte, S. (2000) Global commodity chain analysis and the French filière approach: comparison and critique, *Economy and Society*, 29: 390–417.

Ramesh, M. (1995) Economic globalization and policy choices: Singapore, *Governance: An International Journal of Policy and Administration,* 8: 243–260.

Rapkin, D.P. and Avery, W.P. (eds) (1995) *National Competitiveness in a Global Economy.* Boulder, CO: Lynne Rienner.

Redding, S.G. (1991) Weak organizations and strong linkages: managerial ideology and Chinese family business networks, in G. Hamilton (ed.), *Business Networks and Economic Development in East and South East Asia.* Hong Kong: Centre for Asian Studies, University of Hong Kong. Chapter 3.

Reed, H.C. (1989) Financial centre hegemony, interest rates and the global political economy, in Y.S. Park and N. Essayyad (eds), *International Banking and Financial Centres.* Boston, MA: Kluwer Academic. Chapter 16.

Reich, R.B. (1990) Who is us? *Harvard Business Review,* January–February: 53–64.

Reich, R.B. (1991) *The Work of Nations.* New York: Alfred A. Knopf.

Reich, S. (1989) Roads to follow: regulating direct foreign investment, *International Organization,* 43: 543–584.

Reich, S. (1996) 'Manufacturing' investment: national variations in the contribution of foreign direct investors to the US manufacturing base in the 1990s, *Review of International Political Economy,* 3: 27–64.

Rennstich, J. (2002) The new economy, the leadership long cycle and the nineteenth K-wave, *Review of International Political Economy,* 9: 150–182.

Richardson, J.D. (1990) The political economy of strategic trade policy, *International Organization,* 44: 107–135.

Rifkin, J. (1995) *The End of Work: The Decline of the Global Labour Force and the Dawn of the Post-Market Era.* New York: Putnam.

Roberts, S. (1994) Fictitious capital, fictitious spaces: the geography of offshore financial flows, in S. Corbridge, R. Martin and N.J. Thrift (eds), *Money, Power and Space.* Oxford: Blackwell. Chapter 5.

Roche, E.M. and Blaine, M.J. (2000) Telecommunications and governance in multinational enterprises, in M.I. Wilson and K.E. Corey (eds), *Information Tectonics: Space, Place, and Technology in an Electronic Age.* Chichester: Wiley. Chapter 5.

Rodan, G. (1991) *The Political Economy of Singapore's Industrialization.* Petalang Jaya: Forum Books.

Rodríguez-Pose, A. and Arbix, G. (2001) Strategies of waste: bidding wars in the Brazilian automobile sector, *International Journal of Urban and Regional Research,* 25: 134–154.

Rodrik, D. (1999) *The New Global Economy and Developing Countries: Making Openness Work.* Washington, DC: Overseas Development Council.

Rowthorn, R. and Ramaswamy, R. (1997) De-industrialization: causes and implications, *IMF Working Paper.*

Rugman, A.M. (1981) *Inside the Multinationals.* London: Croom Helm.

Rugman, A. (2000) *The End of Globalization.* London: Random House.

Ruigrok, W. and Tulder, R. van (1995) *The Logic of International Restructuring.* London: Routledge.

Rupert, M. (2000) *Ideologies of Globalization: Contending Visions of a New World Order.* London: Routledge.

Sachs, J. (1997) Nature, nurture, and growth, *The Economist,* 14 June: 19–22.

Sadler, D. (1995) National and international regulatory framework: the politics of European automobile production and trade, in R. Hudson and E. Schamp (eds), *Towards a New Map of Automobile Manufacturing in Europe.* Berlin: Springer-Verlag. Chapter 2.

Sampson, G.P. (ed.) (2001) *The Role of the World Trade Organization in Global Governance.* Tokyo: United Nations University Press.

Sassen, S. (2001) *The Global City: New York, London, Tokyo,* 2nd edn. Princeton, NJ: Princeton University Press.

Savary, J. (1995) The rise of international co-operation in the European automobile industry: the Renault case, *European Urban and Regional Studies,* 2: 3–20.

Saxenian, A. (1994) *Regional Advantage: Culture and Competition in Silicon Valley and Route 128.* Cambridge, MA: Harvard University Press.

Sayer, A. (1986) New developments in manufacturing: the just-in-time system, *Capital & Class,* 30: 43–72.

Sayer, A. and Walker, R. (1992) Beyond Fordism and flexibility, in A. Sayer and R. Walker, *The New Social Economy.* Oxford: Blackwell. Chapter 5.

Schary, P.B. and Skjøtt-Larsen, T. (2001) *Managing the Global Supply Chain,* 2nd edn. Copenhagen: Copenhagen Business School Press.

Schoenberger, E. (1997) *The Cultural Crisis of the Firm.* Oxford: Blackwell.

Schoenberger, E. (1999) The firm in the region and the region in the firm, in T. Barnes and M. Gertler (eds), *The New Industrial Geography: Regions, Regulation and Institutions.* London: Routledge. Chapter 9.

Schoenberger, E. (2000) The management of time and space, in G.L. Clark, M.P. Feldman and M.S. Gertler (eds), *The Oxford Handbook of Economic Geography.* Oxford: Oxford University Press. Chapter 16.

Schonberger, R.J. (1982) *Japanese Manufacturing Techniques: Nine Hidden Lessons in Simplicity.* New York: Free Press.

Schumpeter, J. (1943) *Capitalism, Socialism and Democracy.* London: Allen & Unwin.

Scott, A.J. (1995) The geographic foundations of industrial performance, *Competition & Change,* 1: 51–66.

Scott, A.J. (1998) *Regions and the World Economy: The Coming Shape of Global Production, Competition, and Political Order.* Oxford: Oxford University Press.

Scott, A.J. and Angel, D. (1988) The global assembly operations of US semiconductor firms: a geographical analysis, *Environment and Planning A,* 20: 1047–1067.

Scott-Quinn, B. (1990) US investment banks as multinationals, in G. Jones (ed.), *Banks as Multinationals.* London: Routledge. Chapter 5.

Seidler, E. (1976) *Let's Call it Fiesta.* London: Patrick Stephens.

Sen, A. (1999) *Development as Freedom.* Oxford: Oxford University Press.

Sheard, P. (1983) Auto production systems in Japan: organizational and locational features, *Australian Geographical Studies,* 21: 49–68.

Shepherd, G. (1983) Textiles: new ways of surviving in an old industry, in G. Shepherd, F. Duchene and C. Saunders (eds), *Europe's Industries: Public and Private Strategies for Change.* London: Pinter. Chapter 2.

Siegel, S. (1980) Delicate bonds: the global semiconductor industry, *Pacific Research,* 11: 1–26.

Simonis, U.D. and Brühl, T. (2002) World ecology – structures and trends, in P. Kennedy, D. Messner and F. Nuscheler (eds), *Global Trends and Global Governance.* London: Pluto Press.

Sit, V.F.S. and Liu, W. (2000) Restructuring and spatial change of China's auto industry under institutional reform and globalization, *Annals of the Association of American Geographers,* 90: 653–673.

Sklair, L. (1989) *Assembling for Development: The Maquila Industry in Mexico and the United States.* London: Unwin Hyman.

Sklair, L. (ed.) (1995) *Sociology of the Global System,* 2nd edn. Hemel Hempstead: Prentice Hall/Harvester Wheatsheaf.

Sklair, L. (2001) *The Transnational Capitalist Class.* Oxford: Blackwell.

Smelser, N. and Swedberg, R. (eds) (1994) *The Handbook of Economic Sociology.* Princeton, NJ: Princeton University Press.

Smith, D.M. (1981) *Industrial Location: An Industrial-Geographical Analysis,* 2nd edn. New York: Wiley.

Stalk, G. Jr and Hout, T.M. (1990) *Competing Against Time: How Time-Based Competition is Reshaping Global Markets.* New York: Free Press.

Stalk, G., Evans, P. and Shulman, L.E. (1998) Competing on capabilities: the new rules of corporate strategy, *Harvard Business Review,* March–April: 57–69.

Stallings, B. (ed.) (1995) *Global Change, Regional Response: The New International Context of Development.* Cambridge: Cambridge University Press.

Sternquist, B. (1998) *International Retailing.* New York: Fairchild.

Stewart, T.A. (1997) *Intellectual Capital.* London: Nicholas Brealey.

Stiglitz, J.E. (1999) The World Bank at the millennium, *The Economic Journal,* 109: F577–F597.

Stiglitz, J.E. and Yusuf, S. (eds) (2001) *Rethinking the East Asia Miracle.* New York: Oxford University Press.

Stopford, J.M. and Strange, S. (1991) *Rival States, Rival Firms: Competition for World Market Shares.* Cambridge: Cambridge University Press.

Storper, M. (1992) The limits to globalization: technology districts and international trade, *Economic Geography,* 68: 60–93.

Storper, M. (1995) The resurgence of regional economies, ten years later: the region as a nexus of untraded interdependencies, *European Urban and Regional Studies*, 2: 191–221.

Storper, M. (1997) *The Regional World: Territorial Development in a Global Economy*. New York: Guilford.

Storper, M. and Walker, R. (1984) The spatial division of labour: labour and the location of industries, in L. Sawers and W.K. Tabb (eds), *Sunbelt/Snowbelt: Urban Development and Regional Restructuring*. New York: Oxford University Press. Chapter 2.

Strange, S. (1986) *Casino Capitalism*. Oxford: Blackwell.

Strange, S. (1996) *The Retreat of the State: The Diffusion of Power in the World Economy*. Cambridge: Cambridge University Press.

Strange, S. (1998) *Mad Money*. Manchester: Manchester University Press.

Sturgeon, T.J. (2001) How do we define value chains and production networks? *IDS Bulletin*, 32, 3: 9–18.

Sturgeon, T.J. (2002) Modular production networks: a new American model of industrial organization, *Industrial and Corporate Change*, 11.

Swyngedouw, E. (1997) Neither global nor local: 'glocalization' and the politics of scale, in K.R. Cox (ed.), *Spaces of Globalization*. New York: Guilford. Chapter 6.

Swyngedouw, E. (2000) Elite power, global forces, and the political economy of 'glocal' development, in G.L. Clark, M.P. Feldman and M.S. Gertler (eds), *The Oxford Handbook of Economic Geography*. Oxford: Oxford University Press. Chapter 27.

Taplin, I.M. (1994) Strategic reorientations of US apparel firms, in G. Gereffi and M. Korzeniewicz (eds), *Commodity Chains and Global Capitalism*. Westport, CT: Praeger. Chapter 10.

Taylor, M.J. and Thrift, N.J. (eds) (1986) *Multinationals and the Restructuring of the World Economy*. London: Croom Helm.

Taylor, P.J. (1994) The state as container: territoriality in the modern world-system, *Progress in Human Geography*, 18: 151–162.

Taylor, P.J. (2001) Being economical with the geography, *Environment and Planning A*, 33: 949–954.

Terpstra, V. and David, K. (1991) *The Cultural Environment of International Business*. Cincinnati, OH: South-Western Publishing.

Thoburn, J. and Howell, J. (1995) Trade and development: the political economy of China's open policy, in R. Benewick and P. Wingrove (eds), *China in the 1990s*. London: Macmillan. Chapter 14.

Thrift, N.J. (1990) Doing regional geography in a global system: the new international financial system, the City of London and the South East of England, 1984–1987, in R.J. Johnston, J. Hauer and G.A. Hoekveld (eds), *Regional Geography: Current Developments and Future Prospects*. London: Routledge. pp. 180–207.

Thrift, N.J. (1994) On the social and cultural determinants of international financial centres: the case of the City of London, in S. Corbridge, R. Martin and N.J. Thrift (eds), *Money, Power and Space*. Oxford: Blackwell. Chapter 14.

Thrift, N.J. (1996) *Spatial Formations*. London: Sage.

Tobin, J. (1984) Unemployment in the 1980s: macroeconomic diagnosis and prescription, in A.J. Pierre (ed.) *Unemployment and Growth in the Western Economies*. New York: Council for Foreign Relations. pp. 79–112.

Toffler, A. (1971) *Future Shock*. London: Pan.

Townsend, A.M. (2001) Network cities and the global structure of the Internet, *American Behavioral Scientist*, 44: 1697–1716.

Tudor, G. (2000) *Rollercoaster: The Incredible Story of the Emerging Markets*. London: Pearson.

Turner, A. (2001) *Just Capital: The Liberal Economy*. London: Macmillan.

Turner, R.K., Pearce, D. and Bateman, I. (1994) *Environmental Economics: An Elementary Introduction*. Hemel Hempstead: Harvester Wheatsheaf.

Turok, I. (1993) Inward investment and local linkages: how deeply embedded is 'Silicon Glen'? *Regional Studies*, 27: 401–417.

Tyson, L.D. (1993) *Who's Bashing Whom? Trade Conflict in High-Technology Industries*. Washington, DC: Institute for International Economics.

UNCTAD (1993) *World Investment Report, 1993: Transnational Corporations and Integrated International Production*. New York: United Nations.

UNCTAD (1994) *World Investment Report, 1994: Transnational Corporations, Employment and the Workplace*. New York: United Nations.

UNCTAD (1995) *World Investment Report, 1995: Transnational Corporations and Competitiveness.* New York: United Nations.

UNCTAD (1996a) *Trade and Development Report, 1996.* New York: United Nations.

UNCTAD (1996b) *World Investment Report, 1996: Investment, Trade, and International Policy Arrangements.* New York: United Nations.

UNCTAD (2000a) *The Competitiveness Challenge: Transnational Corporations and Industrial Restructuring in Developing Countries.* New York: United Nations.

UNCTAD (2000b) *World Investment Report, 2000: Cross-Border Mergers and Acquisitions and Development.* New York: United Nations.

UNCTAD (2001) *World Investment Report, 2001: Promoting Linkages.* New York: United Nations.

UNCTAD (2002) *Trade and Development Report, 2002: Developing Countries in World Trade.* New York: United Nations.

UNCTC (1988) *Transnational Corporations in World Development: Trends and Prospects.* New York: United Nations.

UNDP (1999) *Human Development Report, 1999.* New York: Oxford University Press.

UNIDO (1980) Export Processing Zones in developing countries, *UNIDO Working Paper on Structural Change,* 19.

United Nations (1996) *An Urbanizing World: Global Report on Human Settlements.* New York: Oxford University Press.

United Nations Centre for Human Settlements (2001) *Cities in a Globalizing World: Global Report on Human Settlements, 2001.* London: Earthscan.

United Nations Population Division (2001) *World Population Prospects: The 2000 Revision.* New York: United Nations.

United Nations World Commission on the Environment and Development (1987) *Our Common Future.* Oxford: Oxford University Press.

US Department of Commerce (2000) *Digital Economy 2000.* Washington, DC: US Department of Commerce.

Vernon, R. (1966) International investment and international trade in the product cycle, *Quarterly Journal of Economics,* 80: 190–207.

Vernon, R. (1971) *Sovereignty at Bay: The Multinational Spread of US Enterprises.* New York: Basic Books.

Vernon, R. (1974) The location of economic activity, in J.H. Dunning (ed.), *Economic Analysis and the Multinational Enterprise.* London: Allen & Unwin. pp. 89–114.

Vernon, R. (1979) The product cycle hypothesis in a new international environment, *Oxford Bulletin of Economics and Statistics,* 41: 255–268.

Vernon, R. (1998) *In the Hurricane's Eye: The Troubled Prospects of Multinational Enterprises.* Cambridge, MA: Harvard University Press.

Villareal, R. (1990) The Latin American strategy of import substitution: failure or paradigm for the region?, in G. Gereffi and D. Wyman (eds), *Manufacturing Miracles: Paths of Industrialization in Latin America and East Asia.* Princeton, NJ: Princeton University Press. Chapter 11.

Voltaire (Arouet, F.-M.) (1947) *Candide* (translated J. Butt). London: Penguin.

Wade, R. (1990a) *Governing the Market: Economic Theory and the Role of Government in East Asian Industrialization.* Princeton, NJ: Princeton University Press.

Wade, R. (1990b) Industrial policy in East Asia: does it lead or follow the market?, in G. Gereffi and D.L. Wyman (eds), *Manufacturing Miracles: Paths of Industrialization in Latin America and East Asia.* Princeton, NJ: Princeton University Press. Chapter 9.

Wade, R. (1996) Globalization and its limits: reports of the death of the national economy are greatly exaggerated, in S. Berger and R. Dore (eds), *National Diversity and Global Capitalism.* Ithaca, NY: Cornell University Press. Chapter 2.

Walker, R.A. (1999) Putting capital in its place: globalization and the prospects for labour, *Geoforum,* 30: 263–284.

Warf, B. (1989) Telecommunications and the globalization of financial services, *Professional Geographer,* 41: 257–271.

Warf, B. (1995) Telecommunications and the changing geographies of knowledge transmission in the late 20th century, *Urban Studies,* 32: 361–378.

Warf, B. (2001) Segueways into cyberspace: multiple geographies of the digital divide, *Environment and Planning B: Planning and Design,* 28: 3–19.

Warf, B. and Purcell, D. (2001) The currency of currency: speed, sovereignty, and electronic finance, in T.R. Leinbach and S.D. Brunn (eds), *Worlds of E-Commerce: Economic Geographical and Social Dimensions*. Chichester: Wiley. Chapter 12.

Webber, M.J. and Rigby, D.L. (1996) *The Golden Age Illusion: Rethinking Postwar Capitalism*. New York: Guilford.

Weidenbaum, M. and Hughes, S. (1996) *The Bamboo Network: How Expatriate Chinese Entrepreneurs are Creating a New Economic Superpower in Asia*. New York: Free Press.

Weiss, L. (1998) *The Myth of the Powerless State: Governing the Economy in a Global Era*. Cambridge: Polity Press.

Wells, L.T. Jr (ed.) (1972) *The Product Life Cycle and International Trade*. Boston, MA: Harvard Business School.

Whitley, R.D. (1992) *Business Systems in East Asia: Firms, Markets and Societies*. London: Sage.

Whitley, R.D. (1999) *Divergent Capitalisms: The Social Structuring and Change of Business Systems*. Oxford: Oxford University Press.

Williams, K., Haslam, C., Williams, J. and Cutler, T. (1992) Against lean production, *Economy and Society*, 21: 321–354.

Williamson, O.E. (1975) *Markets and Hierarchies*. New York: Free Press.

Wills, J. (1998) Taking on the CosmoCorps? Experiments in transnational labour organization, *Economic Geography*, 74: 111–130.

Womack, J.R., Jones, D.T. and Roos, D. (1990) *The Machine that Changed the World*. New York: Rawson Associates.

Wong, K.Y. and Chu, D.K.Y. (eds) (1995) *Coordination in China: The Case of the Shenzhen Special Economic Zone*. Hong Kong: Oxford University Press.

Wong, K.Y., Lau, C-C. and Li, E.B.C. (eds) (1988) *Perspectives on China's Modernization*. Hong Kong: Chinese University of Hong Kong.

Wood, A. (1994) *North–South Trade, Employment, and Inequality*. Oxford: Oxford University Press.

World Bank (1995) *World Development Report, 1995: Workers in an Integrating World*. New York: Oxford University Press.

World Bank (1997) *World Development Report, 1997: The State in a Changing World*. New York: Oxford University Press.

World Bank (2000) *World Development Report, 2000: Entering the 21st Century*. New York: Oxford University Press.

World Bank (2001) *World Development Report, 2001: Attacking Poverty*. New York: Oxford University Press.

Wrigley, N. (2000) The globalization of retail capital: themes for economic geography, in G.L. Clark, M.P. Feldman and M.S. Gertler (eds), *The Oxford Handbook of Economic Geography*. Oxford: Oxford University Press. Chapter 15.

WTO (1996) *Annual Report, 1996*. Geneva: World Trade Organization.

Yao, S. (1997) The romance of Asian capitalism: geography, desire and Chinese business, in M.T. Berger and D.A. Borer (eds), *The Rise of East Asia: Critical Visions of the Pacific Century*. London: Routledge, Chapter 9.

Yeung, H.W-c. (1998) The political economy of transnational corporations: a study of the regionalization of Singaporean firms, *Political Geography*, 17: 389–416.

Yeung, H.W-c. (ed.) (1999a) *The Globalization of Business Firms from Emerging Economies*, 2 vols. Cheltenham: Edward Elgar.

Yeung, H.W-c. (1999b) Regulating investment abroad: the political economy of the regionalization of Singaporean firms, *Antipode*, 31: 245–273.

Yeung, H.W-c. (2000) The dynamics of Asian business systems in a globalizing era, *Review of International Political Economy*, 7: 399–433.

Yeung, H.W-c. (2002) *Entrepreneurship and the Internationalization of Asian Firms*. Cheltenham: Edward Elgar.

Yeung, H.Y-c., and Olds, K. (eds) (2000) *Globalization of Chinese Business Firms*. London: Macmillan.

Yeung, H.Y-c., Poon, J. and Perry, M. (2001) Towards a regional strategy: the role of regional headquarters of foreign firms in Singapore, *Urban Studies*, 38: 157–183.

Yeung, Y-m. and Li, X. (2000) Transnational corporations and local embeddedness: company case studies from Shanghai, China, *Professional Geographer*, 52: 624–635.

Yeung, Y.-m. and Lo, F.-C. (1996) Global restructuring and emerging urban corridors in Pacific Asia, in Y.-m. Yeung and F.-C. Lo (eds), *Emerging World Cities in Pacific Asia*. Tokyo: United Nations University Press. Chapter 2.

Yoffie, D.B. (ed.) (1993) *Beyond Free Trade: Firms, Governments, and Global Competition*. Boston, MA: Harvard Business School Press.

Yoffie, D.B. and Milner, H.V. (1989) An alternative to free trade or protectionism: why corporations seek strategic trade policy, *California Management Review*, 31: 111–131.

Young, D., Goold, M., Blanc, G., Bühner, R., Collis, D., Eppink, J., Kagono, T. and Jiménez, G. (2000) *Corporate Headquarters: An International Analysis of their Roles and Staffing*. London: Pearson.

Zanfei, A. (2000) Transnational firms and the changing organization of innovative activities, *Cambridge Journal of Economics*, 24: 515–542.

Zook, M.A. (2001) Old hierarchies or new networks of centrality: the global geography of the internet content market, *American Behavioral Scientist*, 44: 1679–1696.

Zukin, S. and DiMaggio, P. (eds) (1990) *Structures of Capital: The Social Organization of the Economy*. Cambridge: Cambridge University Press.

INDEX

The following abbreviations have been used in this index:

EU   European Union
NIEs   Newly Industrialized Economies
TNCs   transnational corporations

Key words are in bold

ABB, 221
accountancy firms, 441
Acer, 265
acid rain, 595
acquisition, 106, 107, 207, 252, 278
    in automobile industry, 373, 375–6
    in distribution industries, 485
    in retailing, 500–1
    in semiconductor industry, 422, 432
    in textiles industry, 341
    see also **mergers**
adjustment, 544–6
advertising, 97, 98
aerospace industry, 170
affiliated companies, 229, 248, 268, 269–70
Africa,
    foreign investment in, 60, 66, 67
    migration flows, 520
    sub-Saharan, 46, 513, 514
    urban-rural contrasts, 559
AFTA (ASEAN Free Trade Agreement), 148, 150, 158–9
ageing population, 517–18, 550
agriculture, 568, 570
    employment in, 555, 559
aid programmes, 573, 584
air transportation, 91, 93
    regulation/deregulation, 482, 483
Amazon.com, 478, 480
AMD, 404, 428
analog devices, 405
ANCOM (Andean Common Market), 150
Angola, debt in, 573
anti-globalization protests, 7, 317, 324, 578–9, 591
APEC (**Asia-Pacific Economic Cooperation Forum**), 148, 159–61
Apparel Industry Partnerships, 334
Argentina, 39, 49
    automobile production, 372
    debt in, 573
    European retailers in, 498, 501
    regional trading arrangements, 148
ASEAN (Association of South East Asian Nations), 71, 158
Asia, 41, 559
    automobile industry, 359, 362
    firms in international strategic alliances, 258

Asia, *cont.*
    foreign investment, 71
    migration flows, 520, 521
    services trade, 43
    textiles and garments industries, 323, 324, 326, 327
    urban growth, 558
    values, 569
    see also East and South East Asia
assembly industries, 355, 356, 362, 364, 366, 376, 408
asset-oriented production, 210–12
Aston Martin, 373
AT & T, 482
atmospheric damage, 28–9, 595–6
Australia,
    and APEC, 159, 160, 161
    automobile industry, 362
    foreign investment, 67
    trade, 41
Austria,
    accession to EU, 151
    trading agreements, 146
automobile industry, 311, 355
    acquisition and mergers in, 373–5, 377, 383
    corporate strategies, 372–85
    demand, 362–3
    global shifts in, 357–62
    government policies, 355, 369–72, 383, 384–5
    Japanese plants, 173, 358, 371, 380–1, 388–9, 390–2
    production, *see* **production of automobiles**
    regional networks, 386–97
    supplier relationships, 366–9, 388–9, 390–2
    technological change, 364–5
    trade tensions, 355, 370–1, 380
Automobile Pact 1965, 390

B2B (business-to-business), 478, 504
B2C (business-to-consumer), 478, 504
balance of payments, 282
Bangladesh,
    child labour, 594
    income from labour migration, 561
Bank for International Settlements, 583
Bank of Tokyo-Mitsubishi, 452
banking, 439–40, 441
    acquisition and merger, 450–3
    decentralization, 467–8
    diversification, 457–8
    in Europe, 454, 468
    in Japan, 450–2, 454, 456

banking, *cont.*
    in South Korea, 181
    transnationalization of, 453–7
bar codes, 476
bargaining between TNCs and states, 304–6, 308–11, 312
Beijing, 462
Belgium,
    automobile industry, 378, 386
    garments industry, 326
    technological leadership, 89
    trading agreements, 146
    transnationality index, 221
    women in labour market, 526
Benetton, 344–5, 502
Bhopal, 286
biotechnology, 170
BMW, 373, 377, 383
    in Thailand, 396
    in USA, 393
Bolivia, trading arrangements, 148
Boo.com, 503
branch plant economy, 298
Brazil, 49, 247
    automobile industry, 359, 362, 367–8, 372, 378, 381, 383–4
    debt levels, 573
    European retailers in, 498, 501
    foreign investment, 68, 247
    income distribution, 567
    industrialization, 178
    trading arrangements, 148
Bretton Woods, 35, 581–2
British Telecom, 482
broadband communications, 96, 102
Brunei, and ASEAN, 158
Bulgaria,
    accession to EU, 154
    garments industry, 343
business, social accountability of, 166
business firms, 17, 21, 463
    influence of foreign TNCs on, 286
business services, 42, 187
buyer-driven production chains, 18, 19, 264–5, 319

cable and satellite broadcasting, 98
cabotage, 483–4
call centres, 492–4
Cambodia,
    and APEC, 148
    and ASEAN, 158
Canada, 221, 290
    and APEC, 160
    automobile industry, 359, 361, 378, 385, 390
    foreign investment in, 60, 62, 65, 67
    and NAFTA, 155, 156, 157
    textiles and garments industries, 348

622